# SUBSURFACE EXPLORATION STRATIGRAPHY
by Burr A. Silver

The Institute for Energy Development
Oklahoma City, Oklahoma

**SUBSURFACE EXPLORATION STRATIGRAPHY**

Printed in 1983 with permission of the author by:

© The Institute for Energy Development
P. O. Box 19243
Oklahoma City, Oklahoma 73144

FOR:

The Institute for Energy Development INSTITUTE
"Subsurface Exploration Stratigraphy"

All rights reserved. No part of this book may be reproduced or transmitted in any form or by any means, electronic or mechanical, including photocopying, recording or by any information retrieval system, without permission in writing from the author and from The Institute for Energy Development.

Printed at IED Press, Inc.
Oklahoma City, Oklahoma
UNITED STATES OF AMERICA

ISBN Number: 0-89419-254-X

*IN MEMORY OF
MY MOTHER*

*AUGUSTA MAE SILVER*
*August 1920-June 1982*

# ABOUT THE AUTHOR . . .

## BURR A. SILVER

Burr A. Silver is President of Olympic Exploration and Production Company and a petroleum industry consulting geologist. He is a member of the faculties of several major oil companies' training centers and he is the instructor for four IED Exploration institutes.

Dr. Silver's past experience includes positions with Cities Service Research, Jersey Production Research Company, and Humble Oil and Refining Company. He served as a member of the geology faculties at the University of Oklahoma and Arizona State University.

Dr. Silver has B.S. and M.S. degrees in geology from Baylor University and a Ph.D. in geology from the University of Washington. He is author and co-author of numerous technical papers and books. He is a member of the American Association of Petroleum Geologists, the Society of Economic Mineralogists and Paleontologists, and the Canadian Association of Petroleum Geologists.

# CONTENTS

| | | |
|---|---|---|
| | **PREFACE** | ix |
| | **ACKNOWLEDGEMENTS** | x |

## PART I: EXPLORATION TECHNOLOGY

| | | |
|---|---|---|
| 1 | Proper Use of Subsurface Data | 3 |
| 2 | Pre-Entrapment History of Petroleum | 20 |
| 3 | The Stratigraphic Framework: A Fundamental Prerequisite to a Successful Exploration Program | 40 |

## PART II: EXPLORATION FOR SANDSTONE RESERVOIRS

| | | |
|---|---|---|
| 4 | Exploration for Continental Sandstone Reservoirs | 61 |
| 5 | Exploration for Delta and Fan Delta Reservoirs | 91 |
| 6 | Exploration for Interdeltaic Sandstone Reservoirs | 123 |
| 7 | Exploration for Offshore Marine Sandstone Reservoirs | 142 |
| 8 | Diagenesis of Potential Sandstone Reservoirs | 165 |

## PART III: EXPLORATION FOR CARBONATE RESERVOIRS

| | | |
|---|---|---|
| 9 | Overview of Carbonate Fundamentals | 179 |
| 10 | Carbonate Depositional Environments and Stratigraphy | 190 |
| 11 | Diagenesis of Potential Carbonate Reservoirs | 225 |
| 12 | Reservoirs and Carbonate Traps | 233 |
| | **REFERENCES** | 279 |
| | **ADDITIONAL READING** | 291 |
| | **APPENDIX I** | 295 |
| | **APPENDIX II** | 311 |
| | **INDEX** | 329 |

# PREFACE

Development of an accurate physical stratigraphic framework in a prospective area is one of the most important tasks of a petroleum explorationist. Poor application of the principles of stratigraphy to exploration is probably one of the most important single reasons that a prospect is unsuccessful. Without a valid stratigraphic framework, the explorationist cannot (1) map the proper horizon(s) that will accurately depict structure, (2) correctly interpret the facies patterns, and (3) accurately interpret the timing relationship between the formation of a potential trap and the generation, expulsion, and migration of hydrocarbons.

Early stratigraphers were general geologists. They had excellent backgrounds in paleontology and sedimentology and they understood the complex interplay between tectonics, sea-level changes and stratigraphy. Amadeus W. Grabau could be considered the "type stratigrapher" of the early 1900's. Starting in the 1920's, Twenhofel, Milner, Krumbein, and Pettijohn began to divide stratigraphy into a number of subdisciplines. As earth science became more qualitative, it became unfashionable to call oneself a stratigrapher. Universities began cutting back on the courses they offered in the field and few graduates recognized the need for a sound stratigraphic framework that is fundamental to a data base, whether it is geochemical, geophysical, or sedimentologic.

The reunification of stratigraphy by Sloss and Wheeler began quietly in the late 1950's. Their approach to solving stratigraphic problems was readily adaptable to the reflection seismogram, particularly with the advent of Common-Depth-Point data acquisition and processing techniques. The modern approach to stratigraphy, popularized by Vail, Mitchum, and Todd, to name a few, permits the explorationist to integrate all forms of surface and subsurface data in order to map unconformity bounded sequences. Delineation of sequences enhances the explorationist's ability to focus his or her attention on specific stratigraphic intervals that potentially contain source rocks that have passed through the generation window, carrier beds, and reservoir rocks. Interpretation of pathways of migration, facies relationships, and the relative timing of generation of hydrocarbons and the formation of potential trap are also facilitated by the delineation of sequences.

It is the purpose of this book to review (1) the basic principles of stratigraphy that are applicable to petroleum exploration, (2) the facies models of the important potential reservoir and source rocks, and (3) numerous case histories that demonstrate the successful application of the sound stratigraphic principles to exploration. There is little data in this book that has not been published previously. However, data does not find oil or gas. "Data" has never been listed as the originator of a map or cross section. Petroleum is discovered by individuals who systematically approach the development of a prospect with the creative interpretation of data. It is therefore hoped that the systematic and creative approach to exploration that is interlaced throughout this book will be useful to the inexperienced as well as the experienced explorationist.

January, 1983
Burr A. Silver

# ACKNOWLEDGEMENTS

This book would not have been possible to write without the benefit of ideas from numerous educators, researchers, and explorationists. My fundamental approach towards subsurface exploration stratigraphy was obtained from Harry Wheeler while a graduate student of his at the University of Washington. When I joined Humble Oil & Refining Company (Exxon U.S.A.) in Midland, Texas, the sequence approach "dovetailed" nicely with the work headed up by Peter Vail and Robert Mitchum at Esso Production Research (Exxon). My approach towards clastics is highly influenced by Phil Shannon (Exxon) and Rufus LeBlanc (Shell). My exploration approach to carbonates was influenced by Robert Terriere (Cities Service Research), Robert Mitchum (at that time with Jersey Production Research), and Robert Todd (Exxon). Modification of the early influences of the above individuals occurred over many years of field observations of both modern and ancient clastic and carbonate rocks.

The philosophy of exploration that is interlaced throughout this book was obtained in part from numerous peers in Exxon, U.S.A. As an independent geologist, my approach to exploration has been influenced by the joy of discoveries and the pain of dry holes.

This book would not have been cost effective without the liberal use of illustrations originally published by the American Association of Petroleum Geologists, Society of Economic Paleontologists and Mineralogists, and the Canadian Association of Petroleum Geologists. Use of these illustrations is greatly appreciated.

# PART I:
# EXPLORATION TECHNOLOGY

*1 Proper Use of Subsurface Data*
*2 Pre-Entrapment History of Petroleum*
*3 The Stratigraphic Framework: A Fundamental Prerequisite to a Successful Exploration Program*

# 1
# Proper Use of Subsurface Data

## INTRODUCTION

During the generation of a prospect, it is typically necessary to deal with data that was not only accumulated by many individuals, but that was also developed over a span of time during which a change in the technology of obtaining the data may have occurred. It is therefore necessary for the explorationists to evaluate exploration data and, in many cases, selectively weigh the importance of the data.

Many variables affect the quality of subsurface data. Some of these variables include: the quality and experience level of the individual who made original measurements, accuracy of the individual or individuals who transcribed the original data, and the stage of technology when the data was generated. Some of the variables that affect the quality of exploration data are reviewed for your edification.

## ROCK SAMPLES

### WELL CUTTINGS

Variables that affect the quality of cuttings include: quality of the mud system, the degree of caving, the accuracy of lag time determination, sharpness and type of drill bit, weight on the drill bit, and rate of bit rotation in relation to circulation rate. The quality of the mud system is by far the most important variable. During the drilling of a wildcat, a percentage log may be prepared by the well-site geologist or mudlogger, and after mechanical logs have been run in the well, an interpretative log can be prepared. Both logs use the standard set of symbols depicted in Figure 1.1.

### CORES

The homogeneity of cored strata, degree of fractures, sharpness and type of core barrel, degree of bedding, and lithology of the cored interval control the quality of cores. Porosity, permeability, water saturation, hydrocarbon content, and lithology can be determined from a core (Fig. 1.2).

## MUD LOGS

The quality of a mud log is influenced by the quality of the mud system, lag time determinations, and the accuracy of the logging equipment. Modern mud logs (Fig. 1.3) are generally reliable. They include a wealth of data, namely drill time, gas shows, sample shows, drill-stem tests, side-wall tests, and lithology.

## DRILLSTEM TESTS (DST)

Variables that control the quality of a DST include: (1) quality of the mud system (invasion of mud-filtrate into the potential reservoir can produce serious formation damage), (2) construction of the borehole, (3) mechanics of the DST (if two packers are required, this adds risk, Figure 1.4), and (4) kind and amount of load fluid (if an overbalanced load system results in invasion into the formation and the load fluid chemically reacts with formation water, severe formation damage may result). Data derived from a DST may be extremely useful to the explorationist during the openhole evaluation (Fig. 1.5). This data includes: (1) initial bottomhole pressure (IBHP), (2) initial flow pressure (IFP), (3) final flow pressure (FFP), (4) initial shut-in pressure (ISP), (5) final shut-in pressure (FSP), and (6) amount and type of fluid recovered. This data can be used to interpret porosity and permeability of the reservoir, its production characteristics, and its potential size. Pressure and fluid (or gas) data may also be useful to refine correlations of potential reservoir rocks.

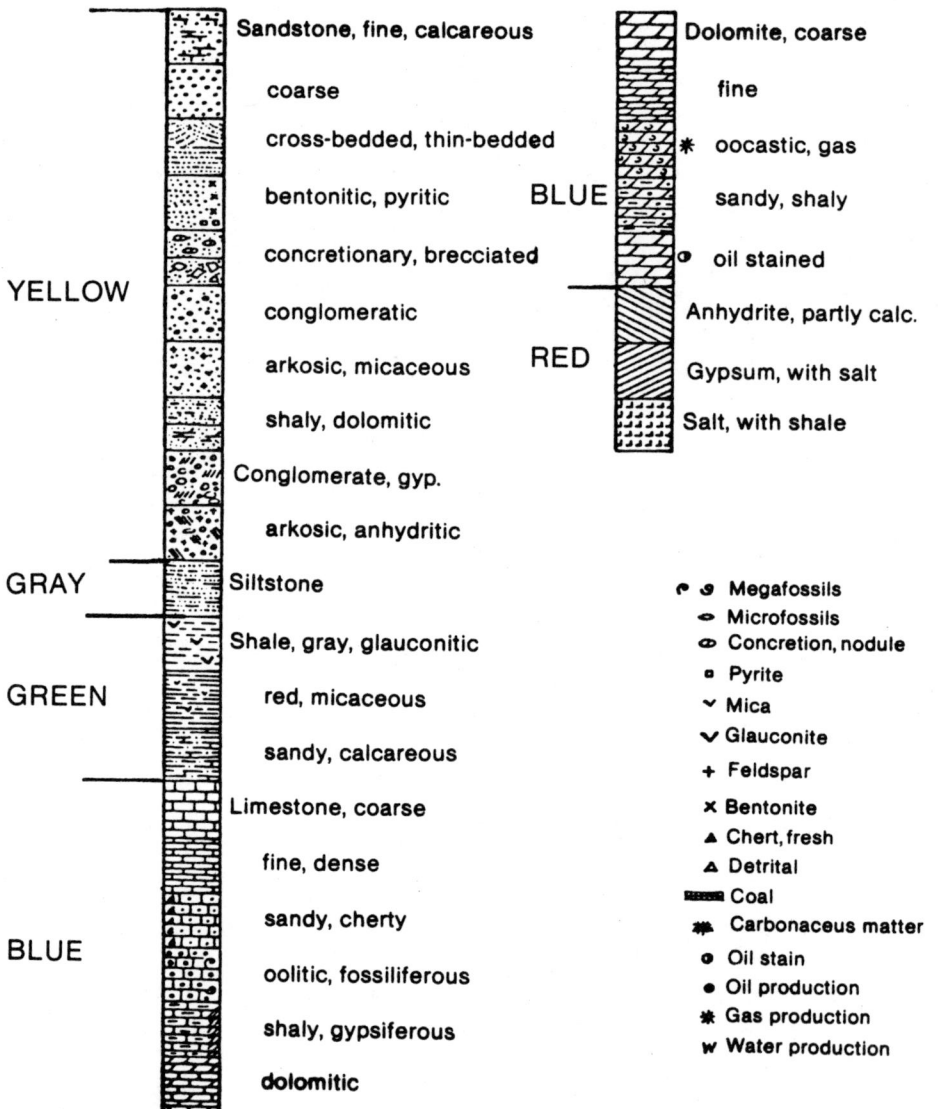

**Figure 1.1:** *Standard lithologic symbols (Haun and LeRoy, 1958).*

CORE LABORATORIES, INC.
Petroleum Reservoir Engineering
DALLAS, TEXAS

## CORE ANALYSIS RESULTS

Company: EXXON COMPANY, U.S.A.  Formation: ___  File: 2106-0349C
Well: #1 SCOTT PAPER CO. G. U. 33  Core Type: DIAMOND  Date Report: 3/14/73
Field: WILDCAT  Drilling Fluid: ___  Analysts: GAL/FJC
County: ESCAMBIA  State: ALABAMA  Elev.: ___  Location: SECTION 33-2N-7E

Lithological Abbreviations

| SAND-SD  SHALE-SH  LIME-LM | DOLOMITE-DOL  CHERT-CH  GYPSUM-GYP | ANHYDRITE-ANHY  CONGLOMERATE-CONG  FOSSILIFEROUS-FOSS | SANDY-SDY  SHALY-SHY  LIMY-LMY | FINE-FN  MEDIUM-MED  COARSE-CSE | CRYSTALLINE-XLN  GRAIN-GRN  GRANULAR-GRNL | BROWN-BRN  GRAY-GY  VUGGY-VGY | FRACTURED-FRAC  LAMINATION-LAM  STYLOLITIC-STY | SLIGHTLY-SL/  VERY-V/  WITH-W/ |

| SAMPLE NUMBER | DEPTH FEET | PERMEABILITY MILLIDARCYS | POROSITY PER CENT | RESIDUAL SATURATION PER CENT PORE ||  INTRP | SAMPLE DESCRIPTION AND REMARKS |
|---|---|---|---|---|---|---|---|
| | | | | OIL | TOTAL WATER | | |

CORE NUMBER 3, 15028-15121; CUT 93 FEET - RECOVERED 28 FEET.
15028-15093 LOST CORE.

|  | 15093-03 |  |  |  | * | ANH SAL | |
|  | 15103-07 |  |  |  | * | ANH | |
| 1 | 07-08 | .03 | 4.6 | 0. | 62.3 (6) | DOL LMY | NO ODR MINERAL |
| 2 | 08-09 | .09 | 8.3 | 1.3 | 59.4 (6) | LM DOL | NO ODR NO FLU |
| 3 | 09-10 | .04 | 8.3 | 0. | 51.3 (6) | DOL LMY | NO ODR NO FLU |
| 4 | 10-11 | .01 | 10.3 | 0. | 56.0 (6) | DOL LMY | NO ODR NO FLU |
| 5 | 11-12 | .08 | 8.7 | 0. | 59.1 (6) | DOL LMY | NO ODR NO FLU |
| 6 | 12-13 | <0.01 | 3.2 | 3.5 | 69.4 (6) | DOL LMY | NO ODR NO FLU |
| 7 | 13-14 | <0.01 | 2.6 | 0. | 67.3 (6) | LM DOL | NO ODR NO FLU |
|  | 14-15 |  |  |  | * | ANH | |
| 8 | 15-16 | <0.01 | 1.9 | 0. | 81.6 (6) | LM | NO ODR NO FLU |
| 9 | 16-17 | <0.01 | 3.5 | 3.2 | 76.1 (6) | LM | NO ODR NO FLU |
| 10 | 17-18 | <0.01 | 1.9 | 0. | 58.6 (6) | LM | NO ODR NO FLU |
| 11 | 18-19 | <0.01 | 2.9 | 0. | 73.9 (6) | LM | NO ODR NO FLU |
| 12 | 19-20 | .01 | 5.2 | 0. | 64.7 (6) | LM | NO ODR NO FLU |
| 13 | 15120-21 | .01 | 5.9 | 1.8 | 75.1 (6) | LM | NO ODR NO FLU |

* NO ANALYSIS BY REQUEST OF CLIENT
(6) LOW PERMEABILITY

**Figure 1.2:** *Typical core analysis report. (Courtesy of Core Laboratories)*

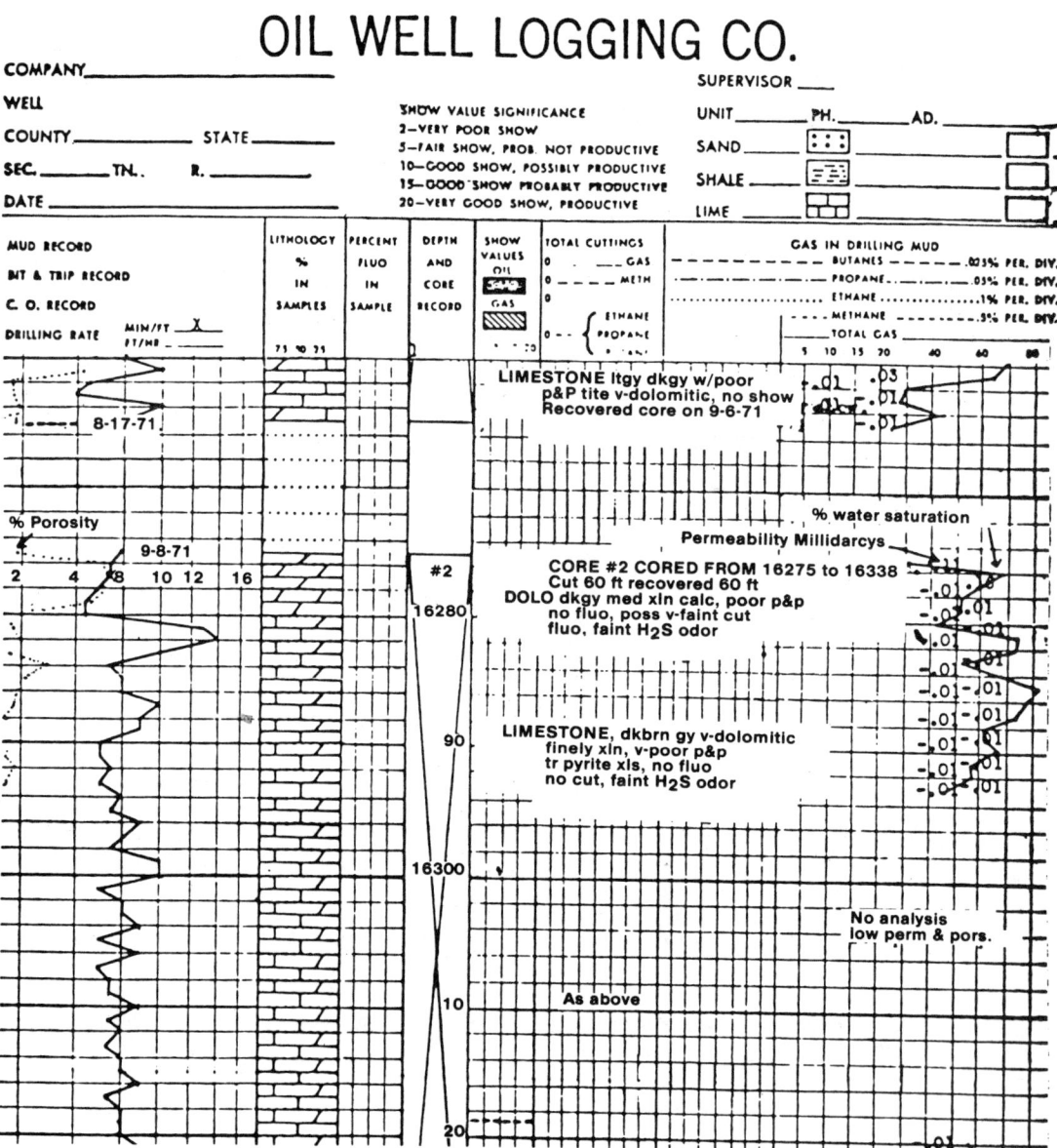

Figure 1.3: *Typical mud log.*

# PROPER USE OF SUBSURFACE DATA

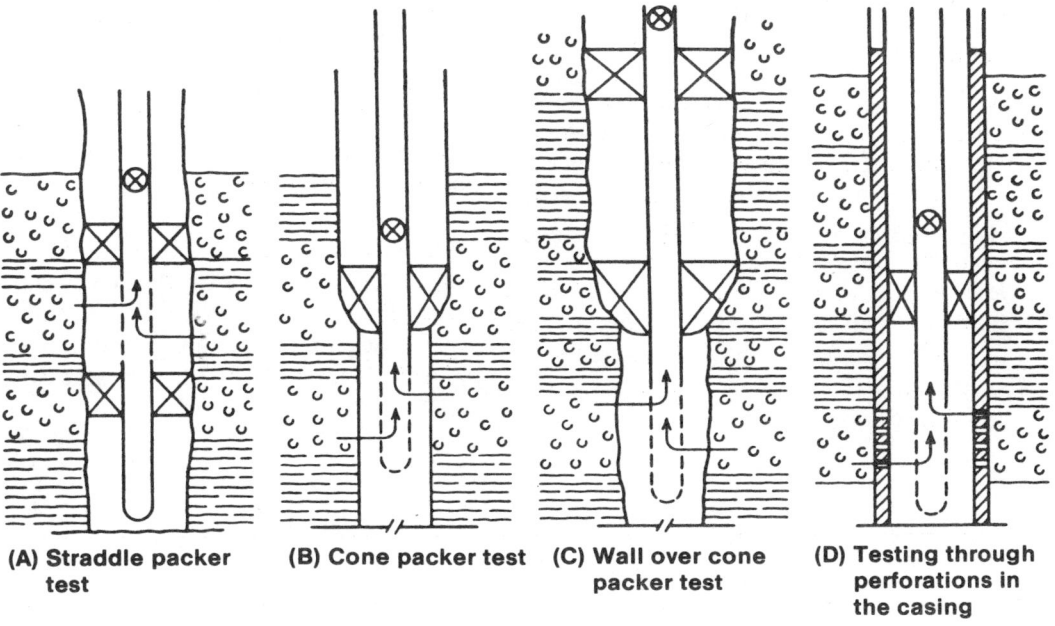

**Figure 1.4:** *Types of drill-stem-test tools (Kirkpatrick, 1954).*

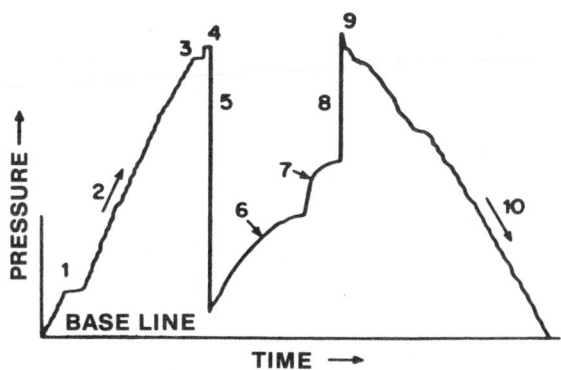

1. Putting water cushion in drill pipe
2. Running in hole
3. Hydrostatic pressure (weight of mud column)
4. Squeeze created by setting packer
5. Opened tester, releasing pressure below packer
6. Flow period, test zone producing into drill pipe
7. Shut-in pressure, tester closed immediately above packer
8. Equalizing hydrostatic pressure below packer
9. Released packer
10. Pulling out of hole

**Figure 1.5:** *Data and procedure for running drill-stem-tests (Kirkpatrick, 1954).*

# MECHANICAL LOGS

## INTRODUCTION

Mechanical logs are by far the most abundant source of subsurface geological data used by explorationists. Variables that affect the quality and character of the most common mechanical logs are depicted in Table 1.1. Each mechanical log responds to one or more specific properties of a rock that are related to its diagenetic and depositional history. These logs may also be variably affected by the construction of a borehole and/or the properties of the drilling fluid.

## INFLUENCE OF AUTHIGENIC MINERALS ON MECHANICAL LOGS

Authigenic minerals, especially clays in a sandstone, cause a decrease in true resistivity. The high surface area of diagenetic clays is conducive to current flow and can cause a zone to appear "wet" when in fact it is either potentially productive or a permeability barrier. A small volume of diagenetic clay can cause a significant reduction in permeability even though it is characterized by a moderate to well developed SP *(Almon and Schultz, 1979).*

**TABLE 1.1**

### VARIABLES THAT AFFECT MODERN MECHANICAL LOGS

| LOG | ROCK PROPERTIES | FLUID PROPERTIES | MECHANICAL |
|---|---|---|---|
| Spontaneous Potential | Lithology<br>Bed thickness<br>Grain size | Gas<br>Salinity of pore fluid | Rate logged<br>Scale changes<br>Salinity of drilling mud |
| Resistivity | Porosity & permeability<br>Bed thickness | Hydrocarbon type<br>Salinity of formation waters | Rate logged<br>Borehole size |
| Gamma-ray | Lithology (particularly clay minerals)<br>Grain size | None | Borehole size |
| Neutron | Porosity | Gas (attenuates) | Borehole size<br>Neutron source |
| Sonic | Lithology<br>Porosity | Gas (attenuates) | Borehole size<br>Position of tool in hole<br>Rate logged |
| Density | Clay content | Oil has little effect since average density close to unity. Gas may produce high porosity determinations. | Borehole size (negligible for hole less than 10" in diameter)<br>Logging speed must be constant |
| Dip Meters* | Lithology<br>Bedding (particularly cross bedding) | Little | Depth of invasion |

*Early logs were 3-SP curves; therefore, they were affected by the same variables as SP logs.

Gamma ray logs (GR) are affected adversely by high percentages of U, Th, and K. K may be common in diagenetic clays derived from potassium feldspar. It may be abundant in illite and mixed layered illite/smectite. This is usually not abundant in authigenic clays but may occur in arkosic sandstones. U is absorbed on clays at cation-exchange-sites.

Like GR logs, neutron logs can be attenuated by authigenic clays. Neutron logs essentially measure hydrogen. Consequently, large amounts of hydrogen bearing minerals, such as gypsum, zeolites, and authigenic clays will attenuate a Neutron log *(Pitman, 1980)*.

Sonic logs are also influenced by authigenic clays. This family of mechanical logs measure the interval transit time of compressional sound waves. Because some authigenic minerals, notably dolomite and chlorite, have anomalously high or low interval transit times, porosity measurements based on the sonic log may be incorrect.

## CLASSIFICATION OF SP AND GR CURVES

Assuming that the SP and GR curves are not influenced by mechanical and/or fluid properties, they are typically continuous grain-size profiles through clastic rocks *(Allen, 1975)*. Six hypothetical SP (or GR) curves are recognized (Fig. 1.6). The smooth curve with abrupt upper and lower contacts (Fig. 1.6A) indicates that the sand body is homogeneous with no significant vertical change in grain size. The serrated curve (Fig. 1.6B) depicts interbedded sand and shale, the sands being similar in grain size. Figure 1.6C represents a smooth, funnel-shaped curve that indicates increasing grain size upward. The basal contact of the sand is gradational with the shale but its upper contact is abrupt. In contrast, Figure 1.6D is serrated and funnel-shaped; this indicates that the sands are interbedded with shales and that the sands are increasing in grain size upward. The smooth, bell-shaped curve (Fig. 1.6E) depicts a sand that is characterized by decreasing grain size upward, a sharp basal contact, and gradational upper contact. In contrast, the serrated, bell-shaped curve (Fig. 1.6F) indicates interbedded sands and shales and the sand beds become finer-grained upward *(Jageler and Matuszak, 1972)*.

## NONSEISMIC GEOPHYSICAL DATA

Only about three percent of the exploration geophysical budget is invested in the acquisition of magnetic and gravity data. Little information regarding subsurface stratigraphy can be gained from this data; therefore, no further discussion is warranted.

## SEISMIC DATA

### INTRODUCTION

The ultimate goal of the explorationist is to develop a direct hydrocarbon locating tool that will delineate commercial hydrocarbon-bearing traps prior to drilling a wildcat. Such a tool may never be developed, and until it is, explorationists must use all the technology possible to indirectly locate areas that are favorable for the commercial production of hydrocarbons.

In order to interpret the variables that control the generation, expulsion, and trapping of hydrocarbons, explorationists measure and record mineral density, fluid density, porosity, and fluid saturation of potential reservoirs. The explorationists who specialize in geology measure and record resistivity, self-potential, radioactivity, hydrogen content, and electron density, and display this data on wireline logs. Geophysicists measure and record velocity data from which they can determine bulk density, acoustic impedance, and reflection coefficients. These measurements are graphically displayed by a seismic tract. The seismic record has excellent horizontal resolution, but vertical resolution varies with depth. In contrast, the wireline log has excellent vertical resolution, but only implies a lateral distribution of strata within an unconformity-bounded sequence.

The geophysicist and the geologist rely on their respective graphic displays to make three categories of interpretation; namely rock, fluid, and geometry. Rock interpretations include lithology type and reservoir potential. The determination of the presence of oil, gas, and/or water is of prime concern to all explorationists. Geometric interpretations include shape of potential reservoir bodies, structure, facies, bed thickness, and the sequence of lithologies. Finally, the explorationist must be able to place these variables into a stratigraphic

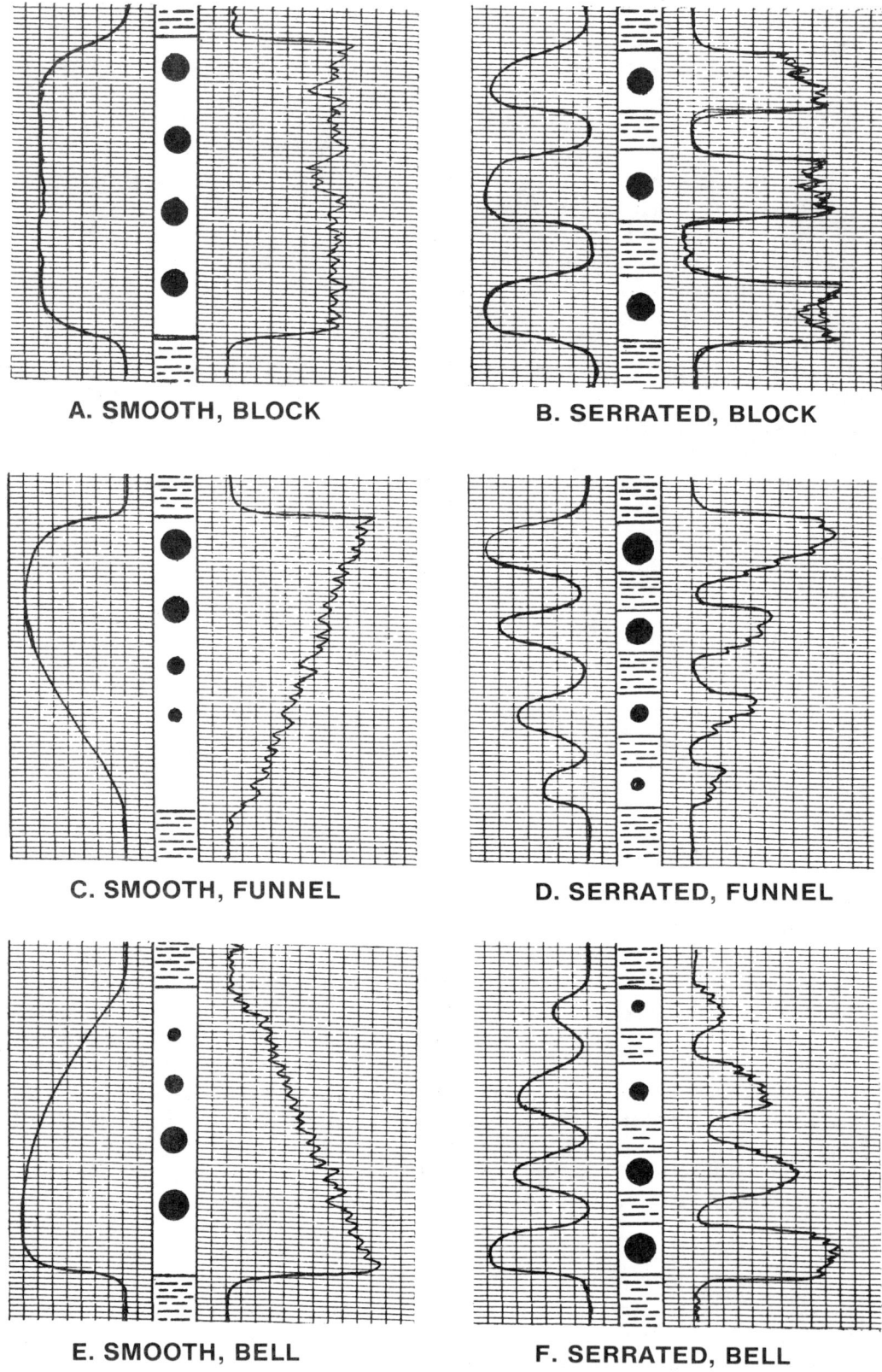

**Figure 1.6:** *Typical end members of SP and GR curves. The hypothetical SP or GR curve is to the left and its counterpart to the right is the hypothetical resistivity curve.*

# PROPER USE OF SUBSURFACE DATA

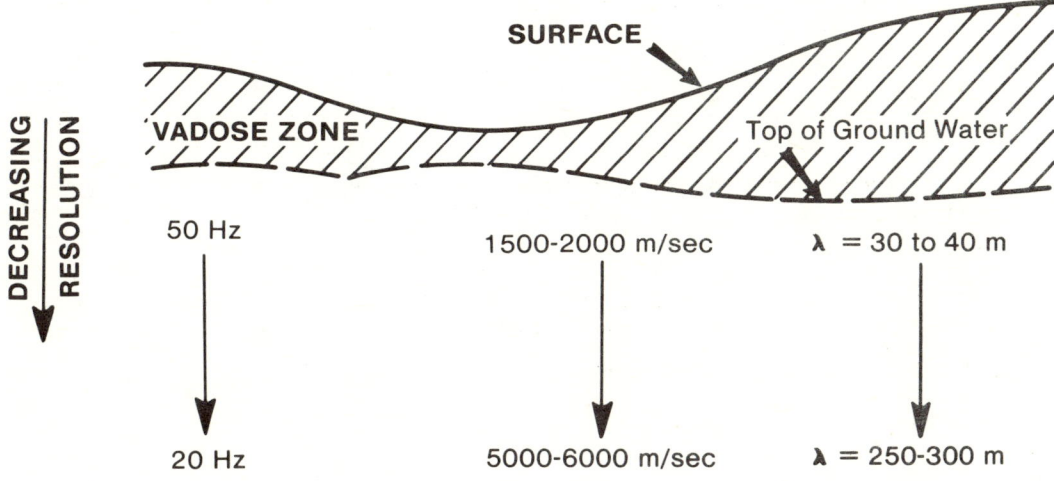

**Figure 1.7:** *Depth of weathering profile, thickness of vadose zone, rock type at surface and depth, and thickness of reflector influences resolution.*

framework that will permit the interpretation of the relative timing of generation, expulsion, migration, and entrapment of hydrocarbons.

## LIMITATIONS OF SEISMIC DATA

**Hazards to Consider** Many modern regional seismic sections so strongly resemble a well-correlated subsurface geologic cross section that many explorationists, particularly geologists, are tempted to interpret them in the same manner as they would interpret a cross section. This can be a very dangerous mistake that can lead to erroneous interpretations and the drilling of needless dry holes. Some of the hazards of interpreting seismic data include: noise, wavelength, reflections, and wavefronts.

**Noise** *Noise* can be most simply defined as any unwanted signal that represents a disturbance which does not indicate any geologic source. Noise includes microseisms, tape-modulation, harmonic distortions, shot-generated noise, etc. Noise can be caused by geologic differences in the site of the energy source. These differences include rock type at the surface, weathering profile, thickness of the vadose zone, and karst topography (Fig. 1.7). The same geologic variables at the site of each geophone can cause noise. Sheriff *(1977)* presents a review on the statistical techniques to attenuate noise in a seismic section.

**Wavelength** Wavelength is to a geophysicist as spacing of electrodes is to a petrophysicist. The geologic features to be mapped must be larger than the wavelength; wavelength is related to velocity multiplied by the period of velocity divided by frequency. Because velocity increases with depth, wavelength increases with depth (Fig. 1.7). The only variable that can be controlled is frequency. Resolution can be improved by either recording and processing a broader band of frequencies (Fig. 1.8) or by using a higher frequency source (Fig. 1.9).

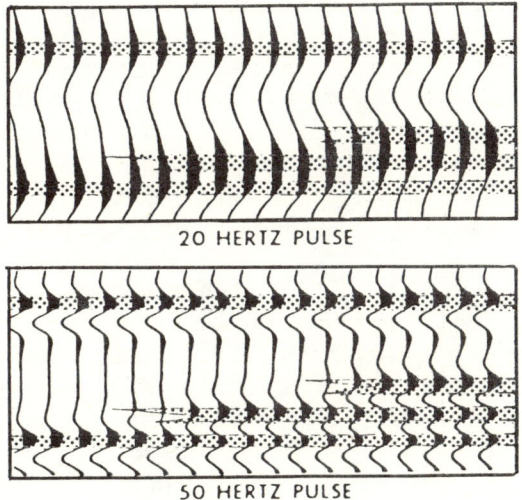

**Figure 1.8:** *Resolving ability of 50 Hz and 20 Hz pulse (Vail and others, 1977).*

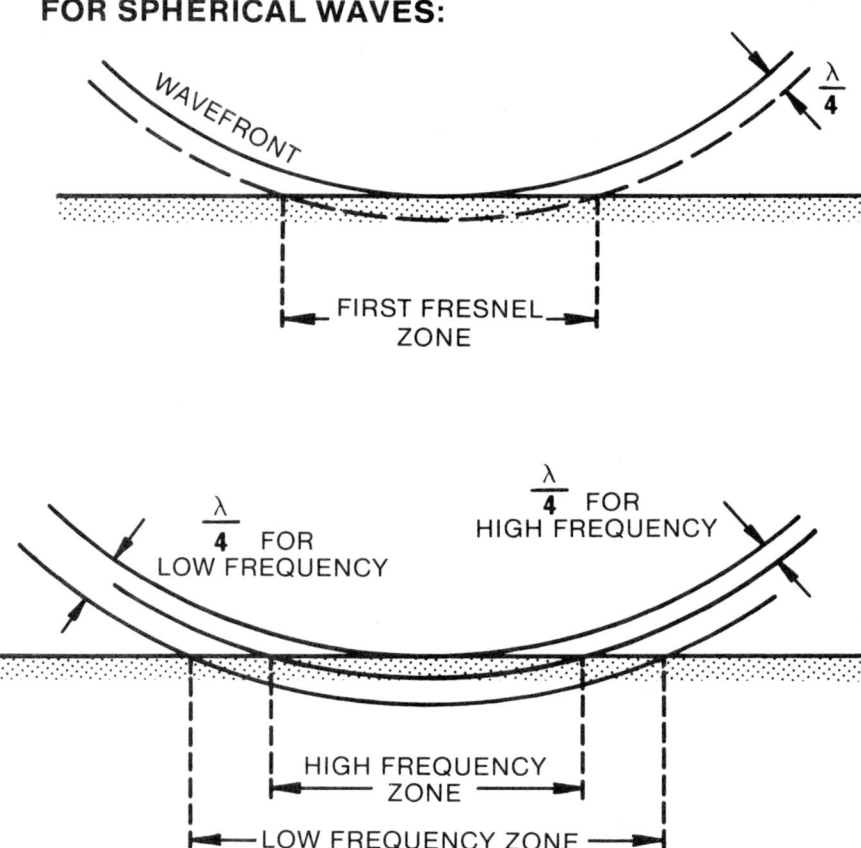

**Figure 1.9:** *Seismic waves should be viewed as wavefronts rather than as thin lines of energy (Sheriff, 1977).*

**Reflections** Most reflections represent the interference composites that result from four or more velocity interfaces. Reflection constancy generally is due to bedding constancy which changes gradually in a horizontal direction *(Sherrif, 1977)*. Consequently, a single reflection commonly depicts several reflecting horizons, especially in geologic provinces that are characterized by interbedded sand and shale lithologies. In these type geologic situations, the reflections change their character slowly until they gradually "die out."

The character of a reflection, therefore, is a clue to the nature of the reflecting horizon(s). If the character is constant, it commonly represents a single reflector; if it is ever-changing, it probably represents a composite of reflectors.

**Wavefronts** A common mistake made by many explorationists is the viewpoint that seismic waves are thin vectors of energy that travel along a ray path, and that each energy vector reacts independently of the adjacent energy vector. This erroneous viewpoint leads to the idea that a reflection represents only a point on the reflector. In practice, the reflection represents a substantial surface.

It is more accurate to consider a seismic wave as a wavefront that expands in size and decreases in energy as it moves away from the point of origin. As the wavefront reaches a velocity or density interface, part of it will be reflected (Fig. 1.9). The point of contact that produces the reflection is called the *first Fresnal Zone*. Thus reflection represents a two-dimensional area on the reflector rather than a point. This can be a hazard in attempting to precisely map a fault, but this is an advantage to the explorationists who are attempting to interpret stratigraphy because the overlapping nature of the reflections adds continuity.

## ORIGIN OF REFLECTIONS

Physical surfaces in rocks that produce reflections are primarily bedding surfaces and unconformities that are characterized by velocity contrasts *(Vail and others, 1977)*. Both surfaces subdivide geologic time into two parts; namely, post-surface and pre-surface. Therefore, seismic reflections can be used to interpret (1) post-depositional structural deformations and thickness changes, (2) geologic time correlations, and (3) delineation of depositional units. From these interpretations, the depositional environment, paleobathymetry, burial history, and geologic history can be inferred. To date, however, rock type cannot always be determined directly from reflection correlation patterns *(Vail and others, 1977)*.

Two types of physical surfaces occur in the geologic record: *bedding surfaces* and *unconformities*. Each type can produce seismic reflections if sufficient velocity and/or density contrasts occur along the physical surface. Bedding planes (or surfaces) represent a change in the depositional sequence that can be caused by a change in the material being deposited and/or a change in the depositional energy. Bedding planes also can be caused by periods of nondeposition and cross lithologic units. For example, bedding planes in sands of a channel-mouth-bar deposit extend into the shales deposited in the pro-delta facies equivalent to the bar. However, as stated previously, only those bedding surfaces that are characterized by significant velocity or density contrasts will produce seismic reflections.

Unconformities are surfaces that represent a break in the continuity of sedimentation. Major unconformities are represented by periods of erosion; minor unconformities may represent periods of nondeposition. Both types of unconformities separate younger rocks from older rocks. Unconformities, if they are characterized by velocity or density contrasts, will produce seismic reflections.

## TIME-STRATIGRAPHIC SIGNIFICANCE OF SEISMIC REFLECTIONS

Many primary seismic reflections are generated by stratal interfaces that represent time surfaces rather than by boundaries that correspond to units (formation, members, etc.) which are defined on the basis of lithologic content. Many lithostratigraphic units cross time surfaces because they are frequently composited from segmented outcrop and/or subsurface data (principally boreholes) where bedding planes cannot be continuously traced *(Vail and others, 1977)*.

Vail and others *(1977)* conclusively demonstrated the time-stratigraphic implications of primary seismic reflections in several areas where closely-spaced well data can be compared to seismic lines. One example from their work is presented here for evaluation.

**Stratigraphic Framework** Figure 1.10 is a stratigraphic cross section located along the crest of a plunging anticline (Fig. 1.11). Correlations were focused within the shale intervals because these low energy deposits are more aerially extensive. The datum is a regional electric log marker. A number of resistivity curve markers were correlated from well 6 to well 1; however, only a unconformity and markers 8, 10, and 15 are illustrated.

The interval from the unconformity to marker 8 is dominated by sandstone that thins significantly from well 6 toward well 1 as it onlaps the erosional surface. It is obvious that the sand in well 6 is older than the sand in well 1. Therefore, it can be described as a time-transgressive sand. Shale represents the interval between markers 8 and 10 in well 6. The interval becomes increasingly sandy toward well 1. Little horizontal change occurs in the interval between markers 10 and 15 in all of the wells. Bedding is relatively uniform.

**Velocity Relations** In order to evaluate the seismic response to the geologic relationships illustrated on the stratigraphic cross section, it is necessary to determine lateral velocity variations *(Vail and others, 1977)*. Continuous velocity logs are available for wells 5 and 1. The stratigraphic framework illustrated on Figure 1.11 was used to correlate the CVL data in wells 5 and 1. The dashed line on Figure 1.12 is a lithologic correlation that represents the top of sand in both wells. There is no physical surface that corresponds with the dashed line.

The velocity derived from the CVL data varies with lithology *(Vail and others, 1977)*. Lower velocities correspond, in general, with

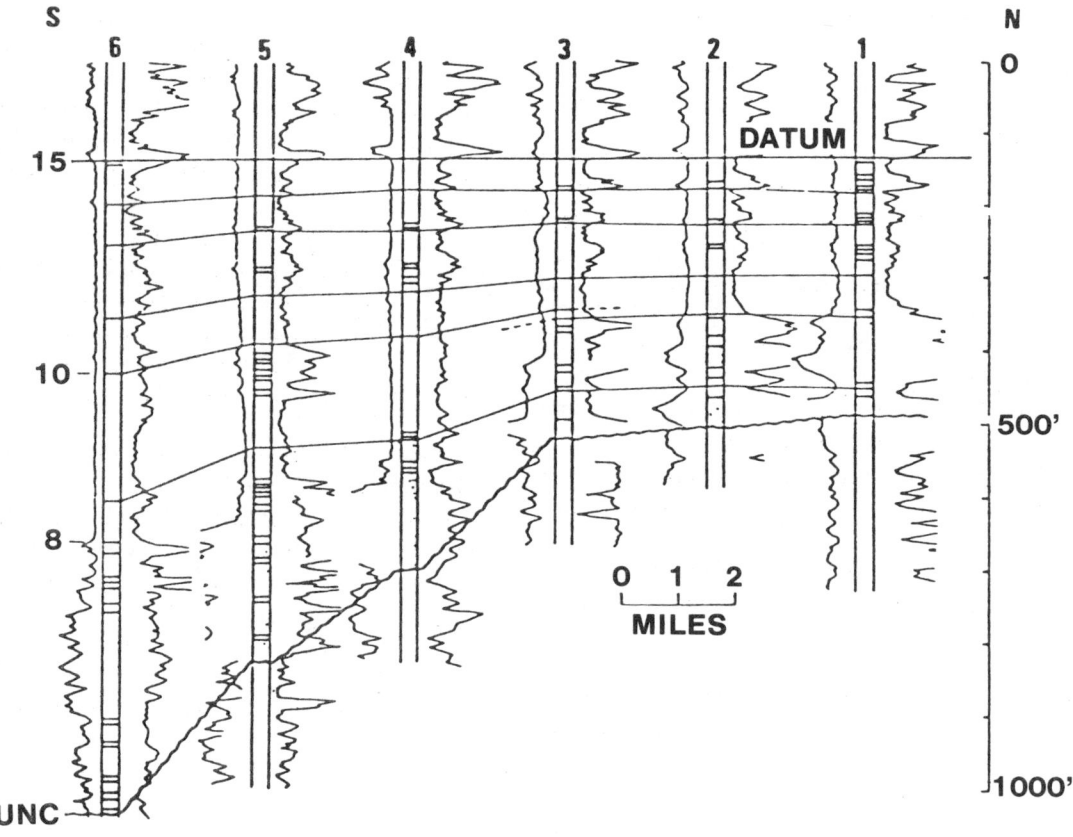

**Figure 1.10:** *Stratigraphic cross section showing major electric log correlations. Tertiary example from South America (Vail and others, 1977).*

shales and range from about 6,600 to 8,200 ft/sec. Medium velocities range from 8,200 to 10,000 ft/sec. and correspond to siltstone. Sandstones represent the highest velocities (10,000 to 12,700 ft/sec.). The gradual decrease in velocity from well 5 to well 1 is attributed to a corresponding increase in the percentage of shale in the total interval, shallower depths, and lower compaction percentages.

A synthetic seismic section was generated by a 20-Hz sine-wave input pulse (Fig. 1.13). Velocity data was obtained from Figure 1.12. The synthetic section is a normal frequency presentation and shows the time-stratigraphic relation of the seismic reflections and the pattern correlation lines. The unconformity, and correlation lines 8, 10, and 15 have been drawn on the section as are wells 6 through 1. In well 6, correlation line 8 corresponds to the top of the sandstone and the corresponding reflection onlaps and terminates against the unconformity southwest of well 1. Moreover, correlation line 10 is a resistivity kick at the top of well 1 that occurs 250 feet above the top of the sand in well 6. Although the reflection changes character from well 1 to 6, it does parallel the correlation marker. No diagonal reflector that corresponds to the dashed line on Figure 1.12 occurs because there is no continuous physical surface that corresponds to this line.

Perhaps a review of a fundamental concept of geology is appropriate to help clarify the above. When geologists draw facies lines on a cross section, it is often forgotten exactly what the "lines" represent because most cross sections have great vertical exaggerations. If the sand in well 6 was considered to be a facies of the sand in well 1, a jagged facies line would be drawn from the top of the sand in well 1 to the top of the sand in well 6 (Fig. 1.10). However, Figure 1.10 has a vertical exaggeration of nearly 280:1 and each of the "jags" of the

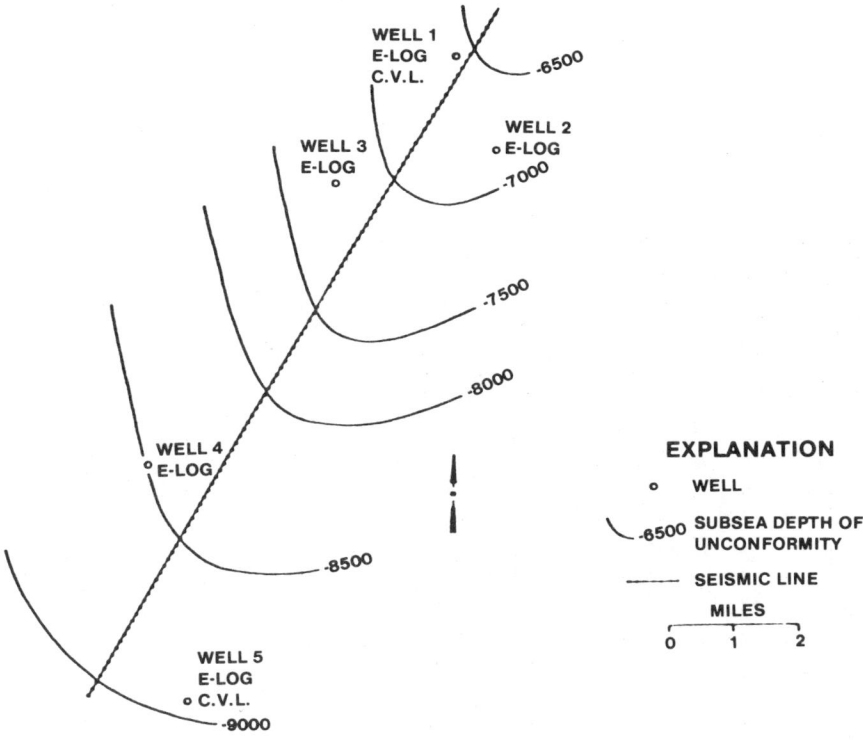

**Figure 1.11:** *Index map showing structure on the unconformity and location of wells used in Figure 4 (Vail and others, 1977).*

facies line would represent thousands of feet laterally. Because the reflection does not represent a point, but rather a tangent of a wave, the reflection follows the bedding plane (tongue of the facies) into the facies equivalent (in this case, shale).

**Seismic Section** Figure 1.14 is an outdated but adequate single-fold variable density film magnetic-tape seismic section. The line is located near wells 1 through 5 (Fig. 1.11). The location of each well is plotted at the top of the section. Plotted on the section is the unconformity and correlation lines 8, 10, and 15. The reflection just beneath marker 8 corresponds to the top of the sand in wells 6 through 4. Note that this reflection terminates against the unconformity just southwest of well 3. The reflection just beneath correlation horizon 10 corresponds to the top of the sandstone in well 1. The reflection in well 1 diminishes southwestward. It also decreases in amplitude. There is no reflector that corresponds to the dashed line on Figure 1.11. In other words, the reflectors in the outdated single-fold VDF seismic line correspond to bedding planes (time surfaces) rather than facies (lithostratigraphic units).

**Synthetic Seismograms** Synthetic seismograms *(Peterson and others, 1955 and Wuenshel, 1960)* and their comparison to both local and regional geologic models support the thesis that reflections commonly follow depositional time surfaces rather than facies boundaries. For example, Figure 1.13 illustrates reflections from a series of sandstone beds that are shaling out in an easterly direction. Because seismic traces are commonly 50 to 250 feet apart, the adjacent traces usually record the same reflecting sequence. Therefore, since the horizontal continuity is constant, the reflections will mainly parallel the bedding planes. On the other hand, a geologist working with wellbore data will likely map only the top of each sandstone because the wells are generally much farther apart. Only in areas where bottomhole pressures or other reservoir data is available will the geologist be able to differentiate the individual sands. Therefore, a top-of-

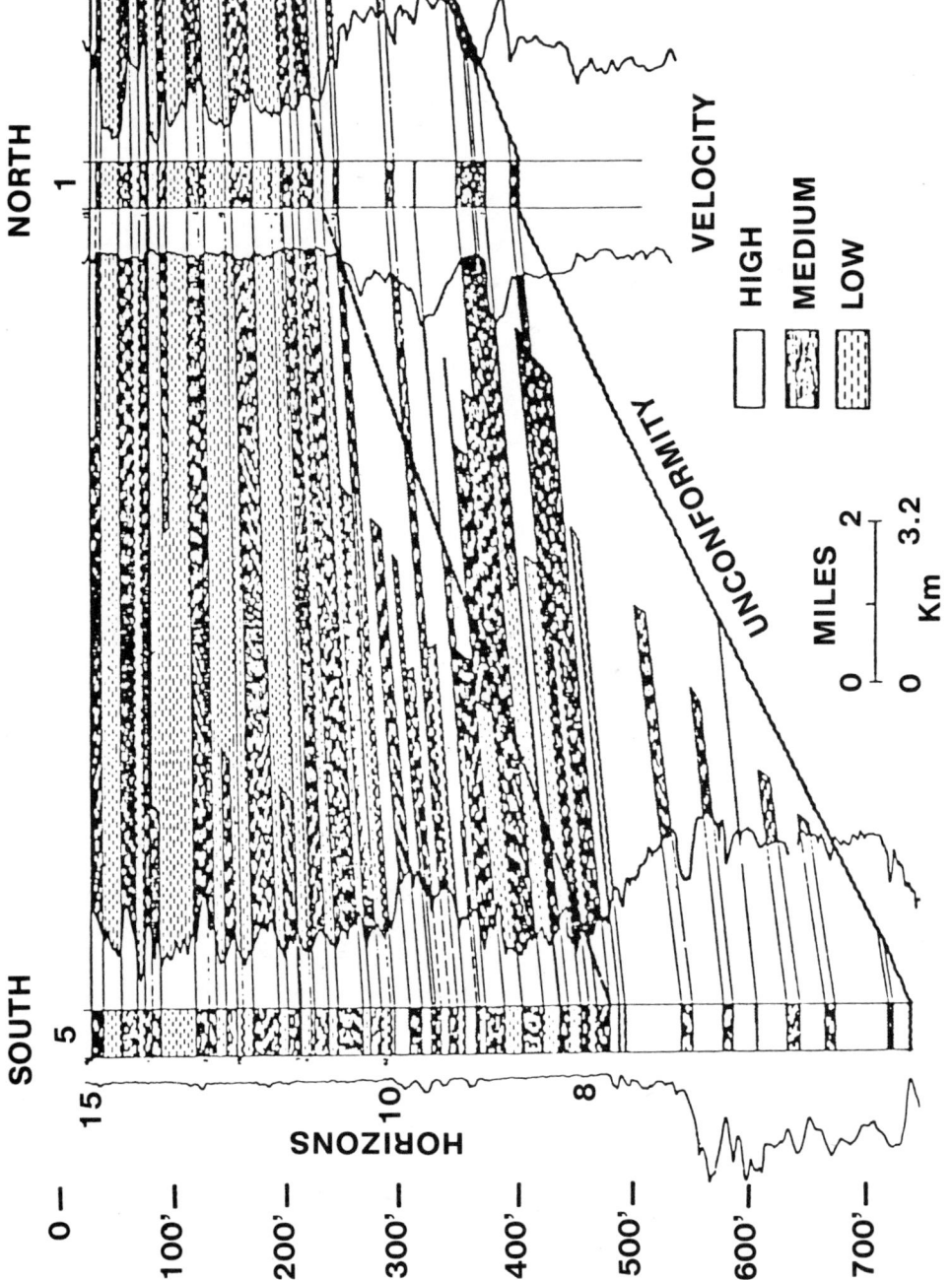

**Figure 1.12:** *Velocity cross section based on correlations illustrated in Figure 4 (Vail and others, 1977).*

sands map based on mechanical log data may cross time lines, and the resultant map will not be useful for depicting structure.

## SCOUT TICKETS

Early scout tickets may contain gross errors. Modern scout tickets generally are accurate as to location of the well, date spudded, date completed, what logs were run, status completion status, and if completed, where the well was perforated, and what was its initial production. Core and other test data are also normally accurate. Data that is typically not accurate include formation tops and structural data, such as where a fault occurs in a well, etc. A typical scout ticket is illustrated in Figure 1.15.

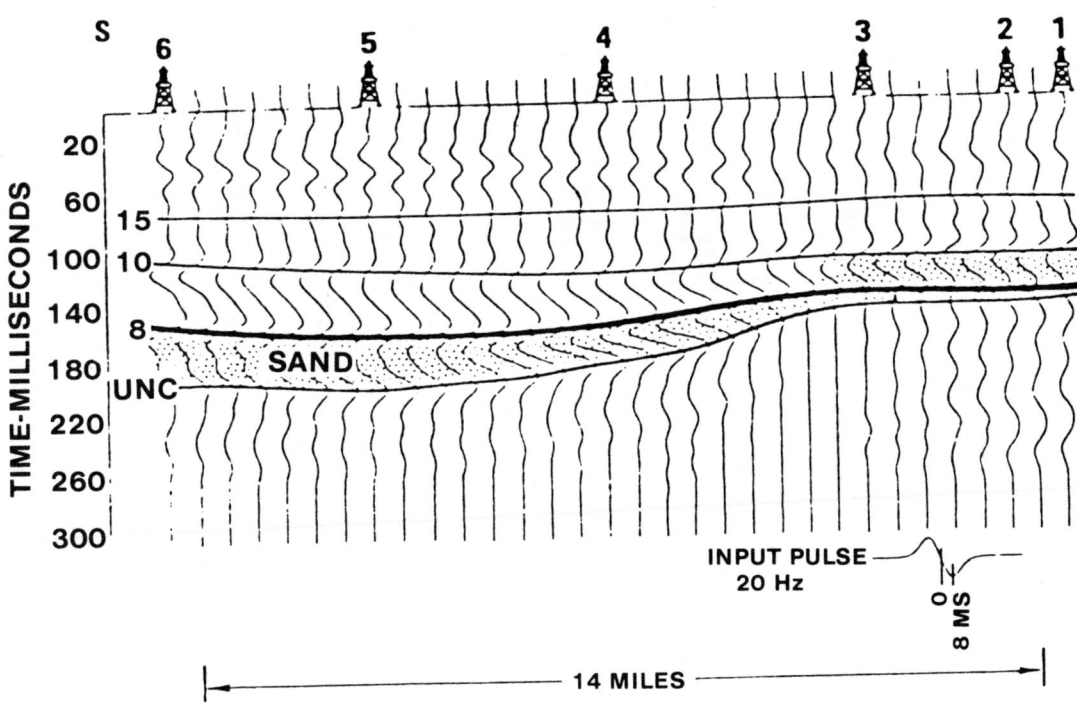

**Figure 1.13:** *Synthetic seismic section based on 20 Hz sine wave that was derived from data illustrated on Figures 5 and 6 (Vail and others, 1977).*

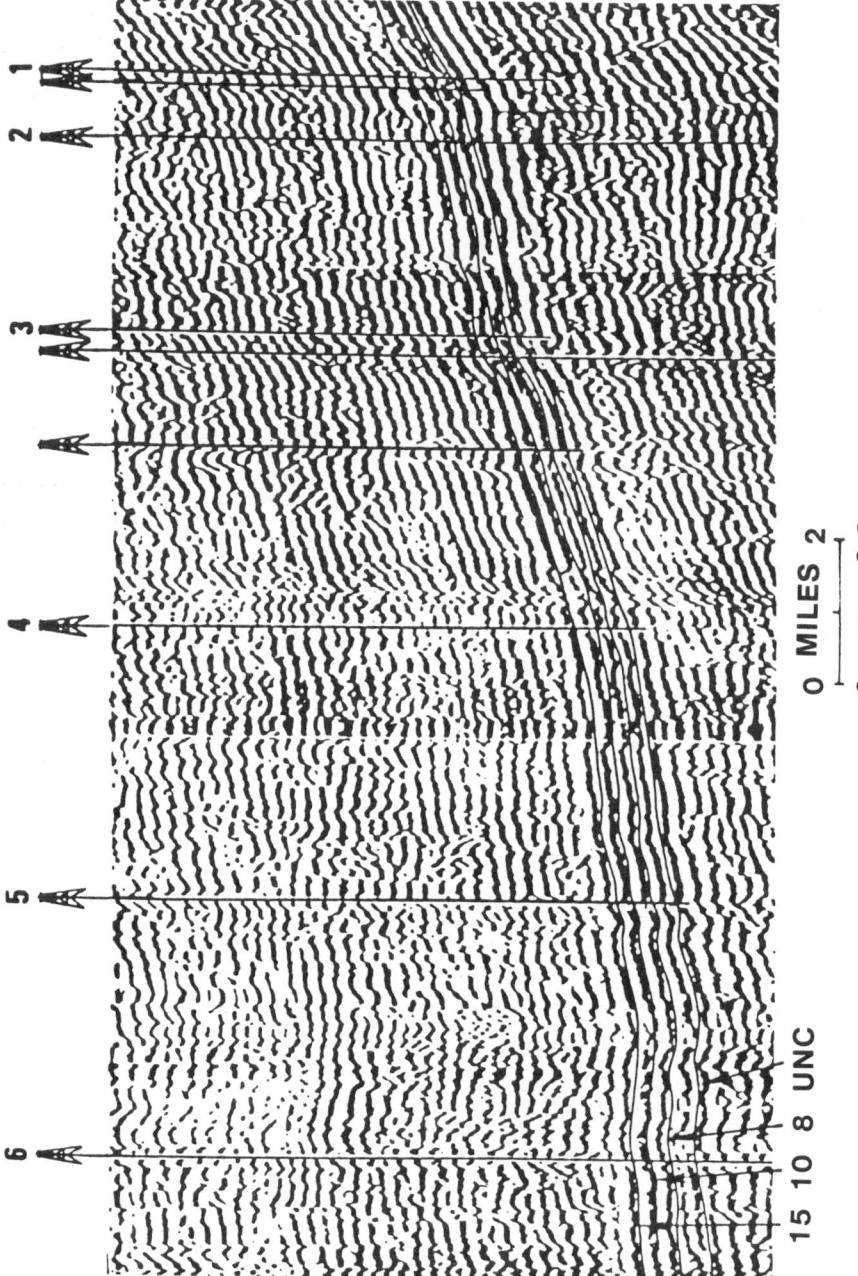

**Figure 1.14:** *Reflections follow time surfaces rather than facies (Single fold VDF seismic section in vicinity of Figure 4, Vail and others, 1977).*

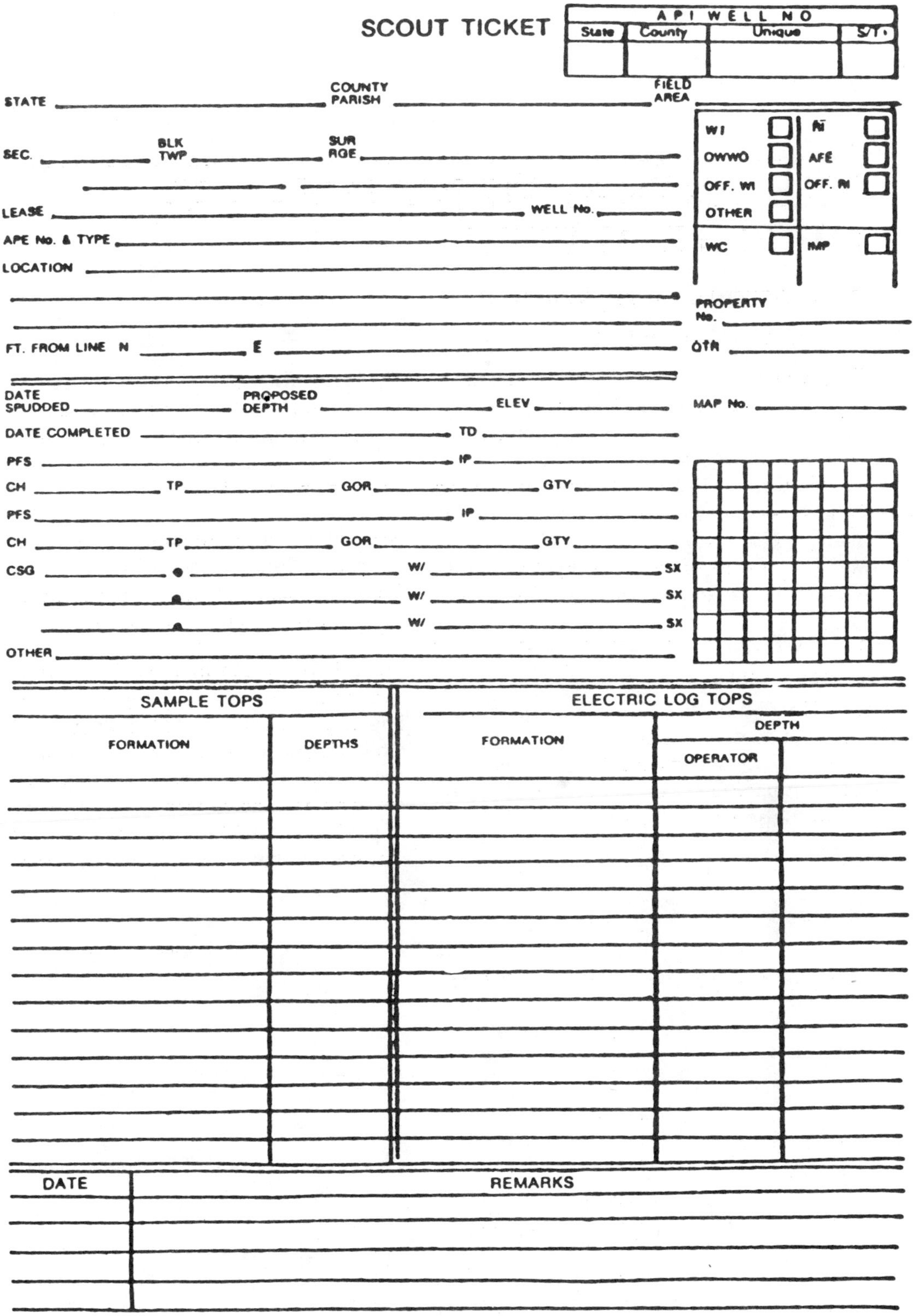

# PROPER USE OF SUBSURFACE DATA

**Figure 1.15:** *Typical form for a scout ticket.*

# 2

# PRE-ENTRAPMENT HISTORY OF PETROLEUM

## COMPOSITION OF PETROLEUM

### GENERAL COMPOSITION

Petroleum is a naturally occurring, complex mixture of hydrocarbons which may be either gas, liquid, or solid depending upon its own unique composition and the pressure and temperature at which it is confined. Petroleum usually contains between 11-15% hydrogen and 82-87% carbon. It generally contains various amounts of paraffins, cycloparaffins, and aromatics, and one or more impurities.

### STRUCTURE OF PETROLEUM

**Paraffins** The general formula for paraffins is $C_nH_{2n+2}$. Nearly all crudes contain some paraffins, particularly the more volatile (low boiling point) constituents. Some of the more common paraffins and their chemical formulas are illustrated in Figure 2.1.

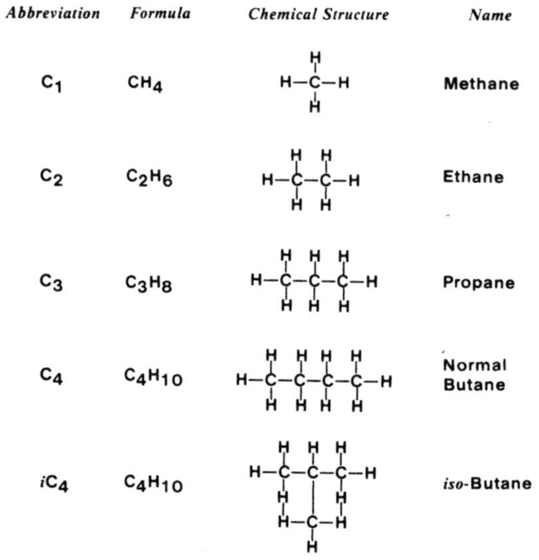

**Cycloparaffins** Cycloparaffins (napthenes) are represented by the general formula $C_nH_{2n}$ and have a ring structure. Some of the more common cycloparaffins include: cyclopropane and cyclobutane (Fig. 2.2).

**Aromatics** The aromatics or benzenes are represented by the general formula $C_nH_{2n-6}$. They are chemically active and complex and the simplest member is benzene.

**Impurities in Petroleum** Petroleum may contain a wide variety, as well as amounts, of impurities. Sulphur and $H_2S$ are perhaps the most widely feared impurities because they produce sour petroleum which is not only an environmental hazard but also adds to the cost of production and refining. $CO_2$, N, and $O_2$, are other common impurities in oil and gas.

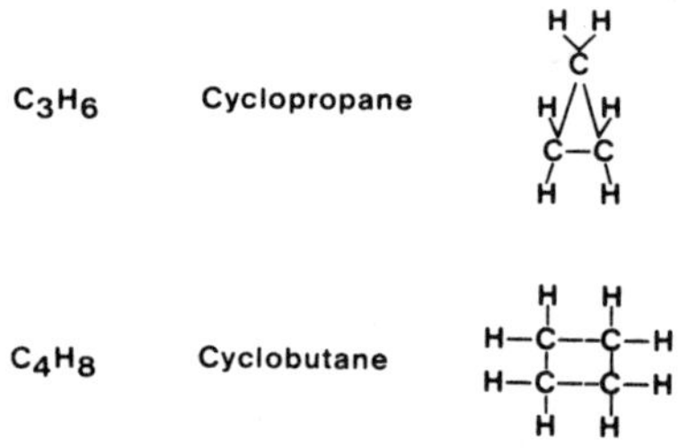

| Abbreviation | Formula | Chemical Structure | Name |
|---|---|---|---|
| $C_1$ | $CH_4$ | H–C–H (with H above and below) | Methane |
| $C_2$ | $C_2H_6$ | H–C–C–H | Ethane |
| $C_3$ | $C_3H_8$ | H–C–C–C–H | Propane |
| $C_4$ | $C_4H_{10}$ | H–C–C–C–C–H | Normal Butane |
| $iC_4$ | $C_4H_{10}$ | branched structure | iso-Butane |

**Figure 2.1:** *Structure of Paraffins (Gatlin, 1960).*

| | | |
|---|---|---|
| $C_3H_6$ | Cyclopropane | (ring structure) |
| $C_4H_8$ | Cyclobutane | (ring structure) |

**Figure 2.2:** *Structure of Cycloparaffins (Gatlin, 1960).*

### TYPES OF CRUDES

The composition of petroleum is the basis for classifying it into three types — paraffin, asphalt, and mixed. Paraffin base petroleum is composed chiefly of the isomers of paraffin; when distilled, it leaves a residue of wax. The crudes from Pennsylvania and Appalachia are generally paraffin base.

Asphalt base crudes are composed chiefly of cycloparaffins (napthenes). When distilled, these crudes leave a solid asphalt residue. Most of the California and Gulf Coast crudes are asphalt base petroleums. The mixed-base crudes are a combination of the paraffin and asphalt base petroleums. The mid-continent oils are generally mixed-based.

## PRECURSORS OF PETROLEUM
### EVIDENCE FOR ORGANIC ORIGIN

**Major and Minor Constituents** The major constituents of petroleum, carbon, and hydrogen require geochemists to conclude that petroleum is derived from plant and animal material. However, the best evidence for a biologic source for petroleum is derived from its minor constituents — porphyrin pigments and nitrogen. Porphyrin pigments are derived from the red coloring matter of blood (hemin) as well as the green coloring (chlorophyll) of plants. It has been demonstrated that vegetable porphyrins ($C_{32}H_{35}N_4COOH$ and $C_{32}H_{36}N_4$) derived from chlorophyll are more plentiful than animal porphyrins ($C_{32}H_{34}N_4$ and $C_{32}H_{36}COOH$) derived from hemin *(Thriebs, 1936)*. Nitrogen is the essential component of amino acids ($CH_2NH_2COOH$). The ratio of organic matter is 1.6 times greater than carbon and 24 times greater than nitrogen. The continuous chain of occurrence of nitrogen from living matter to organic matter in sediments implies an organic source material.

**Optical Activity** Most petroleums are optically active; that is, they can rotate the plane of polarization of light. This characteristic has been attributed to the cholesterin of animal material and phytosterin of plants in petroleum. The degree of optical activity is dependent upon the molecular weight fractions of the petroleum. Maximum activity occurs at a molecular weight of about 400 *(Oakwood et al., 1952)*.

### AQUATIC PLANTS

Aquatic plants are perhaps the most likely group of plants which may be a source of petroleum. Algae, marine fungi, and bacteria have the highest potential. These plants are most abundant in the zone of photosynthesis. Benthonic (bottom dwellers) are restricted to shallow-clear water and are, therefore, concentrated in coastal waters. Planktonic forms (floaters) are most prolific in the upper few meters of the ocean. Coastal waters contain about 50 times more plants than the open ocean.

Marine algae offer some of the most promising source material for petroleum *(Hackford, 1932)*. Blue and green algae are the most common forms. These algae can convert $CO_2$, $H_2O$ and sunlight into $O_2$ and carbohydrates. Diatoms, a siliceous shelled or frustuled algae, may be one of the principal sources for petroleum *(Anderson, 1926)*. It has been estimated that from 5% to 50% of the volume of diatoms consists of oil globules *(Mann, 1916)*. Organic matter derived from Mangrove roots and *Thalassia* have been recognized in modern carbonate sediments *(Sawyer, 1980)*.

### ANIMALS

Mega and micro fauna contain 2% to 40% organic matter. Some of the more common organisms considered to be important source material include corals (4-8% O.M.), mollusks (6-20% O.M.), and fish (35-80% O.M.). Microorganisms, especially foraminifera, may be significant contributors to the source material. The foraminifera are most abundant in shallow waters, but are also common in open marine waters.

### NONMARINE SOURCE MATERIAL

Terrestrial derived plant and animal material may be transported into a marine environment by streams. Modern streams contain from 3.25% (Danube River) to 59.9% (Uruguay) organic material *(Clark, 1924)*. Humus substances are the dominant organic material in modern rivers. These include humic acid ($C_{20}H_{10}O_6$), geic acid ($C_{20}H_{12}O_7$), and ulmic acid ($C_{20}H_{14}O_6$). This material is derived by the slow decomposition of the lignins in peat and is most abundant in soils charged with decaying vegetation.

**Influence of Source Material and Age on Type of Hydrocarbons** There is evidence that the kind of source material and its environment of preservation influences the type of hydrocarbon to be generated. Paraffinic crudes are typically produced from sands that are associated with shales that accumulated in nonmarine or less-than-normal marine salinities. Typically these crudes are Devonian to Pliocene in age and they are low in sulfur. Asphaltic crudes are produced from sands and carbonates that are associated with marine deposited source rocks. They range from Cambrian to Pliocene and contain variable amounts of sulfur. Mixed based crudes contain nonmarine and marine source material. They range in age from Devonian to Pliocene and as expected, contain variable amounts of sulfur.

Data accumulated by CITCO Research, suggest that source material influences the generation of oil or gas (Fig. 2.3). Algal and amorphous marine kerogen typically favors the generation of oil whereas woody debris and coal favor the generation of gas. Inertinite, regardless of its abundance, has little hydrocarbon generating capacity.

**PRESERVATION–THE KEY QUESTION**

In order for plants and animals to be preserved in the rock record, three variables must be satisfied. They are (1) low depositional energy, (2) reduction beneath the sediment-water interface, and (3) rapid sedimentation. Low depositional energies favor rapid accumulation of the low specific gravity organic matter and the limited currents and/or wave action reduces the supply of oxygen. Deposition of organic matter in an oxygen-rich environment results in the complete decomposition of the organic matter into $CO_2$ and $H_2O$. Preservation can only occur in an oxygen deficient environment. In this case, carbohydrates ($CH_2O$) partially react with oxygen to form $xCO_2$ and hydrocarbons ($yCH_2$). This involves the partial reduction of part of the carbohydrates. The oxidation-reduction is actually a series of complex reactions which must occur in a reducing environment. Rapid burial is necessary in order to reduce the consumption of the organic matter by reducing bacteria. These bacteria require heat and sunlight and therefore burial of the organic matter by rapid sedimentation favors preservation.

**SOURCE ROCKS**

EARLY SOURCE BED CONCEPTS

James Hutton in 1795 was one of the first geologists to consider the concept of source

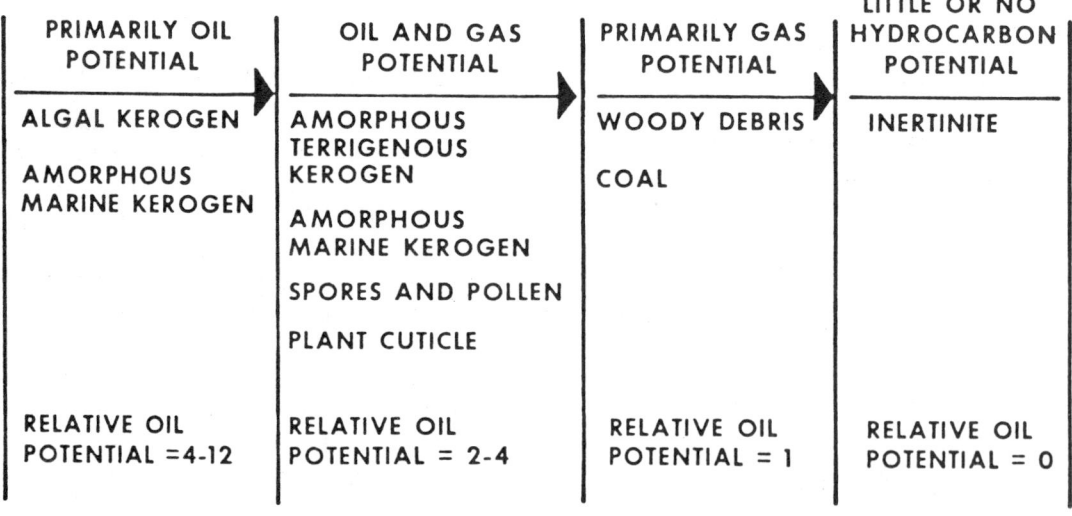

**Figure 2.3:** *Hydrocarbon generation potential of the important organics that are commonly preserved in source rocks (CITCO publication, 1979).*

beds. Alexander Winchel in 1865 was the first to coin the term *"source rock."* He considered a source rock to be one which is deposited together with organic material which may be transformed into hydrocarbons.

## REQUISITES FOR SOURCE BEDS

A source bed must be deposited proximal and contemporaneous to abundant plant and/or animal communities. It must accumulate in a low energy environment; otherwise, the organic material will be "flushed" from the environment. The aqueous environment must be characterized by an oxidizing condition to facilitate active biotic growth and reproduction. However, beneath the sediment-water interface, a reducing environment is necessary in order to facilitate preservation of the organic material. Reasonably rapid sedimentation rates are required in order to cover the organic material which, in turn, will aid preservation.

## FAVORABLE AREAS FOR ACCUMULATION

**Sea Margin**  Sea margins are perhaps the most favorable areas for accumulation of source rocks. The adjacent continent is an excellent supply of sediment; the environment is conducive to organic life and anaerobic conditions are common at the sediment-water interface.

**Deltas**  Deltas are low, nearly flat, alluvial tracts of land deposited at or near the mouth of a major river. Commonly, the resultant deposits are characterized by a triangular shape. Rapid sedimentation and subsidence result in a thick accumulation of both source rocks and potential reservoir rocks.

The extensive interfluvial (river) environments are dominated by floodplain, marsh, and swamp deposits which contain abundant terrestrial and marine plant debris as well as marine animal remains. Terrestrial derived humus material is concentrated on the deltaic plain. Deltaic deposits may be the most prolific source bed rocks *(Rainwater, 1963)*.

**Silled Basins**  A silled basin is partially separated from the open sea by land or submarine barriers. It is connected to the sea by a sill or threshold that is shallower than the basin itself. These basins are effective source bed environments because below the depth of the sill, the basin waters are stagnant *(Weeks, 1952)*. For example, waters of the Gulf of Riga, located adjacent to the Baltic Sea, are characterized by reducing conditions which protect the organic matter from bacterial oxidation and possible scavenger animals. Thus, the preservation of the abundant terrestrial and marine plant material is enhanced.

**Barred Basin**  A barred basin is similar to a silled basin, but the obstruction between the basin and open sea is more restricted. If the barred basin is located in an arid climate, evaporation in the basin equals or exceeds inflow of water from the open sea. Thus, the basin sediments are dominated by evaporites. This environment is favorable for the preservation of organic matter. The adjacent open sea may be the source of organic matter. For example, the Gulf of Kara Bogaz is separated from the Caspian Sea. The latter is teeming with plankton which may be swept into Kara Bogaz where it can be preserved. Thus, barred basins are sites which are supplied with organic matter derived from continuous plankton inflow from the open sea that is adequate for petroleum genesis. They are also sites favorable for the preservation of organic material because aerating currents are minimal below the bar *(Woolnough, 1937)*.

Barred basins located in humid climatic belts contain the normal facies of silt and clay. They are generally poor sites for the preservation of organic material.

**Carbonates**  Although carbonate rocks are directly or indirectly sourced by organisms, many petroleum geochemists do not feel that carbonate rocks are good source rocks for petroleum. This is because the highly-oxygenated waters required to support the rich biologic community that produced the carbonate sediment have been considered detrimental to the preservation of the organic matter. Some early workers *(Gehman, 1962 and Hunt, 1967)* however, questioned this assumption. Research in the southwestern shelf of Puerto Rico *(Sawyer, 1980)* demonstrated that, even in high energy environments such as reefs, 0.2% to 1.6% organic matter is preserved in carbonate sediments.

The southwestern margin of Puerto Rico is characterized by an insular shelf that extends southward 3 to 7 kilometers from the island. The shelf margin consists of a dormant barrier reef that is submerged in 20 meters of water. The shelf contains three distinct structurally-controlled reef trends that are separated by barren interreef troughs (Fig. 2.4). Concentration of organic carbon and lipid material is generally low in the fore reef and outer shelf environments where oxidizing conditions prevail (Fig. 2.5). The low depositional energies in the inner shelf province promote rapid sedimentation and the accumulation of large quantities of organic matter *(Sawyer, 1980)*.

A different influence on the amount and type of organic matter in sediment occurs at Enmedio reef (Fig. 2.4). In general, reef apron and back reef slope sediments contain greater amounts of organic matter than do sediments from other reef environments (Fig. 2.6). Organic carbon in fore reef sediments range from 0.3% to 0.7%. Reef crest and reef flat sediments contain 0.3% to 0.8% organic carbon. Reef apron and back reef slope environments contain 0.4% to 1.0% organic carbon *(Sawyer, 1980)*.

Reducing conditions occur in the reef flat and reef apron of Enmedio reef. Depletion of interstitial sulfate occurs within one meter of the surface in both these environments. Carbon isotopic evidence indicates that a transfer of organic matter from certain organisms to the sediments occurred in these reducing environments. *Thalassia, Dictyota, Laurencia,* and *Zooanthus sociatus* all show evidence of contribution organic carbon and, in particular, lipid material to the sediment. Where *Halimeda* and coral growth was dominant, no significant transfer of organic matter was documented *(Sawyer, 1980)*.

**Inland Lakes** Although most of the world's petroleum reservoirs are thought to be sourced by marine deposits, there are several areas where major accumulations have been attributed to nonmarine deposits. An excellent example is Lake Maracaibo, Venezuela. The lake has existed since Mid-Cretaceous time ($\pm 120$ million years ago) and presently occupies about one-fourth of the basin. The concentration of organic matter depends upon the relative rates of sedimentation with the accumulation of organic material. For example, sedimentation rates in the center of the lake are low; consequently, most of the organic matter is oxidized. This is in contrast to the margin of the lake, where sedimentation rates are rapid, which, in turn, covers the organic matter before it is oxidized *(Redfield, 1958)*.

**Phosphorites** Phosphorites are sedimentary rocks which are composed chiefly of phosphate minerals; for example, pyromorphite ($Pb_5CL[PO_4]_3$). Most commonly, phosphorites are a bedded marine deposit composed of micritic fluorapatite in the form of laminae, pellets, oolites, nodules, and skeletal fragments. Generally, they accumulate on the continental shelf (for example, Southwest Africa and off the Peruvian Gulf), but some ancient phosphorites are thought to represent estuarine environments.

In some cases, phosphorites may be a better source rock than shales *(Powell and Snowdon, 1976)*. The average organic carbon content is between shales and carbonates. Generally, phosphorites contain more organic matter that is soluble in organic solvents, more asphalt, and more complex saturated hydrocarbon fraction than shale. Source of the hydrocarbon is attributed to microorganisms. Phosphorites commonly have a high nitrogen content (70.2%).

# GENERATION
## INTRODUCTION

**The Problem** No modern petroleum liquid has been found to date. Moreover, organic matter in modern sediments consists of complex, resistant, and semisolid mixtures of hydrocarbons. These recent hydrocarbons contain less proportion of diphatic hydrocarbons in extract and light hydrocarbons ($C_4$ and $C_8$) than crude oil. No simple aromatics (benzenes or naphthalenes) have been detected in modern sediment hydrocarbons. The odd-carbon ratio is much higher for modern hydrocarbons than crudes. The question then is — how does the very different recent sediment hydrocarbons transform to petroleum?

**Suggested Transformation Processes** Several decades of research have been focused on this problem. Such processes as radioactive bombardment *(Beers, 1945)* and catalytic reactions *(Fash, 1944)* have received a great deal of attention in the past. Present research

# PRE-ENTRAPMENT HISTORY OF PETROLEUM

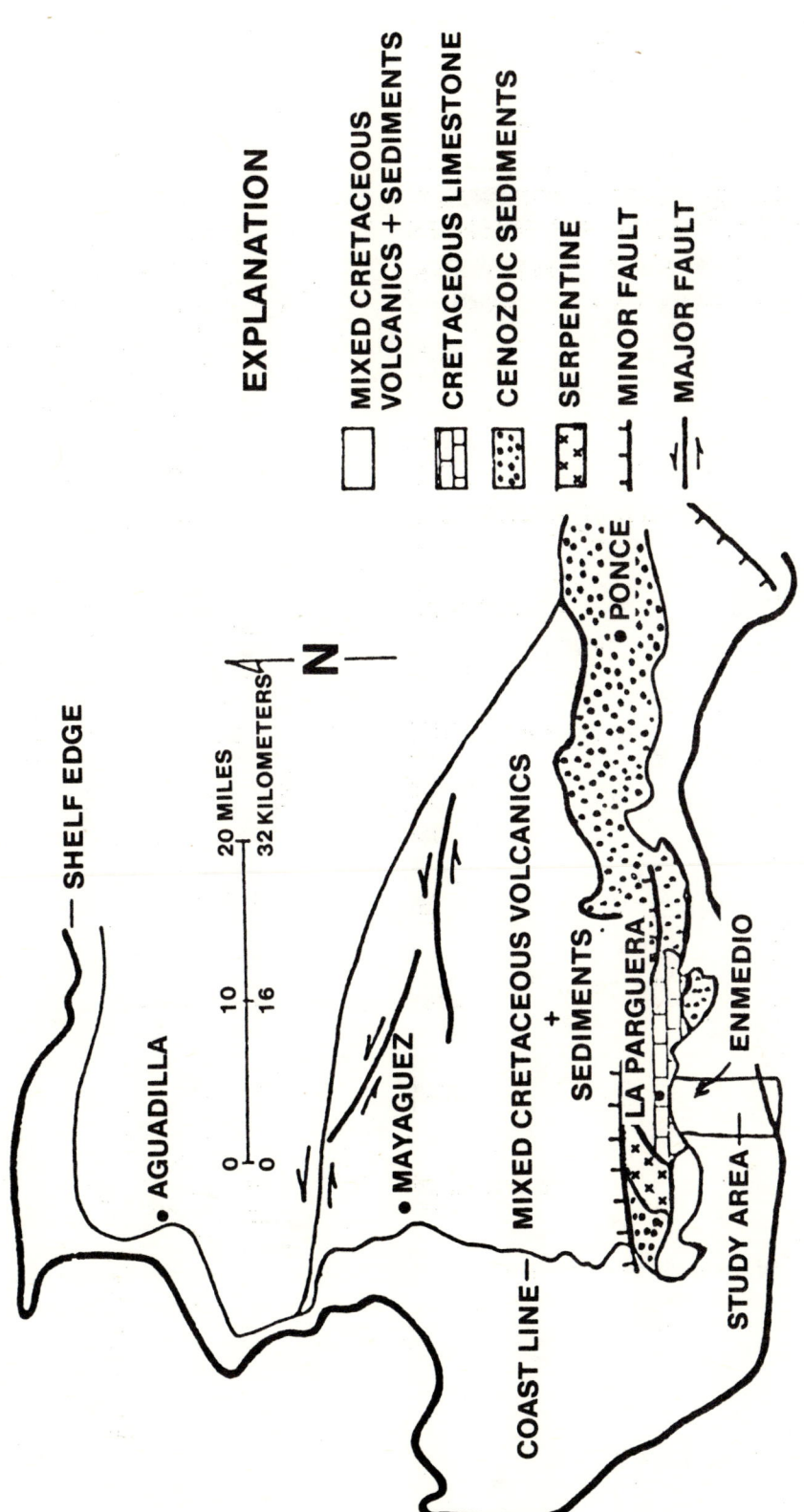

**Figure 2.4:** *Southwestern Puerto Rico (Sawyer, 1980).*

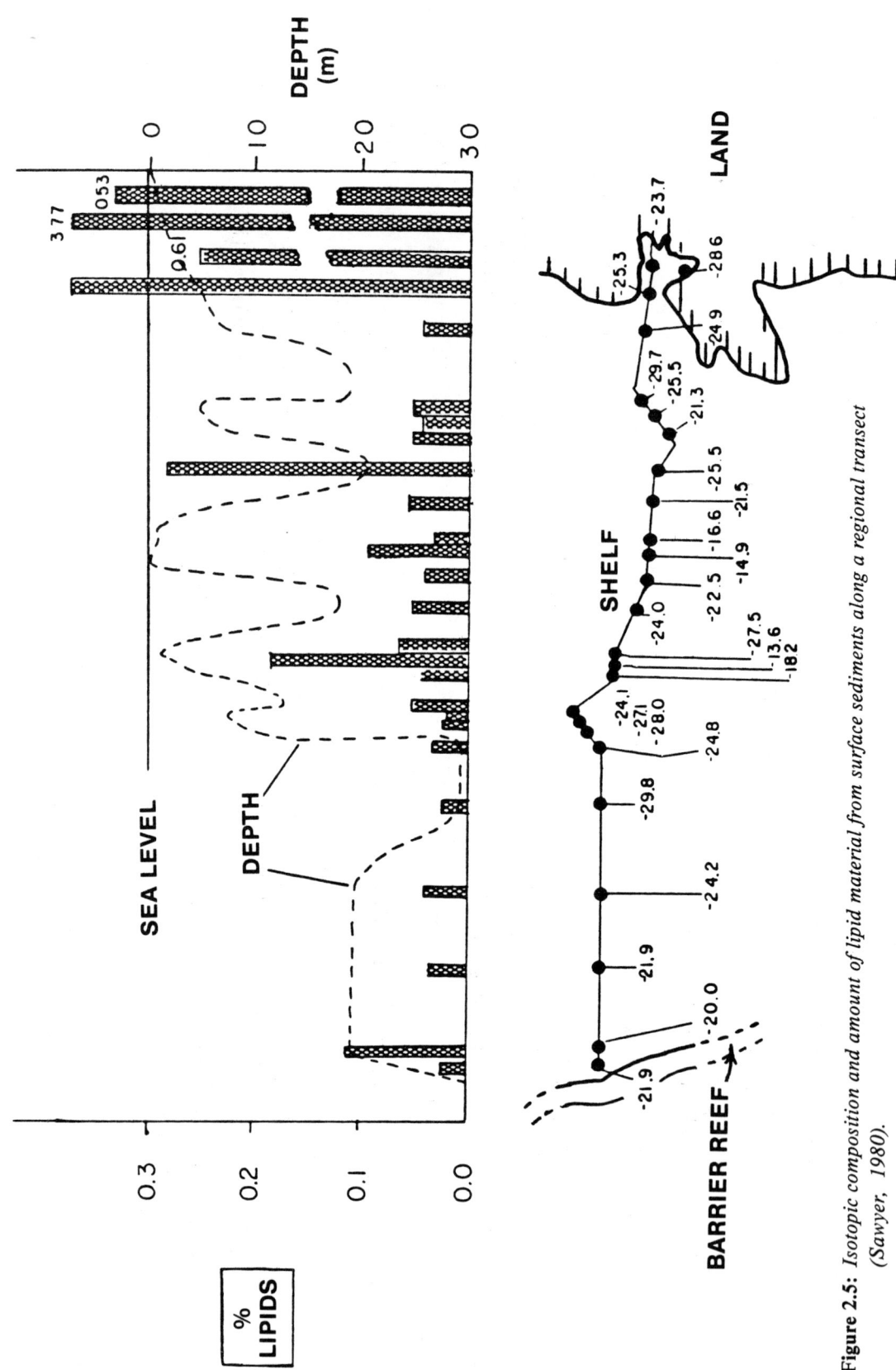

**Figure 2.5:** *Isotopic composition and amount of lipid material from surface sediments along a regional transect (Sawyer, 1980).*

# PRE-ENTRAPMENT HISTORY OF PETROLEUM

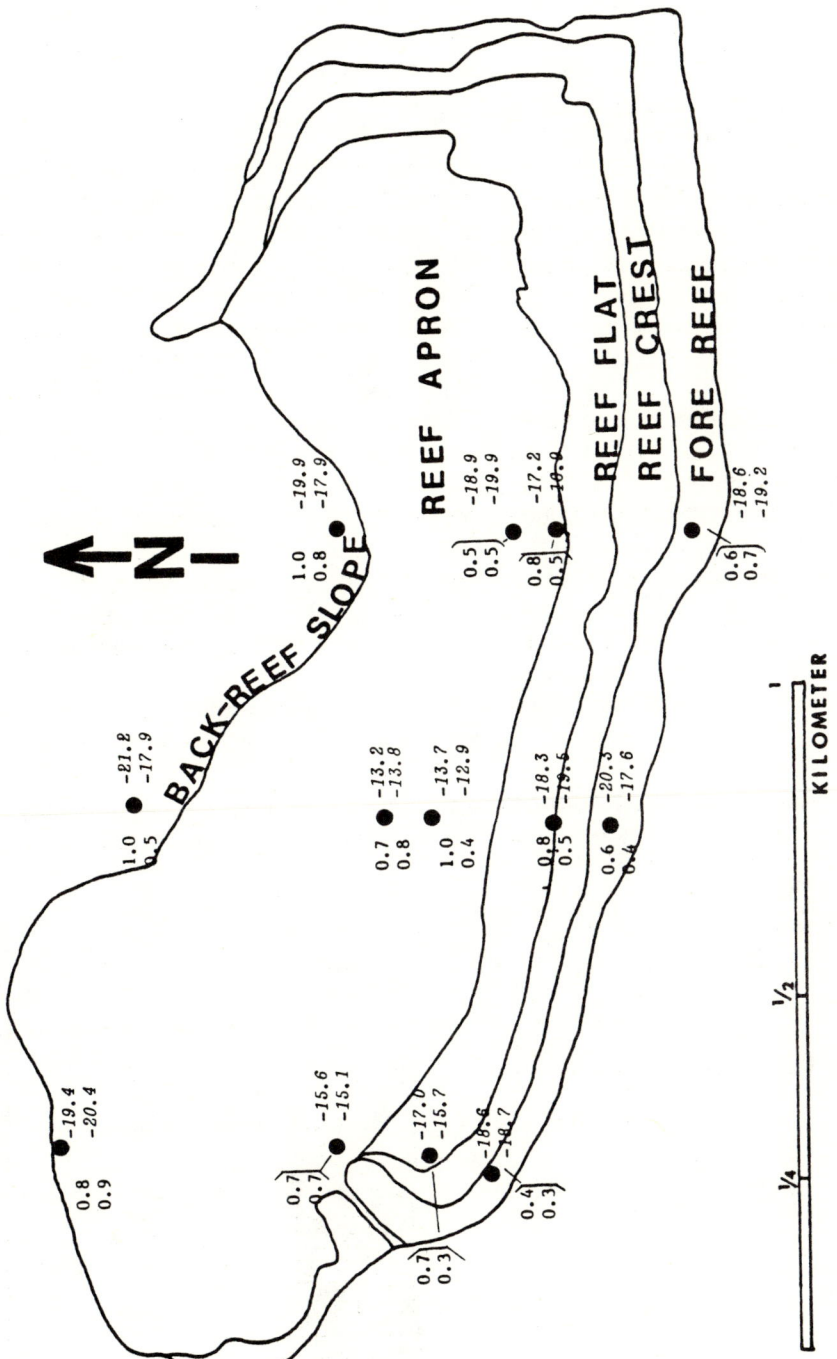

**Figure 2.6:** *Diagram depicts organic carbon content and isotopic composition of organic carbon from surface and subsurface sediments from Enmedio Reef. Values to the left of the sample location reflect the TOC for the surface sediment (top) and the subsurface sediment (bottom). Their respective isotopic compositions appear to the right of the sample location (Sawyer, 1980).*

is focused on the interaction of temperature and time as the main variables to be considered for the transformation process.

## THERMOCHEMICAL GENERATION

**Quantity of Organic Matter** The percent of organic matter in limestones and shales varies from .01% to 32%. The mean percent of organic matter in limestones is about .25%, whereas it is about 1.2% for shales. Gehman *(1962)* has demonstrated that a potential source rock must contain about 0.5% organic matter. However, this number is dependent upon the thickness and aerial extent of the source rock. Higher values are necessary if the source rock is thin.

**Carbonization** Recent kerogen (organic matter) shows little or no alteration. The original structure, pigmentation, and transparency is preserved. In many cases, a paleobiologist can recognize the source of the kerogen. Late diagenetic kerogen shows marked alteration (Fig. 2.7). The texture and color of the kerogen is eliminated as well as its transparency (Fig. 2.8). The process for these changes is thought to be due to increased temperature and pressure.

During carbonization, the residue products of kerogen shift toward 100% carbon (Fig. 2.9). Volatiles (for example, hydrocarbons, $CO_2$, and water) are driven off; oxygen and hydrogen percents are reduced and carbon percent is increased. The least stable bonds are broken first, and the most stable bonds broken last.

**Thermochemical Process** The empirically derived thermochemical concept explains the temperature-time relationship responsible for the generation of petroleum. The theory is based on the assumption that thermal cracking reactions are responsible for the generation of petroleum. Figure 2.10 illustrates that for every 10°C increase in temperature, the reaction rate is about doubled; or, stated another way, the reaction time is halved. Also note that reaction yield varies linearly with time, but exponentially with temperature. Therefore, a 40 million (40M) year old source rock must have been characterized by a temperature of at least 80°C, whereas a 80M year old rock requires a peak generation temperature of only 70°C.

Support for the thermochemical concept is empirically based on the time-temperature

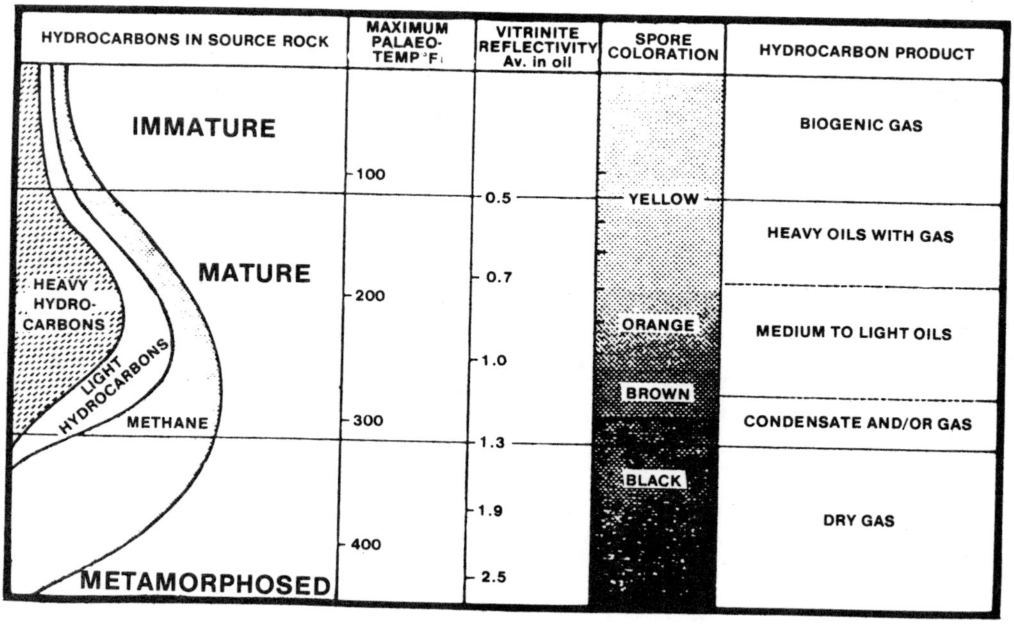

**Figure 2.7:** *Vitrinite and spore coloration criteria for determining stage of hydrocarbon generation (CITCO publication, 1979).*

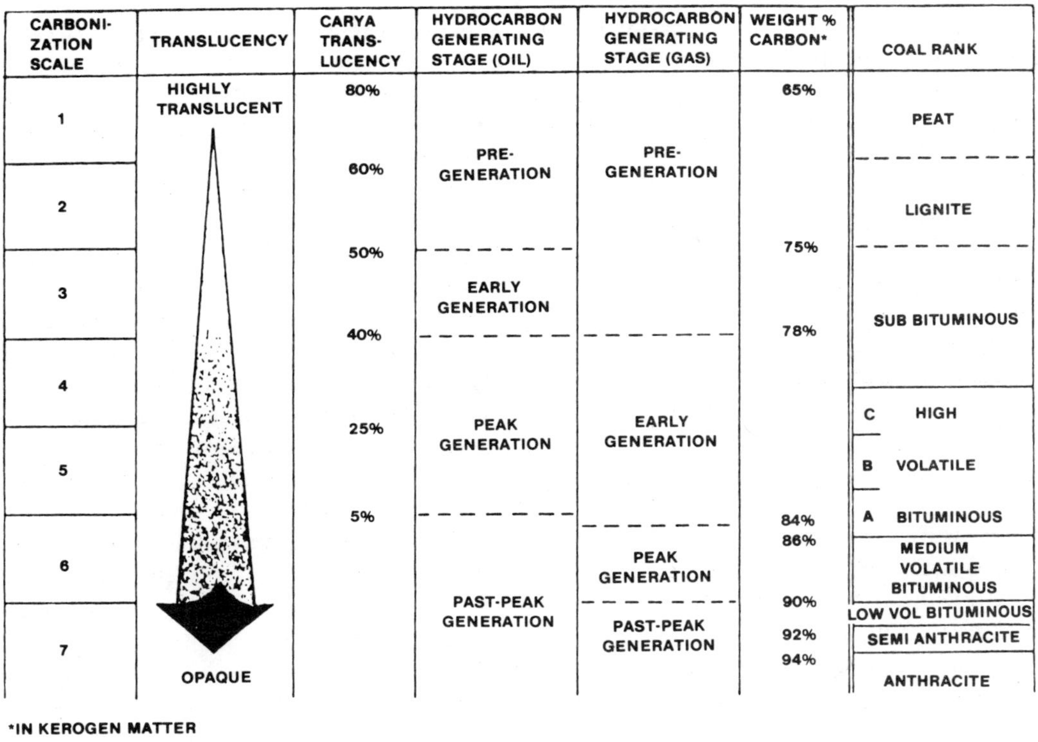

| CARBONI-ZATION SCALE | TRANSLUCENCY | CARYA TRANS-LUCENCY | HYDROCARBON GENERATING STAGE (OIL) | HYDROCARBON GENERATING STAGE (GAS) | WEIGHT % CARBON* | COAL RANK |
|---|---|---|---|---|---|---|
| 1 | HIGHLY TRANSLUCENT | 80% | PRE-GENERATION | PRE-GENERATION | 65% | PEAT |
| 2 | | 60% | | | | LIGNITE |
| 3 | | 50% | EARLY GENERATION | | 75% | |
| 4 | | 40% | | | 78% | SUB BITUMINOUS |
| 5 | | 25% | PEAK GENERATION | EARLY GENERATION | | C HIGH<br>B VOLATILE<br>A BITUMINOUS |
| 6 | | 5% | | PEAK GENERATION | 84%<br>86% | MEDIUM VOLATILE BITUMINOUS |
| 7 | OPAQUE | | PAST-PEAK GENERATION | PAST-PEAK GENERATION | 90%<br>92%<br>94% | LOW VOL BITUMINOUS<br>SEMI ANTHRACITE<br>ANTHRACITE |

*IN KEROGEN MATTER

**Figure 2.8:** *Translucency of kerogen is an indicator of stage of generation (Grayson, 1972).*

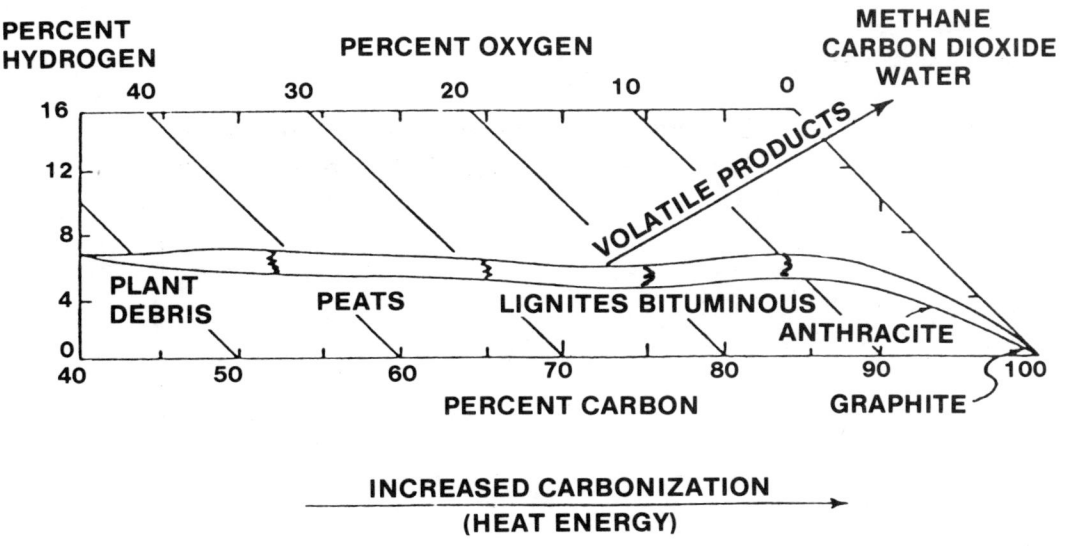

**Figure 2.9:** *Evolution of volatile products during coalification (Himus, 1951).*

relationships required to thermally destroy liquid hydrocarbons. The maximum temperature known to exist for crudes of all ages is depicted in Figure 2.11. Note that maximum known temperature for Cambrian crudes is about 290°F, whereas only gas has been found in Cambrian age reservoirs. Miocene gas has been produced from reservoirs with a temperature of 390°F, which would completely destroy the gas if that temperature existed for 400-500M years. The boundary between oil-condensate-gas and condensate-gas approximate the peak generation curve depicted on Figure 2.10.

## EVALUATION OF SOURCE BEDS

There are four important techniques for evaluating source rock potential that are currently used by the industry. They are (1) organic carbon, (2) hydrocarbon gas analysis, (3) elemental analysis, and (4) visual kerogen analysis. The inorganic carbon is removed (*i.e.*, $CaCO_3$) and the weight percent organic carbon is determined. The result is an approximation of the disseminated organic matter which, in turn, indicates hydrocarbon generating capability of the rock. The hydrocarbon gas analysis may be done on the drilling site. The quantity and composition of gas derived from cuttings and core material is often a guide to the type of hydrocarbons that might be derived from the source rock.

The elemental composition of the kerogen fraction can be determined; the percent elemental carbon in kerogen indicates the generation stage if the carbon has not been oxidized. A cross-plot of H/C and percent elemental carbon is useful in predicting not only the source rock potential of a unit, but also the phase of the hydrocarbon (oil, condensate, or gas).

Perhaps the simplest method of evaluating the source potential of a rock is by observing the color and texture of kerogen. The color of kerogen indicates the state of diagenesis. Highly translucent kerogen suggests pregenerations, whereas opaque kerogen indicates post-peak generation (Fig. 2.8). Structured kerogen suggests possible oil generation, whereas amphous kerogen indicates gas (Fig. 2.3).

Experience has shown that there are typical imperical differences in mechanical log response to organics in carbonates and shales. Organic rich shales typically have higher gamma ray and lower SP levels than non-organic rich shales. They are frequently more dense and consequently the borehole is less "washed out" through organic rich shales than through their "lean" counterparts as indicated by the caliper log. Commonly, organic rich shales have a higher resistivity than organic poor shales. Organic rich carbonates also have higher gamma ray and lower SP response than organic lean carbonates. Density and acoustic readings through organic rich carbonates are lower and higher than their organic "lean" counterparts.

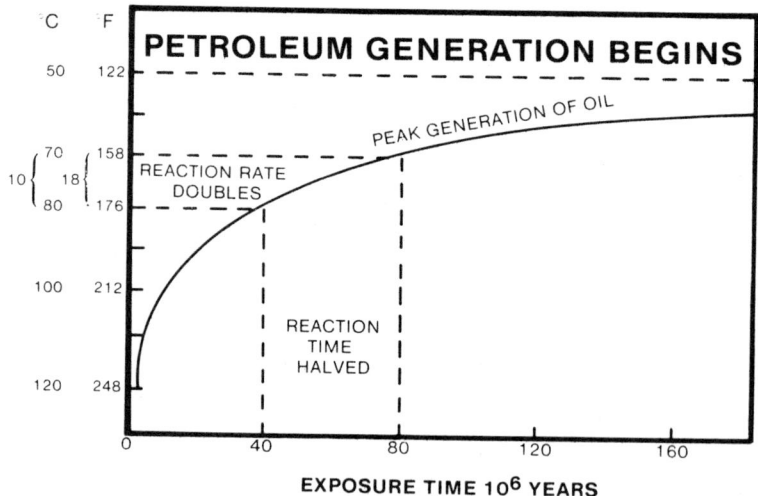

**Figure 2.10:** *Thermochemical concept to explain conversion of kerogen to hydrocarbons.*

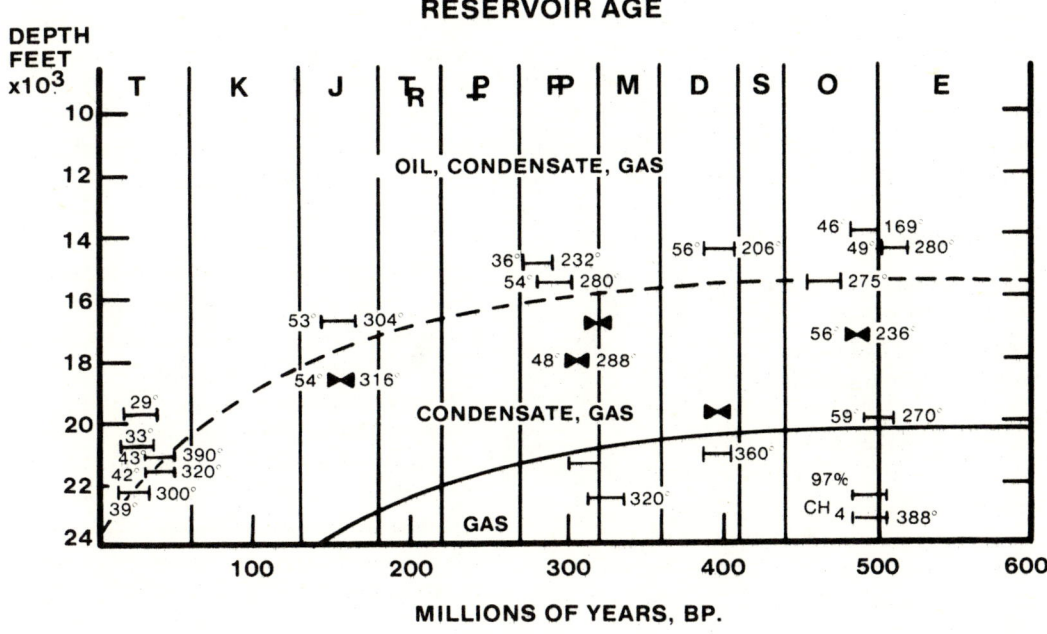

**Figure 2.11:** *Data base for thermochemical concept.*

## EXPULSION OF HYDROCARBONS

### INTRODUCTION

The removal of petroleum (or protopetroleum) from the source rock into a carrier bed is called *expulsion*. Subsequent movement from the carrier bed into a reservoir rock is called *migration*. Several decades of research have been focused on expulsion with special emphasis on the mechanism and timing of expulsion. Four major processes of expulsion have been given particular attention — compaction, capillary pressures, hydrologic gradients, and generation.

### COMPACTION

A decrease in volume corresponds to a loss of interstitial porosity. Pore size reduction results in closer packing and realignment of grains. The loss in pore volume may cause the expulsion of pore fluid.

Porosity is described by $\phi = \phi_o e^{-cz}$, where $\phi$ is porosity, $e$ is $1.42 \times 10^{-3}$ per meter, $\phi_o$ is initial porosity, $e$ is the base of Napierian log, and $z_e$ is equilibrium depth. Integrating this formula with compaction curves, a porosity versus depth graph can be constructed (Fig. 2.12). Porosity is also in part dependent upon the hydrostatic pressure of a particular geologic province. A $\delta$ value of 0 implies a geostatic pressure whereas a $\delta$ of 1 is the normal hydrostatic pressure.

The amount of pore fluid forced from the compacted rock can be determined. If $V_l = V_i - V_c$ where $V_l$ = volume of liquid loss, $V_i$ = initial volume, and $V_c$ = compacted volume, the $(1 - \phi_o)$ = volume of the rock. Compaction will cause a decrease of $\phi/(1 - \phi_o)$. Therefore, the compaction from an initial porosity $\phi_o$ to a porosity $\phi$ is given by the expression:

$$\frac{\phi(1 - \phi_o)}{1 - \phi}$$

$$q = \phi_o - \frac{\phi}{1 - \phi}(1 - \phi_o) = \frac{\phi_o - \phi}{1 - \phi}$$

Consequently, expulsion of pore fluid begins with burial on the lithosphere surface and continues until burial ceases and stress equilibrium is achieved.

### CAPILLARY ACTION

Capillary pressures are considered by some geochemists as a possible process for expulsion. Cook *(1923)* discovered that water moved from large to small pores and, conversely, oil moved from small to large pores. Capillary action probably is not an effective expulsion process except along the boundary between a carrier bed and the source rock; the latter would have to be saturated with oil in order to expect replacement on a wide scale.

## GENERATION AND EXPULSION

**Introduction** Compaction cannot be the major process for expulsion. Philippi *(1965)* has demonstrated that in a normal thermal basin significant generation does not occur until the source rock is buried by at least 8000 feet of sediment. At this depth, most of the pore fluid has been squeezed out of the source rock (Fig. 2.12). Consequently, compaction and the resultant "flushing" of the source rock occur prior to generation. Therefore, other processes of expulsion must account for expulsion.

**Stages of Compaction** Hedberg *(1959)* has delineated several compaction stages. Stage 1 compaction occurs from the surface to about 1200 feet. It is characterized by a nearly linear reduction of porosity. Stage 2, on the other hand, is characterized by continued porosity reduction down to about 13,500 feet. Recent work has shown that the lower 3,000 feet of this stage may contain similar porosities in the Gulf Coast. Nearly 67% of the total reserved in the Gulf Coast correspond to this interval of compaction.

**Pore Size and Compaction** Sands are composed of quartz grains which are commonly rounded, whereas platy clay grains constitute the building blocks of shales. The surface area of a shale grain is about 80,000 times greater than a sand grain. The resultant pore geometry of sands and shales are therefore, greatly different. Pores in sands are commonly voids with accurate borders, whereas shale pores are tabular shaped. Compaction reduces shale pore size, but at or near the boundary of Stages 1 and 2, shale pore size remains constant. The average pore size is larger than water, methane, and liquid hydrocarbon molecules.

Shale pore water is frequently absorbed because the water is charged (polar); therefore, the water is tightly held to the pore wall. The hydrocarbon molecules generated within the shale are not absorbed because they are nonpolar. Thus, the net effect of absorption can be thought of as a tendency of the shale to retain pore water, but not hydrocarbon molecules.

**Explusion Mechanisms Related to Generation** At least two mechanisms are not related to porosity loss that may account for expulsion. They include volume increase of pore fluid due to generation and thermal expansion of shale pore water.

During hydrocarbon generation, the remnant kerogen plus the generated hydrocarbons result in 4½% increase in volume. This would amount to about 40 BOPAF. The increase in volume would preferentially drive the nonpolar generated hydrocarbons from the shale pore. The pore water tends to remain due to the absorption of the polar-water to the wall of the pore. Thermal expansion of free water increases about ±0.5% with each 1,000 feet of additional burial in Stage 2. If pore water acts similarly to liquid water, then it follows that the expansion may be a mechanism to "flush" the generated hydrocarbon from the source rock pore. Diffusion of light hydrocarbons may be another important process of expulsion *(Leythaeuser and others, 1982).*

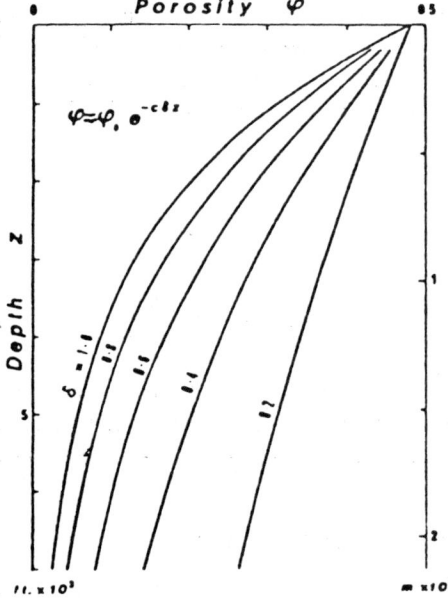

**Figure 2.12:** *Porosity—depth of burial relationship (Chapman, 1973).*

# MIGRATION

## INTRODUCTION

Petroleum must be expelled from the source rock into a carrier bed and concentrated in traps in order to be commercial. The process of concentration is referred to as *migration*. Most explorationists concede that there is a great deal more to learn about migration than perhaps any other aspect of the pre-entrapment history of hydrocarbons.

## GEOLOGIC FRAMEWORK

Water and petroleum coexist in nearly every commercial field. This may be manifested by an oil-water, gas-water, or gas-oil-water contact, or (more subtly) by a thin film of water on the grain substrates. Therefore, any theory for migration must explain the association of water and petroleum.

The extreme variability of petroleum reservoirs negates a simple theory for migration. Petroleum reservoirs range in lithology from fracture igneous rocks to various carbonate and clastic rocks that may be Precambrian to Pliocene in age. Like lithology and age, the porosity and the permeability of reservoir rocks are highly variable. Porosity values range from 5% to 40%, and one millidarcy to several darcies of permeability have been recorded. The temperature of producing fields ranges nearly 400%, and pressures vary from essentially zero to over 1,000 atmospheres.

Trapping mechanisms are highly variable, which, like the above parameters, complicate any theory for explaining migration.

## SHORT MIGRATION vs. LONG MIGRATION

**Short Migration** A significant group of explorationists feel that migration is not an important process. They would cite point-bar deposits (that are encased in an envelope of flood plain silts and shales and that are not in fault communication with a carrier bed) as having received the contained petroleum directly from the source rock. In this example, expulsion from the source rock into the reservoir rock was the only process of petroleum migration. Similar arguments can be made for carbonate bioherms and reefs that are in an envelope of nonporous and permeable carbonates and/ or evaporites.

**Long Migration** Strong arguments can be developed to demonstrate that petroleum does migrate on a regional scale. These include the following: (1) Petroleum could not be produced if it did not migrate through the reservoir to the wellbore. (2) Oil seeps are common. (3) Many reservoirs produce more hydrocarbons than could be generated within the area of the trap.

An excellent example of significant secondary migration occurs on the Central Basin Platform of the Midland Basin, West Texas. Chuber and Rodgers *(1968)* recognized seven types of crude oils that are produced from various Wolfcampian and Pennsylvanian carbonate and sandstone reservoirs. Each of the seven crudes is defined on the basis of similar paraffin-napthene ($V_n/V_p$) and carbon isotope ($\delta^{13}C$) values. Group I was subdivided into two subgroups: Pennsylvanian reservoirs on the Eastern Shelf, and Pennsylvanian reservoirs located on the Central Basin Platform (Fig. 2.13). The Central Basin Platform crudes contain mixed $V_n/V_p$ and $\delta^{13}C$ ratios which are similar to Ellenburger (Ordovician) crudes. This mixing is attributed to a tertiary stage of migration of the older oils. The remigration was caused by destruction of the traps during early Wolfcampian.

**Limited Vertical Migration** There is a great deal of information supporting the concept that secondary vertical migration of petroleum is limited from a carrier bed through a nonporous unit into a second carrier bed or reservoir rock. Most of the data to support this observation is derived from "multipay" fields (Fig. 2.14). Most multipay fields are characterized by erratic reservoir pressures in the same position of the structure. Water sands commonly occur between producing reservoirs. Frequently, no two reservoirs in the field will have identical $V_n/V_p$ and $\delta^{13}C$ ratios or gas/oil ratios. A gas reservoir often will be sandwiched between two reservoirs with very low gas/oil ratios. If vertical migration were a significant process, the above examples of heterogeneity would be replaced by an upward and predictable increase in gas/oil ratios and decreasing $V_n/V_p$ ratios.

## CARRIER BEDS

Most petroleum geologists believe that signifi-

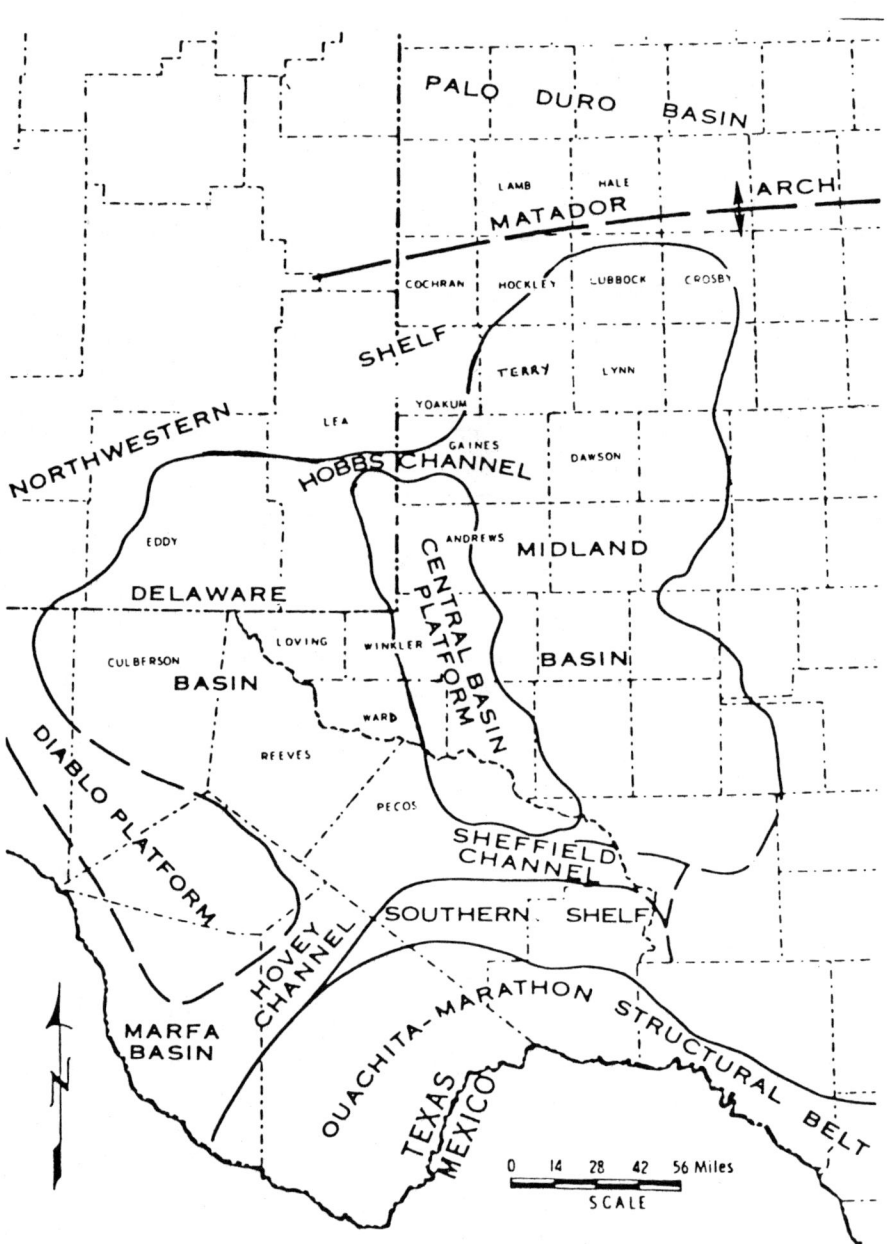

**Figure 2.13:** *Index map of Permian Basin, southwestern New Mexico, and West Texas, U.S.A. (McKee, et al., 1967).*

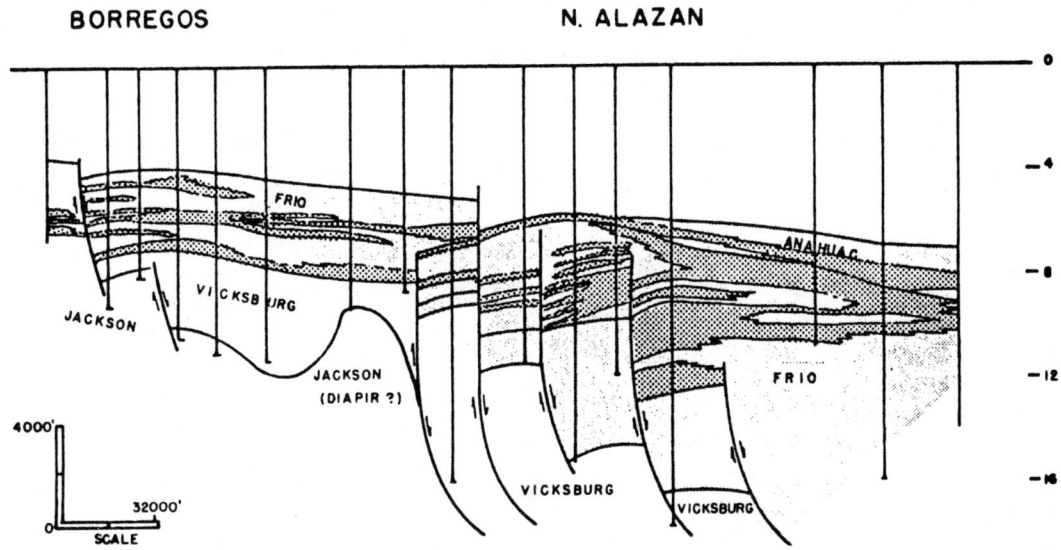

**Figure 2.14:** *Schematic cross section of Borregas and North Alazan fields, south Texas. A 15,000 foot well would encounter over 200 individual sands of which only 40 to 50 are productive.*

cant lateral migration is the rule rather than the exception. Therefore, mandatory to this concept is the presence of a porous and permeable medium through which the petroleum can migrate. Rich *(1939)* was the first to coin the term "carrier beds" to explain long-distance migration in the East Texas field. Carrier beds may be sheet sands (such as the Dakota group, Tensleep, or Simpson) that are regionally extensive, porous, and permeable, and are in regional contact with source rocks. Carbonates can also be carrier beds. For example, the Arbuckle limestone of Oklahoma and the Trenton limestone of the Cincinnati Arch are thought to be excellent carrier beds. Carrier beds may be the reservoir rock in the trap area.

## GATHERING GROUND

Long-distance migration requires a second concept called a *gathering area*. Hunt *(1862)* and Phinney *(1891)* postulated that oil is mobilized and concentrated from a considerable area. Wegemann *(1911)* was first to coin the term "gathering ground," and Mather and Mehl *(1919)* referred to the same as the "drainage area."

A gathering area is simply an area in which petroleum is expelled into a carrier bed from the source rock. It is located down-dip from a trap, and the paleo axis of dip reversal is considered to be the limits of a gathering area. In the classic East Texas field, the gathering area is bound on the west and the north by structural reversal, on the east by the Sabine Uplift, and on the south by the Angelina-Caldwell flexure (Fig. 2.15). It is almost 50 miles wide and 140 miles long.

## CONTINUOUS PHASE THEORIES

**Oil Cannot Migrate as Individual Droplets**
Early writers thought in terms of movement of individual oil droplets passing from pore-to-pore. Oil in a dispersed state cannot move through a water-wet sand (much less a fine-grained shale) because the forces created by natural hydrodynamic gradients are thousands to millions of times too small to overcome forces created by surface tension of petroleum *(Levorsen, 1958)*. This concept of individual oil droplet migration must be abandoned in favor of continuous phase migration.

**Aqueous Solubility of Petroleum** Price (1976) reported the following controls on the solubility of petroleum in water.

1. An increase in temperature causes a gradual increase in aqueous solubility of up to 100°C. Above 100°C, a more drastic change takes place due to a solution mechanism (Fig. 2.16).
2. Increase in salinity causes a decrease in aqueous solubility of petroleum-forming hydrocarbons (Fig. 2.17).

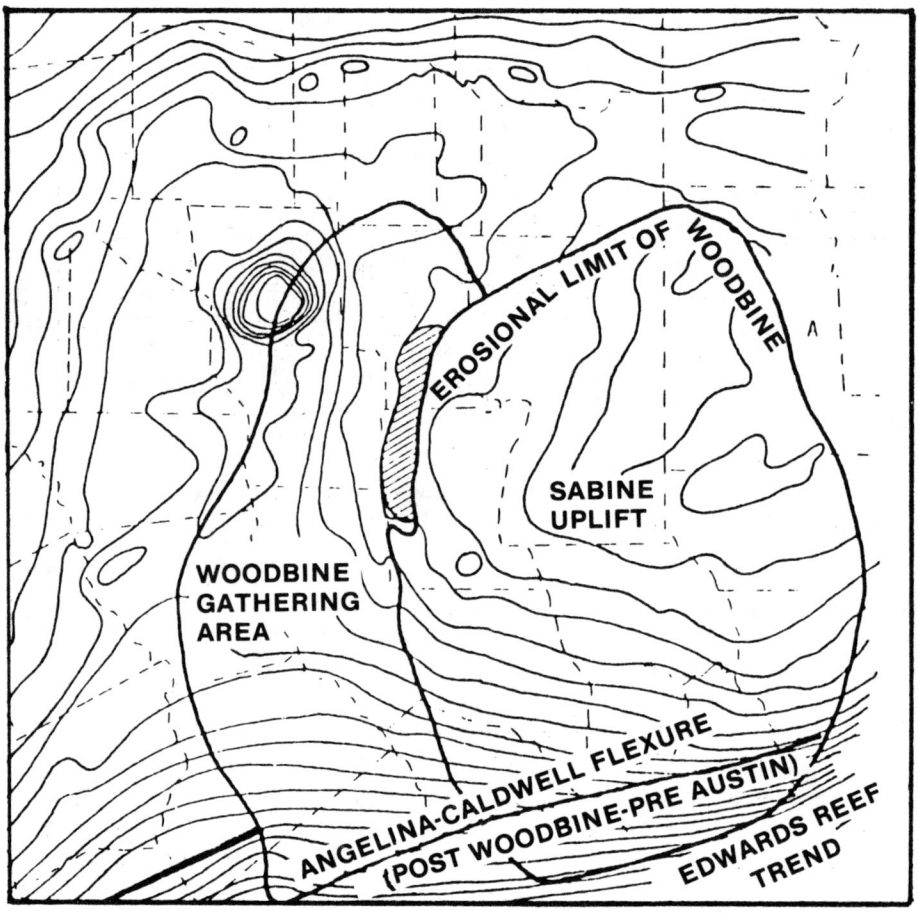

**Figure 2.15:** *The gathering area for the giant East Texas Field is over 7,000 square miles.*

3. Increase in salinity and decrease in temperature would cause the exsolution of hydrocarbons from the aqueous phase in natural systems.

Although the graphs illustrate that hydrocarbons have a very low solubility in water, the solubility may be sufficient to move large amounts of petroleum if sufficient quantities of water are available. Various fractions of hydrocarbons have very different solubilities. Aromatics are most soluble, naphthenes are intermediate, and paraffins are least soluble (Fig. 2.18).

Several mechanisms which enhance the solubility of hydrocarbons have been proposed. Accommodation, the suspension of minute colloidal particles in water, can increase the apparent solubility of hydrocarbons. Peake and Hodgson *(1967)* pointed out that *n*-alkanes can be accommodated in distilled water as colloidal particles in amounts that are much greater than their solubility. Cordell *(1973)* reported that polar organic molecules may form small colloidal aggregates called *"micelles."* The hydrocarbonlike volume in the center of the micelle can incorporate other organic molecules, thus enhancing their apparent solubility.

**Exsolution Mechanisms** Exsolution literally means "to come out of solution." If two oil-saturated groundwaters of different temperatures mix, or if a single groundwater is raised closer to the lithosphere surface, the solution temperature will be lowered. At the lower temperature, oil will be exsolved (Fig. 2.19). Mixing along the migration path of differing salinity water will result in an intermediate salinity and can cause exsolution (Fig. 2.20). Therefore, secondary migration of

# PRE-ENTRAPMENT HISTORY OF PETROLEUM

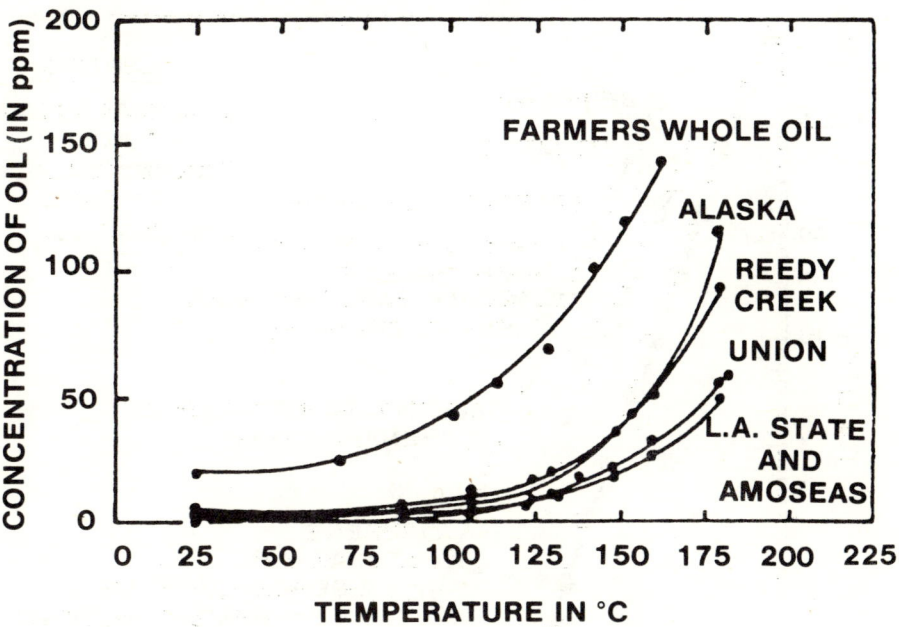

**Figure 2.16:** *Solubility of hydrocarbons as influenced by temperature (Price, 1976).*

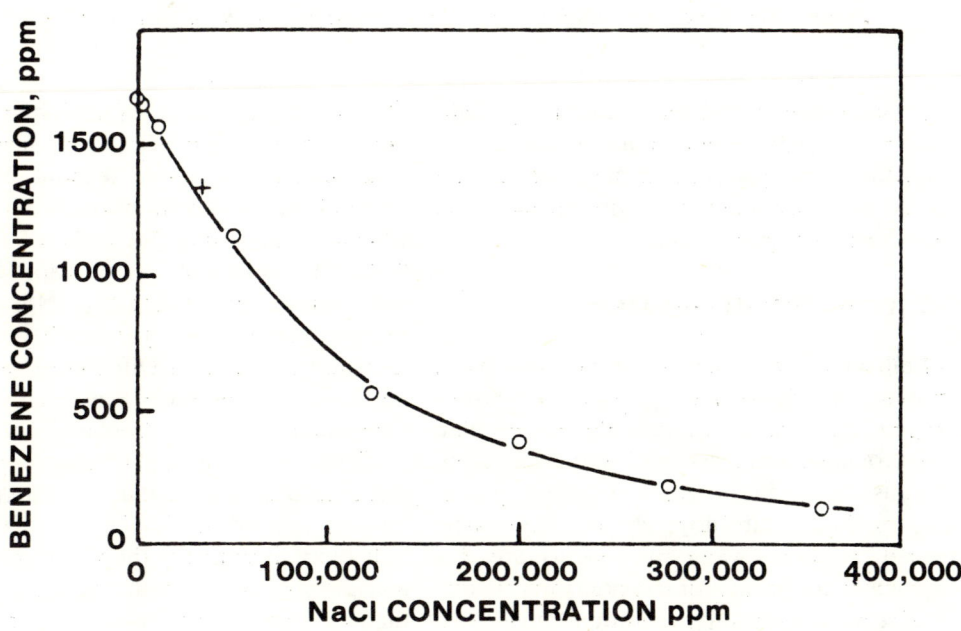

**Figure 2.17:** *Solubility of hydrocarbons as influenced by salinity (Price, 1976).*

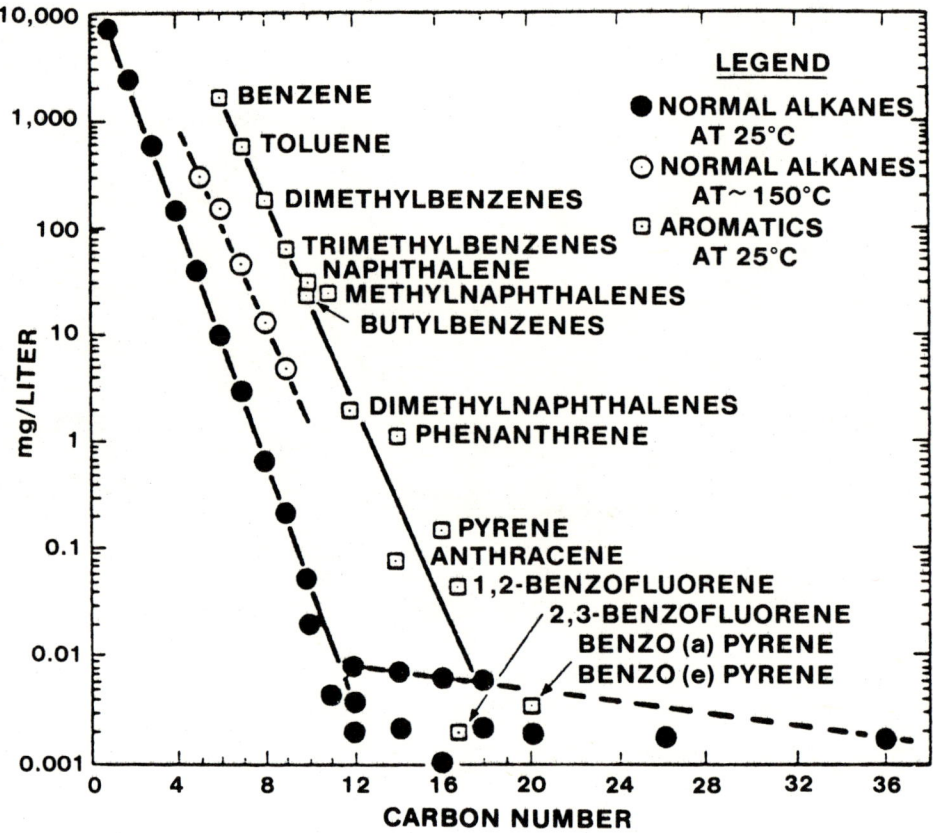

**Figure 2.18:** *Relative solubility of various hydrocarbon fractions (McAuliffe, 1978).*

petroleum may be related to normal groundwater movement; and capillary attraction, buoyancy, and gas-solution mechanisms are active processes only after the oil has been exsolved from groundwater.

## SECONDARY MIGRATION

Mobilization of crude oils after they have been trapped and reservoired is termed *secondary migration*. Regional tilting is the process that most frequently triggers secondary migration. Because the oil and/or gas is in a single phase, it is probable that buoyancy and capillarity (aided by groundwater movement) are the dominant mechanisms of secondary migration.

Effects of secondary migration range from the simple relocation of hydrocarbons with minor chemical changes to complex chemical changes and relocation. Regional tilting and/or uplift can lower the pressure in a gas cap which, in turn, may force the oil downward below the "spill point" of the trap. The oil first will migrate out of the trap. Normally, the original gas cap is not entirely removed by secondary migration. Differential entrapment of the crude produces gas-filled reservoirs downdip from oil-filled reservoirs (Fig. 2.21).

Complex chemical changes are possible in the crude during secondary migration. Crudes may be oxidized and biodegraded during remobilization. Frequently, they have a lower API and contain more sulfur and less aromatics than their parent crude in the original reservoir.

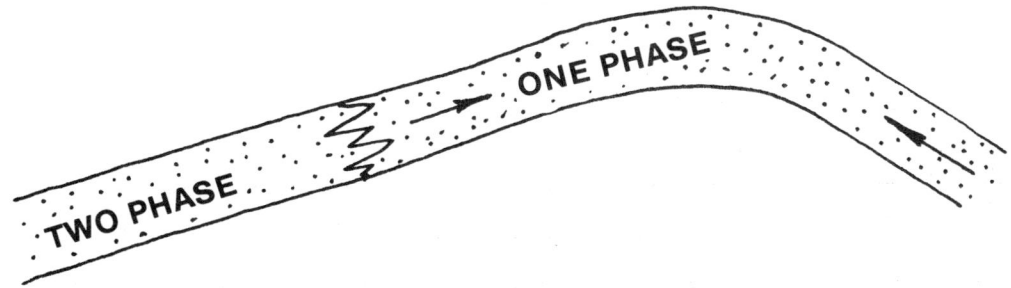

**Figure 2.19:** *Exsolution may be produced by upward migration that causes a reduction of temperature, therefore, exsolution.*

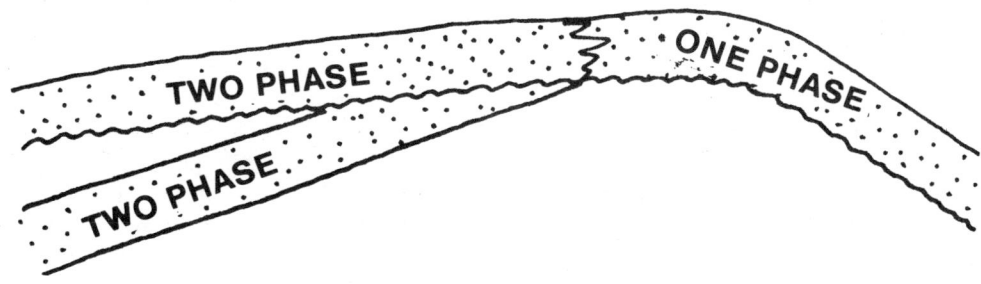

**Figure 2.20:** *Exsolution may be caused by mixing of oil-saturated ground waters of different salinities.*

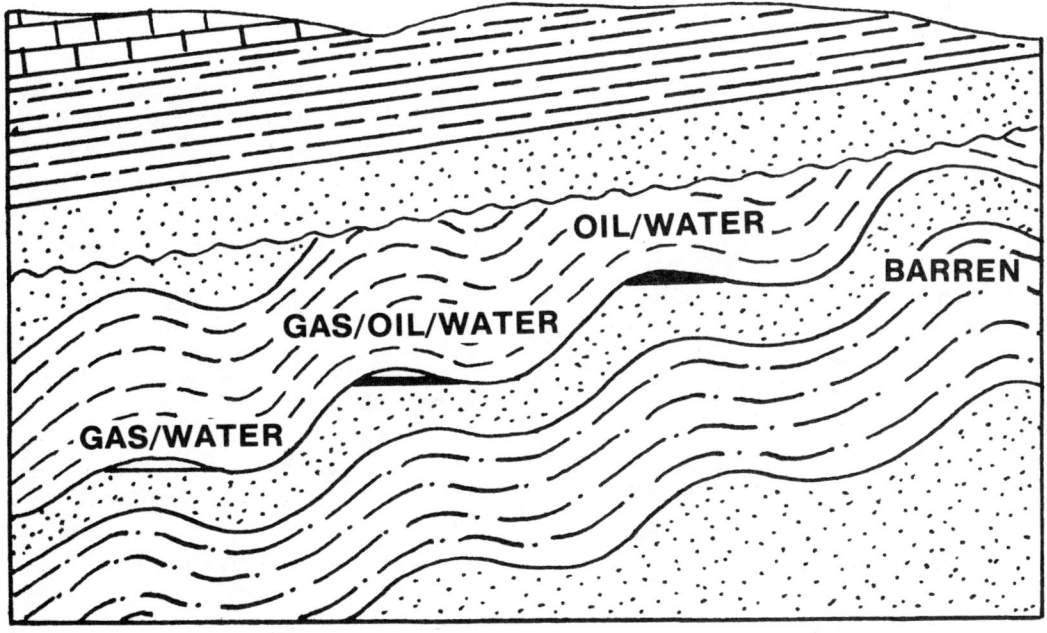

**Figure 2.21:** *Secondary migration may result in differential entrapment of hydrocarbons.*

# 3

# The Stratigraphic Framework: A Fundamental Prerequisite to a Successful Exploration Program

## INTRODUCTION

Modern subsurface exploration stratigraphy is based on data derived from surface information, wireline logs, seismic data, paleontology, and sedimentology. This data set must be incorporated into an understanding of process-response of global sea level changes and global tectonics in order to develop a stratigraphic framework. A valid stratigraphic framework is a fundamental prerequisite for correct evaluation of a basin and for generation of prospects. An accurate stratigraphic framework also provides the explorationist with the ability to predict the occurrence of source rocks and their depth of burial, and permits the delineation of carrier beds, migration paths, and the vertical and lateral distribution of potential reservoir rocks. Correct mapping of structures and interpretation of their ages is dependent upon accurate selection of stratigraphic horizons that approximate time surfaces.

## TYPES OF STRATIGRAPHIC UNITS

### INTRODUCTION

Experienced, as well as novice explorationists, have been introduced to a "family" of nomenclature that is used to describe stratigraphic units. In order to ensure a uniform understanding of terms used in this publication, a review of the terms that are accepted by the Code of Stratigraphic Nomenclature *(American Commission of Stratigraphic Nomenclature)* is appropriate.

### TIME-STRATIGRAPHIC UNITS

The Commission defines a time-stratigraphic unit as a ". . . subdivision of rock considered solely as the record of a specific interval of geologic time." Time-stratigraphic units do not normally contain unconformities and they are typically bounded by synchronous surfaces. The delineation of time-stratigraphic units requires the recognition of surfaces of contact that divide sedimentary deposits of one portion of geologic time from those of another. All points on the surface bounding a time-stratigraphic unit lie on sedimentary laminae of identical age, called *synchronous surfaces.*

Underlying this concept of layering is the idea that these synchronous surfaces are bedding planes or surfaces (Fig. 3.1A). These surfaces form during brief periods of nondeposition or change in the depositional energies. This belief is documented by observations of beaches, varves, ash falls, and fluvial deposits, to name a few. Comparison of these recent observations with ancient rocks illustrate identical bedding surfaces that are synchronous.

Such terms as *Era, Period,* and *Series* are standard time-stratigraphic terms. The *periods* are the fundamental units of geochronology *(Stratigraphic Code, Articles 29 and 37).*

### ROCK-STRATIGRAPHIC UNITS

A rock-stratigraphic unit is a ". . . subdivision of the rocks in the earth's crust distinguished and delimited on the basis of lithologic characteristics," according to the Commission. The unit shows the general distribution of rock types and it does not typically convey any information on rock layering or time. A rock-stratigraphic boundary is defined by observable differences in lithology and the lithologic change may cross layering and therefore time (Fig. 3.1B).

# THE STRATIGRAPHIC FRAMEWORK

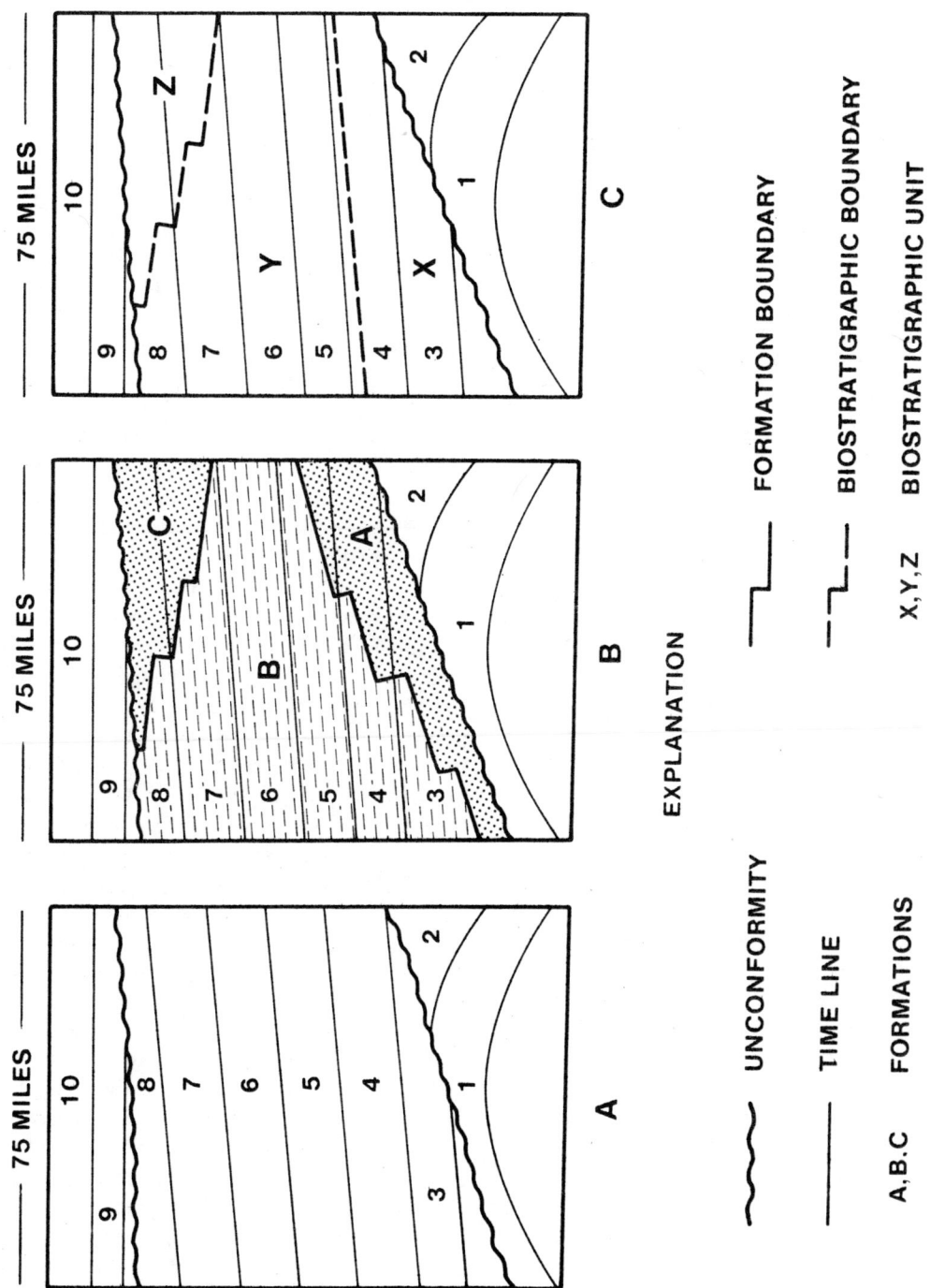

Figure 3.1: *Types of stratigraphic units used in exploration.*

Typical rock-stratigraphic terms include *group, formation,* and *member.* Each unit is subject to an arbitrary cut-off where the unit is no longer representative of the characteristics of the unit at its type section.

## BIOSTRATIGRAPHIC UNITS

The Commission defines a biostratigraphic unit as "... a body of rock strata characterized by its content of fossils contemporaneous with the deposition of the strata." Because fossils, both fauna and flora, reflect both environmental adaptations and evolutionary changes, biostratigraphic units record both time and environment. A biostratigraphic unit may either parallel or cross rock-stratigraphic and time-stratigraphic units.

Terms that describe biostratigraphic units include *zone, subzone,* and *zonule.* Note that the biostratigraphic units are also subjective to arbitrary cut-off (Fig. 3.1C).

## SEQUENCES: THE FUNDAMENTAL MAPPABLE UNIT

**Fundamentals of a Sequence** A sequence is an unconformity bounded unit *(Sloss, 1963; Wheeler, 1963).* The unconformities must be pan-continental and therefore associated with epeirogeny and/or eustatic changes in sea level rather than orogeny. Basal rocks of a sequence onlap an erosional surface during relative sea-level rise. Normally, the basal strata are clastics; however, carbonates are typical basal deposits of several Paleozoic sequences. Generally, basal strata of a sequence are time transgressive, and are more widely preserved than are the upper strata. If preserved, the upper part of a sequence depicts an offlap pattern that is most frequently deposited during sea-level stillstand. Sedimentary patterns around the globe suggest that sea-level drop is very rapid and little, if any, sediment is deposited during this time interval *(Mitchum and others, 1977).* Lithologies are varied, but evaporites, coastline, and continental deposits are most common. The thickness of the offlap sediments is generally less than that of the onlap interval, because of differential preservation.

Three kinds of missing geologic time are recorded in an unconformity; namely, the time represented by strata that were removed by erosion, the time during which erosion occurred, and the time during which nondeposition and onlap occurred. The hiatus represented by the unconformity is variable. It may range from very short intervals to hundreds of millions of years.

**Difficulties in Delineating Sequences** Geologists recognize six to eight pan-continental unconformities. These pan-continental unconformities are not easily recognized at all localities. The center of a craton is characterized by many unconformities, whereas deposition along continental margins is more continuous. Little difficulty is encountered where the unconformity is angular or a non-conformity. But disconformities, especially those that are characterized by little or no erosional topography, are sometimes impossible to recognize at a single locality. An excellent example of this problem occurs in southwestern Oklahoma. In northeastern Oklahoma, basal Kaskaskia rocks are in obvious unconformable contact with Sauk rocks of the Arbuckle limestone; but in southwestern Oklahoma, basal Kaskaskia rocks are in paraconformable contact with Tippecanoe strata represented by Hunton limestone. If geologic mapping is restricted to the southwestern part of the state, the explorationist would not recognize the pre-Kaskaskia unconformity and, therefore, miss much of the key history of the area related to possible generation and migration. Perhaps one of the important keys to a successful exploration program is that a knowledge of the regional geology of any exploration play will result in improving the success ratio of new discoveries.

**Practical Time-Stratigraphic Unit** Sequences are a practical time-stratigraphic unit. Strata in the sequence form a natural genetic unit of rocks that was deposited in a definite time interval. Time datum variance of rocks at the base of the sequence is expected; consequently, strata in the sequence may represent different time intervals in different parts of the area of deposition. However, strata of the sequence are everywhere younger than strata below the basal unconformity, and older than strata above the upper bounding unconformity. Rocks within the sequence may consist of

# THE STRATIGRAPHIC FRAMEWORK

different lithologies, but they probably had a common post-depositional history within a given geologic province.

**Geometric Patterns of Sequences**

*Lower Patterns: Baselap* is a progressive lapout of initial horizontally deposited sediments against the lower boundary of a sequence. *Onlap* and *downlap* are two important types of baselap. Downlap is the basinward termination of initially inclined strata onto a horizontal or inclined surface (Fig. 3.2). Deltaic front, mass transport, fore-reef, and fore-bank deposits are common depositional facies that form downlap stratal surfaces. *Onlap* is the landward termination of initially horizontal stratal surfaces onto an inclined surface of greater inclination. *Base-concordance* is used to describe those situations where basal strata of a sequence are parallel to the surface boundary of the sequence. Onlap, downlap, and base-concordance may occur during global sea level rise, fall, or stillstand.

*Upper Patterns:* Only two important types of upper depositional patterns are recognized. They are *toplap* and *concordance*. *Toplap* is characterized by initially inclined strata that terminate against the upper boundary of a sequence in a basinward direction (Fig. 3.2). If the stratal surfaces are parallel to the upper boundary, they are said to be *concordant*. Post-depositional erosion produces erosional truncation, which is the most common upper pattern of a sequence.

## DEPOSITIONAL SIGNIFICANCE OF SEDIMENTARY PATTERNS

### ONLAP

A relative rise in sea level causes a relative rise in base level which, in turn, permits the coastal aggradation and onlap of sediments. Low to moderate influx of terrigenous clastics during relative sea level rise is represented by successive stratal surfaces that depict coastal onlap (Fig. 3.3). Nonmarine coastal sediments are subordinate in thickness and areal extent to littoral deposits. Sand content is dependent upon the source area. Petroleum source rocks may be rare because sedimentation rates are slow. Slow sedimentation rates are not conducive to rapid burial of organics which is a factor in preserving organic material. Sands in the littoral environments will be relatively thin, but they may form sheet-like sands that are proximal to the basal unconformity. These sands are typically well sorted and may be excellent carrier beds as well as reservoir rocks.

A low influx of terrigenous clastics into a marine environment characterized by a tropical to subtropical environment favors the deposition of carbonate sediments. Figure 3.4 illustrates barrier reef (windward) and oolite

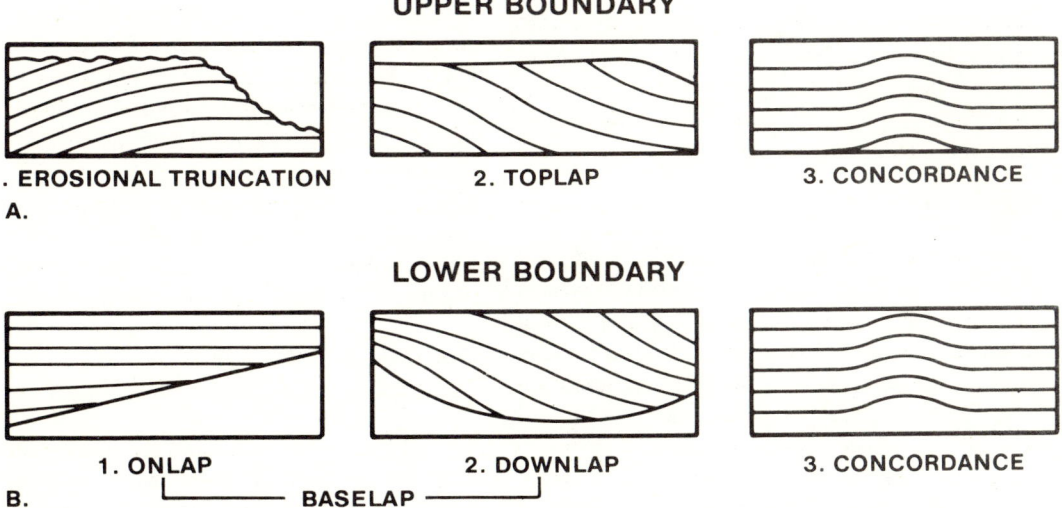

**Figure 3.2:** *Geometric patterns within a sequence (Mitchum and others, 1977).*

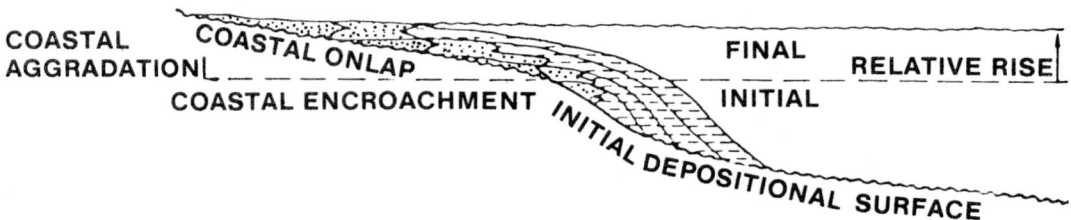

**Figure 3.3:** *Low to moderate influx of terrigenous clastics during sea level rise causes coastal onlap (Vail and others, 1977).*

bank (leeward) deposits on the shelf margin that prograde basinward, but coastal onlap will occur on the shelf. Potential reservoir rocks may occur in these facies or in the shelf facies, especially if the latter facies were dolomitized.

A high terrigenous influx of clastics during sea level rise results in extensive nonmarine coastal deposits (Fig. 3.5). Littoral deposits prograde seaward forming a relatively narrow, but thick belt of sediments. If the biologic community is abundant, accumulation of petroleum source rocks is favored as a result of rapid sedimentation rates. Potential reservoir rocks of the littoral environment will be in an envelope of potential source rocks.

Stratal surfaces formed in an environment characterized by a balance of terrigenous influx with relative sea level rise are similar to high influx of clastics, except where the shoreline remains relatively stationary. This produces a very narrow belt of littoral sediments that may be stacked (Fig. 3.6).

## TOPLAP

Coastal toplap indicates relative stillstand of sea level. Sea level stillstand inhibits a relative rise in base level on the shelf (sea level is not always base level). With no relative in base level, nonmarine coastal and/or littoral deposits cannot aggrade. Moderate to high influx of terrigenous clastics causes sediment by-passing and toplapping of coastal and littoral sediments across a topographically featureless shelf (Fig. 3.7). Accumulation of potential petroleum source rocks is commonly restricted to the shallow marine deposits just basinward of the prograding littoral environments.

## INFLUENCE OF SEA-LEVEL CHANGES ON SEDIMENTARY PATTERNS

Figure 3.8 illustrates the trends in the rise and fall of sea level during Phanerozoic time. Analysis of the global sea level changes reveals that sea level rises are gradual. Where detailed data exists, sea level rise is more accurately depicted by a relatively rapid rise followed by a relatively long period of stillstand. After maximum sea level has been reached during a given cycle, a fairly significant amount of geologic time is represented by sea level stillstand.

In contrast to pulses of sea level rise followed by significant periods of stillstand, sea level fall is generally a very rapid process (Fig. 3.8). A seaward shift in onlap is produced by the fall. Erosion typically removes much of the underlying coastal sediments, as well as sediments deposited during the previous interval of sea-level stillstand. The unconformity produced by erosion is generally concordant and frequently very difficult to identify.

## MAPPING SEQUENCES

### INTRODUCTION

Perhaps one of the most challenging decisions made by an explorationist when assigned a new area is what to map. Most frequently, this decision is made for us by our predecessors because it is commonly easier to update pre-existing maps than to "generate" new ones. In

# THE STRATIGRAPHIC FRAMEWORK

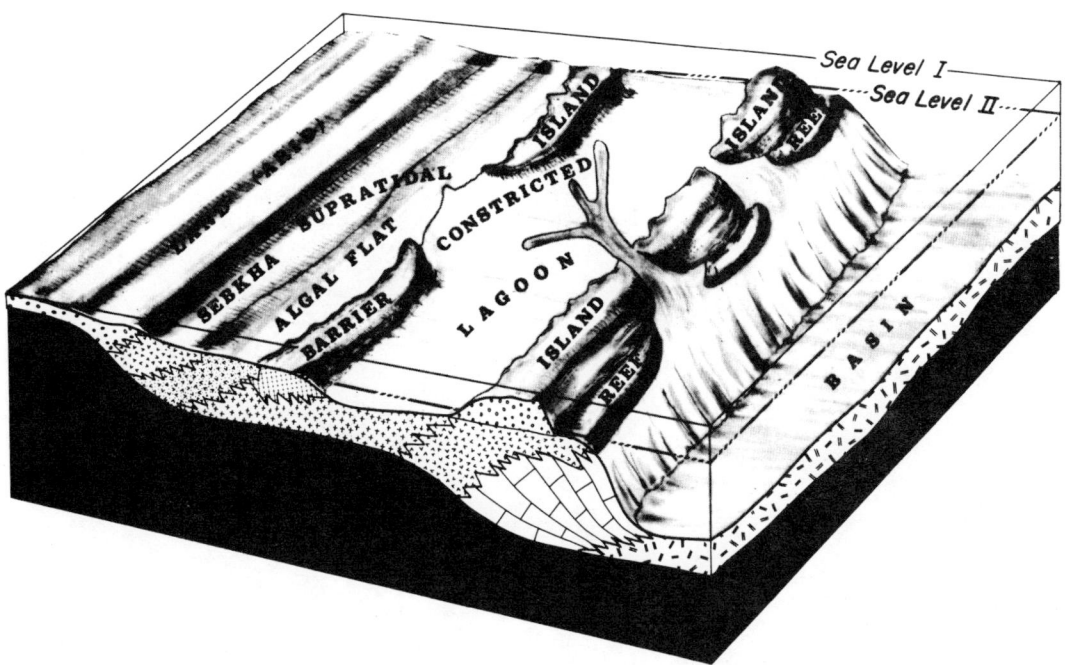

**Figure 3.4:** *Low influx of terrigenous clastics in a tropical to subtropical climate favors accumulation of carbonate sediments (Silver and Todd, 1969).*

most cases, regional structural maps are built on a reservoir rock that is productive throughout an area or basin. Most sandstone and carbonate reservoirs depict varying amounts of depositional topography and most cross time surfaces. Therefore, these rock stratigraphic units are not always valid indicators of structure. The technical success of an exploration program is highly influenced by the selection of mappable units that not only accurately depict structure, but also correctly delimits coeval rock bodies.

## GEOLOGIC TECHNIQUE

### Procedure

1. Gross time-stratigraphic correlations can be facilitated by reducing the mechanical logs to 1 inch equals 200 feet. Reduction via a copier, for example-Xerox, is sufficient because exact reduction is not necessary for these general correlations. Better quality copies can be obtained by overlying the blue-line log with a yellow transparency.
2. Construct a network of cross sections, preferably parallel and perpendicular to depositional strike (parallel to interpreted coast line).
3. Correlate low depositional energy deposits, *i.e.*, shales, siltstones. In contrast to sands and carbonate grainstones and boundstones, shales and siltstones are generally more continuous and they are more reliable time surfaces.
4. Draw all possible correlation lines between adjacent wells. Numbering each correlation line enhances the organization of the section.
5. Analyze the resultant "pattern correlations" for subtle changes in thickness.
6. Mark the position of all correlation line terminations that are related to the thickness variations.
7. The terminations should depict onlap or downlap patterns.
8. Draw in the sequence boundaries below the onlap or downlap zones and above their truncation zones.

The initial step in selecting valid mappable horizons within a sequence involves the determination of detailed correlations. At least four different types of data are used for delineating time-stratigraphic units. They include (1) marker beds, (2) biostratigraphic facies, (3) lithogenetic facies, and (4) seismic profiles.

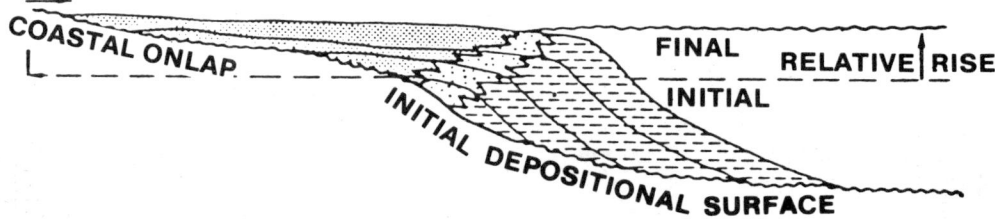

**Figure 3.5:** *High influx of terrigenous clastics during sea level rise results in coastal onlap (Vail and others, 1977).*

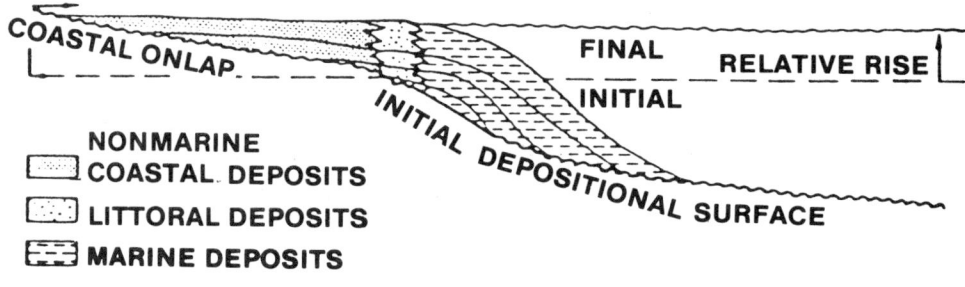

**Figure 3.6:** *Balance of terrigenous clastic during sea level rise results in coastal onlap, but a stationary shoreline (Vail and others, 1977).*

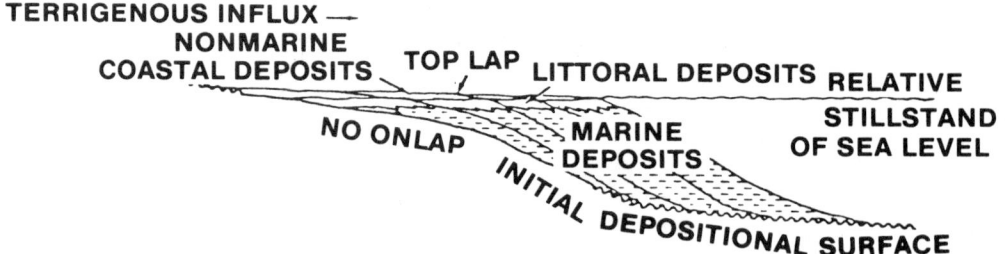

**Figure 3.7:** *Toplapping geometric patterns are produced by sea level stillstand (Vail and others, 1977).*

# THE STRATIGRAPHIC FRAMEWORK

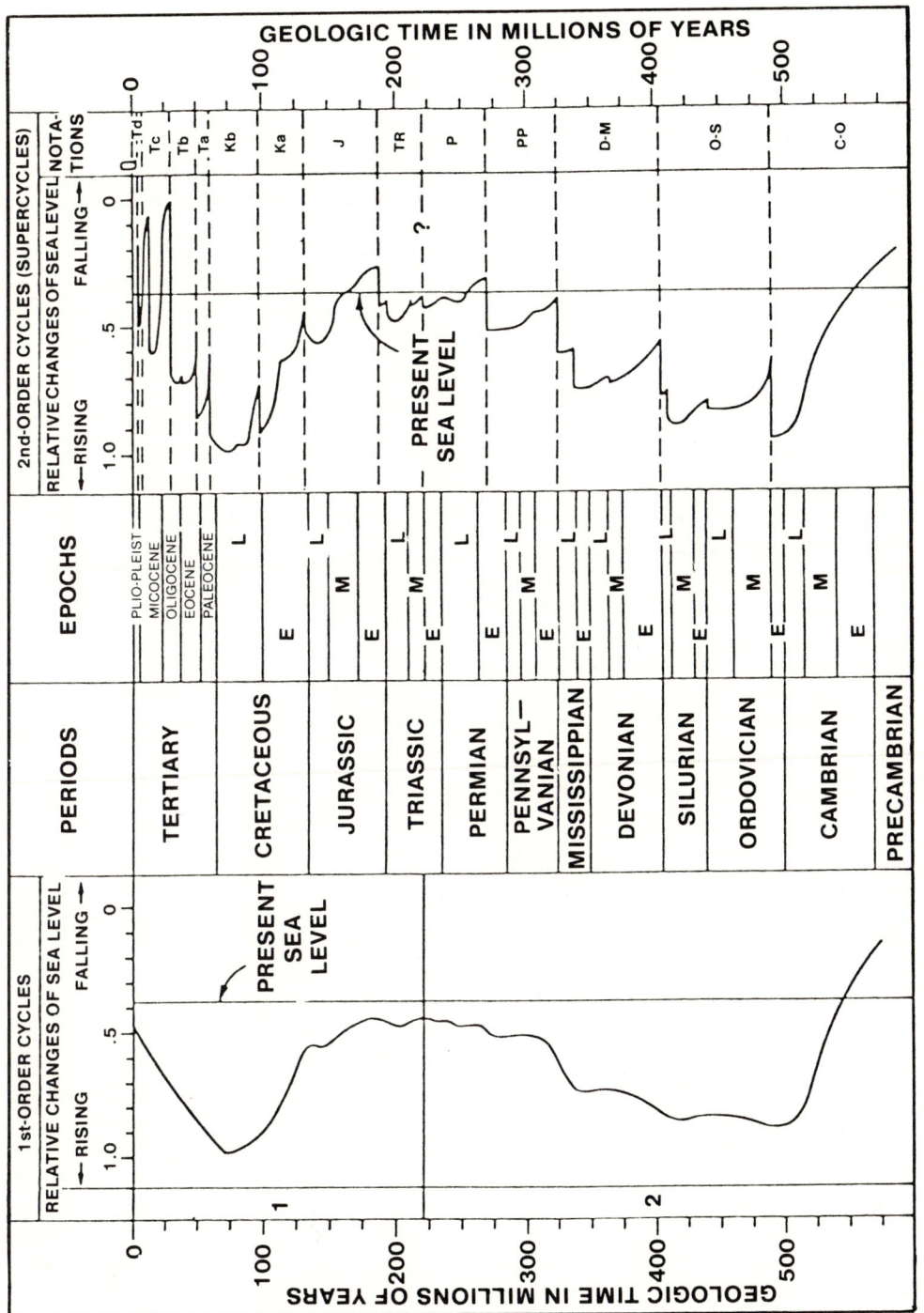

**Figure 3.8:** *Global cycles of relative changes of sea level during Phanerozoic time (Vail and others, 1977).*

**Marker Beds** Marker beds are the paramount correlations used by explorationists for stratigraphic studies within a basin. Marker beds may be a regionally consistent (1) lithologic unit or bed, (2) faunal unit, or (3) mechanical log response that is easily recognized. Typically, the best marker bed occurs within a shale interval and is identified by one or more of the resistivity curves (Fig. 3.9). Commonly, the markers are thin individual beds or groups of beds that show close agreement with time surfaces that are defined by time-indicative fossils or bentonites. Increased validity of the markers is indicated by a number of markers that depict parallel patterns.

**Faunal Markers** Significance of time-indicative fossils in exploration not only depends upon rapid changes in the species, but also their size. Most subsurface sample data comes from cuttings. Therefore, the time-indicative fossils must be micro in size. The micro-fossils are the most practical means of placing the gross stratigraphic section of a basin into the global time scale, although seismic data may be more practical, particularly in non-drilled basins. Faunal markers are rarely precise enough for the detailed mapping required for the generation of prospects, but they are useful for development of a physical stratigraphic framework.

**Lithogenetic Rock Units** Lacking marker beds and fauna markers, particularly in areas characterized by cyclic depositional patterns and/or depositional topography, the explorationist is hard pressed to develop a time-stratigraphic framework. In such circumstances, the explorationist must use Walther's Law and attempt, through detailed analysis, to reconstruct the depositional environment and correlate rock bodies that are lithogenetically related. For example, Wells 2 and 3 (Fig. 3.9) contain two thick sands that do not have any apparent physical similarities. However, they can be correlated on the basis of lithogenetic rock bodies. Well 1 penetrates deltaic swamp and marsh shales, Well 2 penetrates a deltaic plain sand, and Well 3 penetrates a deltaic front sand and prodeltaic shales. Walther's Law states that in a continuously deposited sequence of strata, the vertical distribution of rocks is a manifestation of the lateral distribution of strata. Consequently, the pattern correlation lines depicted on Figure 3.9 support the thesis that the two sands are correlative and lithogenetically equivalent.

## GEOPHYSICAL TECHNIQUE

Configuration of time surfaces within sequences can be determined by "pattern correlation" of reflections. The following procedure is recommended for mapping seismic sequences.

**Procedure**
1. If possible, select a network of depositional strike and dip seismic lines that have been processed on a stratigraphic datum.
2. Analyze the section(s) for changes in thickness.
3. Locate seismic cycle termination within a zone(s) of thickness changes.
4. Place arrows in the direction of cycle terminations and draw in sequence boundaries beneath onlap and downlap zones above the truncated zones.
5. Extend sequence boundaries over the rest of the section wherever cycles and strata are conformable.
6. The above interpretations may be tested and refined by correlation throughout the net of seismic lines.

**Some Pitfalls** There are some pitfalls in using seismic data. Not all cycle terminations or thickness changes are indicative of sequence boundaries *(Sangree and Widmier, 1974)*. Thickness changes on the shelf that are caused by algal mounds or other carbonate buildups do not, of course, indicate a sequence boundary. Thickness changes produced by deltas, barrier reefs, or banks along a shelf margin can locally be misinterpreted as sequence boundaries. Submarine fans, because of their low depositional topography, have been mapped as delineating sequence boundaries.

## DETAILED TIME-STRATIGRAPHIC CORRELATION

### PROCEDURE

After successful delineation of the sequences

# THE STRATIGRAPHIC FRAMEWORK 49

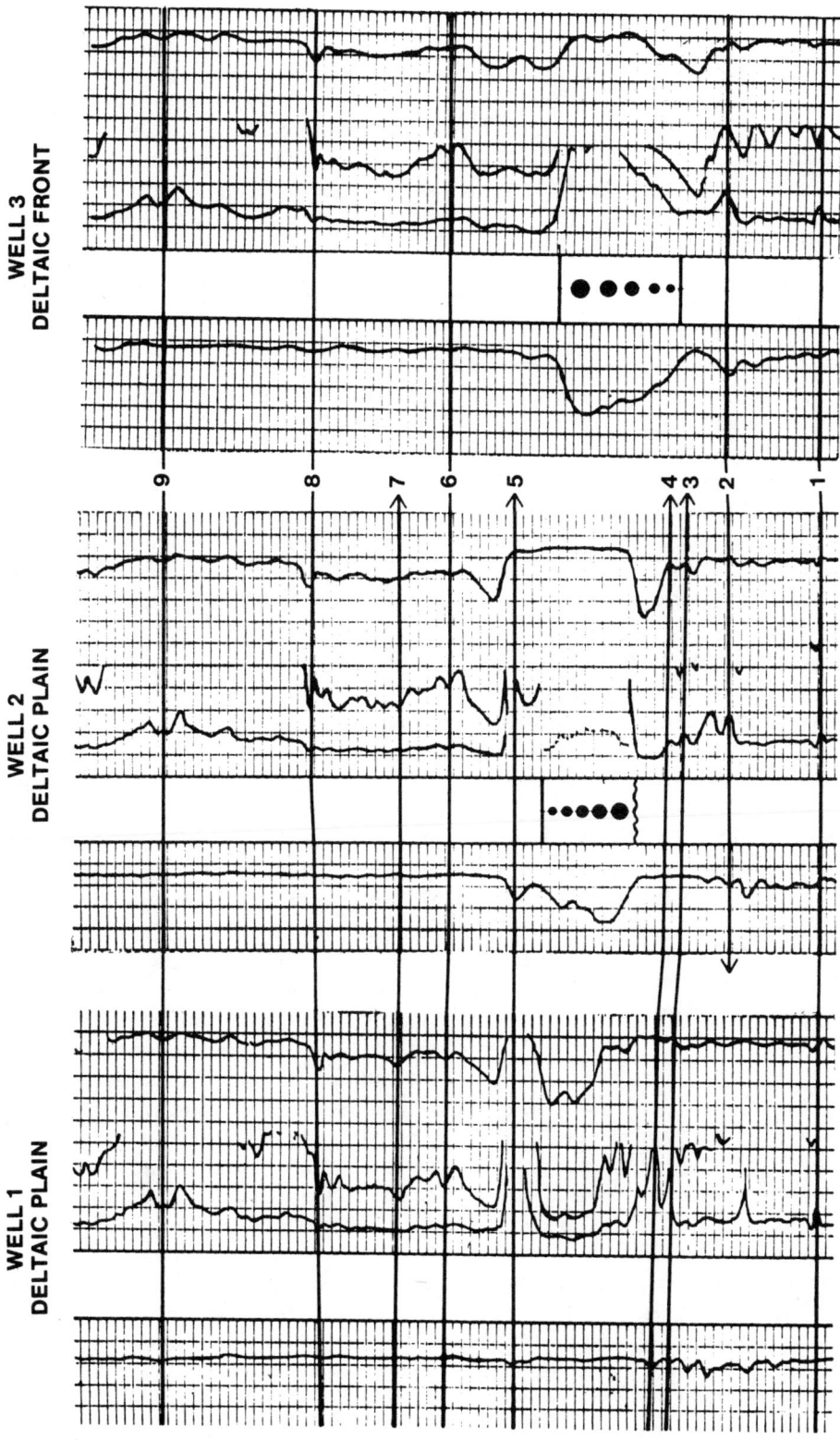

**Figure 3.9:** *Pattern correlations indicate that the distributary sand in Well 2 is the lithogentic equivalent of the channel-mouth bar sand in Well 3.*

are completed, all subsurface (and surface) sections should be correlated in detail. All sample data should be plotted in the center of the mechanical logs for best results. Although mechanical logs frequently contain more detail in thick units, such as shales, sample data may show a lithologic marker not reflected in the mechanical log and the sample data is necessary to accurately depict the rock types encountered in the well.

Construction of a detailed systematic vertical pattern of closely spaced correlation lines between wells is necessary in order to obtain the full benefit from the stratigraphic sections. The vertical spacing of correlation lines should be as close as possible in order to help guide "weaker" or more subjective correlations in more complex stratigraphic intervals. Although a majority of the correlations between adjacent wells may not be continuous between the first and third wells, the pattern will be a guide for adding confidence that the first and third wells are correlated properly (Fig. 3.9).

Particular attention should be given to the detailed resistivity log character in thick shales because experience has demonstrated that these correlations approximate time surfaces. Similarly, thin silts, shales, and anhydrites yield excellent correlations. Correlation of boundaries of thick sandstones and carbonates are not typically time surfaces. Experience has shown that narrow peaks, "valleys," or shoulders of the resistivity curve(s) are the most reliable log characters for detailed correlations. However, the test of a certain curve configuration is to determine whether or not it carries from well to well as part of a general systematic pattern.

## CHECKS ON CORRELATIONS

Two techniques are useful to check correlations. They are tying loops and isopach maps. A network of stratigraphic cross sections should be constructed parallel and perpendicular to the sedimentary strike of the sequence under investigation. Of course, correlations of the common wells in the network of cross sections should be checked to make sure the correlations are consistent. Moreover, each well not on the network of cross sections should be correlated to the nearest cross-section that is located on sedimentary strike with the well. Circular correlation loops of the non-cross section wells should be made continuously to assure consistency of correlations. Numbering of each pattern correlation line facilitates the cross-checking procedures.

Interval thickness maps are useful to help delineate anomalous thicknesses and gradients. Each map anomaly should be checked against the network of stratigraphic cross-sections in order to determine if the map anomaly is real or a correlation "bust."

## SIGNIFICANCE OF STRATIGRAPHIC CROSS SECTIONS

### INTRODUCTION

Correlation lines on a network of stratigraphic cross sections depict depositional, contemporaneous, and post-depositional thickness anomalies. The origin of each anomaly should be carefully determined by cross checking the correlations, and integrating them with faunal, lithologic and structural data. Interpreting the origin of each anomaly is critical to a successful exploration program.

### DEPOSITIONAL ANOMALIES

**Clastic Anomalies** Depositional thickness anomalies can be produced by clastic and carbonate depositional patterns. Thickness anomalies can be produced by sand bodies. Bar sands will produce linear thickness anomalies that are typically narrow (Fig. 3.10). Correlation markers in coeval shales will terminate against the bar. Channel sands will also produce linear thickness anomalies, but unlike bar sands, their shape may vary from linear to crescent. Due to erosion, markers within the coeval shales will be truncated by the channel sand. Individual alluvial fans will produce fan-shaped rock bodies, but coalescing fans will form a linear, wedge-shaped anomaly. Post fan correlation markers will depict onlap patterns. Deltaic front sands typically produce banana-shaped rock bodies. Coeval markers in the prodelta and deltaic plain deposits will terminate against the sand. Post sand markers will depict onlap patterns. Thickness anomalies produced by submarine fans will conform to the topography and axis of the basin. Certain correlation markers within coeval shales should extend through the fan deposited rock body.

Correlation markers in shales above the

# THE STRATIGRAPHIC FRAMEWORK 51

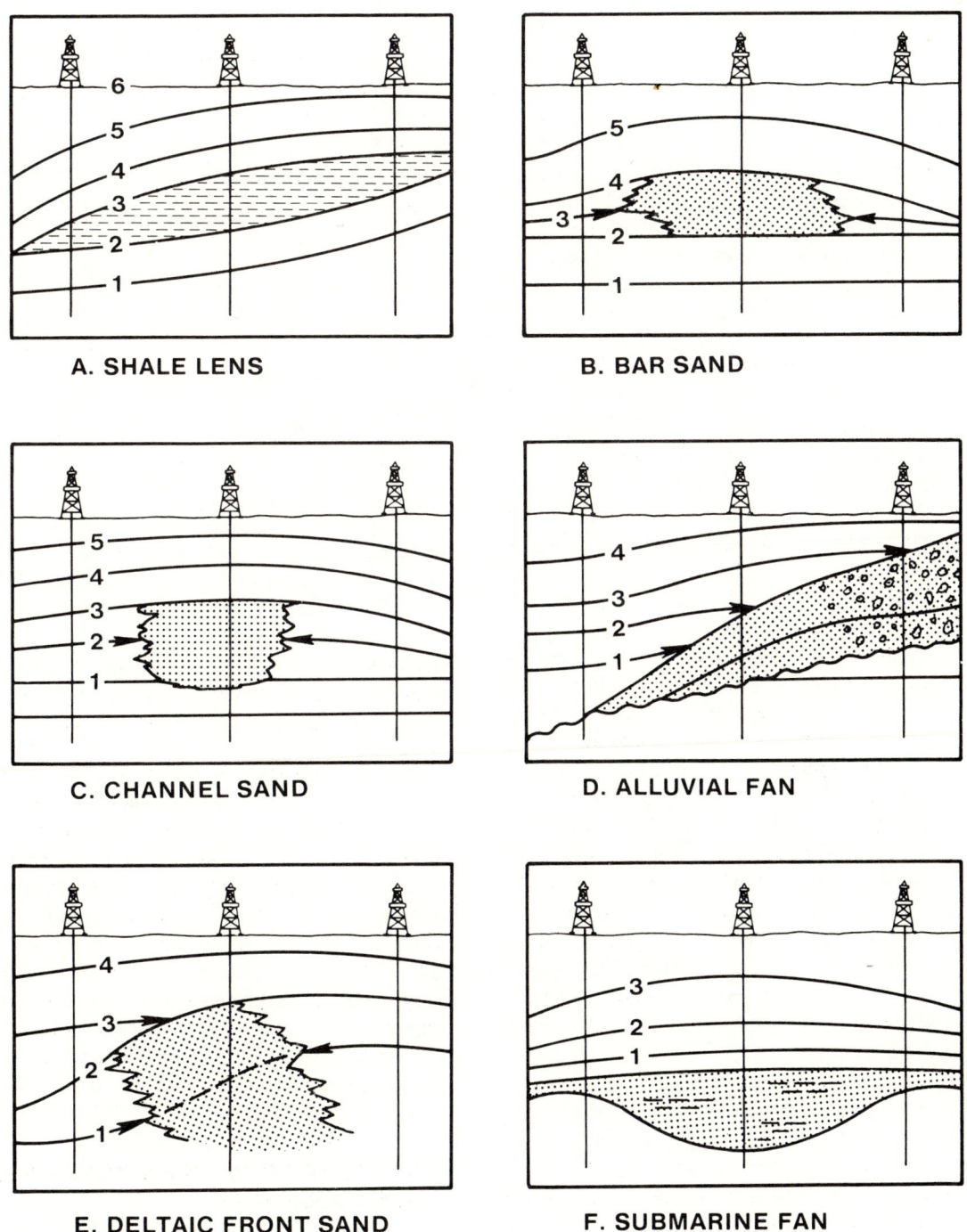

**Figure 3.10:** *Depositional thickness anomalies in clastics. Termination of marker beds indicated by arrow.*

sand bodies are normally convex. This is attributed to differential compaction. Shales, of course, will compact more than sands and, therefore, the markers appear to be draped over the individual sand bodies.

**Carbonate Anomalies** Barrier reefs and barrier banks will produce linear thickness anomalies (Fig. 3.11). Correlation markers within coeval deposits of the reef or bank will terminate against the carbonate buildup. Post reef or bank markers will onlap the buildups. Pinnacle buildups will produce thick, circular anomalies. Coeval correlation markers will terminate against the buildup. Carbonate mounds and shoals will produce minor thickness anomalies. Typically the anomalies will cover a larger area relative to their thickness than pinnacle buildups. Because carbonate sediments are commonly cemented during accumulation, as well as during burial, they compact very little. Consequently, post buildup markers normally drape over a buildup.

**Evaporite Anomalies** Without regional data, it is sometimes difficult to determine the origin of thickness anomalies that are related to evaporite deposits (Fig. 3.12). Both depositional and post-depositional thinning can produce isopach maps that illustrate constant contour spacing. Both processes can produce isolated "pods" of evaporites. However, anomalies produced by solution should be indicated by regionally consistent contour spacing that is abruptly changed in the area of thinning. Correlation markers above the evaporite solution anomaly typically will depict concave downward patterns in the area of the anomaly. In contrast, correlation markers above the depositional anomaly will not be influenced by the anomaly.

**Regional Anomalies** Depositional thickness anomalies may be produced by regional depositional patterns (Fig. 3.13A and B). Onlap and downlap patterns will produce regionally consistent isopach contour spacing that will thin landward and seaward respectively. Onlap patterns are characterized by regionally extensive marker beds that terminate against an inclined erosional surface. Downlap patterns will be denoted by marker beds that converge in a basinward direction with the marker beds extending further seaward in an upward direction.

Draping over a subsiding high can produce thickness anomalies (Fig. 3.13C). Typically, marker beds are regionally extensive and produce excellent patterns. Differential subsidence, either locally or regionally, results in thickness anomalies (Fig. 3.13D). Marker beds converge away from the subsiding area. This anomaly is more common in clastic depositional patterns than in carbonate environments.

## POST-DEPOSITIONAL ANOMALIES

**Truncation and Apparent Anomalies** Post-depositional anomalies can be produced by truncation, apparent thickness changes, and faulting (Fig. 3.14). Truncated anomalies will be characterized by regional thinning with consistent isopach contour spacing. Marker beds will terminate against the erosional surface in order from top to bottom. Apparent thickening may be produced by steep dips. This type of anomaly can be recognized by selecting a datum for a stratigraphic cross section that occurs at the base of the strata that appears to be thickening. The datum should preferably be an unconformity or a regional marker bed just above the unconformity.

**Fault Anomalies** Post-depositional thickness anomalies may be caused by apparent interval thickening or thinning by fault. If a normal fault cuts a wall, apparent thinning will occur; if a reverse fault cuts a wall, apparent thickening will be implied. The anomalies are easily recognized because they are limited to the faulted well. The missing (or duplicate) section can be determined by pattern-correlating the faulted well to non-faulted wells. The faulted well should be ignored when contouring an isopach map that includes the missing (or duplicated) section.

## ADVANTAGES OF STRATIGRAPHIC CROSS SECTIONS

### HIGHLIGHTS POTENTIAL SOURCE AND RESERVOIR ROCKS

The delineation of sequences highlights the stratigraphic position of potential source and reservoir rocks. Because a sequence is differentially preserved, preservation of potential source rocks and reservoir rocks is more likely in the base of a sequence than in its upper part. Furthermore, generation of petroleum is more likely in the deeper buried basal deposits than in the shallower buried upper strata. Once the sequences have been delineated, evaluation of known petroleum occurrences within the study area, or comparing them to other areas of known occurrences and similar geologic situations, the exploration program can be designed to find a specific type of trap.

## OPTIMUM SELECTION OF MAPPABLE UNIT

Stratigraphic cross sections permit the selection of proper mappable units. They help the explorationist to (1) select the most accurate time-stratigraphic correlations that will best depict the type of trap he or she is exploring for, (2) make sure facies changes are recognized, (3) make sure interval thickness changes are not interpreted as facies changes, and (4) understand the vertical distribution of potential source and reservoir rocks.

## ACCURATE DEPICTION OF PALEOTOPOGRAPHY

Stratigraphic cross-sections permit the explorationist to map paleotopography and unconformity surfaces. For example, the amount of thickening caused by onlap is an indication of paleoslope on the erosional surface. This can be an important factor in attempting to explore for paleogeomorphic traps.

## PHYSICAL STRATIGRAPHIC FRAMEWORK AIDS IN INTERPRETING SEISMIC PROFILES

Direct interpretation of seismic profiles is difficult without integrating all subsurface data into the interpretation. Because seismic reflections typically follow layering surfaces, an understanding of the layering surface greatly facilitates interpretation of seismic profiles where well control is scarce or absent.

## DELINEATION OF CARRIER BEDS AND MIGRATION PATHS

Migration of petroleum, whether in a single or dual phase, is largely controlled by regional layering. Detailed physical stratigraphic correlations not only permit the delineation of carrier beds, but by hanging the sections on different unconformities, paleohydrologic gradients can be interpreted.

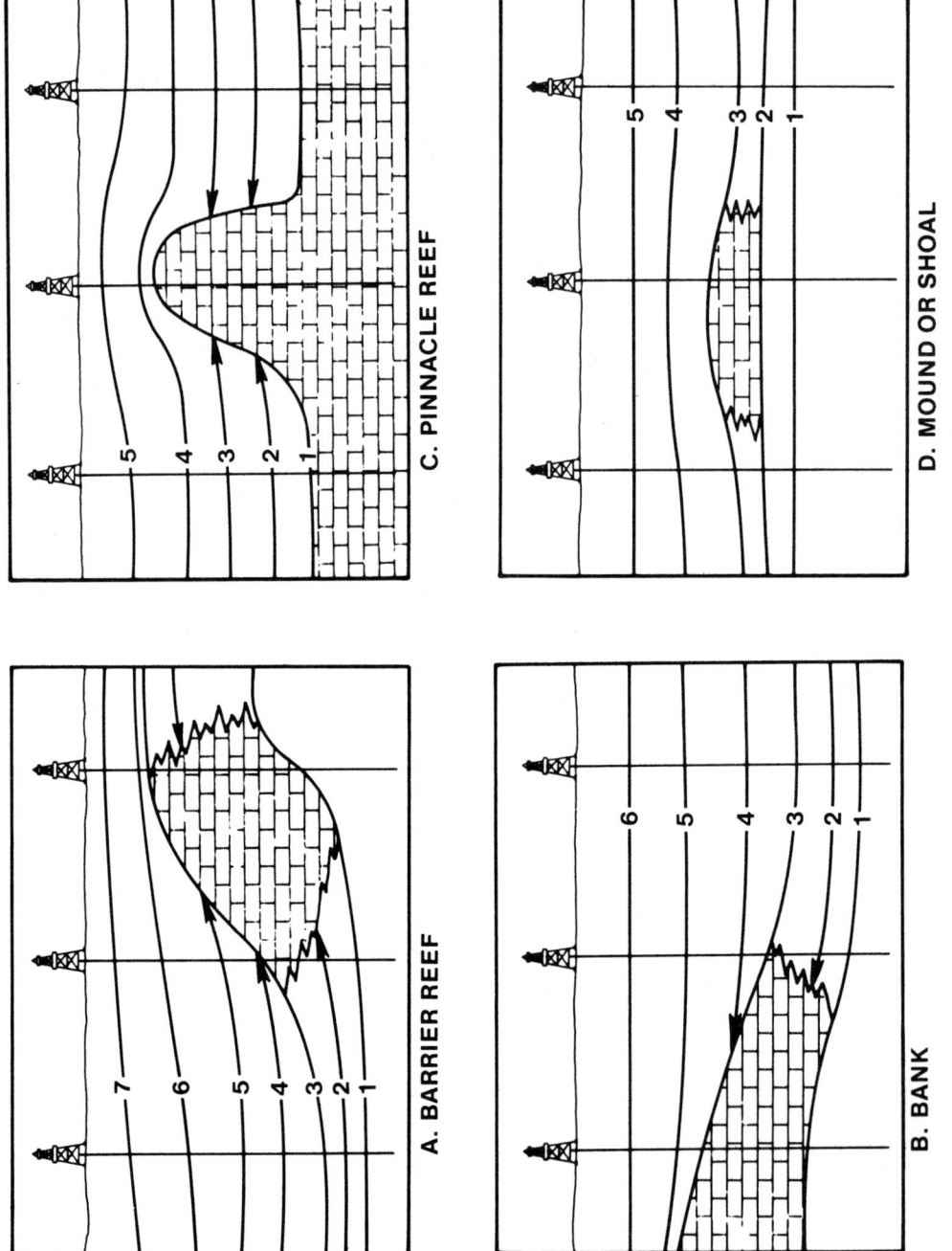

**Figure 3.11:** *Depositional thickness anomalies in carbonates. Termination of marker beds indicated by arrow.*

# THE STRATIGRAPHIC FRAMEWORK

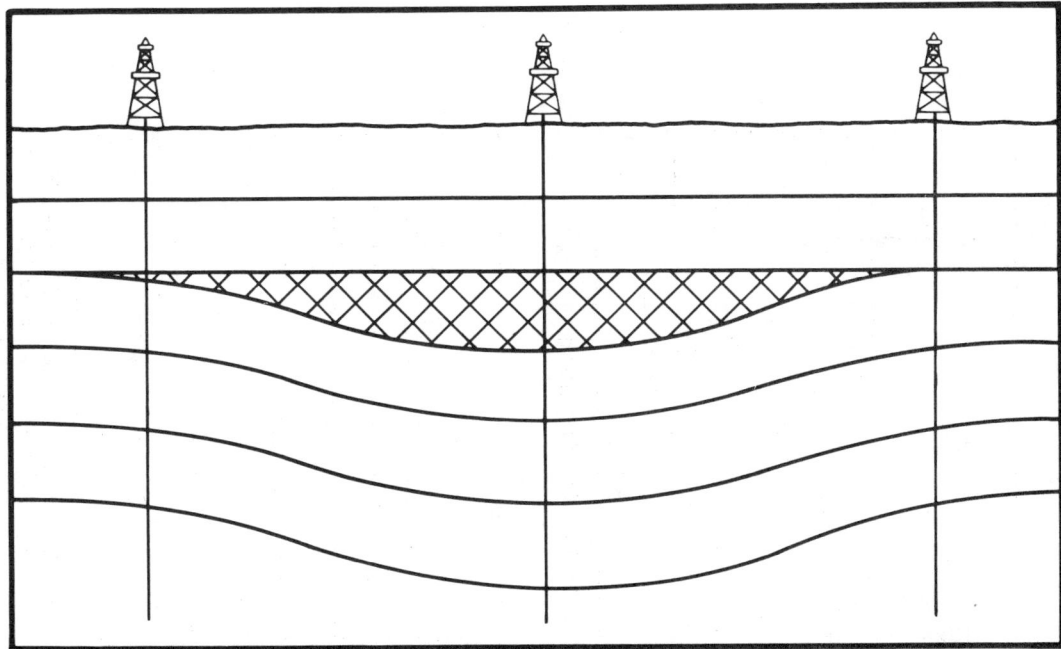

**A. SALT LENS**

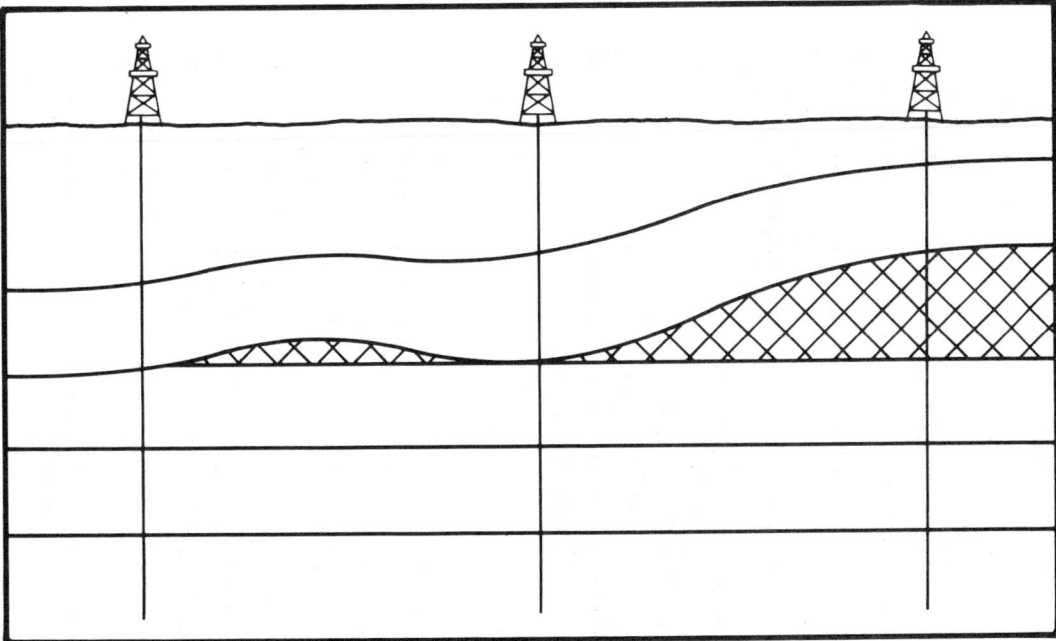

**B. SALT SOLUTION**

**Figure 3.12:** *Salt thickness anomalies can be produced by depositional and post-depositional processes.*

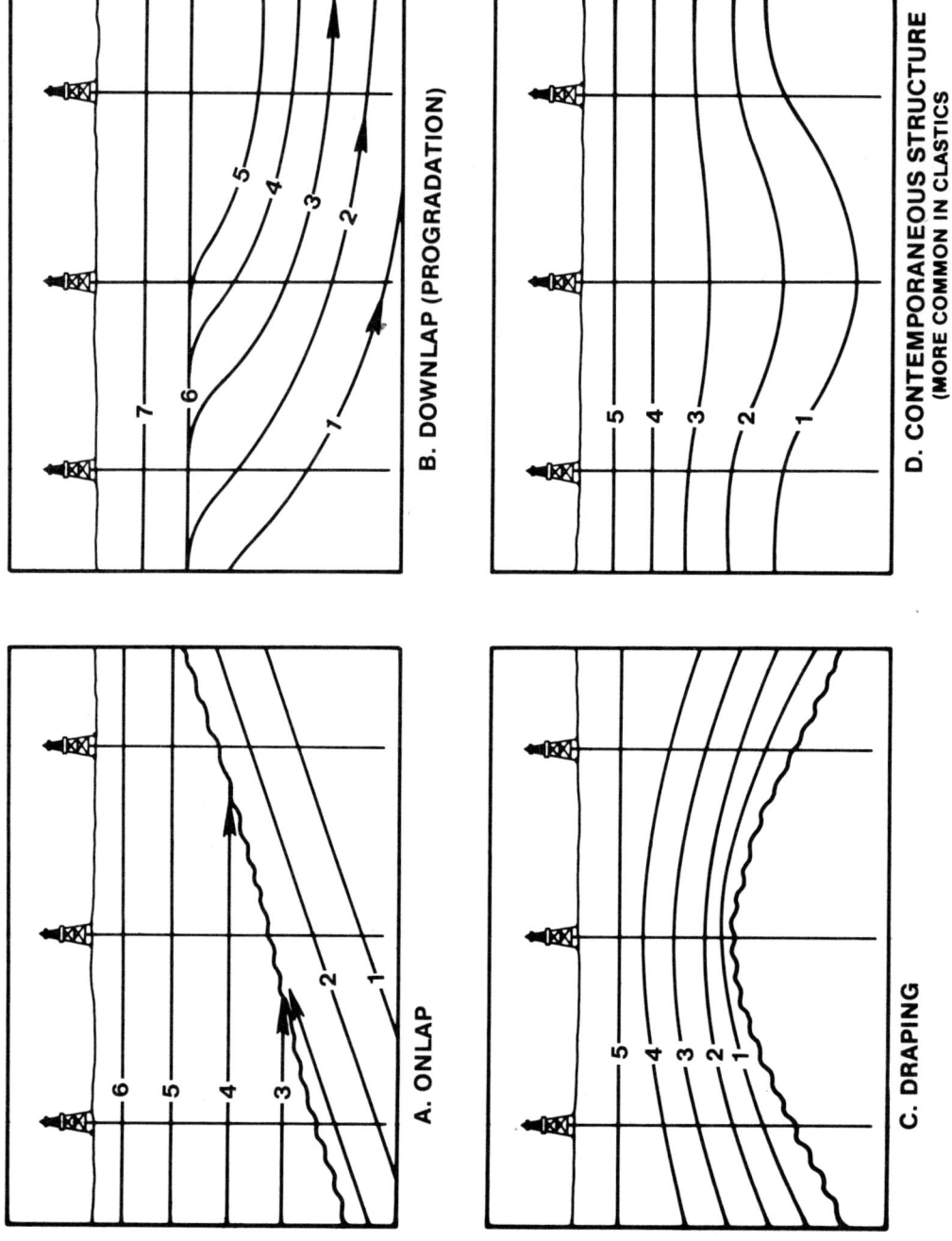

**Figure 3.13:** *Depositional thickness anomalies in clastics and/or carbonates. Termination of marker beds denoted by arrow.*

# THE STRATIGRAPHIC FRAMEWORK

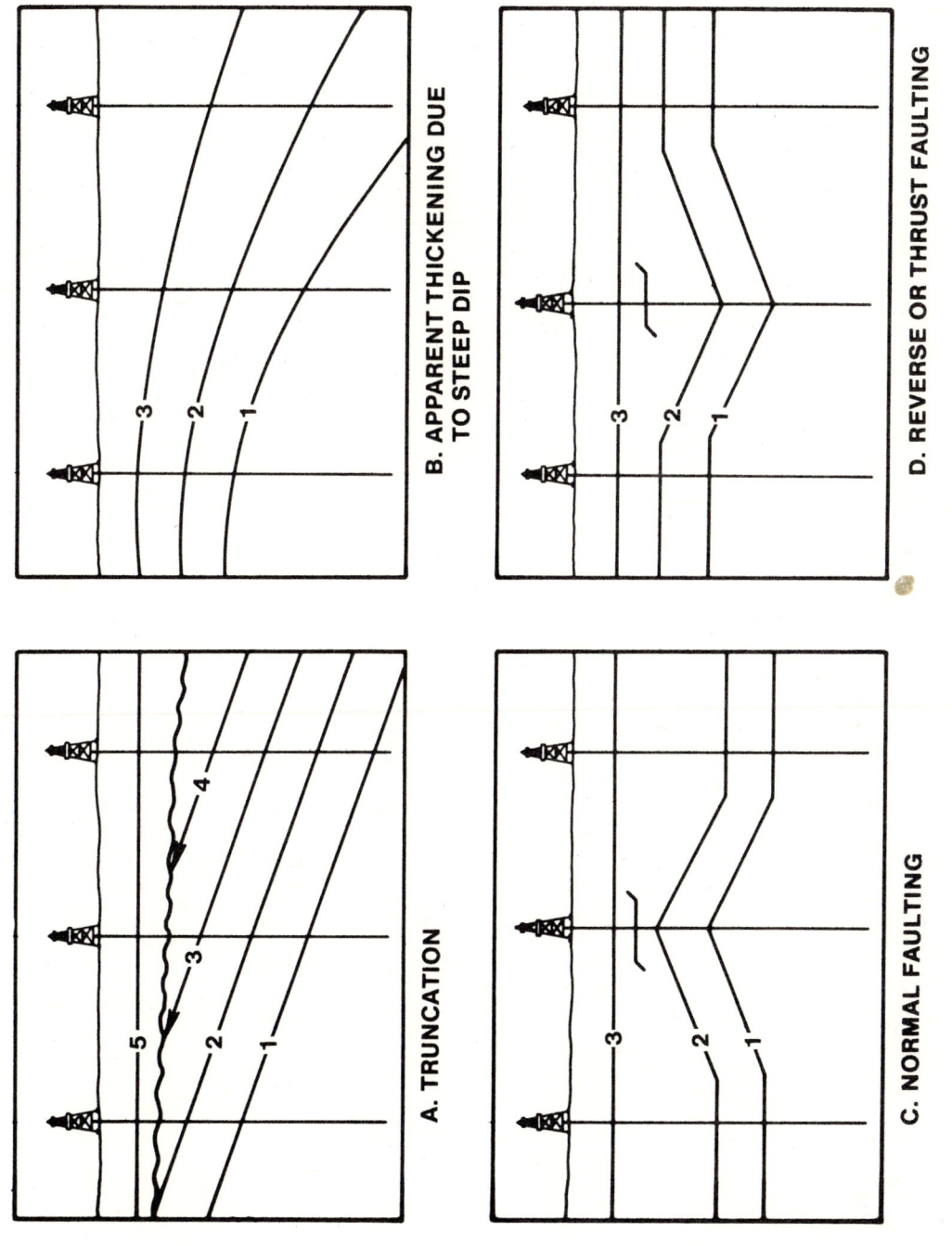

**Figure 3.14:** *Post-depositional thickness anomalies in clastics and carbonates. Termination of marker beds denoted by arrow.*

# PART II:
# EXPLORATION FOR SANDSTONE RESERVOIRS

4  *Exploration for Continental Sandstone Reservoirs*
5  *Exploration for Delta and Fan Delta Reservoirs*
6  *Exploration for Interdeltaic Sandstone Reservoirs*
7  *Exploration for Offshore Marine Sandstone Reservoirs*
8  *Diagenesis of Potential Sandstone Reservoirs*

# 4
# Exploration for Continental Sandstone Reservoirs

## METHOD OF APPROACH

### RESERVOIR - TRAP EVALUATION

A reservoir rock is herein considered to be any rock from which commercial quantities of hydrocarbons can be produced. The delineation of a sedimentary deposited reservoir presents the explorationists with a two-fold problem; namely, location of the reservoir and reservoir quality. Reservoir location involves definition of not only its stratigraphic position but mapping its geometry, extent, and orientation. Success is based on the accuracy of basin analysis studies and the recognition of depositional systems which "filled" the basin. Reservoir quality (porosity and permeability) considerations may be focused almost entirely upon the depositional environment, if generation, expulsion, and migration of hydrocarbons occurred prior to cementation. If, on the other hand, late generation occurred, porosity and permeability will be a function of the diagenetic history of the reservoir. In general, the best sandstone reservoirs, in terms of volume and production rate, are those that contain intergranular porosity. Sandstone reservoirs with a significant amount of microporosity and intragranular porosity generally have low permeability and therefore, low production rates.

Trap considerations are focused on the spatial distribution of the reservoir with respect to the associated seal that produces the trap. Some of the possibilities include: (1) unconformities (Prudhoe Bay Field, Alaska; *Jones and Speers, 1976*), (2) paleotopography (Martin Hills Field, Alberta; *Silver, 1980*), (3) lateral changes due to diagenetic patterns (Lyons Sandstone, Denver Basin, Colorado: *Levandoski and others, 1973*), (4) lateral changes due to facies changes (Bell Creek Field, Montana; *Davies and Berg, 1969*), and (5) lateral and vertical changes due to variations in capillary pressures *(Berg, 1975).*

### GEOLOGY BY ANALOGY

After the exploration team has developed an accurate physical stratigraphic framework, the next task is to develop an understanding of the depositional environment of each potential reservoir rock in the area under investigation. This knowledge will provide the team with the ability to predict the distribution and geometry of the potential reservoir rock(s) as well as afford the team with a feeling for the variations in reservoir quality and continuity. Although the geologist has in the past been solely responsible for this phase of the exploration program, the seismic interpreter is becoming an increasingly important contributor to this task.

Geologists have recorded a wealth of data on the internal and external features that are characteristic of modern sandstone and carbonate deposits. For modern sandstone deposits, grain size, texture, type and scale of sedimentary structures, geometry, and thickness variations are considered diagnostic. In contrast, skeletal content, grain type, fabric, nutrient supply, wind directions, and depositional topography are the most characteristic features of modern carbonate deposits. These sets of data have been cataloged for each modern sandstone and carbonate environment.

The explorationists can use well cuttings and cores to measure the internal features of rock bodies. This data can be plotted on SP and/or GR logs to calibrate these tools for a given suite of rocks and extend the measurements to wells where cuttings and/or core data is not available. Stratigraphic cross sections, isopach maps, and seismic sections are excellent tools to measure the external features of potential sandstone and carbonate reservoirs. The internal and external measurements can be compared to the data derived from modern environments, and by analogy, the exploration team can reconstruct the depositional environment of the rocks in question. Thus, geology by analogy is a powerful technique that must be used by the exploration team.

## SOME PITFALLS

One of the most compelling reasons for studying modern depositional processes is that exploration models may be constructed for use in the interpretation of reservoir rocks. In areas characterized by limited data, the modern models are used by explorationists to predict where, as yet undiscovered, reservoir rocks may be developed. However, rigid attachment to a particular model or dogmatic assertion of its applicability to the ancient may subordinate facts that do not fit the model and produce erroneous interpretations. Furthermore, there are several pitfalls that must be considered prior to the direct application of the modern to the ancient. Some of these are discussed below.

**Snapshot in Time** The modern depositional environment is a snapshot in geologic time. Geologists are able to view the various facies of an area at a given instant, for example, the levee, active channel, point bar, and abandoned channel of a meandering stream. At any instant in time the facies of a modern environment are considered to be in equilibrium with the physical and/or biological parameters that control the facies. However, tectonics, sea level changes, and/or storms can produce disequilibrium and, in time, perhaps the environment will "re-equilibrate" and produce facies patterns that are characteristic of the dominant parameters. However, the question remains, which will be preserved, the "disequilibrium" pattern, the equilibrium pattern, or a combination pattern?

For the explorationists engaged in facies analysis, it is obvious that the lateral facies of a modern environment can be observed directly, whereas the vertical facies must be interpreted or mechanically sampled for observation. In contrast, facies analysis of ancient strata is most frequently based on the vertical relationships of strata, and the lateral facies must be interpreted. According to Walther's Law of Succession of Facies, only those facies that can be observed beside one another can be superimposed. However, this assumes that there are no major breaks in the stratigraphic section and that the environment always obtains equilibrium. Clearly, the succession observed in the modern environment is a snapshot in time and it is most unlikely that the vertical succession of the ancient analogy will display all of the facies developed laterally.

**Scale** The magnitude of a depositional system is, in part, controlled by tectonics, sea level stability, depositional topography, and depositional site. The present day globe is quite unique relative to the Phanerozoic globe. Currently, there is more land surface, relief, life forms, and tectonic activity than geologists interpreted for the Phanerozoic. Consequently, direct application of the present models is sometimes difficult.

**Climate** Climate clearly influences modern depositional environments. Ancient climate regions probably differed from the global patterns that exist today. Furthermore, even if the climatic patterns have not been significantly altered, most explorationists would agree that the continents have passed through several climatic belts during continental drifting. Perhaps climatic variations influence carbonate environments more than clastic environments. Geologists study any given modern carbonate setting in terms of climate, *i.e.*, Persian Gulf vs. Bahamian Platform. In a stratigraphic section, hundreds of meters thick, climatic factors influencing deposition in the lower part of the section may have influenced the upper part, for example, the Smackover of the Florida panhandle, USA. In this case, solution of the older Louann Salt may have influenced depositional patterns within the upper part of the Smackover.

**Tectonics** Tectonic processes highly influence carbonate and clastic environments. Because of global Tertiary tectonic activity, abundant terrigenous debris is being transported to the continental margins producing rapid progradation of interdeltaic and deltaic deposits. This same tectonic activity inhibits the development of epicontinental seas where extensive sheet-like carbonate deposition occurs. Many of our tectonically influenced modern models are difficult to apply directly to the extensive Phanerozoic deposits characterized by limited tectonic activity.

**Global Changes in Sea Level** A cursory look at Figure 3.8 reveals that global changes in sea level have increased not only in frequency, but also in amount, during the Phanerozoic. This, in part, may be misleading because the more recent changes, *i.e.,* Tertiary, are better preserved in the rock record than those of the Cambrian. Of particular note is that the last significant change in global sea level occurred about 5,000 years ago. Has this been sufficient time for modern depositional regimes, particularly carbonate systems, to equilibrate?

**Fossils and Niches Through Time** Life today is easily observed, ecological niches recorded, and the influence on rock types by the interaction of the species cataloged. Interpretation of the effect of extinct species, or organisms that are not preserved in the rock record, (algae, sea grasses, etc.) or species that have changed ecological niches, *i.e.,* Crinoids, on the resultant rock type is, at best, difficult. The modern is a living system, composed of sediment and organisms interacting under varying conditions; the ancient is merely a rock record of the vital activities and processes that prevailed.

**Diagenesis** What of the rock itself? How much information is actually preserved, and how much information has been altered by diagenetic processes? Compaction, cementation, neomorphism, leaching, solution, to name a few, can partially or completely change the mineralogy and/or texture of a sedimentary rock. Consequently, it is frequently difficult to observe the texture and fabric of a sandstone or limestone and attempt to compare it to a modern equivalent for environmental implications.

**Distribution of Data** In the study of modern depositional environments, one can make an almost infinitely close sampling system, especially in shallow, coastal, or continental deposits. Furthermore, the site of sampling is controlled by the collector. In direct contrast, non-seismic subsurface data is irregular and sample density is very low. Even within fields, spacing (10 to 640 acre) represents only a minute fraction of the depositional system. To complicate matters, most wildcats are drilled on the assumption that the drill-site represents an anomalous position within the subsurface, and therefore, the resultant data is highly skewed.

## SUMMARY

Geology by analogy is a valid technique to *begin* the reconstruction of an ancient depositional system. However, while appreciating that many ancient environments can be described by a general modern model, the explorationist must be aware of the variability of facies models through geologic time and that the modern models presented in this, or any other text, are only a starting point. The explorationist must not categorically fit the ancient into the modern, but rather look for differences between them in order to develop more precise ancient models that will permit the accurate extension of ideas into areas of little or no data.

## SEDIMENT MOVEMENT

Sediment transport occurs by one of three processes, rolling (or sliding), saltation, or suspension. In a series of controlled laboratory experiments, Sanborg *(1956)* determined the velocity that is required for water to initiate transport of particles of various grain sizes (Fig. 4.1). Of particular note is that unconsolidated clay and silt-sized particles require essentially the same velocity for initiating transport as does fine-grained sand. This is attributed to the platey nature of clay minerals which make them more difficult to set in motion than spherical grains. If the clay-sized particles are slightly consolidated (for example, a soil horizon on a point bar or partially vegetated tidal flat) the required current to initiate transport is greater than gravel-sized particles.

## FLOW REGIME

As soon as sediment transport begins in a channel, the bed material is shaped into a number of bed forms (Fig. 4.2). The bed form produced in a channel is related to channel depth (and hydraulic radius) and velocity. The latter, of course, in turn, is affected by friction on the bed-load surface and channel walls. Flow regime is also affected by grain size, river stage, and type of stream.

Bed forms may be classified as low flow regime and upper flow regime. Low flow regime structures include ripples, sandwaves, and dunes (megaripples). Upper flow regime forms include planes, antidunes, and chute-spools. Combinations of bed forms are possible; for example, lower flow regime forms may form on the backs of higher flow regime forms. The vertical and lateral change in bed forms within a sand body are strong indicators of the depositional environment of that sand because the changes reflect variations in current velocities.

## DEPOSITIONAL ENVIRONMENTS OF POTENTIAL SANDSTONE RESERVOIRS

Sandstone reservoirs can be grouped into four major environments of deposition (Fig. 4.3). They include continental, interdeltaic, deltaic, and marine. *Continental sands* include those that were deposited as alluvial fans, eolian, and alluvial channels. Alluvial channels include point bar and braided stream deposits. *Interdeltaic sands* include beach, barrier island, and tidal channel deposits. *Deltaic sands* can be subdivided into deltaic plain, deltaic front, and prodeltaic deposits. Point bar, braided streams, crevasse fan, and distributary sand deposits comprise the major deltaic plain reservoirs. Channel-mouth (steam-mouth or distributary-mouth) bars, delta margin islands, and beach ridge sands represent the dominant deltaic front deposited reservoirs. *Marine sands* include offshore shallow marine deposits and deep marine fan sediments.

## BRAIDED RIVER RESERVOIRS

### CHARACTERISTICS OF BRAIDED RIVERS

Braided rivers are normally characterized by large fluctuations in discharge. During low water stage, flow is restricted to channels. Little sand-size sediment is transported during this stage of flow. During rising water conditions, the bars may be covered, and scouring in the channels as well as on bars may occur. Consequently, load fluctuates greatly during waning water conditions. The river normally has a steep to moderate gradient, and each channel may differ in gradient slightly. The channel configuration is anastomosing, or braided, and the channels are separated by bars (Fig. 4.4).

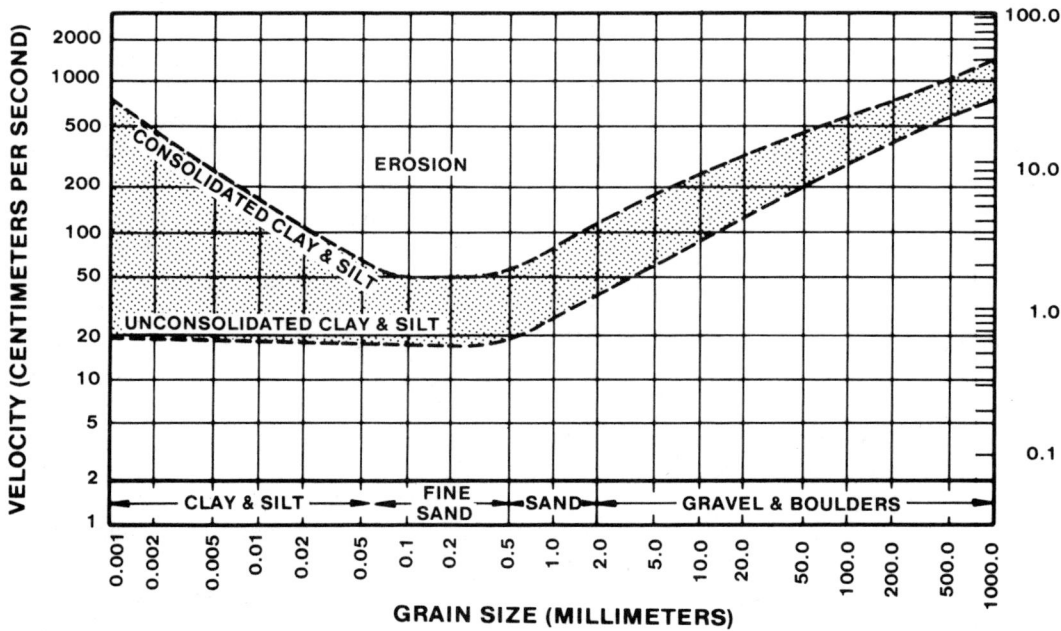

**Figure 4.1:** *Diagrams illustrating the stream power (velocity) that is required to initiate motion of sediment particles in one meter of water (Sandborg, 1956).*

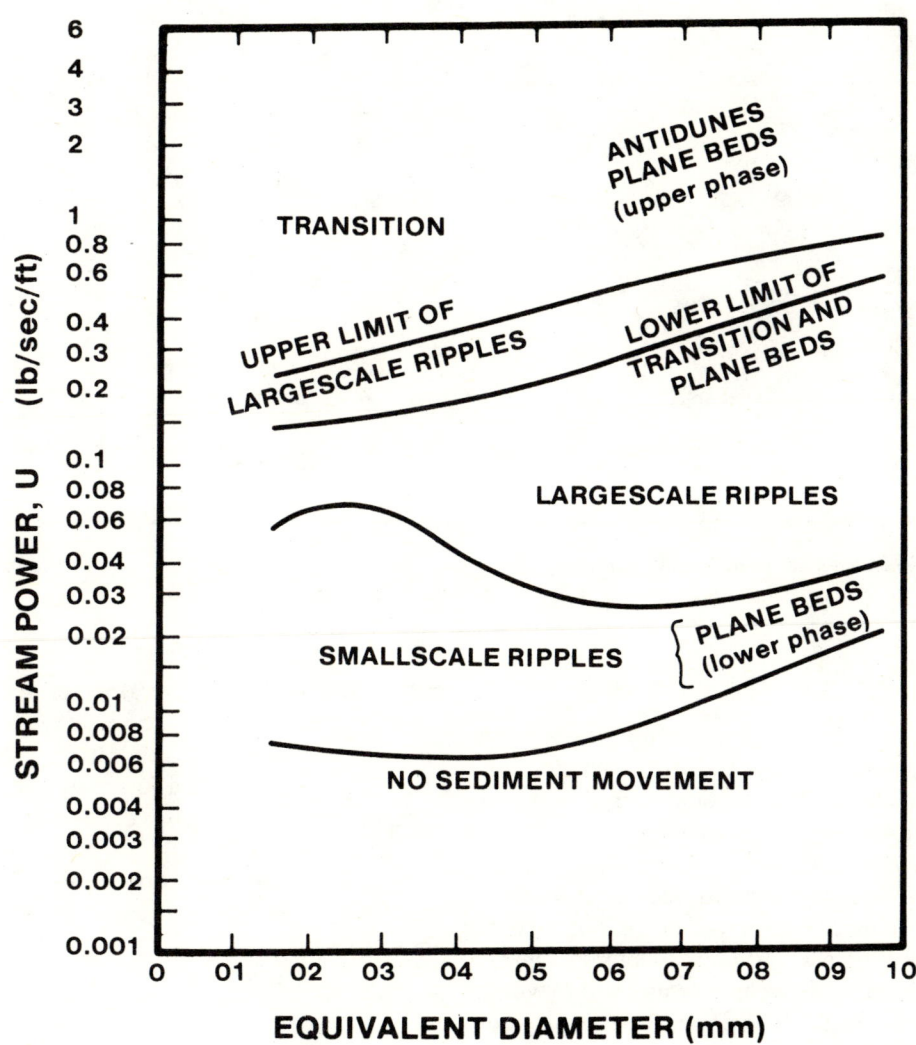

**Figure 4.2:** *The flow regime controls the relationship of grain size and bed forms (Allen, 1969).*

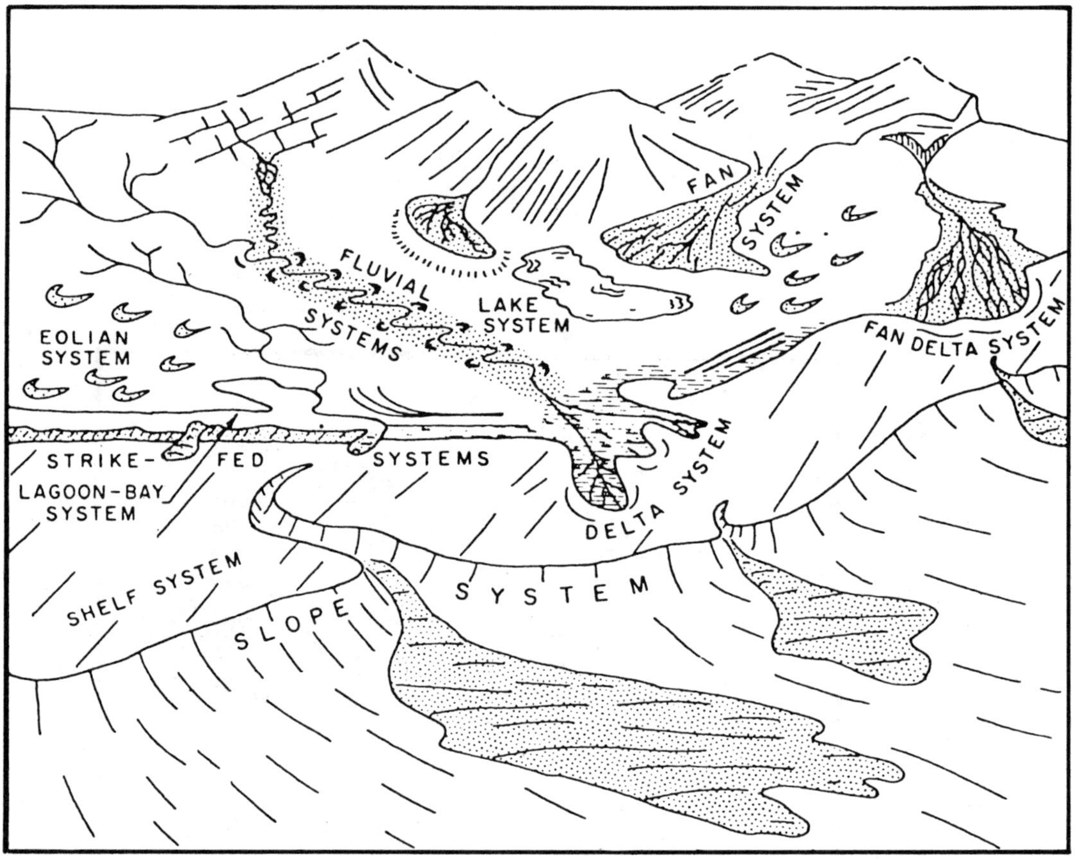

**Figure 4.3:** *Depositional environments of potential sandstone reservoirs. They include continental, deltaic, interdeltaic, and offshore marine (Brown and others, 1973).*

## CHARACTERISTICS OF BAR

**Process of Deposition** Deposition in a braided stream occurs during the waning stages of a flood. Thus, vertical accretion, especially in ephemeral streams, is the dominant process of accumulation. Some lateral accretion of individual bars may occur, but probably this will not be preserved in the rock record. Individual bars will be permanent in perennial streams where they can be stabilized by vegetation (Fig. 4.5). Normally, the bars are destroyed during the flood stage in ephemeral streams (Fig. 4.4).

**Internal Features** Figure 4.6 depicts the internal and external features of a braided stream deposit that may be represented in a core. Deposition during waning flood stage produces upward grading from coarse to fine, with a reduction in sorting. However, this fine-grained set is not normally preserved in the rock record. Gravel to sand-sized material is most common in braided stream deposits, but ranges from cobbles to silt as described in the literature. Festoon cross beds are the dominant sedimentary structure. Commonly they decrease in size upward and may grade into parallel laminae at the top of the last depositional set. Current ripples, particularly in perennial streams, will form in the bottom of the channels, but they are rarely preserved in ancient strata.

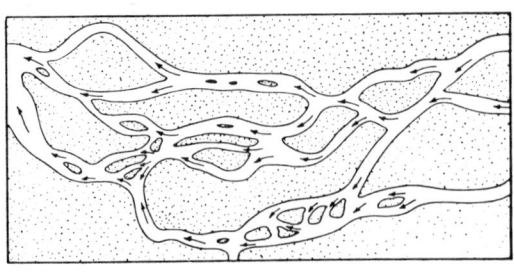

**Figure 4.4:** *Characteristics of braided stream channels.*

# EXPLORATION FOR CONTINENTAL SANDSTONE RESERVOIRS

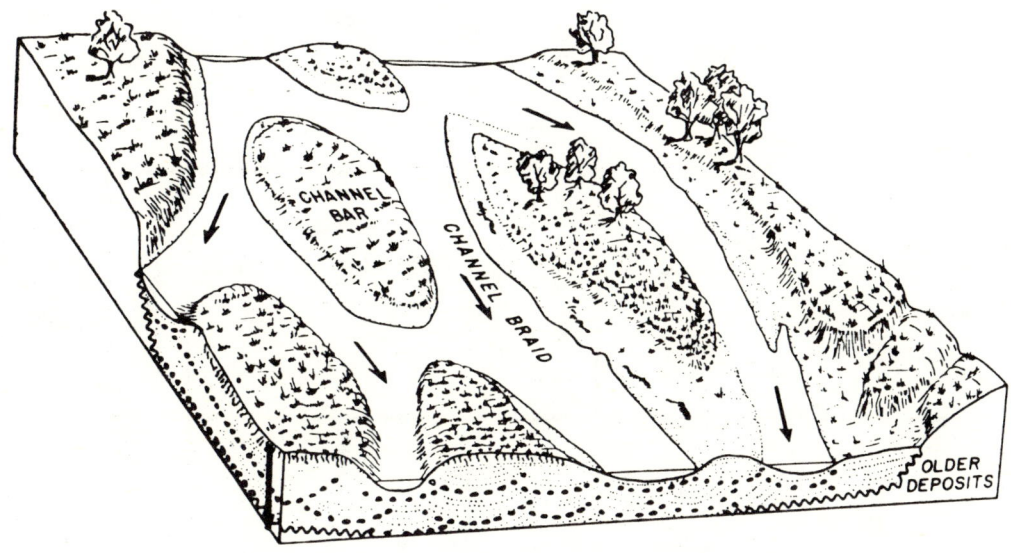

**Figure 4.5:** *Characteristics of braided stream bars.*

**External Features** The lower contacts of a braided stream deposit are sharp and commonly unconformable. If conformable, basal sands may be in contact with marsh, swamp, or flood plain deposits. Upper contacts are sharp and generally conformable with alluvial plain or delta plain. Typically, braided stream deposits are located in alluvial valleys, for example, the Red River, Platte River, or Salt River; or along coastal plain, for example, the San Bernard River. The sands generally form linear bodies that range from 30 to over 200 feet thick.

## RESERVOIR AND STRATIGRAPHIC TRAP POTENTIAL

Vertical and lateral fluid communication within braided stream deposited reservoirs is normally excellent. If the climate is humid, it is possible for the bars to be interbedded with soil horizons which, in turn, will reduce vertical reservoir communication.

The best stratigraphic trap opportunities are at the margin of a braided stream where they cut perpendicular to regional dip or strike across a regional nose. Typically, flood plain or swamp-marsh deposits form the seal to the trap.

## ALLUVIAL FAN RESERVOIRS

### INTRODUCTION

Normally, alluvial fans form at the foot of a steep slope where an abrupt decrease in stream gradient occurs. Commonly, they are located near the base of a mountain range (Fig. 4.3). They form under practically all types of climates and topographic profiles, but they are most common in areas characterized by bold relief and arid climate. Individual wedge-shaped fans may coalesce to produce a linear wedge-shaped rock body.

### PROCESS OF DEPOSITION

Sediment transport occurs by water, sheet flow, and debris flow under some of the highest energy conditions in the clastic reservoir realm (Fig. 4.7). Deposition occurs in channels (braided streams), sheet flows, and debris flows. Channels erode into the fan surface. Sediment from the alpine area is transported through the channels and deposited at the toe of the fan during periods of heavy runoff. As runoff slackens, deposition ensues. When the channels are filled with water during heavy runoff, water spreads out over the surface of the entire fan. Deposition of finer material

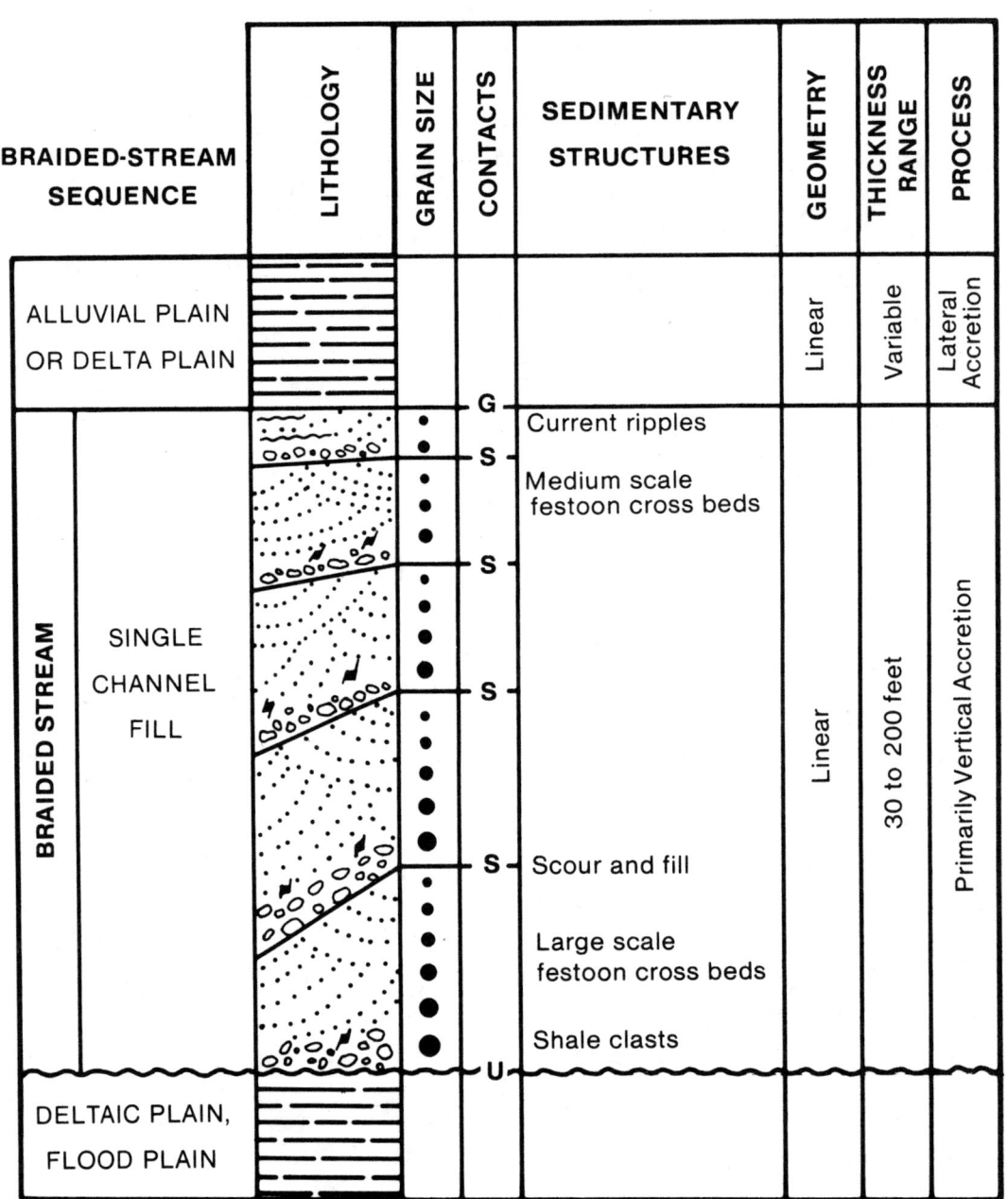

**Figure 4.6:** *Characteristics of a braided stream sequence.*

# EXPLORATION FOR CONTINENTAL SANDSTONE RESERVOIRS

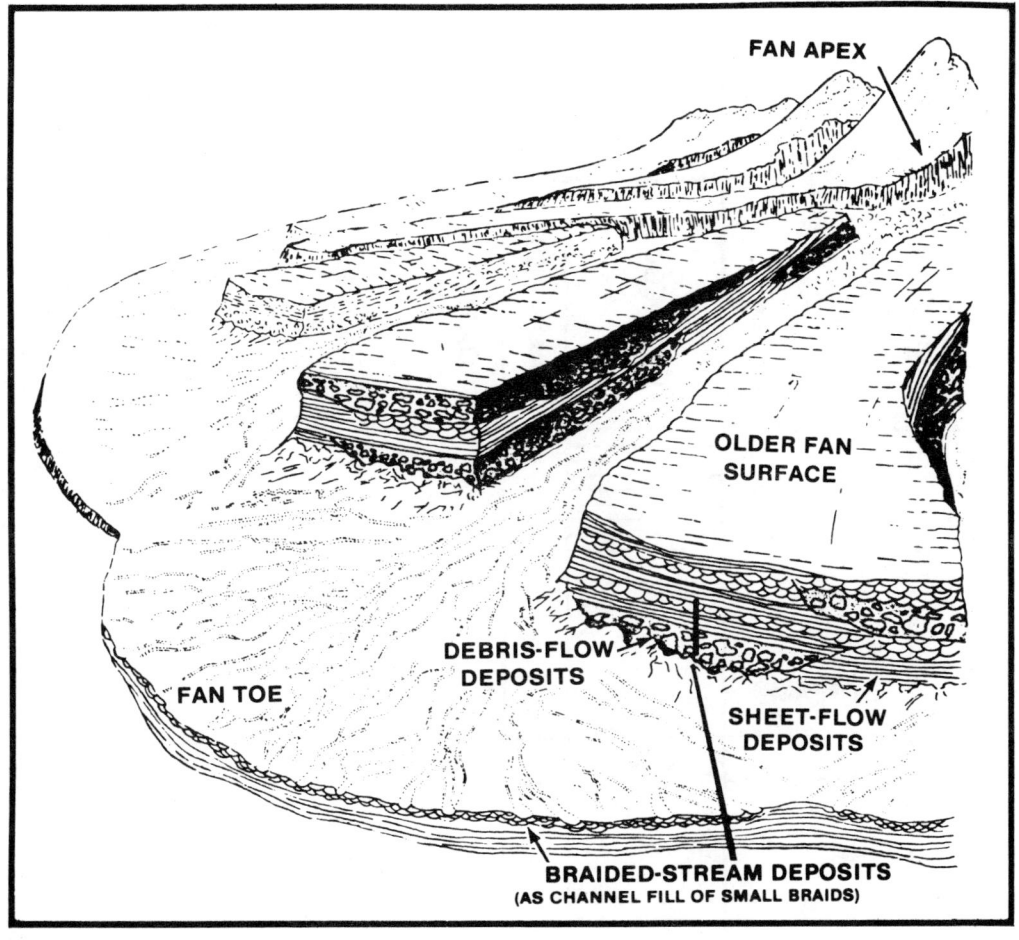

**Figure 4.7:** *Parts of an alluvial fan.*

than that in the channels occurs in a more or less uniform pattern. During torrential floods, viscosity of runoff waters is greater, therefore, large cobbles and boulders may be transported. Vertical and lateral accretion are the dominant processes of deposition.

## INTERNAL FEATURES

The channels typically contain scour surfaces that truncate low angle cross beds (Fig. 4.8). Lower flow regime structures such as coarse to fine-grained bedding and small scale structures are generally not preserved in the rock record because they are eroded by upper flow regime currents. The sheet flow deposits contain low angle scour and fill structures. These deposits may be weakly graded. Pebble imbrication is about the only sedimentary structure formed in the debris flow strata because the grain size is so large.

## EXTERNAL FEATURES

Lower contacts of alluvial fan deposits are normally sharp and probably unconformable; lateral contacts are gradational. Contacts at the toe of the fan may be gradational with the finer-grained, typically evaporite-rich valley facies. The size of individual alluvial fans is controlled by drainage basin area, slope, climate, tectonics, and lithology of the source rocks. Individual fans range in radius from several hundred feet to several tens of miles. Coalescing fans can produce linear, wedge-shaped deposits that are hundreds of miles long. Alluvial fan deposits commonly grade downstream into braided stream or playa-lake deposits. In some areas where mountainous areas are proximal to inland lakes or oceans, alluvial fan deposits are accumulated under both subaerial and subaqueous conditions. Such fans have been termed *Gilbert-type deltas*.

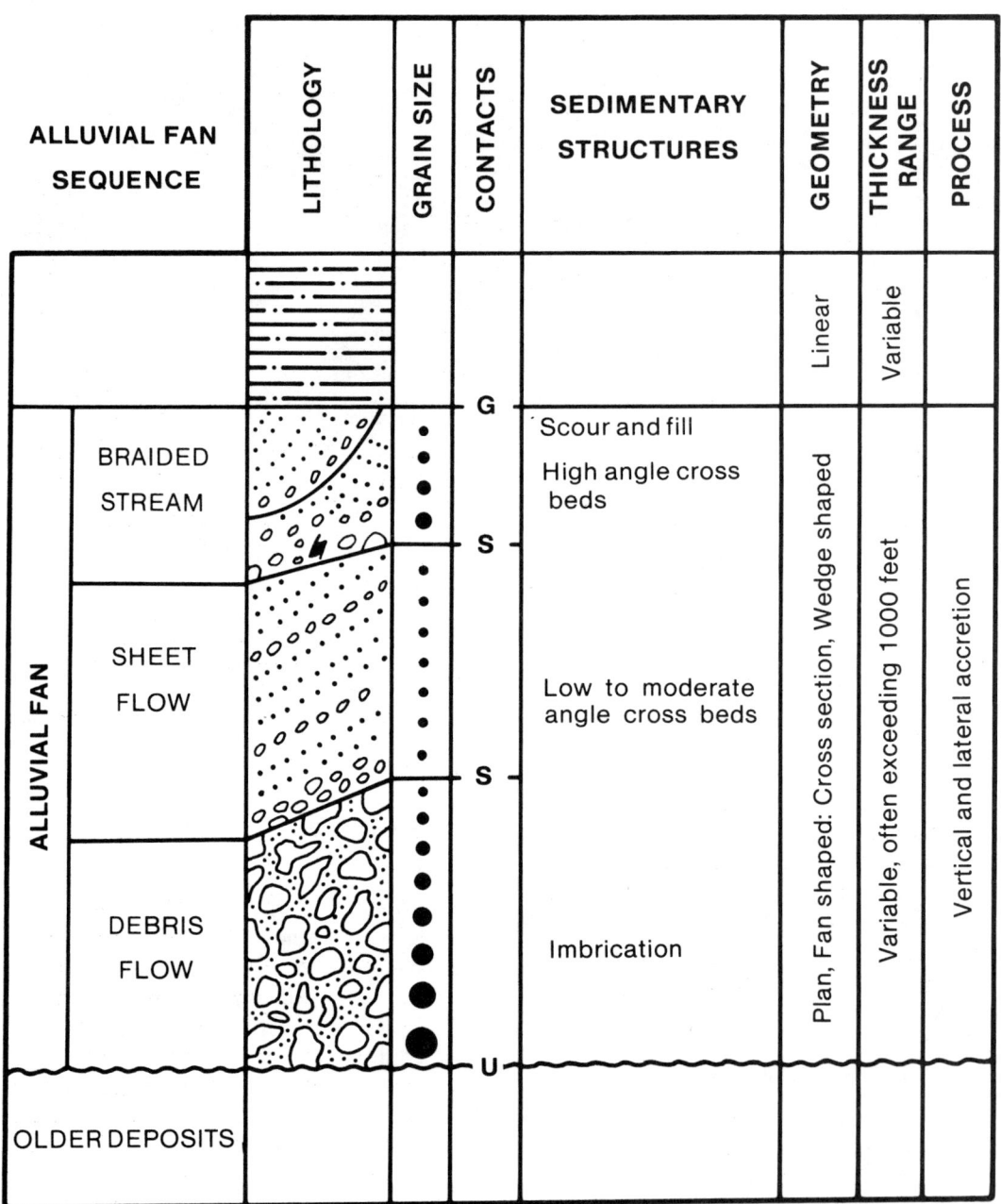

**Figure 4.8:** *Characteristics of an alluvial fan depositional sequence.*

## RESERVOIR AND STRATIGRAPHIC TRAP POTENTIAL

Reservoir characteristics of alluvial fan deposits are normally complex. Maximum reservoir potential occurs in the channel facies. Typically this facies has good to excellent reservoir continuity; however, exceptions do occur, for example, the Quiriquire Field, Venezuela. In this case the channel facies is complexly interbedded with the sheet wash facies. The latter facies provide local seals which, in turn, produces numerous internal stratigraphic traps within the field. Normally, the debris flow facies are highly variable and reservoir quality is controlled by the amount and mineralogy of the clay matrix.

Stratigraphic trap opportunities occur along the margin of alluvial fan systems where it intertongues with fine-grained sediment of marine or lacustrine origin. The latter, of course, is the best environment for potential source rocks to be proximal to the fan deposits.

## EOLIAN SAND RESERVOIRS

### INTRODUCTION

Basic conditions for eolian deposition are a large supply of dry sand and high wind velocity. These conditions most frequently are present along sandy coastlines and in semiarid regions where physical weathering and fluvial sedimentation produce a large quantity of sand. Sources for the sand include alluvial fans, braided streams, meandering streams, and various coastal interdeltaic environments.

### PROCESS OF DEPOSITION

The most common eolian deposition is in the form of sand dunes. Many types of dunes have been recognized and described. Following is a tabulation of some of the types of dunes that have been described in the literature.

1. Foredune ridges or elongate mounds of sand up to a few tens of feet in height, adjacent and parallel with beaches.
2. U-shaped dunes, arcuate to hairpin-shaped sand ridges with the open end toward the beach.
3. Barchans, or crescentic dunes, with a steep lee slope on the concave side, which faces away from the beach.
4. Transverse dune ridges, trending parallel with or oblique to the shore, and elongated in a direction essentially perpendicular to the dominant winds. These dunes are asymmetric in cross profile, having a gentle slope on the windward side and a steep slope on the leeward side.
5. Longitudinal dunes, elongated parallel with wind direction and extending perpendicular or oblique to the shoreline; cross profile is typically symmetric.
6. Blowouts, comprising a wide variety of pits, troughs, channels, and chute-shaped forms cutting into or across other types of dunes or sand hills. The larger ones are marked by conspicuous heaps of sand on the landward side, assuming the form of a fan, mound, or ridge, commonly with a slope as steep as 32° facing away from the shore.
7. Attached dunes, comprising accumulations of sand trapped by various types of topographic obstacles.

### INTERNAL FEATURES

High angle cross beds are the dominant sedimentary structure of dunes. Although ripples are common on the surface of modern dunes, they are rarely preserved in ancient dunes (Fig. 4.9). The sand is well sorted and fine grained. Few shale interbeds occur within the dune sequence.

### EXTERNAL FEATURES

The lower contacts of a dune sequence are normally sharp. Lateral and upper contacts are gradational. Interdeltaic coastal dune fields are typically linear belts of thin sands. In contrast, continental dune fields form regionally extensive sheet sand bodies that frequently exceed 1,000 feet thick. The dune sequence is commonly associated with alluvial fan, braided stream, and coastal interdeltaic deposits. Typical facies pinchouts are broad and regional in nature.

### RESERVOIR AND STRATIGRAPHIC TRAP POTENTIAL

The well sorted, homogeneous continental

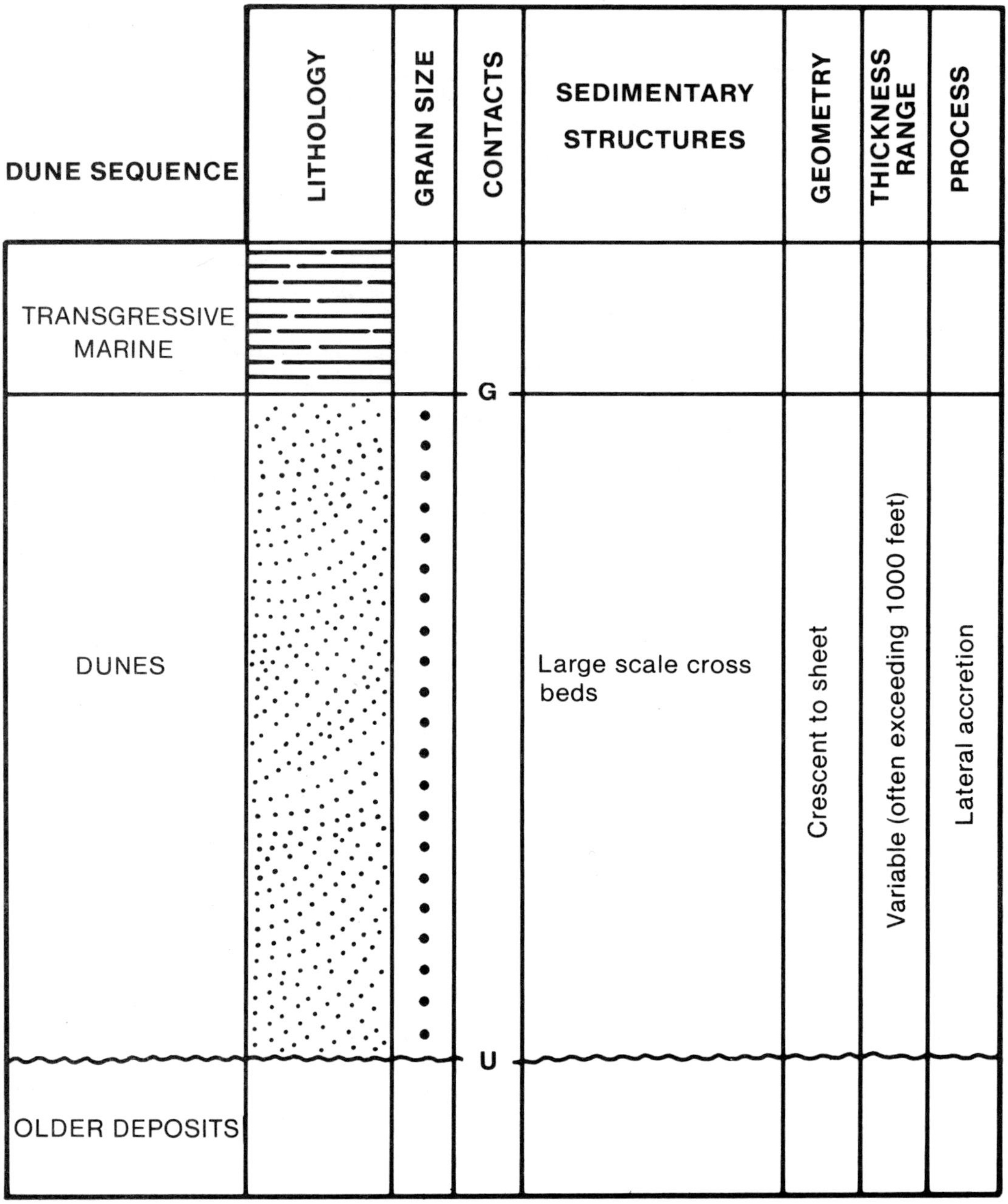

72                                   EXPLORATION FOR SANDSTONE RESERVOIRS

| DUNE SEQUENCE | LITHOLOGY | GRAIN SIZE | CONTACTS | SEDIMENTARY STRUCTURES | GEOMETRY | THICKNESS RANGE | PROCESS |
|---|---|---|---|---|---|---|---|
| TRANSGRESSIVE MARINE | | | G | | | | |
| DUNES | | | | Large scale cross beds | Crescent to sheet | Variable (often exceeding 1000 feet) | Lateral accretion |
| | | | U | | | | |
| OLDER DEPOSITS | | | | | | | |

**Figure 4.9:** *Characteristics of a dune depositional sequence.*

# EXPLORATION FOR CONTINENTAL SANDSTONE RESERVOIRS

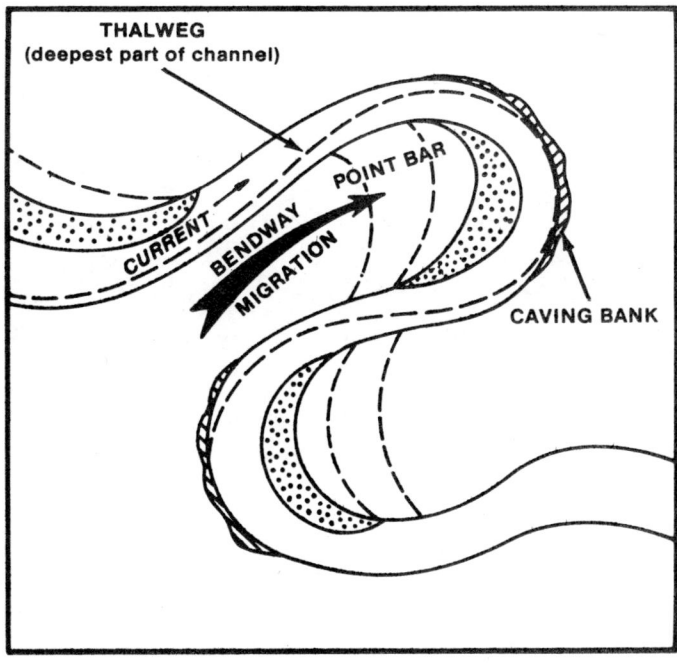

**Figure 4.10:** *Parts of a meandering stream.*

dune sequence normally is characterized by excellent vertical and lateral reservoir continuity. Permeability is best developed parallel to the cross beds. The best trap opportunities occur along regional facies changes where reservoir rocks pinchout into lateral fine-grained (seal) facies. The occurrence of a source rock is one of the major problems in exploring for traps in eolian-deposited sands. An understanding of migration paths is particularly important when exploring for dune deposited reservoirs.

## MEANDERING STREAM RESERVOIRS

### CHARACTERISTICS OF MEANDERING RIVERS

Meandering rivers occur on a continent or on a delta (Fig. 4.3). The river is characterized by moderate discharge fluctuations, low gradient, and moderate sediment load of mixed grain size; it typically occurs in an area of moderate rainfall. The river can be divided into three parts — channel, caving bank, and point bar (Fig. 4.10). Sediment is transported by laminar flow, fluid drag, saltation, and/or suspension.

If a river is flowing across a substratum that is easily eroded, it cannot maintain a straight channel for a distance much greater than about ten times its width. It will meander because the stream constantly undermines the banks and develops a cross section that resembles a banana rather than maintaining the theoretical semicircle cross section. The slightest irregularity in erodibility of the substratum will produce a slight shift in the channel. This in turn produces differential turbulence, and the sidecutting power of the stream is greatly increased in the bends where water turbulences and velocities are greatest. The meanders migrate downstream at more or less constant rates. However, differential downstream migration of meanders can produce cutoffs, or during floods, chute cutoffs may develop which will result in the local abandonment of a channel.

### INTERNAL FEATURES OF POINT BARS

The mode of deposition in meanders produces a natural three-fold zonation within the point bar (Fig. 4.11). The lower zone conforms to deposition in the thalweg (the deepest part of

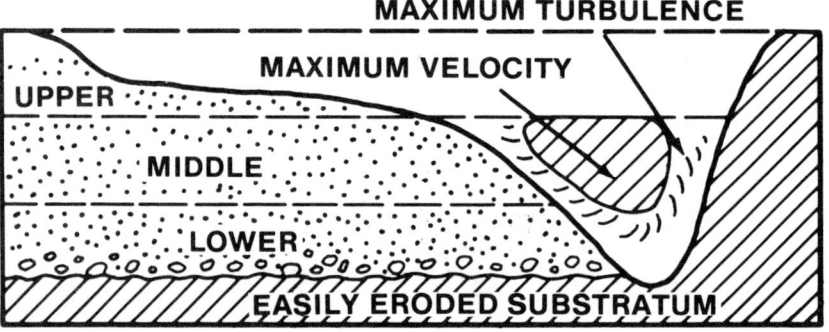

**Figure 4.11:** *Subdivisions of a Point Bar deposit.*

the channel). The middle zone corresponds with low stage of discharge and the upper zone denotes flood stage. The lower zone contains the coarsest fraction of sediment transported by the river (Fig. 4.12). Wood fragments and some shale clasts (clasts derived from soil horizons on the upper zone) are common. Large scale festoon cross beds are the dominant sedimentary structures. Sorting is good to excellent.

The middle zone contains finer-grained sand than that found in the lower zone. Some plant fragments may occur, but they are rare. The large scale festoon cross in the lower zone grade upward to middle scale cross beds. Sorting ranges from good to moderate. Parallel horizontal laminae may occur at the top of the middle zone.

The finest quartz sand carried by the meandering river is deposited in the upper zone of the point bar sequence. The sand may be interbedded with thin shales that represent soil horizons which developed on the upper surface of the bar. Sorting is moderate to poor. Sedimentary structures include even parallel laminae that are interbedded with small scale current rippled zones. Plant fragments are common.

## INTERNAL FEATURES OF ABANDONED CHANNELS

The abandoned channel sequence is zoned like the point bar sequence, but the upper, and in most cases, the middle zone of the point bar will be absent in the abandoned channel sequence (Fig. 4.13). Sedimentary structures and textures are the same for the basal sand in the abandoned channel as that which occurs in the active channel. The abandoned zone, however, contains fine-grained sand, silt and clay, and abundant plant fragments. Peat is not uncommon. This fine-grained sequence is usually rich in organic material and may serve as source and seal for the sand sequence. Laminations and thin bedding are the dominant sedimentary structures.

## EXTERNAL FEATURES

Basal contacts of the point bar and abandoned-channel sequences are sharp to unconformable. Upper contacts are gradational to sharp. Lateral contacts are gradational on the inside of the meander and sharp on the outside (caving bank). Average maximum thickness of point bar sands approaches the maximum depth that the stream will scour during flood stage *(Leopold and others, 1964)*. Depth is a function of discharge, and discharge is the prime factor governing width and meander length. Leopold and others *(1964)* published a series of graphs that permit the prediction of meander size from thickness values of point bar sands (Fig. 4.14). Although discrete point bar deposits are normally crescent-shaped sand bodies, coalescing bars of a meander belt may form linear bodies of sand.

Associated facies include continental and deltaic. If the point bar deposits occur in a continental environment, flood plain and alluvial valley deposits are the normal facies that are in contact with the point bar sands. On the other hand, bay, beach, and lagoonal facies may be associated with point bar sands in an interdeltaic model. Channel-mouth bar, crevasse fan, marsh, and swamp facies are normal rock units that may be in juxtaposition

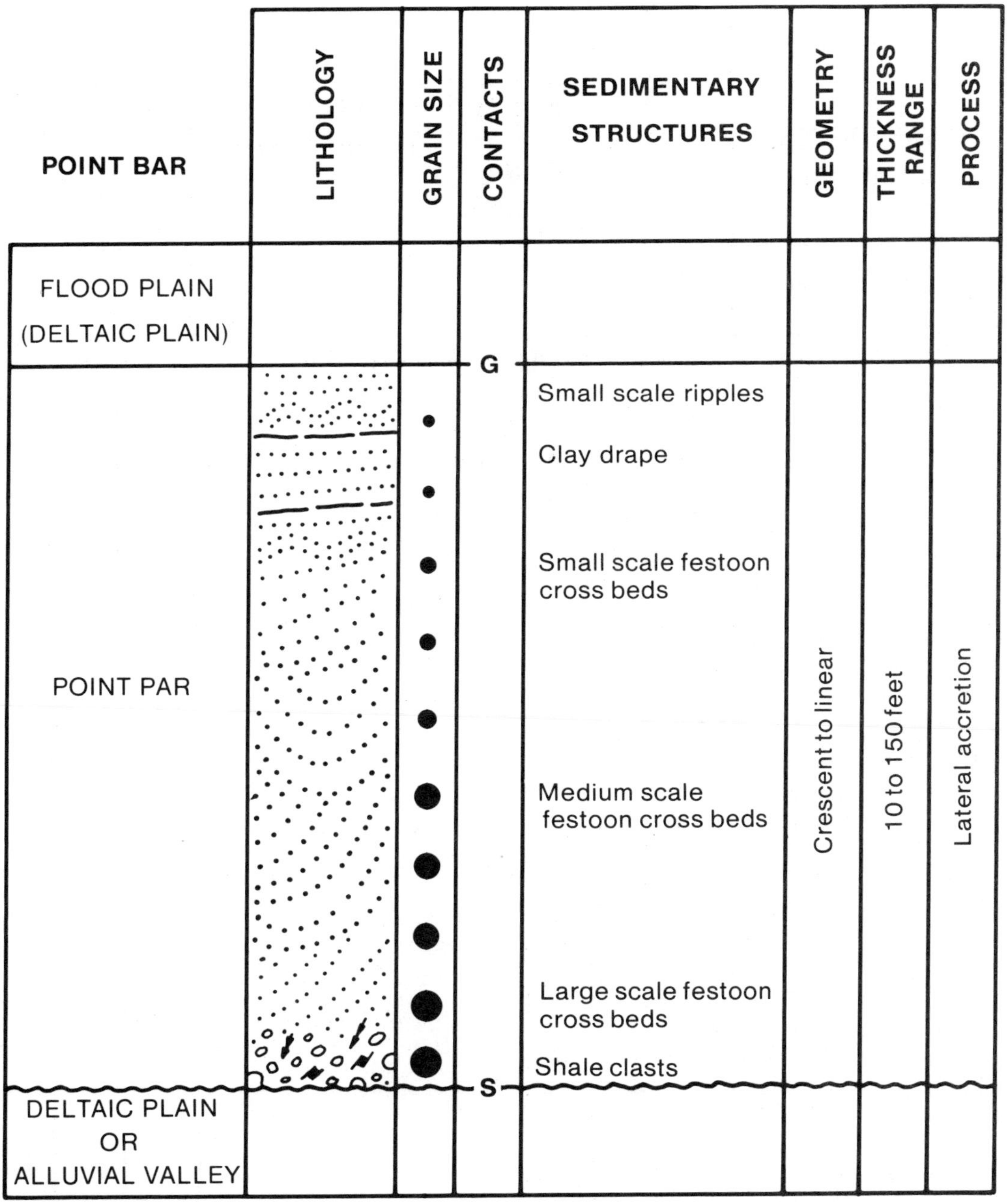

# EXPLORATION FOR CONTINENTAL SANDSTONE RESERVOIRS

| POINT BAR | LITHOLOGY | GRAIN SIZE | CONTACTS | SEDIMENTARY STRUCTURES | GEOMETRY | THICKNESS RANGE | PROCESS |
|---|---|---|---|---|---|---|---|
| FLOOD PLAIN (DELTAIC PLAIN) | | | | | | | |
| POINT PAR | | | G | Small scale ripples<br>Clay drape<br>Small scale festoon cross beds<br>Medium scale festoon cross beds<br>Large scale festoon cross beds<br>Shale clasts | Crescent to linear | 10 to 150 feet | Lateral accretion |
| DELTAIC PLAIN OR ALLUVIAL VALLEY | | | S | | | | |

**Figure 4.12:** *Depositional characteristics of a point bar sequence (Silver, 1980).*

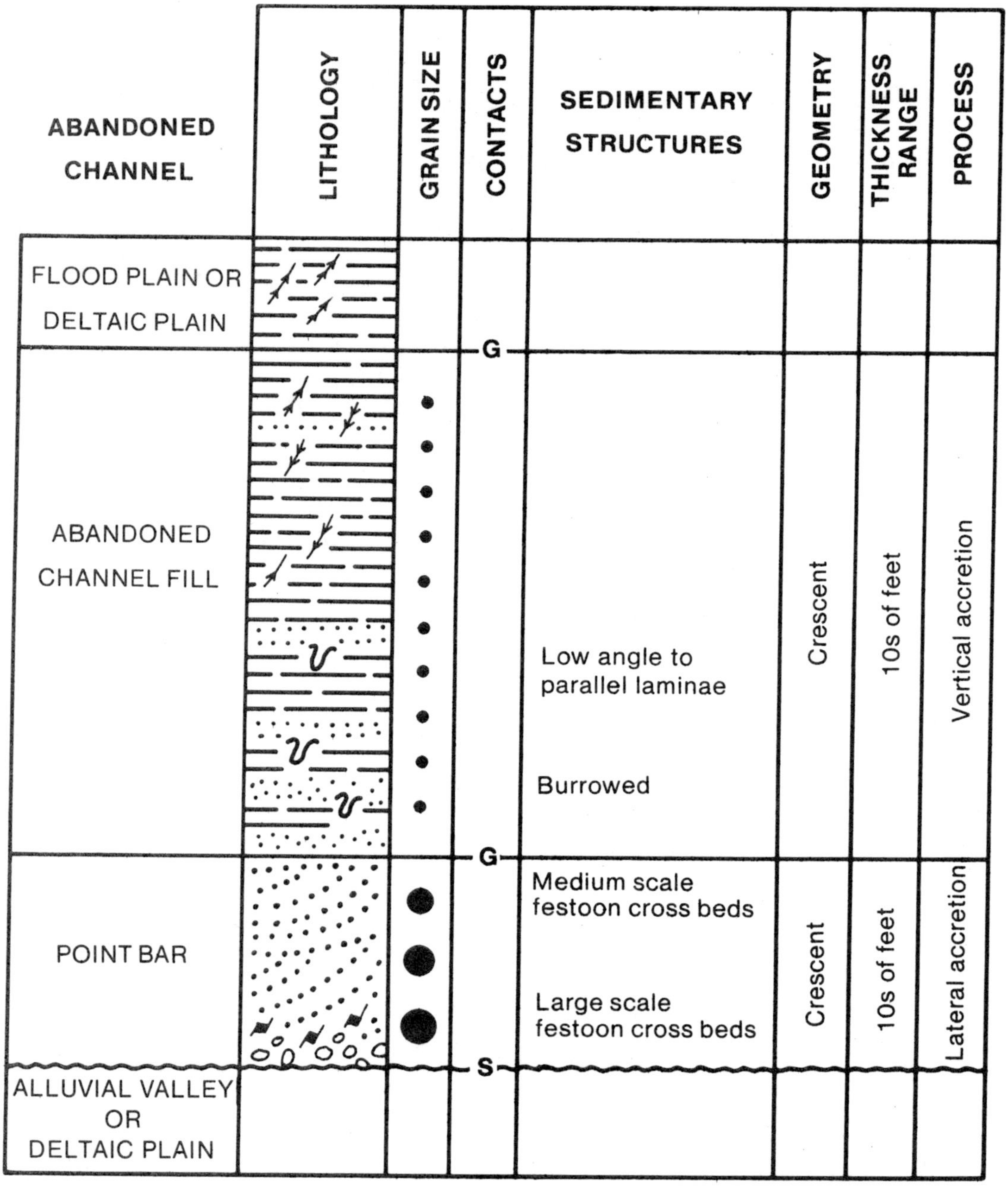

# EXPLORATION FOR SANDSTONE RESERVOIRS

| ABANDONED CHANNEL | LITHOLOGY | GRAIN SIZE | CONTACTS | SEDIMENTARY STRUCTURES | GEOMETRY | THICKNESS RANGE | PROCESS |
|---|---|---|---|---|---|---|---|
| FLOOD PLAIN OR DELTAIC PLAIN | | | G | | | | |
| ABANDONED CHANNEL FILL | | | | Low angle to parallel laminae<br><br>Burrowed | Crescent | 10s of feet | Vertical accretion |
| POINT BAR | | | G<br><br><br>S | Medium scale festoon cross beds<br><br>Large scale festoon cross beds | Crescent | 10s of feet | Lateral accretion |
| ALLUVIAL VALLEY OR DELTAIC PLAIN | | | | | | | |

**Figure 4.13:** *Depositional characteristics of an abandoned-channel sequence.*

# EXPLORATION FOR CONTINENTAL SANDSTONE RESERVOIRS 77

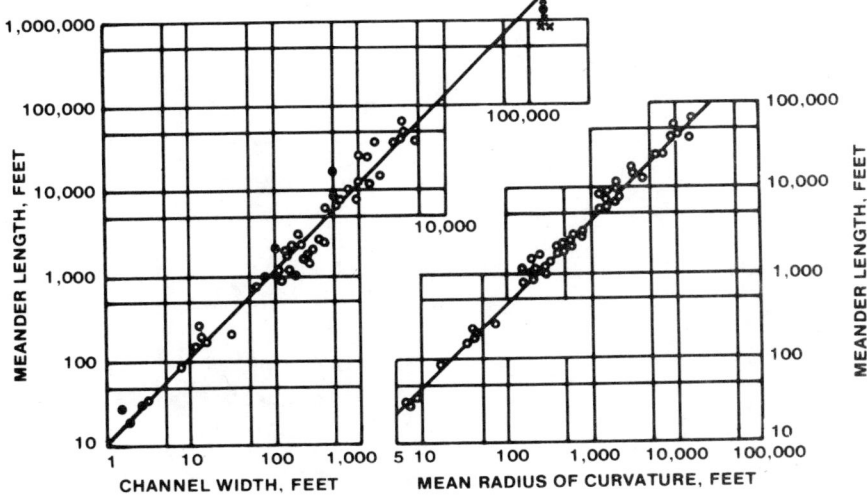

- MEANDERS OF RIVERS AND FLUMES
- × MEANDERS OF GULF STREAM
- • MEANDERS ON GLACIER ICE

**(a)** Relation of meander length to width (A) and to radius of curvature in channels (B).

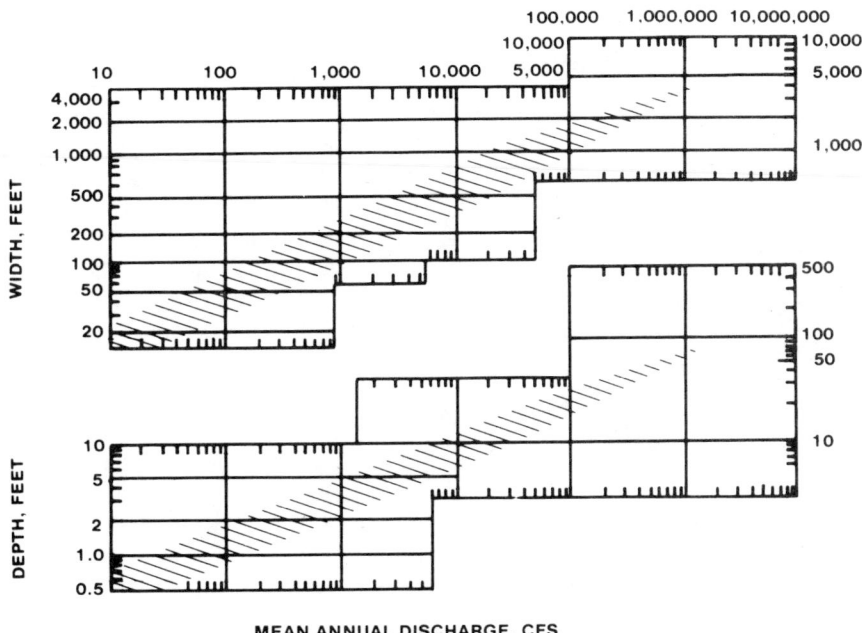

**(b)** Width and depth in relation to mean annual discharge as discharge increases downstream in various river systems.

**Figure 4.14:** *Meander width and thickness of sand are related. For example, a thickness of 20 feet implies a meander width of 700 to 1,500 feet wide (Leopold and others, 1964).*

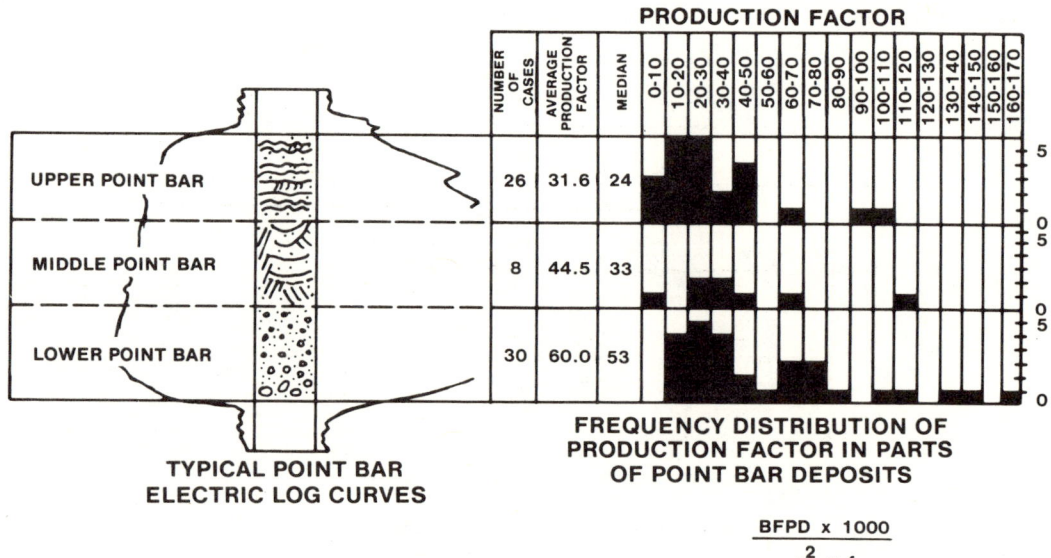

**Figure 4.15:** *Reservoir characteristics of point bar sands, Hough area, Oklahoma (Swanson, 1972).*

with point bar sands in a deltaic environment.

## RESERVOIR AND STRATIGRAPHIC TRAP POTENTIAL

Maximum reservoir quality occurs in the basal zone of the point bar sequence (Fig. 4.15). Although porosity is not significantly influenced by grain size, permeability is and therefore maximum permeability occurs in the basal zone because it is the coarsest and poorest sorted interval in the point bar sequence. Lateral reservoir continuity is good to excellent. In contrast, vertical reservoir continuity is highly variable. Those point bars deposited in a humid climate generally contain abundant soil horizons which diagonally cut through the sand body, drastically reducing vertical communication within the middle and upper zones of the sands.

Potential trap opportunities for individual bars occur where shale-filled oxbows provide the lateral seal, and floodplain shales form the upper seal. Backswamp facies may provide the lateral seal for the edge of a channel belt. Where the meander alluvial valley is entrenched into a non-porous and permeable strata, excellent stratigraphic trap opportunities exist.

## SEISMIC CHARACTERISTICS OF CONTINENTAL FACIES

Continental environments characteristically form sheet or wedge deposits that are thinner than their marine equivalents. Typically seismic patterns are concordant at the top with low angle downlap patterns at the base. Amplitude is variable from high to low with frequent intervals of high amplitude reflectors that rapidly grade to low amplitude reflectors (Fig. 4.16 at 1.0 second interval). Reflections are generally discontinuous, but in non-sand intervals they may be moderately continuous. Locally where fluvial and alluvial sands are dominant, the external form of the seismic reflections will approximate a sheet in which continuity and amplitude of the reflections are high.

If proper data acquisition and processing procedures are employed, individual sands can be mapped. For example, Figure 4.17A illustrates a channel sand in the Mannville Group (Cretaceous), central Alberta. The upper Mannville interval consists of thick but narrow, multistoried, channel sandstones that are bordered by equally thick interbedded siltstones, shales, and coals *(Putnam and Oliver, 1980)*. Dense drilling in the area permits delineation of individual sandstone

EXPLORATION FOR CONTINENTAL SANDSTONE RESERVOIRS 79

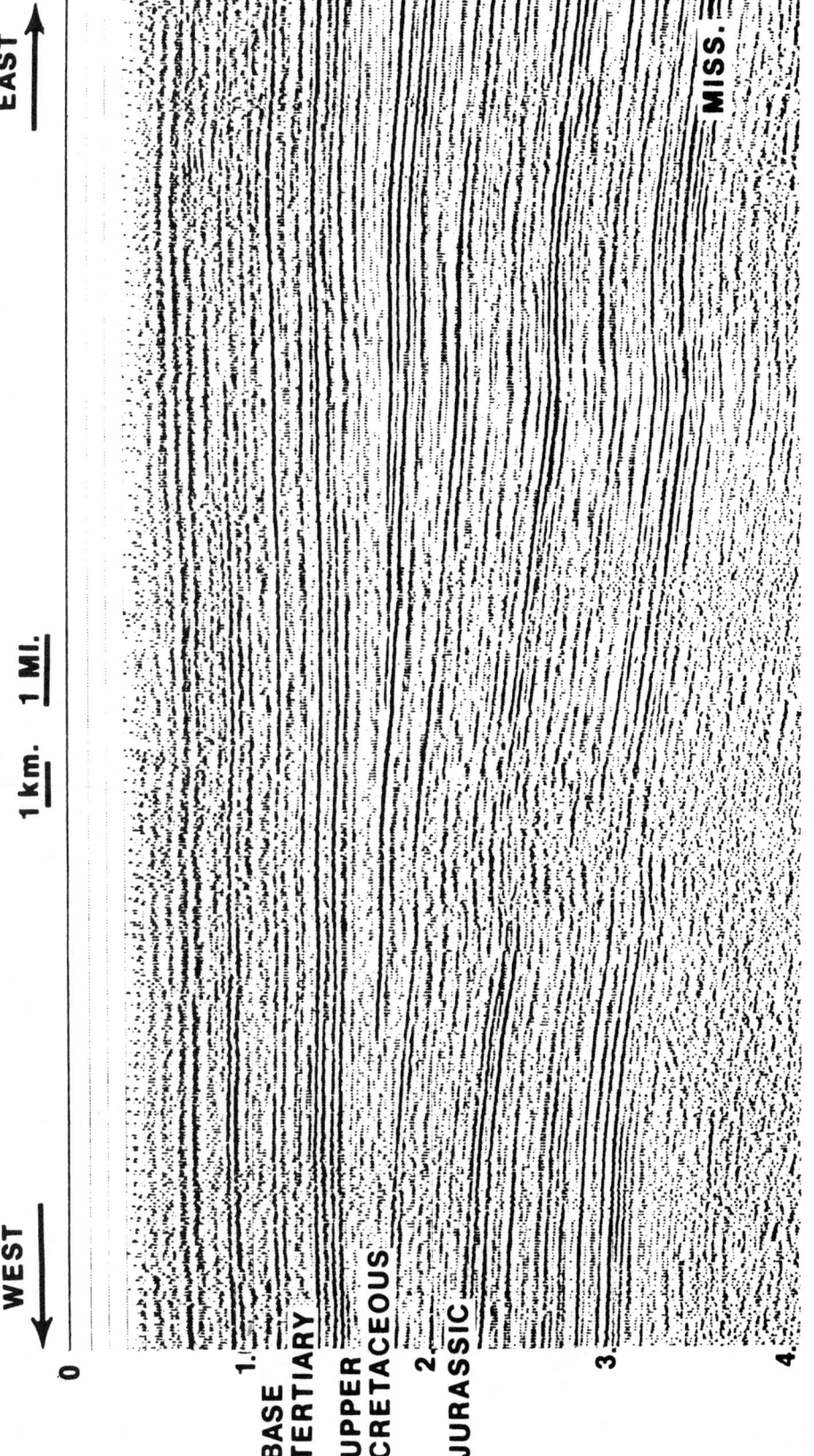

**Figure 4.16:** *Seismic line located in the Green River Basin, Wyoming, illustrates continental and deltaic systems of Jurassic and Middle Cretaceous age (Laing, 1972).*

bodies. Subsurface mapping of the bodies has permitted excellent calibration of seismic data (Fig. 4.17B). Careful study of Figure 4.17A reveals that the reflection on the siltstone-shale-coal sequence changes amplitude across the channel (perhaps due to a "tunning effect") and it is "draped" over the channel sand. Once the seismic data is calibrated, it is an excellent exploration tool for mapping these sandstones.

Other subtle reflection signatures may indicate the presence of thin channel sand reservoirs. For example, careful inspection of Figure 4.18 reveals a change in reflector character above the post Mississippian erosional surface across Red Fork (Pennsylvanian) channel sand production (0.830 sec.). The productive limits of the field corresponds to a lower amplitude Mississippian reflector that attempts to break into two distinct reflectors. Synthetic analysis of the data supports the interpretation and it has been successfully used as an exploration tool to map the channel sands.

# EXAMPLES OF CONTINENTAL SAND RESERVOIRS

## INTRODUCTION

A commercial trap of oil and/or gas requires the simultaneous existence of (1) a reservoir rock (filled with oil and/or gas), (2) an area of low potential in the reservoir, and (3) a seal which is defined as a rock that has sufficiently high capillary entry pressure to restrict oil and/or gas to the low potential area *(Schowater, 1976 and Roberts, 1980)*. Trap types are classified as a function of the mechanism that produces the low potential area. The focus of this book is on stratigraphic traps. Recent classifications of stratigraphic traps include Rittenhouse *(1972)* and Halbouty *(1972)*.

Continental deposited sandstone reservoirs constitute some of the highest quality reservoirs in the sandstone family. For example, the braided stream portion of the Sadlerochit Formation (Jurassic) contains porosities of 25% and permeabilities up to several darcies. It is the principal reservoir at more than 9 billion barrel Prudhoe Bay Field, Alaska *(Morgridge and Smith, 1972)*.

Eolian depositional processes also produce excellent reservoirs. Typically the dune systems have excellent reservoir continuity but exceptions do exist. For example, the Weber Sandstone (Pennsylvanian-Permian) at the Rangely Field, Colorado *(Fryberger, 1979)* and the Tensleep Sandstone (Pennsylvanian) at the Oregon Basin Field *(Morgan and others, 1978)*. Several non-reservoir sandstones occur within the Oregon Basin productive zone. These dolomite and dolomite-anhydrite cemented sandstones are effective permeability barriers that have made it necessary to modify the secondary recovery program in the field.

Point Bar deposits of meandering rivers are probably the most abundant reservoirs deposited on the continent. Harms *(1966)* has documented channel sand production in Nebraska. Traps include both stratigraphic and structural. Berg *(1968)* interpreted production from the Fall River Sandstone (Cretaceous) at the Miller Creek Field, Montana, to be a point bar sand. The trap is produced by the alluvial shale of probable backswamp origin. The seal can also be produced by the substratum rock into which the meandering stream eroded. Such is the case for the Cut Bank Sand (Cretaceous) at Cut Bank Field, Montana *(Blixt, 1941)*. The abandoned channel-fill can also provide the seal for oil and/or gas. Examples of this situation occur at the Cole Creek Field *(Berg, 1968)*, and the South Glenrock Field, Wyoming *(Curry and Curry, 1972)*. Perhaps one of the most frequent trapping mechanisms occurs when the meandering sand belt crosses a structural nose or it is structed by a fault. In excess of 1 billion barrels of oil has been trapped by this mechanism in the Borregas-Seeligson complex of south Texas *(Nanz, 1954)*.

For additional references on examples of non-marine deposited sandstone reservoirs consult Flores and Ethridge *(1981)*.

## BRAIDED STREAM RESERVOIRS - COOK INLET BASIN, ALASKA

The reservoir potential of Tertiary-age braided and meandering stream deposits in the Cook Inlet Basin of Alaska are reviewed by Hayes and others, 1976. Differences in the reservoir potential of the Beluga (braided stream deposits) and the Sterling (meandering stream deposits) formations are related to the depositional environment and diagenetic history of

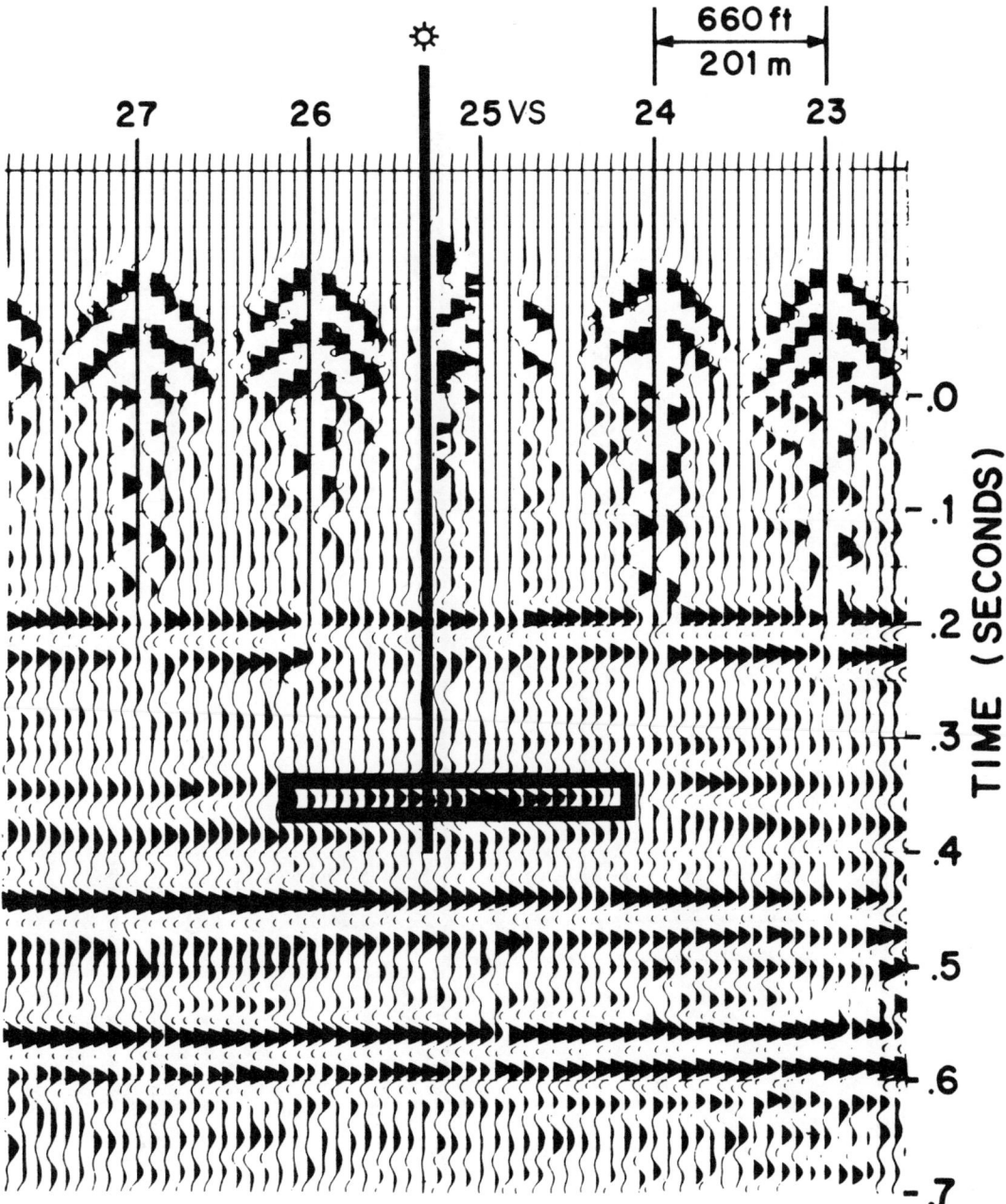

**Figure 4.17A:** *Seismic section showing "bright-spot" across channel sandstone in east central Alberta. Geologic data to support the interpretation is depicted on Figure 4.17B (adapted from Putnam and Oliver, 1980).*

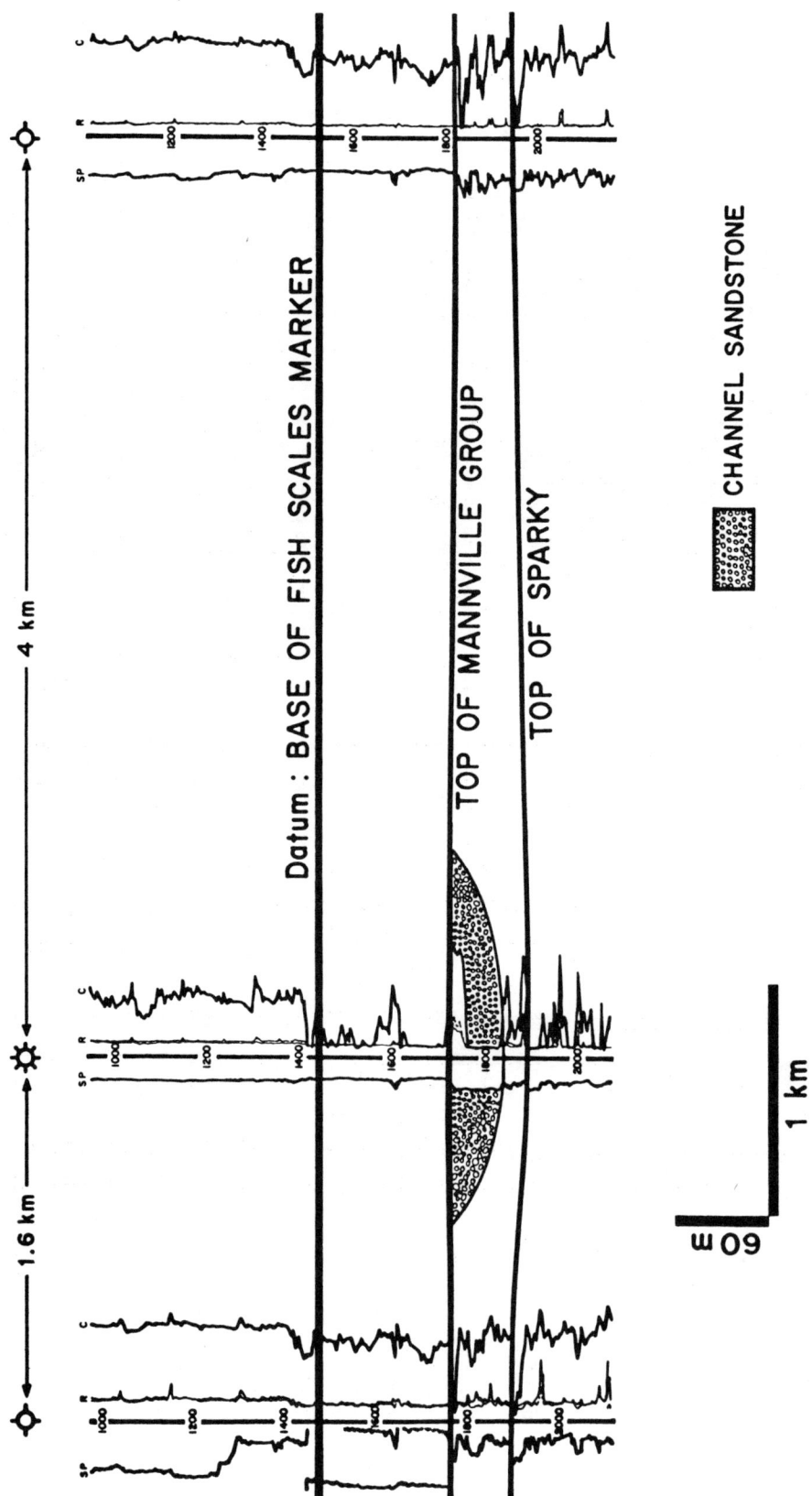

**Figure 4.17B:** *Cross section depicting a channel sandstone in central Alberta. The sandstone is represented by a "bright-spot" on Figure 4.17A between shotpoints 26 and 24 (adapted from Putnam and Oliver, 1980).*

# EXPLORATION FOR CONTINENTAL SANDSTONE RESERVOIRS 83

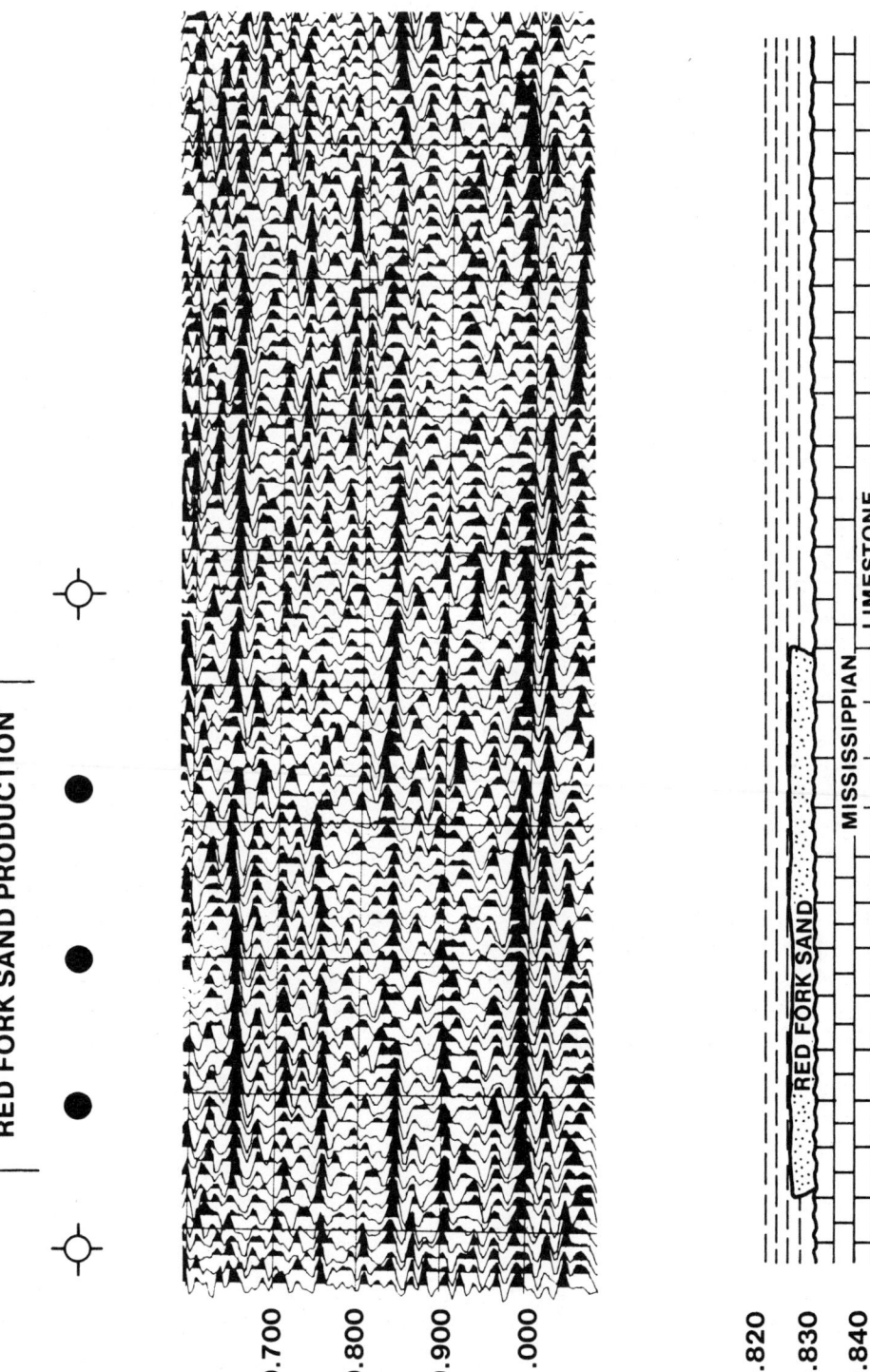

**Figure 4.18:** *Change in reflector character denotes Pennsylvania Red Fork Sand production above the post Mississippi Unconformity (courtesy of Olympia Exploration & Production Co.).*

the two formations. The Sterling Formation was deposited by large meandering streams that produce 30 to 45-foot thick point bar sandstones. The sandstones contain volcanic rock fragments that were dissolved during burial to produce secondary porosity that ranges from 30 to 40%. In contrast, the Beluga Formation was deposited by small, high gradient braided streams that produced lenticular, narrow, 5 to 10-foot thick braided stream sandstones. These sandstones contain metasedimentary rock fragments that were crushed during burial which in turn destroyed most of the primary porosity.

## ALLUVIAL FAN RESERVOIR - QUIRQUIRE FIELD, VENEZUELA

The giant field, Quirquire, is located in northeastern Venezuela on the tectonically unstable northern margin of the Eastern Venezuela Basin. More than 800 million barrels of oil have been produced from alluvial fan deposits since the field's discovery in 1928. The initial discovery was made by drilling downdip from tar and asphalt seeps. The discovery well produced 438 barrels of 16.3 API gravity oil per day *(Borger, 1952)*. Maximum daily production was achieved by 1938 when 241 wells produced an average of 72,000 BOPD. The field covers over 20,000 productive acres and has a maximum productive section of about 1,500 feet.

The main reservoir is the Quirquire Formation. It is a unique, locally developed coarse-grained facies that is part of the aerially extensive Las Piedras Formation (Fig. 4.19). The Quirquire Formation was sourced from the Interior Range that lies just to the north of the field. During the Mio-Pliocene time these mountains were much higher than at present. Alluvial transport of sediment on a southerly direction resulted in the accumulation of over 1,900 feet of coarse-grained, heterogenous sands that fine, but thicken southward to over 4,600 feet at the southern edge of the field. The formation is divided into eight members and the four main productive members have complex oil/water contacts. Each member can be related to a cyclic variation of conglomerate through clay, possibly tied to pulsations of sediment supply as controlled by climate variations. The field is an engineer's nightmare. Because of complex variations in capillarity that are produced by the various facies changes within the field, a simple borehole may have several oil/water contacts within the "pay zone."

A contour map on top of the Quirquire Formation depicts a southeast dipping homocline (Fig. 4.20). The dip is partly depositional although some of it must be due to renewed uplift during post Pliocene time as evidenced by rejuvenated erosional cycle in the Quirquire area. Beneath the erosional surface, upon which alluvial deposits of the Quirquire Formation were deposited, Cretaceous to lower Miocene strata are truncated and intensely deformed (Fig. 4.19).

Source for the Quirquire oil and gas is believed to have been generated from the Miocene La Pica Formation located in the basin to the south of the field *(Borger, 1952)*. The hydrocarbons must have migrated northward into the Quirquire Formation. There it was trapped by a permeability failure that is

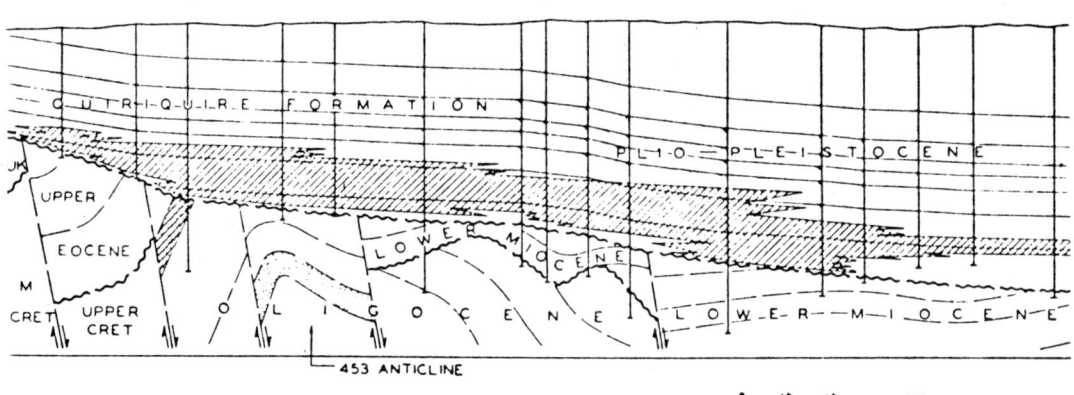

**Figure 4.19:** *North-south cross section of Quirquire Field, Venezuela (Borger, 1952).*

EXPLORATION FOR CONTINENTAL SANDSTONE RESERVOIRS 85

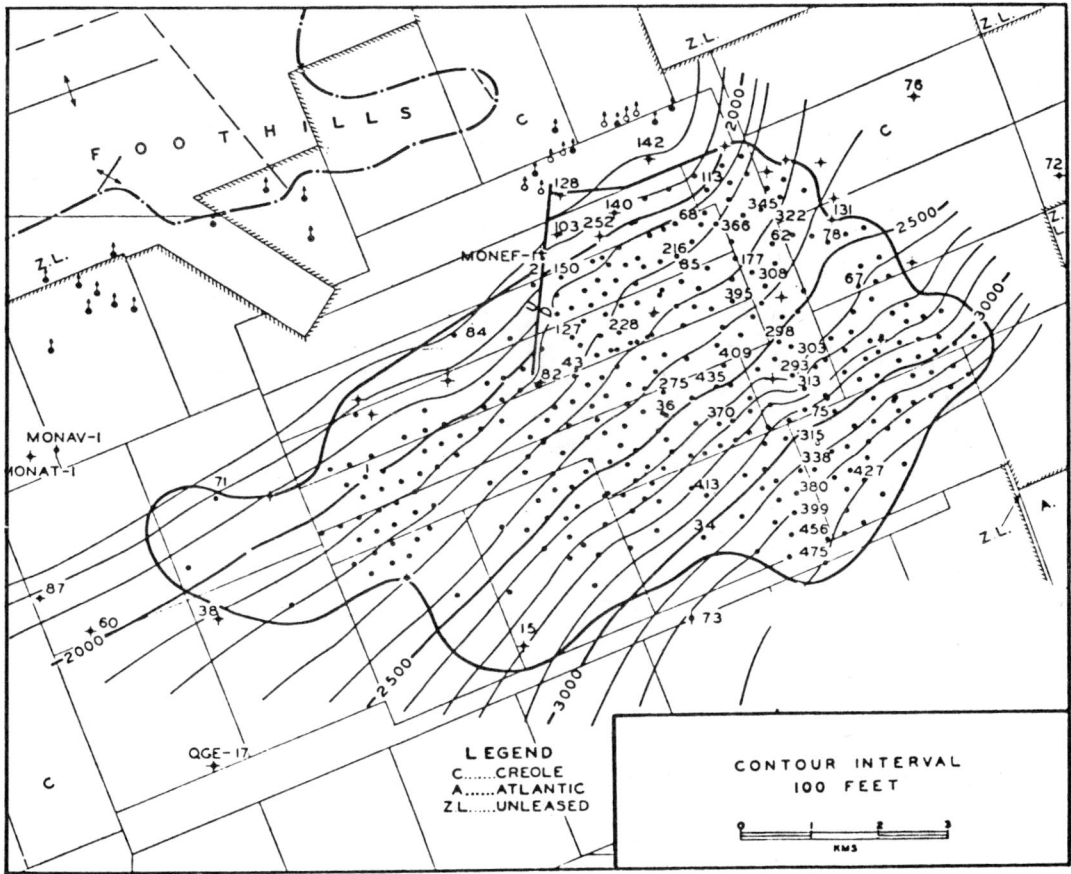

**Figure 4.20:** *Structure map on top of the Eta member, Quirquire Formation, Quirquire Field, Venezuela (Borger, 1952).*

related to depositional patterns but enhanced by the development of a "tar seal" along the northwest margins of the field.

## DUNE RESERVOIR - MEDIA FIELD, NEW MEXICO

The Media Field is located on the southeastern margin of the San Juan Basin, northwestern New Mexico. Although the field is small, less than 1 million barrels of recoverable oil, it is an excellent example of the importance of depositional topography as a possible trapping mechanism. Production from the Entrada Sandstone (Jurassic) was established in 1953. The discovery well produced less than 15,000 barrels of low gravity oil with a high water cut. This discouraged further development of the field until 1969. An offset to the discovery well produced 500 BOPD plus 1,500 BWPD *(Vincelette and Chittum, 1981).*

The discovery well was drilled on a seismically defined Lower Paleozoic closure. Subsequent drilling indicates that the field is located on a northwest plunging anticline as demonstrated by a structural map on top of a marker within the Cretaceous strata (Fig. 4.21). However, a map on top of the Entrada Sandstone depicts at least 85 feet of closure (Fig. 4.22). Vincelette and Chittum, *(1981),* attribute the closure to depositional topography on the Entrada Sandstone. To support their interpretation, Vincelette and Chittum constructed an isopach map from the Cretaceous marker to the top of the Entrada (Fig. 4.23). It depicts about 100 feet of thinning above the field. The thinning could be attributed to pre-marker uplift but the base of the Entrada and the marker are parallel (Fig. 4.24).

The Entrada topography is thought to be

86                                    EXPLORATION FOR SANDSTONE RESERVOIRS

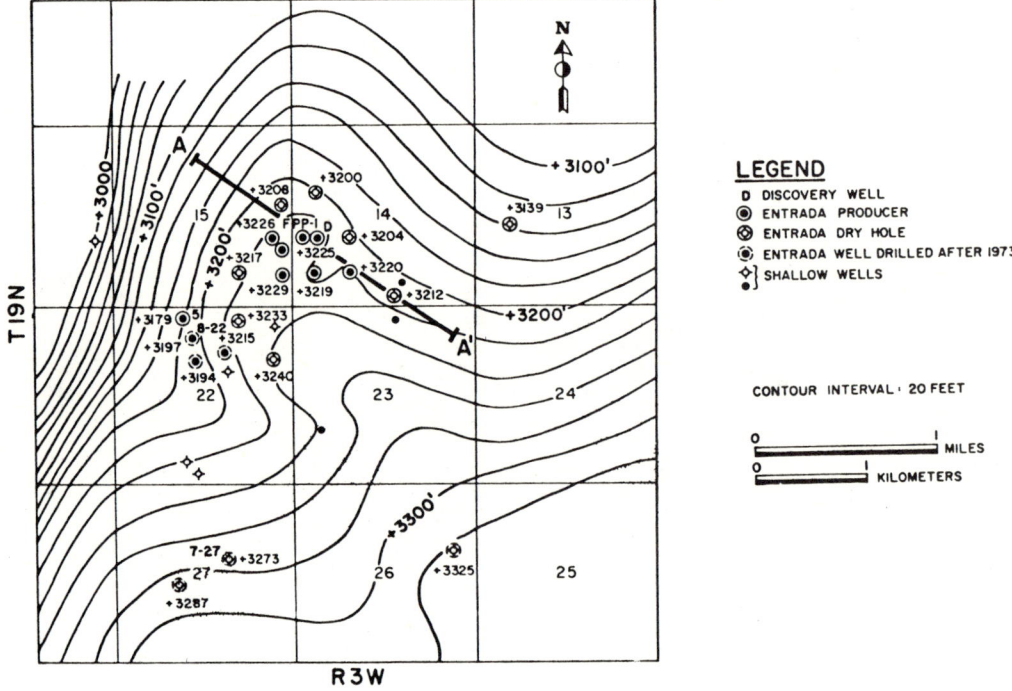

**Figure 4.21:** *Structure map on Cretaceous marker, Media Field, New Mexico (Vincelette and Chittum, 1981).*

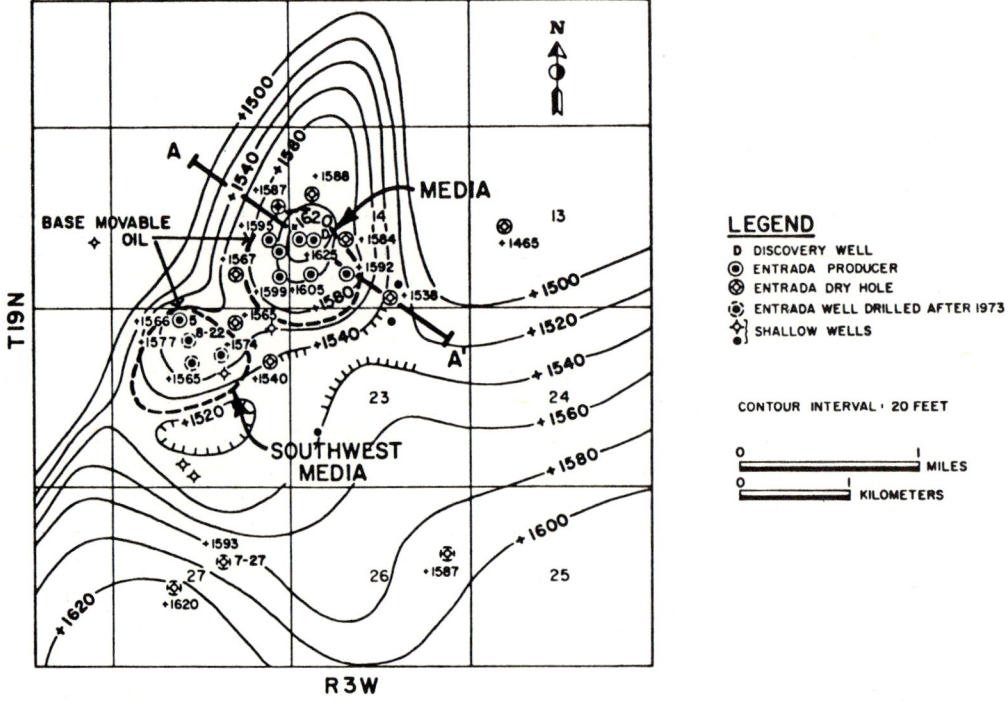

**Figure 4.22:** *Top of Entrada Sandstone, Media Field, New Mexico (Vincelette and Chittum, 1981).*

# EXPLORATION FOR CONTINENTAL SANDSTONE RESERVOIRS

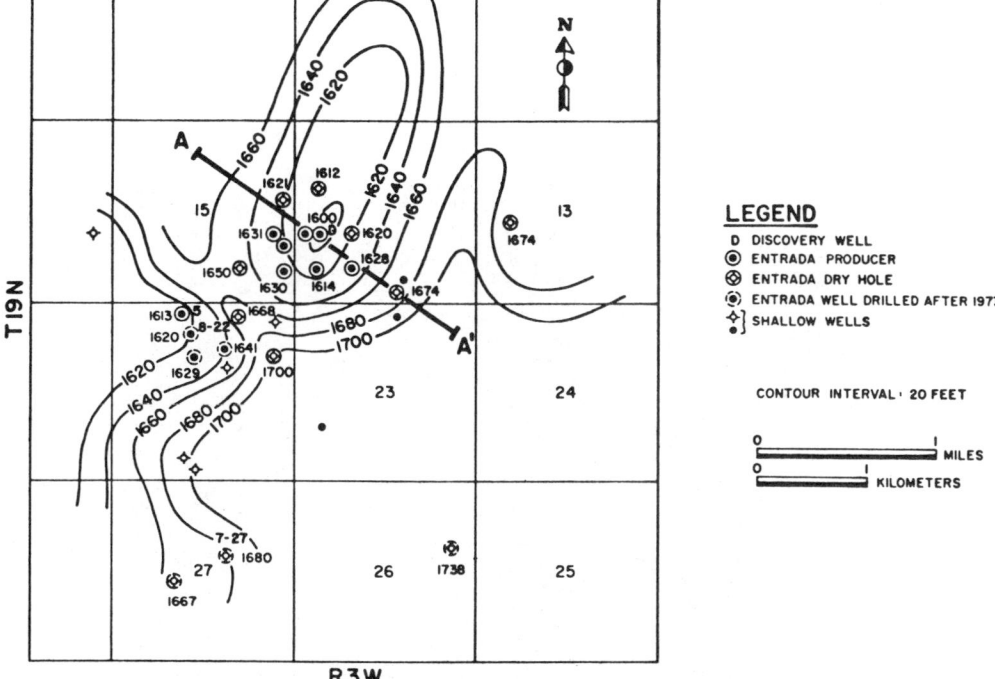

**Figure 4.23:** *Cretaceous marker to Entrada Sandstone isopach, Media Field, New Mexico (Vincelette and Chittum, 1981).*

attributed to eolian depositional processes. Surface data has been used to interpret the Entrada sands as dune deposits. The sands contain large scale eolian cross beds, sorting is excellent, and frosted grains have been reported in the literature *(Tanner, 1974; Stapor, 1972; Anderson and Kirkland, 1960)*. There is evidence, however, that locally the upper sands of the Entrada were reworked by water which in turn would reduce depositional topography of the dunes *(Tanner, 1970)*.

Perhaps it is not worth mentioning, but one possible flaw in the interpretation of Vincelette and Chittum is related to the Todelto formation which overlies the Entrada. The basal member of the Todelto is an organic rich, varved limestone that is 7 to 8 feet thick. It is difficult to explain why this limestone does not reflect the 80 feet or so of interpreted depositional topography by either a facies change or total absence on top of the Entrada structure.

## POINT BAR RESERVOIR - CUT BANK FIELD AREA, MONTANA

Cut Bank Field is located in northwest Montana on the west flank of the Sweetgrass arch. Regional dip is west-southwest at 75 to 100 ft/mile *(Lynn, 1955)*. There is no structural closure on this field that trends parallel to structural strike. The field is about 30 miles long and 10 miles wide.

In the Cut Bank Field area, the lower Cretaceous Kootenai Formation consists of gray, green, and red mudstone, lenticular siltstone, and variable-thickness sandstones. These units were deposited as alluvium in a valley or on a plain. The basal member of Kootenai is the Cut Bank Sandstone. The 40 to 80-foot porous sandstone trends north-south for nearly 70 miles (Fig. 4.25). The Cut Bank Sandstone is a typical point bar deposit. In most places, its stream cut through the Swift Formation and into the underlying marine Rierdon Formation. The Swift Formation, an alluvial siltstone, formed the lateral seal for the hydrocarbons that migrated updip through the Cut Bank Sandstone.

The Cut Bank Sandstone depicts the classic vertical sequence of sedimentary structure and textures that are characteristic of point bar deposits (Fig. 4.21B). Scale of cross beds decrease upward along with average grain size. Mud-cracked shale clasts are present in the basal part of the sand. Numerous shale zones, probable clay drape, occur in the upper part of the sandstone (Fig. 4.26). Porosities range from 12 to 19 %, permeabilities range from 19 to 305md *(Blixt, 1941)*.

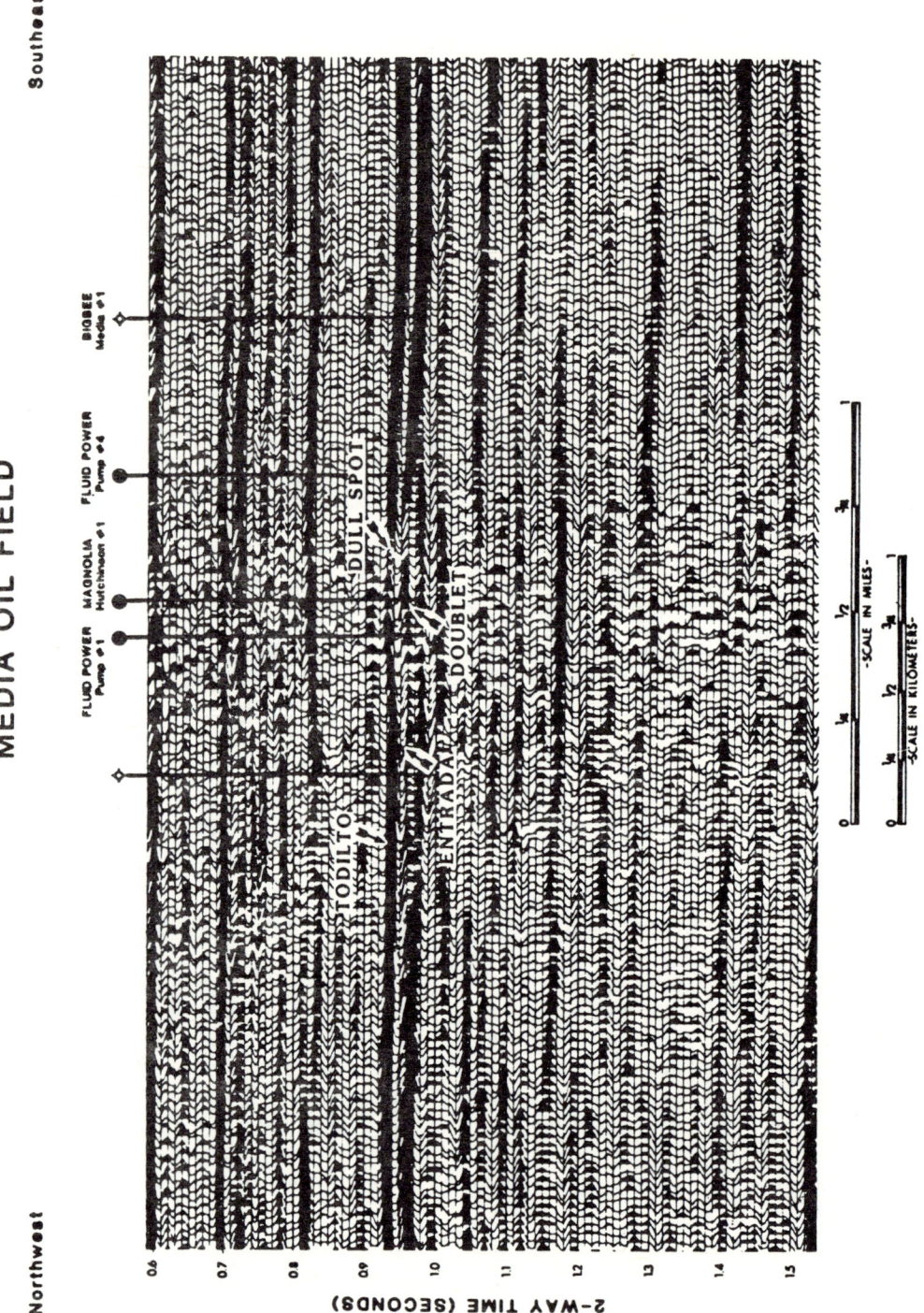

**Figure 4.24:** *Seismic line located approximately on line A-A', Figure 4.23 (Vincelette and Chittum, 1981).*

# EXPLORATION FOR CONTINENTAL SANDSTONE RESERVOIRS

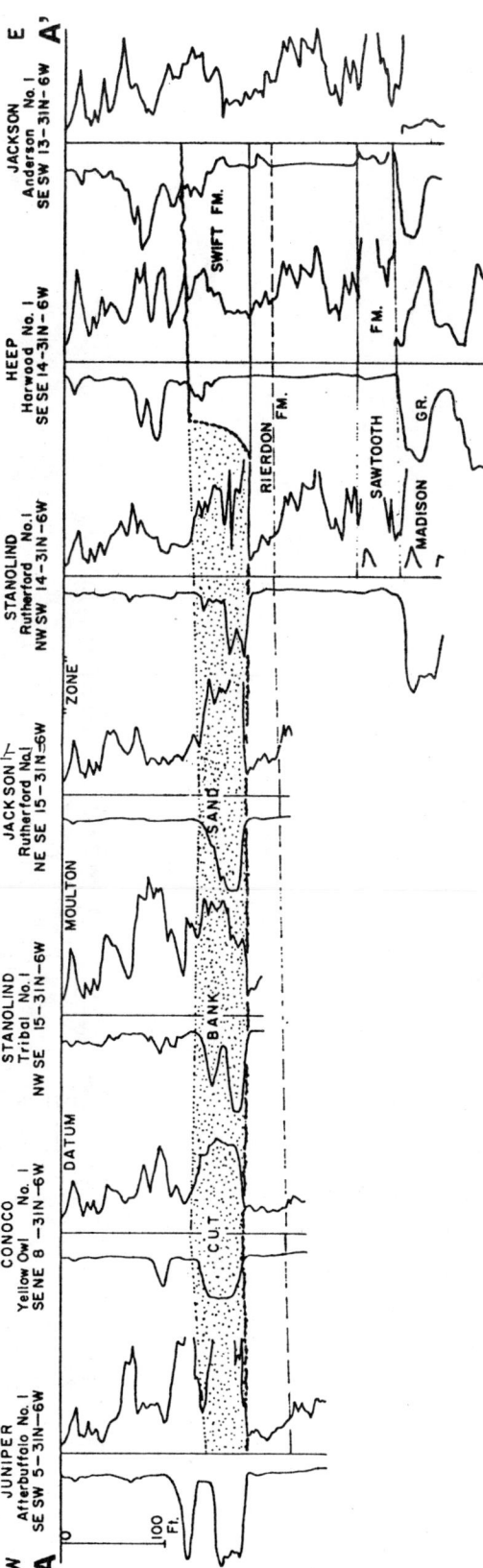

**Figure 4.25:** *West-east cross section across the southern end of Cut Bank Sand trend (Shelton, 1967).*

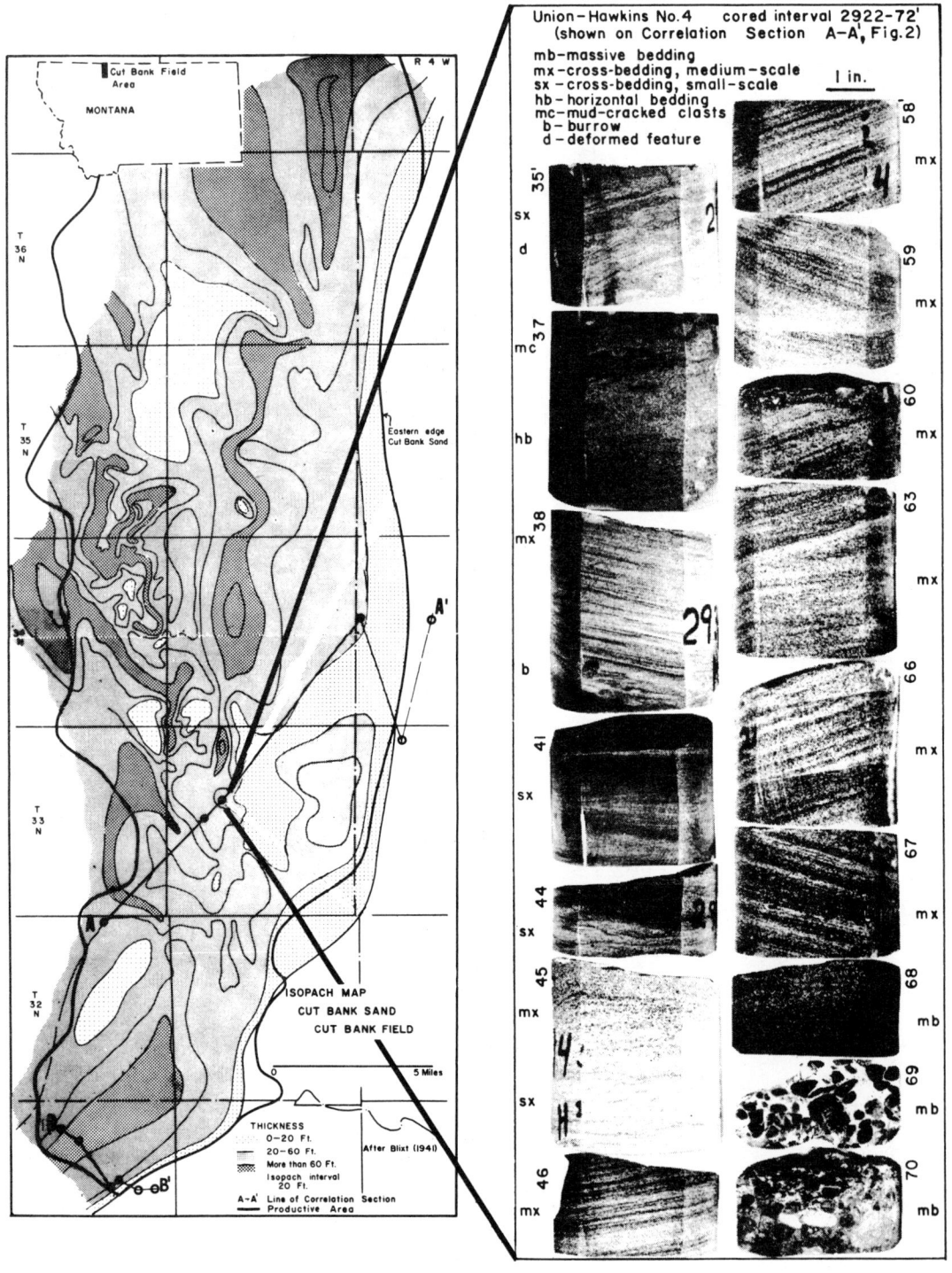

**Figure 4.26:** *Isopach map and core photograph of Cut Bank Sandstone, Montana (Shelton, 1967).*

# 5
# Exploration for Delta and Fan Delta Reservoirs

## CLASSIFICATION OF COASTLINES

Coastal depositional systems include deltaic and interdeltaic environments. Both of these environments are influenced by continuous, seasonal, and/or storm induced marine processes. The characteristics and three-dimensional array of these coastal deposits are determined by five controlling factors *(Hayes, 1977)*. They include: (1) hydrographic regime, (2) sediment supply and source, (3) global tectonic and sea level movements, (4) climate, and (5) geologic history.

Coastal areas have been classified by two very different criteria, namely hydrologic and tectonic. Perhaps the best classification of coastal areas that is based on tectonics was proposed by Inman and Norstrom *(1971)* and modified by Davies *(1973)*. This classification recognizes three major types of coast lines. They are collision coasts, trailing edge coasts, and marginal sea coasts. Marginal sea coastlines are typified by a plate-imbedded coast facing an island arc system. Trailing edge coasts occur along a plate-imbedded coast that faces a spreading zone. It may be a "neo-trailing" coast where a new zone of spreading is separating a land mass; an "afro-trailing" edge coast, where the opposite continental coast is also trailing; or an "amero-trailing" edge coast, where the opposite continental mass is a collision coast.

In general, explorationists prefer a classification of coastlines that is based upon hydrologic factors rather than tectonics. Classification emphasis has been placed on tides because as tidal ranges increase, wave action decreases *(Hayes, 1977)*. Davies *(1964)* proposed a useful classification of coastlines. His classification recognizes three types of coastlines. They are: (1) *microtidal*, tidal range less than 6 feet, (2) *mesotidal*, tidal range 6-12 feet, and (3) *macrotidal*, tidal range greater than 12 feet.

## MODERN RIVERS AND DELTAS

Not all large rivers produce deltas (Table 5.1).

**TABLE 5.1**

Discharge, length, and drainage area of selected rivers

|  | M Ft$^3$/sec. | Length | Drainage m$_2$x1000 | Climate |
|---|---|---|---|---|
| Amazon | 4000 | 3900 | 2231 | Tropic, rain forest |
| Colorado | 23 | 1200 | 260 | Subtropical, desert |
| Danube | 250 | 1400 | 315 | Humid, continental |
| Mackenzie | 300 | 2900 | 697 | Tundra, subarctic |
| Mekong | 400 | 2800 | 310 | Tropical, humid |
| Mississippi | 610 | 2500 | 1244 | Humid, subtropical |
| Niger | 220 | 2600 | 430 | Tropical, rain forest |
| Nile | 100 | 4000 | 1150 | Tropical, desert |
| Volga | 350 | 2200 | 600 | Desert to humid |

The Amazon is filling a large estuary that was produced by a trench of Wisconsin age. The Congo and Columbia Rivers are dumping their bed loads in coastal areas having narrow but steep continental shelves. Rivers that construct deltas have some common characteristics. Typically these rivers empty into a shallow marine environment that normally has a limited tidal range and an extensive shelf. Most delta constructing rivers have a large drainage basin.

## FACTORS THAT CONTROL THE SIZE OF DELTAS

A complex interplay of discharge rate, drainage area, climate, sediment load, and bathymetry of the depositional site control the size of a delta *(Shirely and Ragsdale, 1966)*. Although the Volga and Mackenzie Rivers have about the same discharge rate, length, and drainage area (Table 5.1), the Volga delta is much larger than the Mackenzie delta (Fig. 5.1). The major difference is that the load of the Mackenzie River is much smaller than that of the Volga River because the former is draining a subarctic drainage area. Climate in the drainage area also influences the relative size of the Niger and Mackenzie River deltas. Once again the discharge rate and length of the two rivers are about the same, and in fact the drainage area of the Niger River is about 40% less than that of the Volga River; but the Niger River delta is much larger than the Volga River delta. The difference is that the Niger River is draining a tropical rain forest, whereas the Volga River is draining a desert to humid area.

## FACTORS THAT CONTROL THE SHAPE OF DELTAS

**Water Depth: Mississippi River Delta** The Mississippi River delta is an excellent example of the influence of water depth on deltaic shape (Fig. 5.2). Discharge rates for the Mississippi River are moderate ($197 \times 10^2 \text{ft}^3/\text{yr}$) and the river deposits an average of $516 \times 10^6$ tons of sediment per year into the Gulf of Mexico. About 25% of this load is sand-size. Maximum water depth is 450 feet. Tidal range is about a foot, and the westerly longshore currents are weak to moderate.

The Mississippi delta is characterized by a few long finger-like extensions that are prograding out into deep water near the edge of the continental shelf. Distributary bifurcation is restricted because discharge rates are insufficient to erode fine-grained prodeltaic silt and clay over which the distributary is prograding. Bar finger and channel-mouth-bar deposits are dominant sand-rich facies.

**Strong Longshore Currents: Rhone River Delta** A modern longshore current-dominated delta is examplified by the Rhone River Delta. Discharge rate for the Rhone River is about $22 \times 10^2 \text{ft}^3/\text{year}$ and sediment load is about $45 \times 10^{16}$ tons per year. Sand constitutes about 50% of the load. The delta is building out into about 80 feet of water. Tidal range is only a foot or so. Longshore currents are strong, and transport sand from east to west.

Longshore currents and strong wave action produce an accurate shaped delta. Sand-rich facies are restricted to distributary channel sands and beach ridges along the delta margin.

**Tidal Currents: Niger River Delta** In contrast to Rhone River Delta, the Niger River Delta is dominated by tidal currents. Although the Niger River discharges about three times more water than the Rhone River, its load is only about half (Table 5.1). Sand-size material constitutes about 45% of the $25 \times 10^6$ tons per year that the river carries. Water depth is about 100 feet. Longshore currents are strong, but their direction of flow changes with the seasons. Tides in the Gulf of Guinea mold the blunt deltaic front of this somewhat triangular-shaped, high energy delta. Deep tidal and distributary channels contain the sand facies that cut through extensive mangrove swamps of the deltaic plain facies.

## DEVELOPMENT AND PRESERVATION OF DELTAS

**Initiation** Deltas initiate rapidly once the proper bathymetric conditions are satisfied; most of the modern deltas have developed to their present stage in the last 4,500 to 5,000 years. Sedimentation begins when the stream bed-load material is deposited to form channel-mouth-bars (Fig. 5.3A). This forms a subaqueous sedimentary bulge in the original shoreline.

**Progradation** Progradation initiates at the

# EXPLORATION FOR DELTA AND FAN DELTA RESERVOIRS

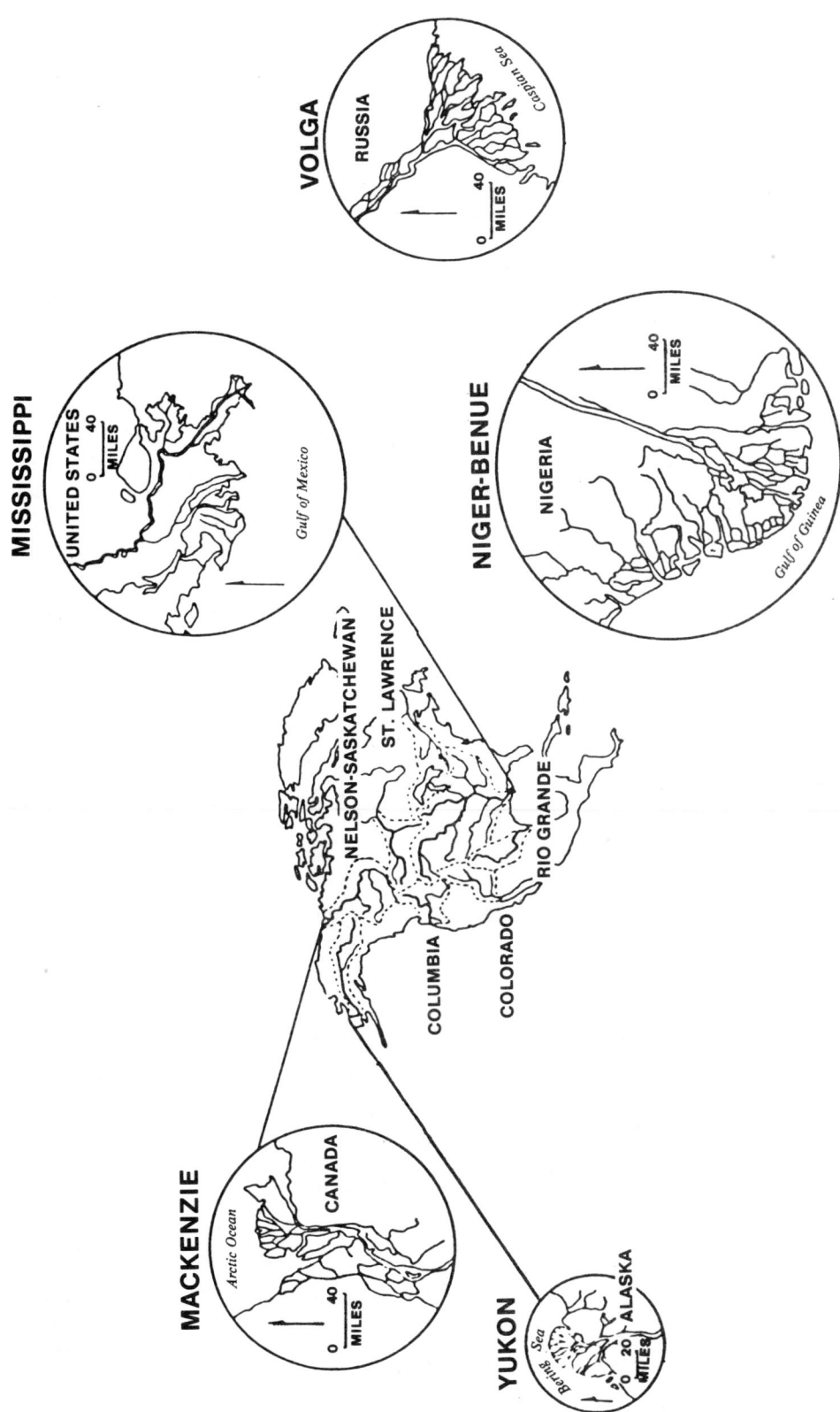

**Figure 5.1:** *Comparative size and shape of several important deltas.*

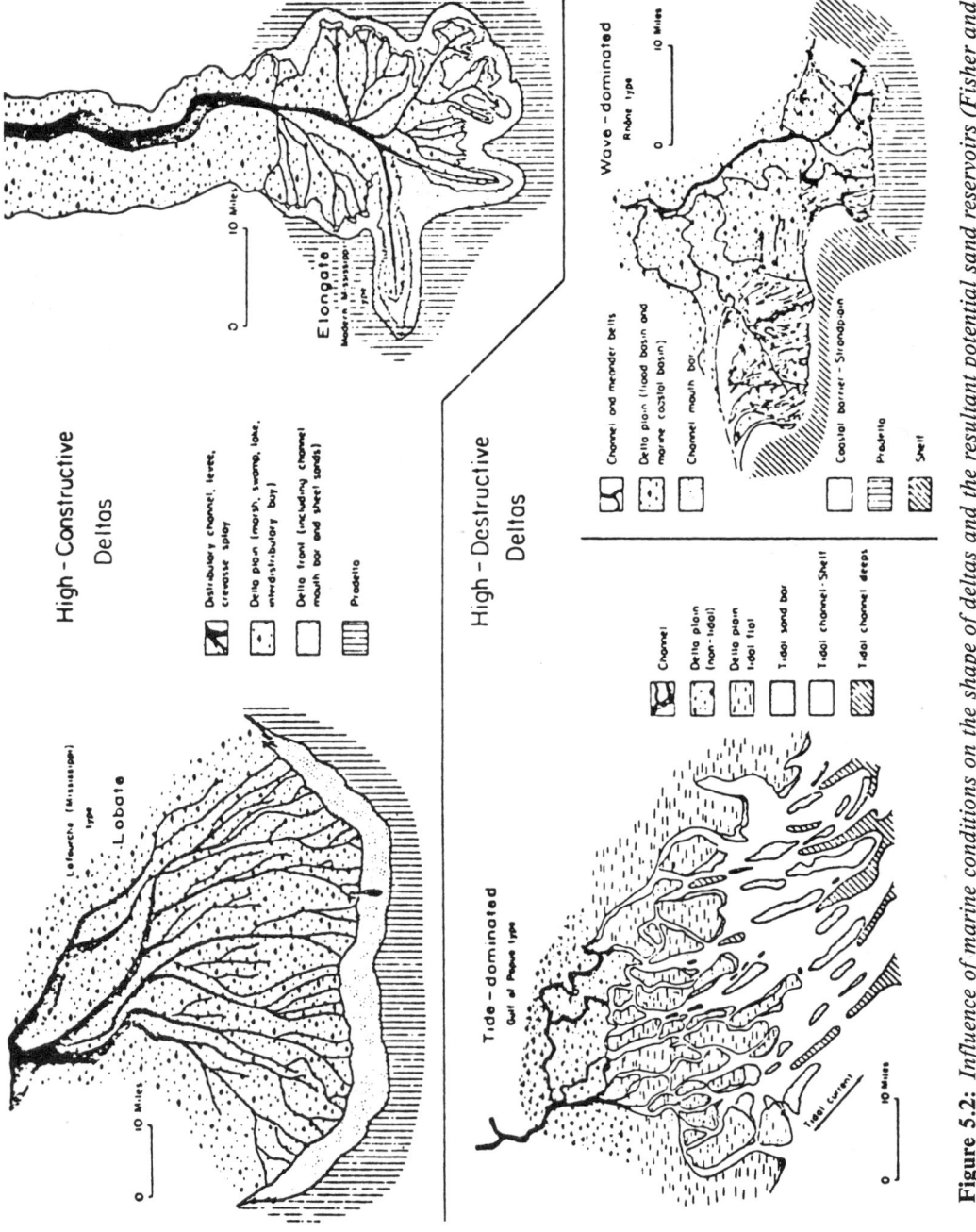

**Figure 5.2:** *Influence of marine conditions on the shape of deltas and the resultant potential sand reservoirs (Fisher and others, 1969).*

original shoreline and shifts the locus of deltaic sedimentation basinward (Fig. 5.3B and 5.3C). The position of the delta reflects the most favorable stream gradient from source to sea. The rate of progradation is a function of rate of subsidence, global sea level change, local tectonics, and competence of the stream. Progradation occurs when rates of sedimentation are greater than subsidence, assuming that sea level is at a standstill.

**Deterioration** Destruction of the delta occurs upon abandonment of a distributary or during a rise in sea level (Fig. 5.4A). If destruction is attributed to the former, subsidence and compaction occur concurrently with deposition long after the terrigenous supply of clastics has been terminated. Marine waves and currents degrade the delta. Channel-mouth-bars may be reworked into barrier bars.

**Preservation** Preservation of deltas can occur by a rise in sea level and/or continued subsidence via compaction after sedimentation has ceased (Fig. 5.4B). Low energy marine deposited silts and clays bury the delta. Much of the deltaic plain facies will be reworked prior to complete burial.

Delta rejuvenation can occur because the subsidence produced by compaction results in a net increase in gradient that in turn can produce a favorable site for the major distributary (river) to reoccupy.

## DELTAIC PLAIN RESERVOIRS

### INTRODUCTION

Deltas have been subdivided into three depositional environments; namely deltaic plain, deltaic front, and prodeltaic. Potential reservoir rocks on the deltaic plain include point bar, braided stream, bar finger, natural levee, and crevasse fan deposits. Deltaic front sands include channel-mouth-bar, delta-margin island, and beach ridge deposits. There are no significant sands that accumulate in the prodeltaic environment.

### POINT BAR AND BRAIDED STREAM RESERVOIRS

Deltas may be constructed by meandering or braided rivers. In both cases, the deposits of these rivers have the same characteristics as those that accumulate in a meandering or braided river in a continental framework. Refer to Chapter Four for a review of these deposits.

### NATURAL LEVEE RESERVOIRS

**Site and Process of Deposition** Natural levees accumulate along the banks of major rivers (Fig. 5.5). They form a narrow linear belt of sand that is sourced by the river during intervals of flooding when sediment-laden waters top the banks and begin unconfined flow. Large quantities of sediment are rapidly deposited laterally to the channels, and finer sediment is deposited successively farther from the channel.

**Internal Features** Silt and clay-sized material is dominant, but it is commonly interbedded with sand-size material which, in many cases, may be larger than the bedload (Fig. 5.6). Natural levee deposits are laminated, some are cross-bedded, and all are highly burrowed.

**External Features** Lower contacts of a natural levee sequence are gradational with floodplain or marsh-swamp deposits. Upper contacts are gradational with floodplain, marsh-swamp, or bay strata. Thickness ranges from a few feet to tens of feet. Length varies tens of miles along the channel trace.

**Reservoir and Stratigraphic Trap Potential** Size, limited reservoir continuity, and low porosity and permeability inhibit the explorationist from actively exploring for natural levee reservoirs. However, it is possible to miss the channel and hit the levee sequence. In general, these reservoirs are limited to salvage operations. The facies change from the natural levee to the finer-grained (seal) facies of the swamp, floodplain, or bay is the best locality for a stratigraphic trap.

### CREVASSE FAN RESERVOIRS

**Site and Process of Deposition** A crevasse is a break in the natural levee or any other stream embankment. A crevasse fan is a miniature delta that constitutes a sheet of sediment that flares outward from a crevasse. The crevasse fan may form on a floodplain,

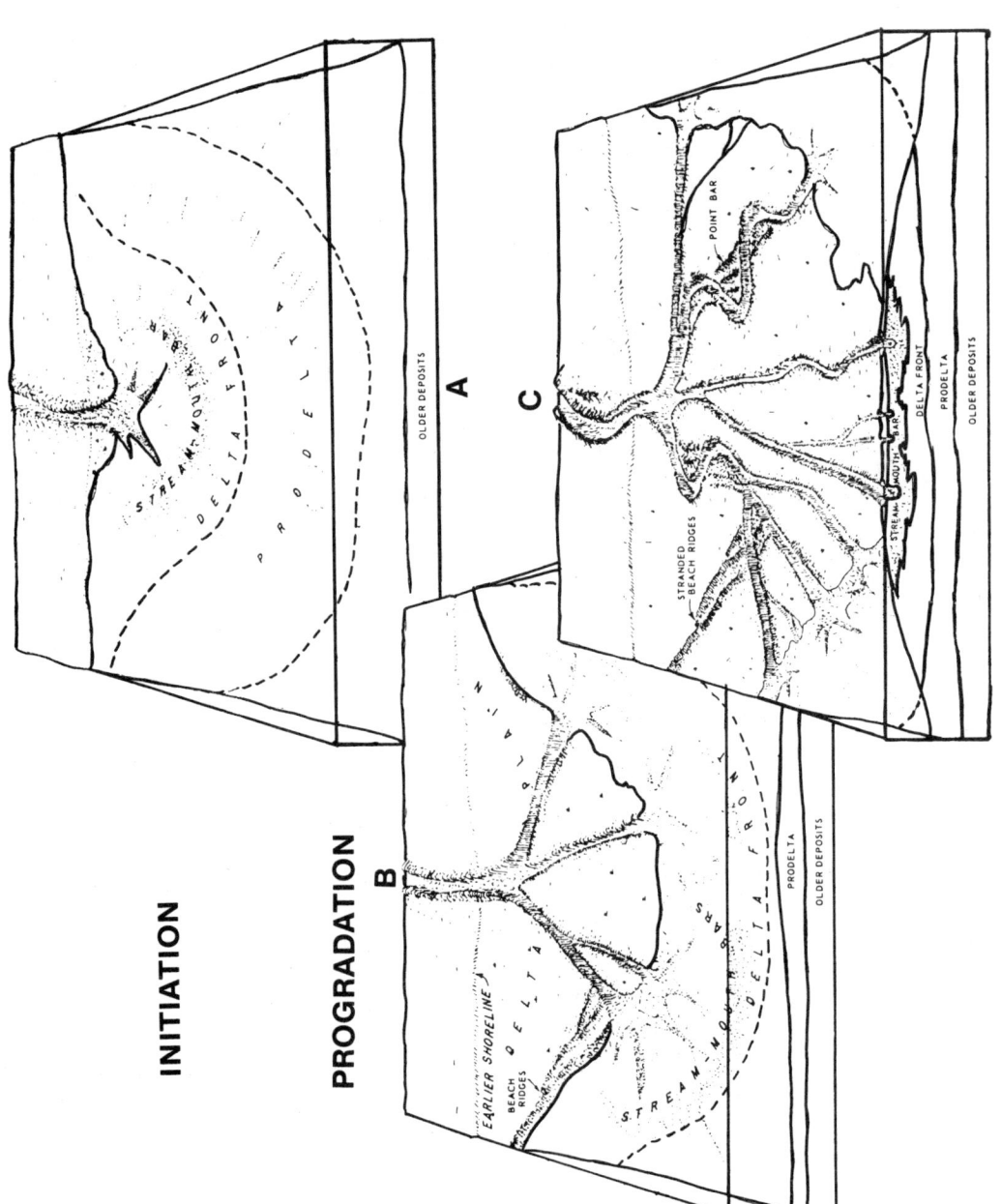

**Figure 5.3:** *Initiation (A) and progradation (B and C) of deltaic complexes.*

# EXPLORATION FOR DELTA AND FAN DELTA RESERVOIRS

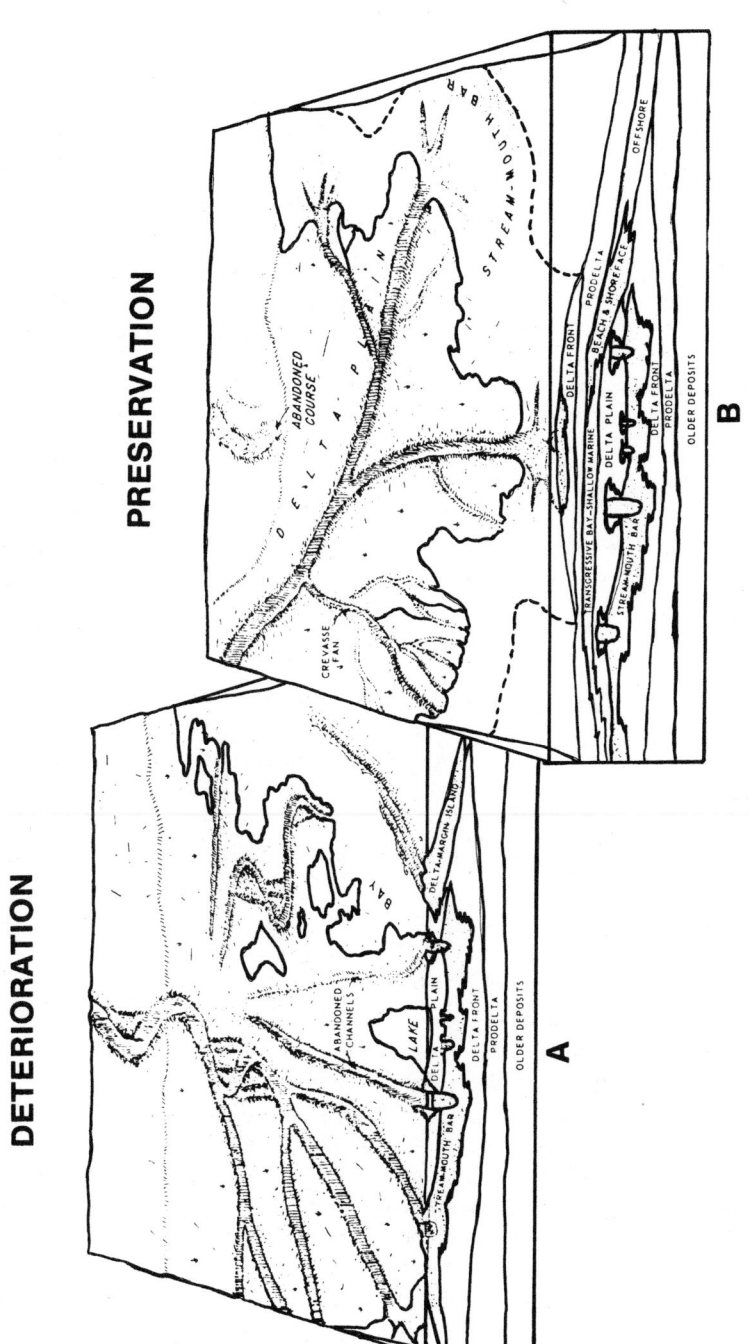

**Figure 5.4:** *Deterioration and preservation of deltaic complexes.*

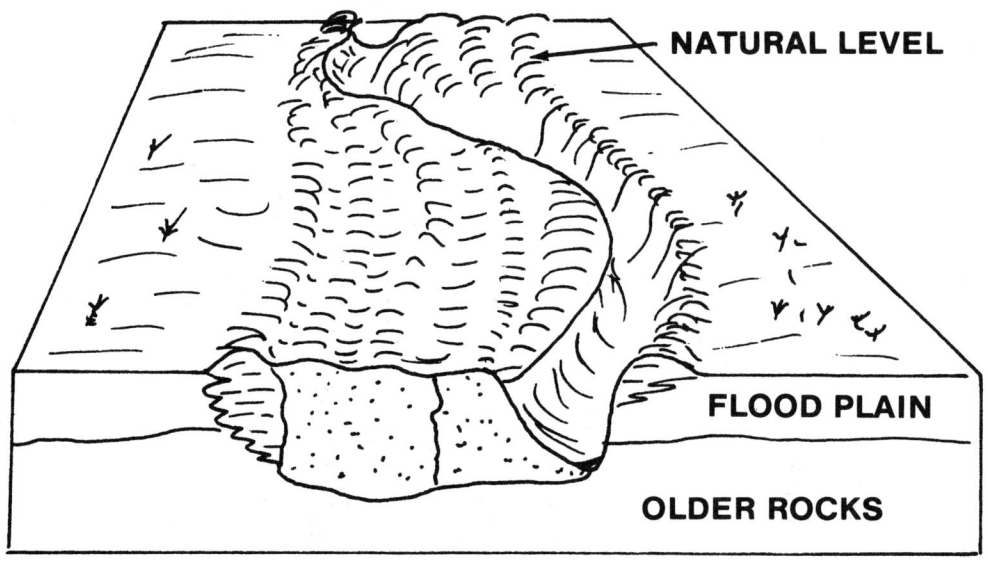

**Figure 5.5:** *Site of deposition and facies relationships of natural levee reservoirs.*

marsh, swamp, or within a bay. Depositional rates are normally rapid.

**Internal Features** Crevasse fan deposits are composed of sand, clay, and silt-sized material (Fig. 5.7). Normally grain size and degree of sorting slightly increases upward. A repeated sequence of cross-bedded, current-rippled sand, and laminated, burrowed, plant-rich interdistributary marsh-swamp deposits characterize a typical crevasse fan sequence.

**External Features** The lower contacts of a crevasse fan are sharp, and in some cases, unconformable with deltaic plain or bay sediments. Upper and lateral contacts are gradational with flood and/or deltaic plain sediments. The fans range in thickness from a few feet to several tens of feet. Individual sands rarely cover more than a few square miles.

**Reservoir and Stratigraphic Trap Potential** Reservoir quality of crevasse fan deposits range from poor to good. Vertical porosity and permeability are better than lateral within a single sand. Consequently, the best stratigraphic trap opportunities are along the margins of the crevasse fan where the coarse-grained fan facies grade into the finer-grained flood-plain or deltaic plain facies.

## DELTAIC FRONT RESERVOIRS
## CHANNEL-MOUTH-BAR RESERVOIRS

**Channel-Mouth Sedimentation** The coarse-grained load of the stream is deposited at the interface between running water and the lower energy marine environment. Under normal conditions, the river is sluggish and the channel is filled with sand-sized material. Channel-mouth sedimentation chiefly occurs during high discharge rates because the stronger currents scour the channel fill and transport it into the subaqueous marine environment. In order for the river to accommodate discharge, the channel widens, which, in turn, produces a typical bell-shaped mouth.

During high water stage, the channel deepens by scouring and marine waves and currents have little effect on sedimentation. Even the channel-mouth-bar may be cut into. The longer the distributary, the lower the distributary gradient which, in turn, facilitates bifurcation. Bifurcation is more common in sand-rich deltas than mud-rich deltas. During low discharge rates, waves and currents effectively winnow the fines leaving a well-sorted channel-(stream)-mouth-bar facies. These sands are typically laminated and rippled. Silt and clay bypass the channel-mouth-bar sands and are transported into deeper water as turbid

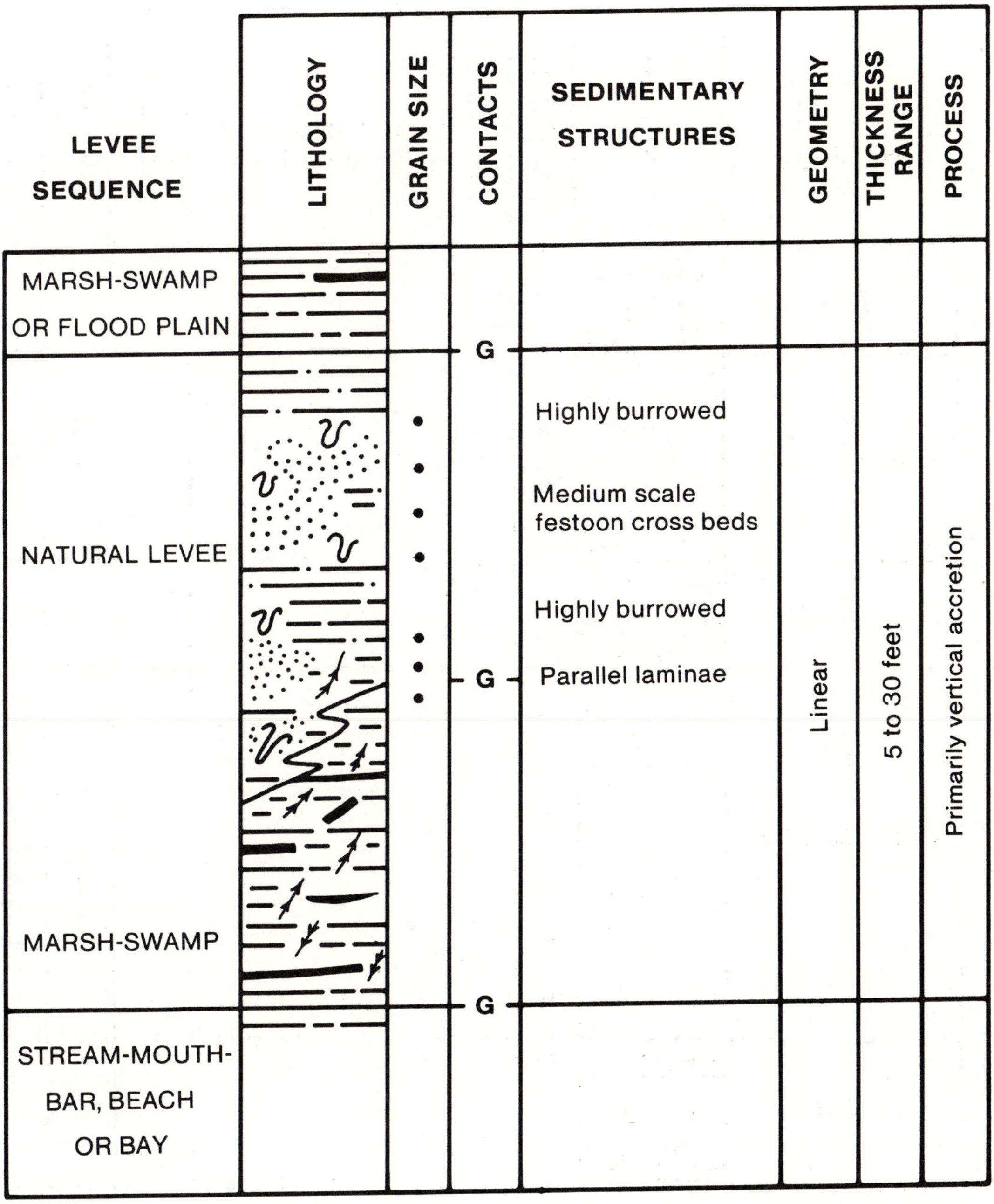

# EXPLORATION FOR DELTA AND FAN DELTA RESERVOIRS

| LEVEE SEQUENCE | LITHOLOGY | GRAIN SIZE | CONTACTS | SEDIMENTARY STRUCTURES | GEOMETRY | THICKNESS RANGE | PROCESS |
|---|---|---|---|---|---|---|---|
| MARSH-SWAMP OR FLOOD PLAIN | | | G | | | | |
| NATURAL LEVEE | | | • • • • • G | Highly burrowed<br><br>Medium scale festoon cross beds<br><br>Highly burrowed<br>Parallel laminae | Linear | 5 to 30 feet | Primarily vertical accretion |
| MARSH-SWAMP | | | G | | | | |
| STREAM-MOUTH-BAR, BEACH OR BAY | | | | | | | |

**Figure 5.6:** *Depositional characteristics of a natural-levee sequence.*

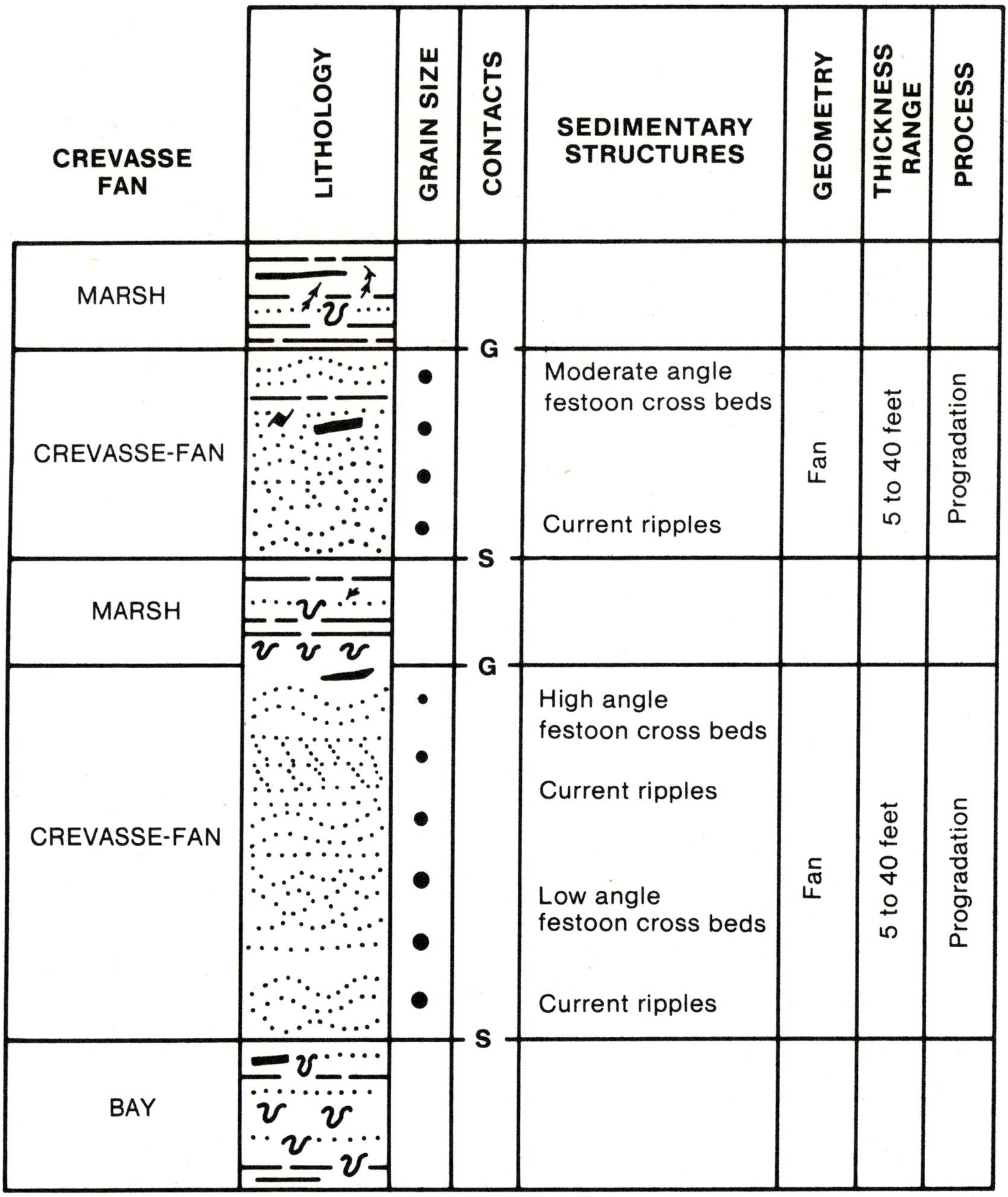

| CREVASSE FAN | LITHOLOGY | GRAIN SIZE | CONTACTS | SEDIMENTARY STRUCTURES | GEOMETRY | THICKNESS RANGE | PROCESS |
|---|---|---|---|---|---|---|---|
| MARSH | | | | | | | |
| CREVASSE-FAN | | ● ● ● ● | G — S | Moderate angle festoon cross beds<br><br>Current ripples | Fan | 5 to 40 feet | Progradation |
| MARSH | | | G | | | | |
| CREVASSE-FAN | | ● ● ● ● ● ● | S | High angle festoon cross beds<br><br>Current ripples<br><br>Low angle festoon cross beds<br><br>Current ripples | Fan | 5 to 40 feet | Progradation |
| BAY | | | | | | | |

**Figure 5.7:** *Depositional characteristics of a crevasse-fan sequence.*

plumes. They are deposited as deltaic front, prodelta facies, and basinal facies (Fig. 5.8).

**Internal Features**  Changes in texture and sedimentary structures within a channel-mouth-bar sequence resemble an upside down point bar (Fig. 5.9). The lower part of the bar contains intercalated fine-grained sand and shale beds. This part of the bar is laminated and burrowed. Scale and abundance of cross beds increase upward from the middle of the bar to its top. Large scale cross beds and some current ripples are the dominant sedimentary structures in the upper part of the bar. Sorting is maximum in these upper sands.

**External Features**  Basal contacts of the bar are gradational with prodeltaic fine-grained sediments. Upper contacts vary from gradational with interbar fine-grained sediments to sharp or unconformable with channel sands. Individual bars are crescent to linear, wedge-shaped deposits, but coalescing bars can form semi-linear rock bodies. Bars that form during periods of significant progradation within intervals of global sea level stillstand may form rectangular-shaped rock bodies. Thickness ranges from a few tens of feet to over 1,000 feet.

**Reservoir and Stratigraphic Trap Potential**  In response to maximum sorting and grain size near the top of a channel-mouth-bar sequence, maximum porosity and permeability occur in the upper part of the bar (Fig. 5.10). The best opportunities for a stratigraphic trap occur along the up-sedimentary dip margins of the bar where it grades into the fine-grained deltaic plain facies.

## DELTA-MARGIN ISLAND RESERVOIRS

**Depositional Processes**  Delta-margin island sands normally are associated with deteriorating deltas. The margin of the delta is subjected to wave and longshore current action. Sands that were originally deposited as deltaic front are reworked and spread along the margin of the deteriorating delta. Commonly, these sands are perched on the old deltaic plain.

**Internal Features**  Basal strata contain interbedded sand, silt, and clays that are typically burrowed (Fig. 5.11). Parallel laminae are typical of the basal deposits, but bioturbation may be so severe that the laminations may have been destroyed. Low angle cross beds and current ripples comprise the other sedimentary structures. Grain size, degree of sorting, and angle of cross beds increase upward. The uppermost part of the delta-margin island sequence is current rippled. Due to the winnowing by marine currents, the upper few feet of the sequence are clean, well-sorted, silt and clay-free sand.

**External Features**  Basal contacts may be sharp or gradational with deltaic plain strata, but more typically with deltaic front deposits. Upper contacts are gradational with offshore deposits. Characteristically, delta-margin island deposits form linear belts of sand that range in thickness up to sixty feet. Normally these deposits are less than two miles wide and may approach ten to fifteen miles in length.

**Reservoir and Stratigraphic Trap Potential**  Good to excellent vertical and lateral porosity and permeability in the upper part of the delta-margin island sands make this sequence an excellent potential reservoir. Intensely burrowed sands in the base of the deposit will have reduced permeability. Maximum opportunities for porosity failure occur along the landward margin of the island facies, where fine-grained seal sediments are in gradational contacts with coarser-grained reservoir strata.

## BEACH-RIDGE RESERVOIRS

**Depositional Processes**  Beach-ridge reservoirs are associated with prograding deltas that were subject to strong longshore currents. As the sand is transported to the mouth of the river, strong longshore currents sweep the sand down current to form laterally accreting beach-ridge deposits.

**Internal Features**  The beach-ridge sequence, like the delta-margin island sequence, can be subdivided into three units — shoreface, foreshore, and backshore (Fig. 5.12). The basal unit (shoreface) contains the finest grain sands that may be intercalated with thin siltstones and/or shale beds. The unit typically is highly burrowed. Low angle cross beds are common, particularly if silt and shale lithologies are absent. Grain size and angle of cross

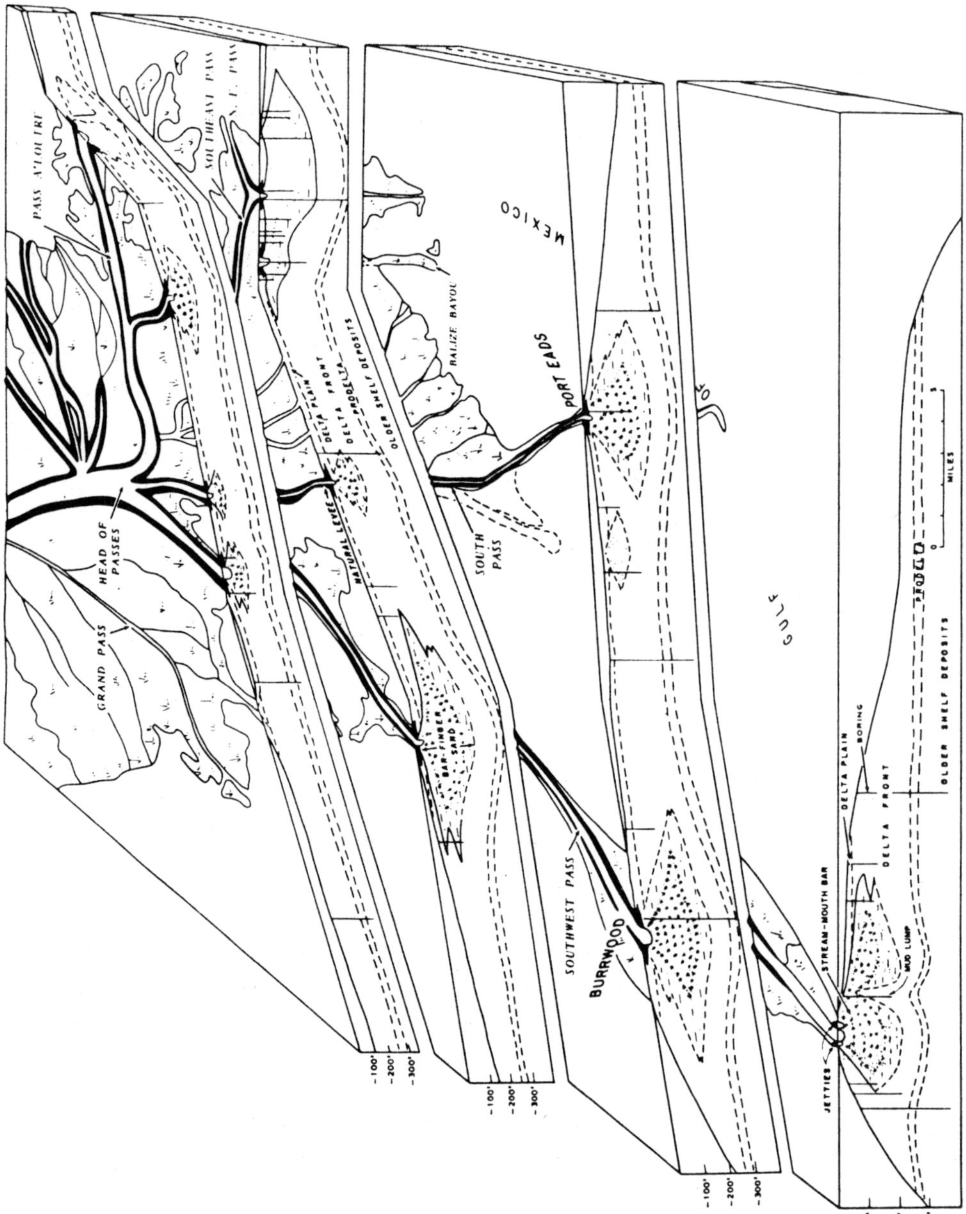

**Figure 5.8:** *Development of distributary channel and channel-mouth-bar deposits, Mississippi River Delta (Fisk, 1961).*

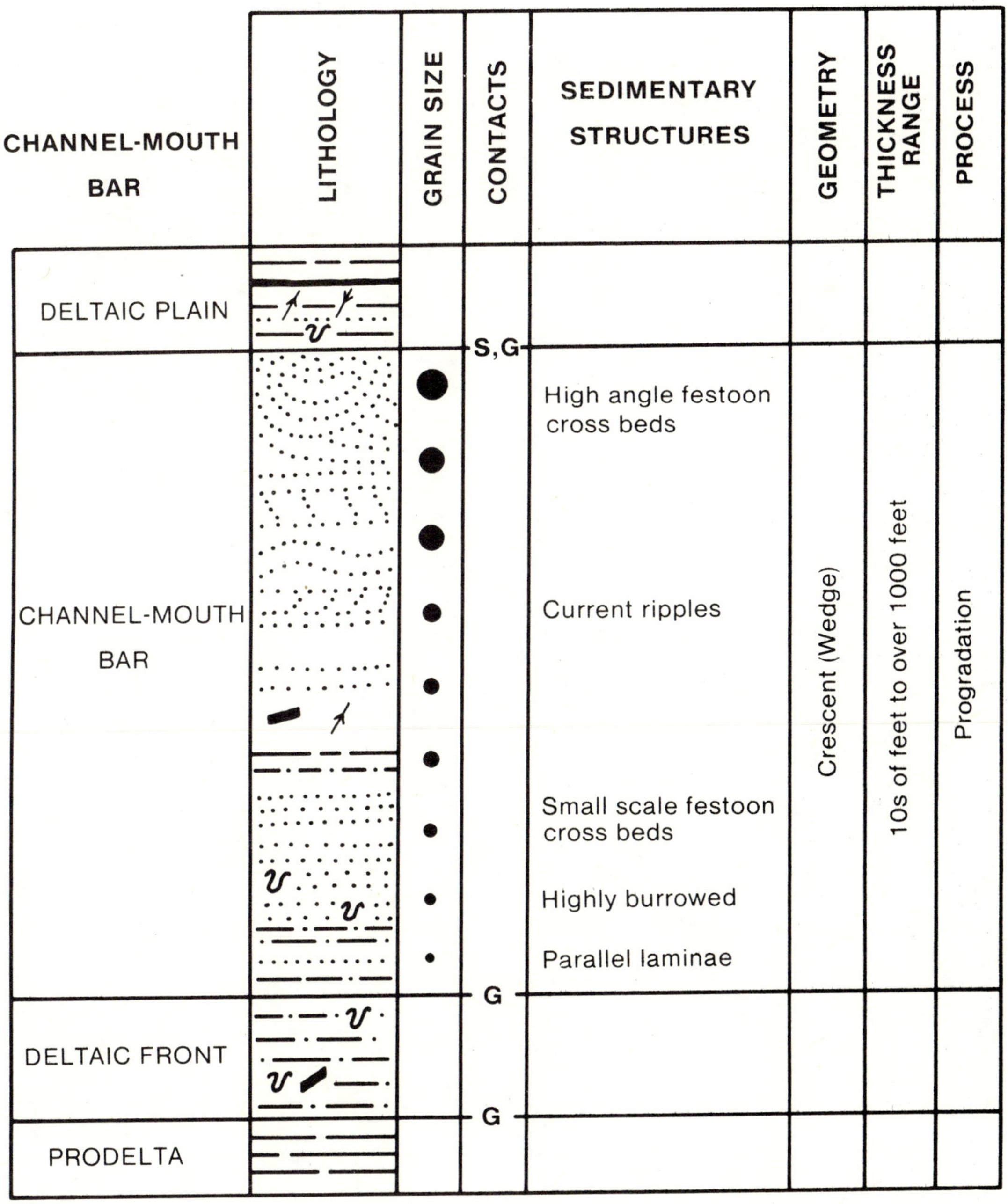

EXPLORATION FOR DELTA AND FAN DELTA RESERVOIRS 103

| CHANNEL-MOUTH BAR | LITHOLOGY | GRAIN SIZE | CONTACTS | SEDIMENTARY STRUCTURES | GEOMETRY | THICKNESS RANGE | PROCESS |
|---|---|---|---|---|---|---|---|
| DELTAIC PLAIN | | | S,G | | | | |
| CHANNEL-MOUTH BAR | | | | High angle festoon cross beds | Crescent (Wedge) | 10s of feet to over 1000 feet | Progradation |
| | | | | Current ripples | | | |
| | | | | Small scale festoon cross beds | | | |
| | | | | Highly burrowed | | | |
| | | | | Parallel laminae | | | |
| DELTAIC FRONT | | | G | | | | |
| PRODELTA | | | G | | | | |

**Figure 5.9:** *Typical channel-mouth-bar sequence (Silver, 1980).*

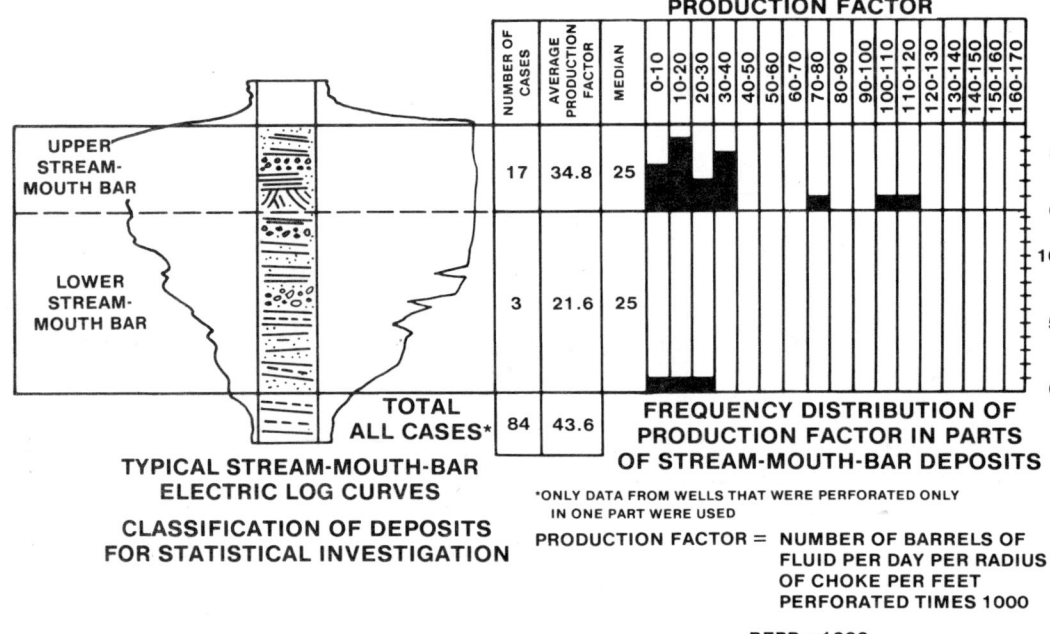

**Figure 5.10:** *Reservoir characteristics of channel-mouth-bar sands, Hough Area, Oklahoma (Swanson, 1972).*

bedding increase upward into the middle unit. Sorting is excellent throughout this unit. The upper unit contains maximum grain size, but sorting is not quite as good as that in the middle unit. Ripples and medium-scale cross beds are dominant sedimentary structures. Burrowing is not common.

**External Features** The lower contacts of the beach-ridge sequence are gradational with prodelta sediments, and upper contacts are gradationally overlain by fine-grained sediments. Coalescing beach-ridge deposits form sheet-like sand bodies that are bioconvex and may obtain thickness in the tens of feet particularly in rapidly subsiding areas. It is frequently difficult to distinguish a beach-ridge from a delta-margin island on the basis of a single core. Geometry and depositional setting are the most important distinguishing features of the two deposits.

**Reservoir and Stratigraphic Trap Potential** Maximum reservoir quality occurs in the middle and upper units of the beach-ridge sequence. Both lateral and vertical continuity in the two units range from good to excellent. Therefore, the beach-ridge sequence is an excellent target for a stratigraphic trap. The gradational coarse-grained facies change with fine-grained seal sediments occurs along the backshore margin of the beach-ridge. The best opportunity for a stratigraphic trap occurs where this facies change cuts across a regional nose or where the change produces an embankment up-dip on regional structure.

## FAN DELTAS

### INTRODUCTION

The term *Fan Delta* was first used by Holmes *(1965)* to describe the Lynmouth delta located on the north shore of the Bristol Channel, England. Fan deltas are essentially alluvial fans that prograde into a standing body of water. Because they normally consist of very coarse-grained terrigenous clastics (potential reservoirs) that are complexly interbedded with coastal deposited (marine or lacustrine) potentially organic rich shales, they are likely

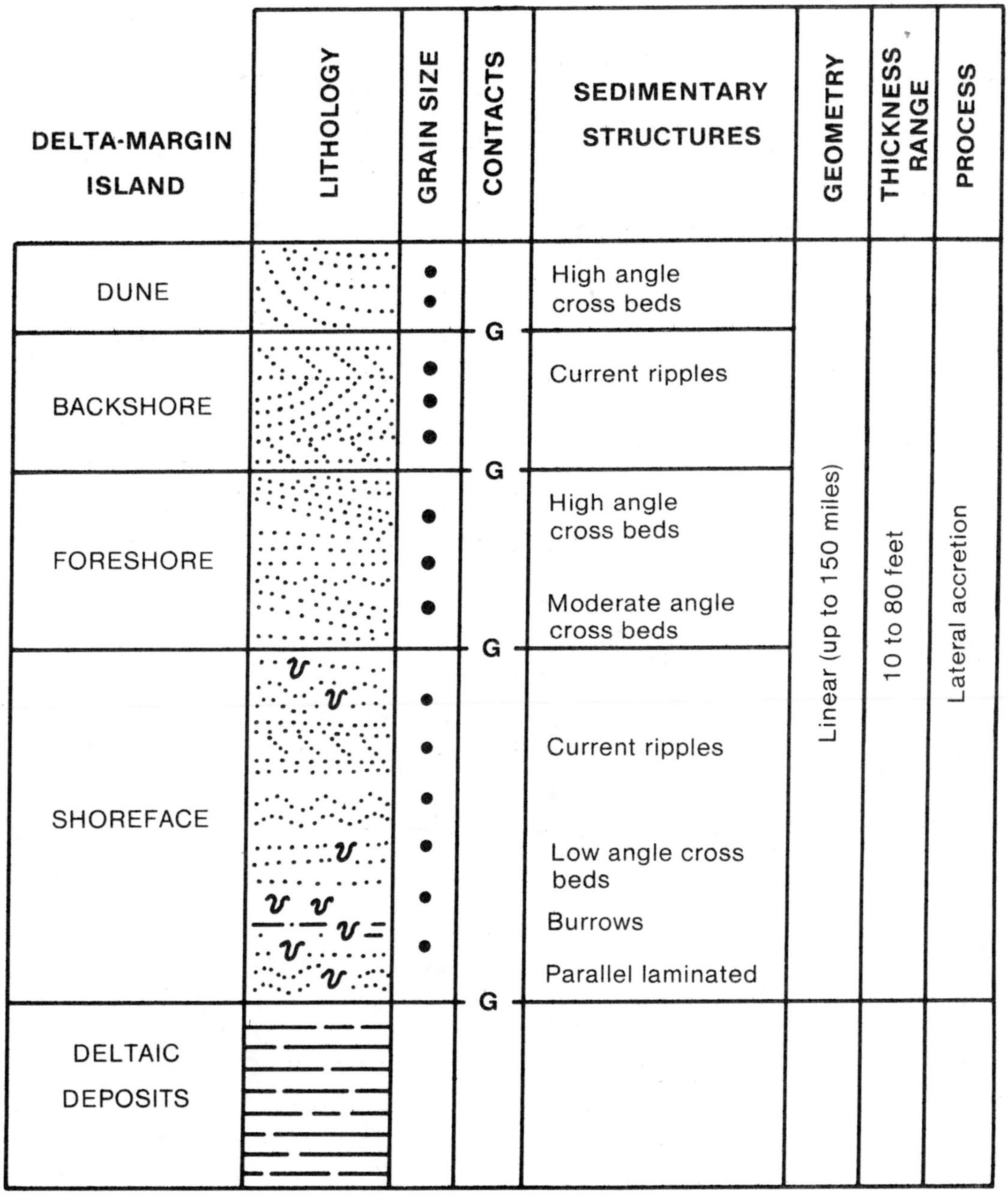

# EXPLORATION FOR DELTA AND FAN DELTA RESERVOIRS

| DELTA-MARGIN ISLAND | LITHOLOGY | GRAIN SIZE | CONTACTS | SEDIMENTARY STRUCTURES | GEOMETRY | THICKNESS RANGE | PROCESS |
|---|---|---|---|---|---|---|---|
| DUNE | | • • | G | High angle cross beds | Linear (up to 150 miles) | 10 to 80 feet | Lateral accretion |
| BACKSHORE | | • • • | G | Current ripples | | | |
| FORESHORE | | • • • | G | High angle cross beds / Moderate angle cross beds | | | |
| SHOREFACE | | • • • • • | G | Current ripples / Low angle cross beds / Burrows / Parallel laminated | | | |
| DELTAIC DEPOSITS | | | | | | | |

**Figure 5.11:** *Depositional characteristics of a delta-margin island sequence.*

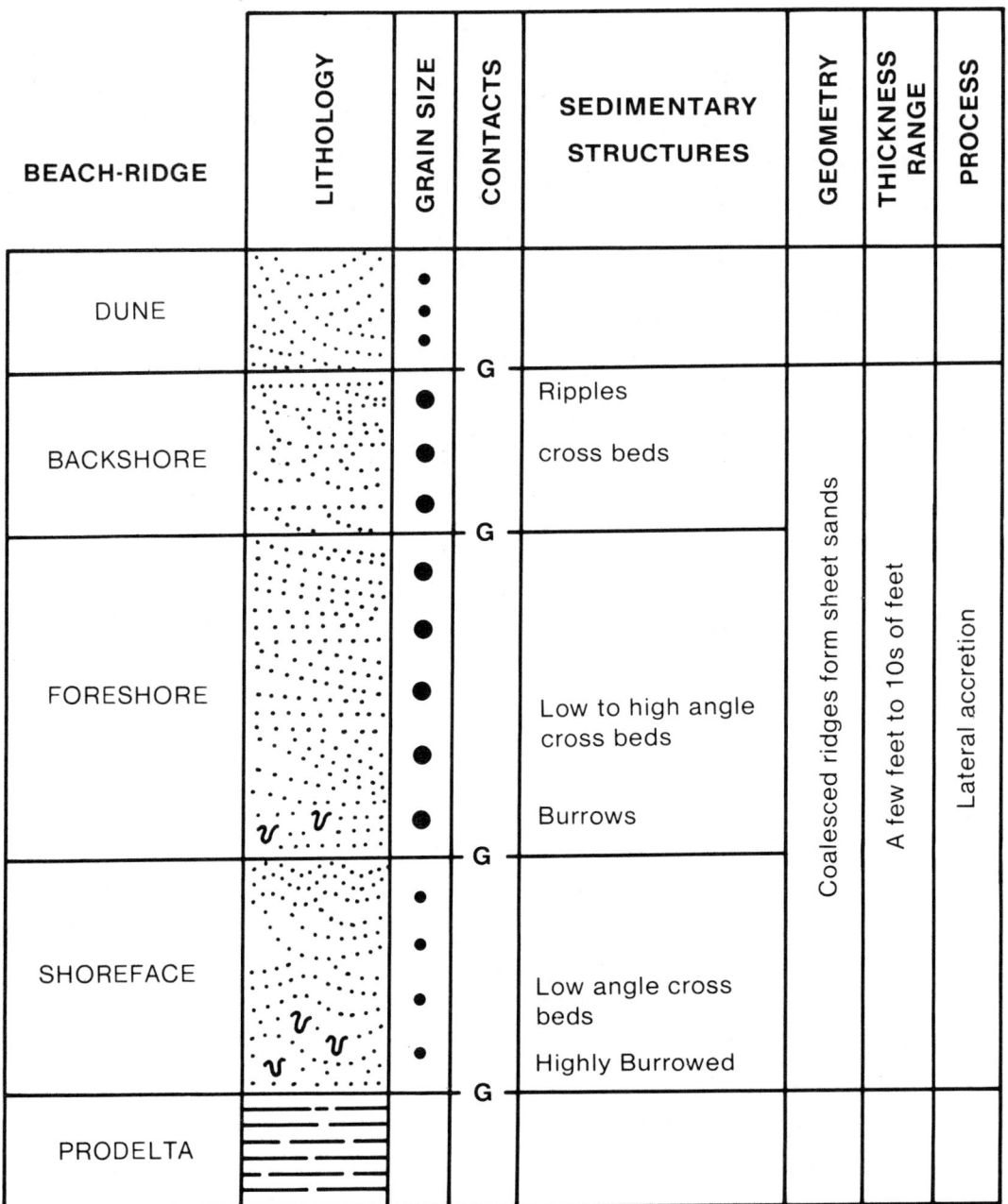

**Figure 5.12:** *Depositional characteristics of a beach-ridge sequence.*

excellent environments for major accumulations of oil and/or gas.

Fan deltas are known to occur along all three major coastlines (Fig. 5.13). The abundance and morphology of the fan deltas are highly controlled by the tectonic character of the continental margin. Fan deltas along trailing edge coastlines include the small fan delta at Lynmouth *(Holmes, 1965)*, Firth River delta, Yukon *(McDonald and Lewis, 1973)*, several small gravel fans along the northwest coast of the Gulf of California *(Thompson, 1968)*, and several deltas of variable size along the shores of the Red Sea and the Gulf of Aqaba *(Friedman and Sanders, 1978)*. These deltas are listed as locations 1-4 on Figure 5.13. McGowan *(1970)* described the only known fan delta to occur along a marginal sea coast (Gum Hollow, locality 5, Fig. 5.13). Large gravel-dominated fan deltas are more typical of collision coastlines. They include the glacial outwash fans along the southeast coast of Alaska *(Boothroyd and Nummdal, 1978,* location 6, Fig. 5.13) and the Yallahs fan delta, southeastern Jamaica *(Wescott and Ethridge, 1980,* location 11, Fig. 5.13). These deltas are formed by subaerial fans that prograde directly into a steeply sloping marine environment (Yallahs fan delta) or fans that prograde onto continental or island shelfs (Alaska east coast fan deltas).

## CHARACTERISTICS OF FAN DELTAS

Fan deltas are typically fan-shaped. In cross section they depict a wedge or prism of coarse detrial sediments that thicken away from the mountain front. Like their alluvial fan counterparts, mineralogically they are immature; sorting and rounding are poor. Mud to grain supported fabrics are common. The depositional sequence should coarsen upward.

Fans that prograde onto steep submarine slopes, such as the Yallahs fan delta, are in effect truncated and only the proximal fan deposits are formed. In contrast, fans that prograde onto shelves will be characterized by a down-fan facies change from horizontally bedded coarse gravels to coarse, cross-bedded

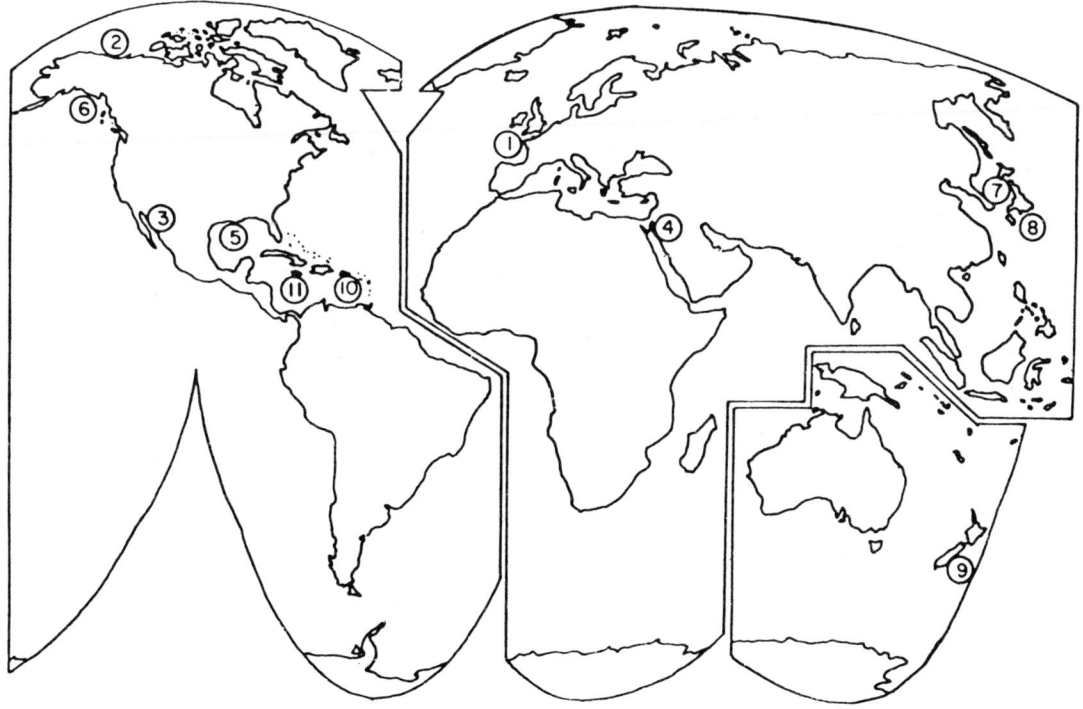

**Figure 5.13:** *Distribution of Holocene fan deltas. Number locations are described in text (Wescott and Ethridge, 1980).*

sandstones. Mid and distal portions of the fan, particularly the cross bedded sandstones, should directly overlay marine lithologies. Post fan deposits and fan thicknesses will be determined by local tectonics and/or global changes in sea level.

## SEISMIC CHARACTERISTICS OF DELTAIC RESERVOIRS

### FLUVIAL - DOMINATED DELTAS

The oblique (tangential), sigmoid, and complex-sigmoid-oblique progradational reflection patterns are representative patterns of a fluvial-dominated deltaic system *(Mitchum and others, 1977)*. The oblique (tangential) seismic character (Fig. 5.14A) is characterized by clinoform reflections that terminate up sedimentary dip by toplap and downdip by downlap. This seismic reflection pattern (SRP) represents a high-energy delta in which the sand-prone deltaic plain is coincident with the upper horizontal reflection event. The clinoform pattern represents the sand-poor prodelta facies. The absence of stacking of horizontal reflections on the deltaic plain indicates that the shelf was stable and/or sea level standstill (Fig. 3.7).

The complex oblique SRP (Fig. 5.14B) is similar to the oblique tangential pattern in that both clinoforms are characterized by oblique patterns that terminate downdip by downlap. However, the complex oblique SRP contains increasingly higher horizontal seismic events on the deltaic plain. This SRP represents a deltaic system that prograded into an actively subsiding shelf *(Berg, 1982)*.

The sigmoid seismic reflection pattern (SRP) has reflections on the deltaic plain that continue across the clinoform and downlap into the lower boundary of the SRP (Fig. 5.14C). This SRP represents a low-energy system in which little if any sand is present. It is probable that this SRP represents an inter sand-rich area of a deltaic system because typically the sigmoid pattern is associated with oblique (tangential) and complex oblique patterns. Normally the sigmoid clinoform reflections are stacked

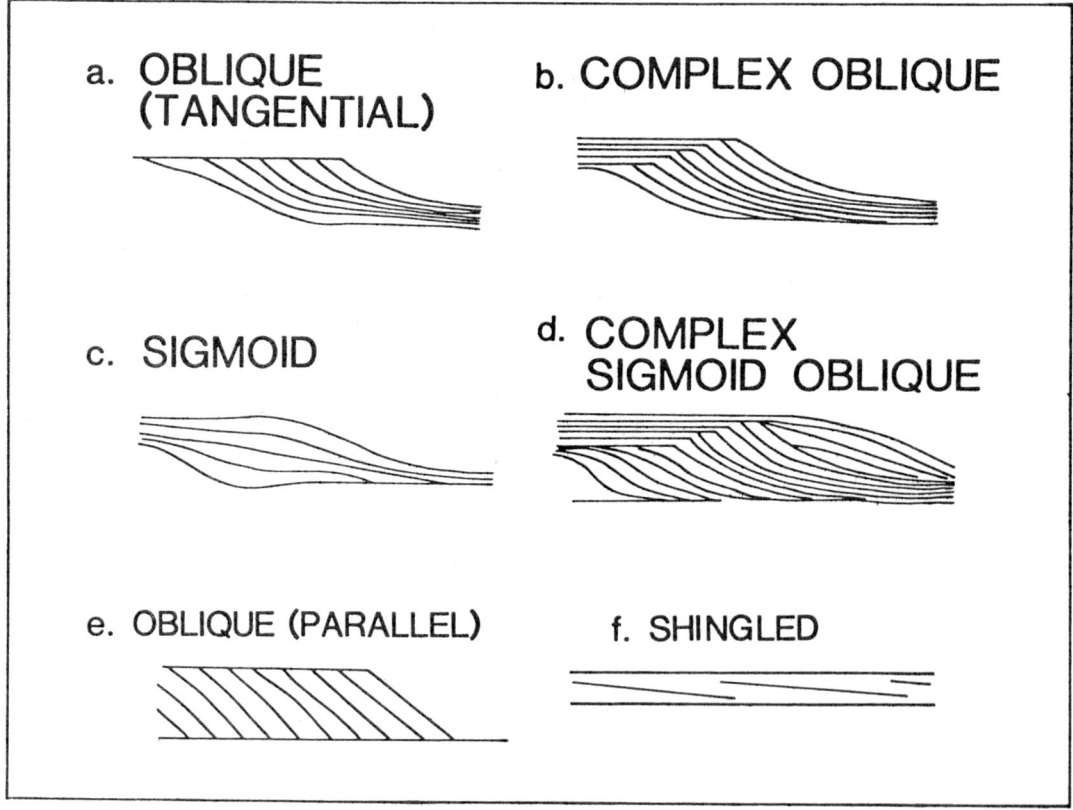

**Figure 5.14:** *Model reflection patterns for fluvial-dominated and wave-dominated deltas (Berg, 1982 and Mitchum and others, 1977).*

which indicates that subsidence of the shelf occurred.

The complex-oblique seismic reflection pattern (SRP) consists of alternating sigmoid and oblique reflection patterns (Fig. 5.14D). This common reflection response depicts a shift from sand-prone to sand-poor environments. The change in influx of sand may be attributed to differential rates of erosion, change in fluvial distributary system, change in rate of subsidence and/or change in global sea level.

Figure 5.15 is a 6-fold seismic section from the North Slope of Alaska. The pre-Cretaceous section dips to the south and it is unconformably overlain by northward prograding Cretaceous deltaic strata. The lower Cretaceous section (2.2 to 2.7 seconds, Fig. 5.15), is characterized by sigmoid reflection patterns. Clinoform reflections terminate downdip by downlap; they are stacked and horizontal on the deltaic plain which implies subsidence on the shelf and/or sea level rise. Sands are rare in this deltaic system. The 2.2 to 1.0 second interval is characterized by clinoform reflections with little, if any, downlap or onlap. They represent progradational "filling" of the basinal by predominantly shales of marine origin.

## WAVE DOMINATED DELTAS

Because progradation of wave-dominated deltas occurs along the deltaic front rather than being concentrated in the distributary channel-mouths, the seismic reflection patterns (SRP) appear to be oblique, parallel, or shingled (Fig. 5.14E and D). In both of these seismic reflection patterns, inclined reflections are terminated at their top and bottom by continuous, variable amplitude, horizontal reflections. Normally it is impossible to separate the deltaic plain from the prodelta environments. Normally, however, potential sandstones occur in the upper part of the interval and they may be coincident with the occurrence of shingled reflections *(Berg, 1982)*. Mitchum and others *(1977)* interpreted shingled SRP to represent progradation into shallow water.

The seismic line illustrated on Figure 5.16 illustrates a wave-dominated delta of Cretaceous age in east Texas. Shingled reflections prograde across the lower Cretaceous shelf and they are terminated at their top and bottom by horizontal and continuous reflections. Resolution of the seismic data is inadequate to map individual sands but these sands form numerous stratigraphic traps, for example, Seven Oaks gas field *(Berg, 1982)*.

A second example of a wave-dominated delta is illustrated on Figure 5.17. Shingled reflections at the 2.7 to 2.9 second interval on the east-central part of the line depict a westward prograding middle Cretaceous deltaic system in the Green River Basin, Wyoming. Like the above Woodbine example, shingled reflections are terminated at their top and bottom by flat continuous reflections. Discontinuous, variable amplitude reflections at the 3.3 to 3.5 second interval on the western part of the line depict Jurassic continental sands and shales. Reflections above 1.0 second on the east side of the line represent various shale and fine-grained sand lithologies of dominantly marine origin.

## EXAMPLES OF DELTAIC RESERVOIRS

### INTRODUCTION

Delta and fan deltaic systems are excellent models for the accumulation of potential reservoir and source rocks. Production from distributary channel sandstones occurs in the Arkoma Basin, Oklahoma *(Busch, 1971)*. The distributary system was part of an early Pennsylvanian river-dominated delta. Most of the reservoirs bear little relation to structure and the trap is produced by facies changes between the coarse-grained distributary sandstones to interdistributary swamp or bay deposited shales.

Wave-dominated deltas provide many of the reservoirs in the Denver Basin, Colorado. Deltaic plain reservoirs include the distributary channel and crevasse-fan deposits of the "J" sandstone at Peoria Field, Colorado *(Land and Weimer, 1978)*. Deltaic front sandstones of the "J" sandstone produce in the Wattenberg Field, Colorado *(Matuszczak, 1976)*. Wave-dominated deltas of the lower Wilcox provide many of the reservoirs in the Gulf Coast *(Fisher and McGowen, 1969)*. Rice *(1980)* summarizes the influence of deltaic and coastal depositional processes on gas production in several north-central Montana fields.

Fisher and McGowen *(1969)* document the only published example of a tidal-domi-

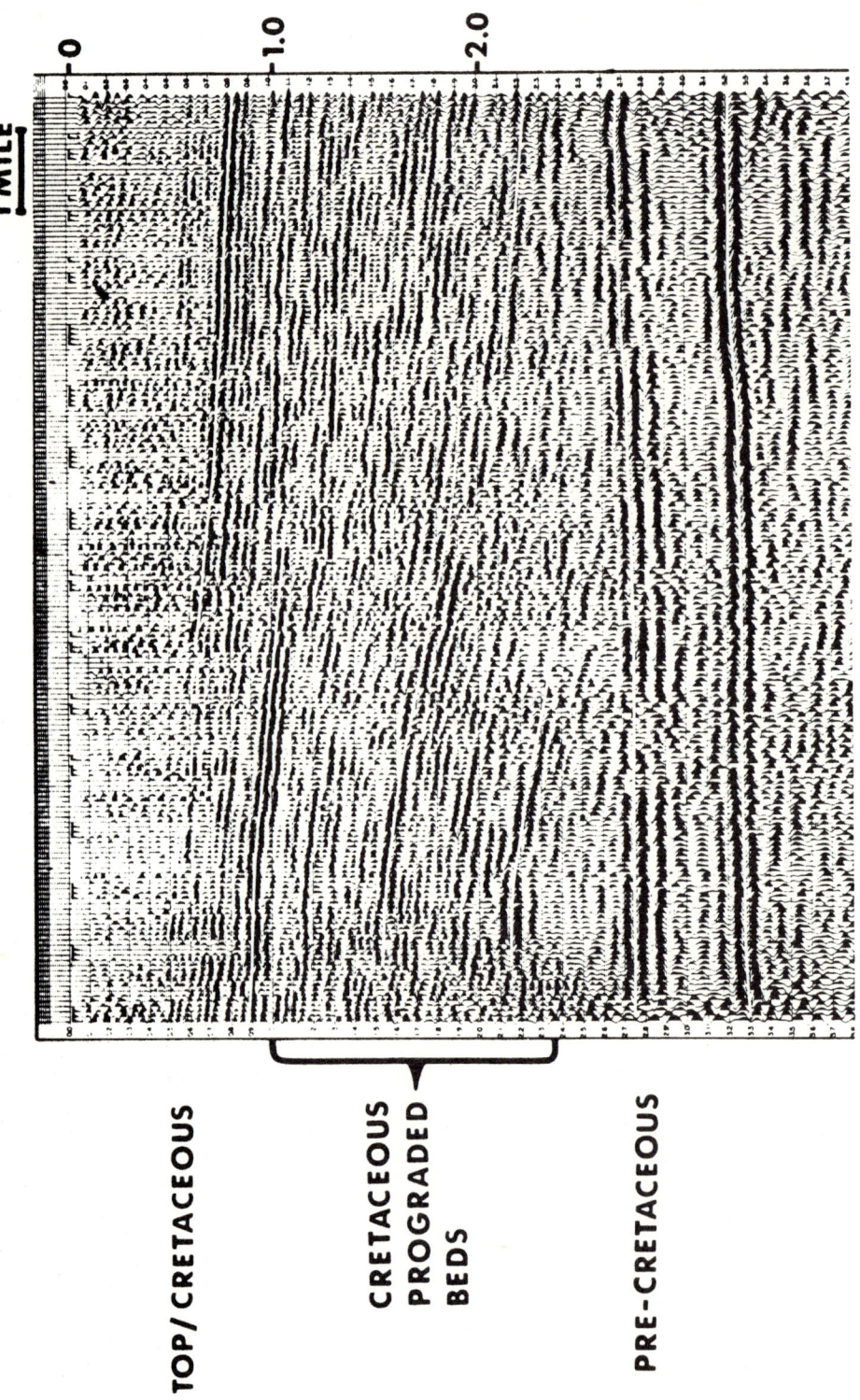

Figure 5.15: *Sigmoid reflections in Lower Cretaceous of the North Slope, Alaska, indicate a fluvial-dominated deltaic system (Laing, 1972).*

# EXPLORATION FOR DELTA AND FAN DELTA RESERVOIRS

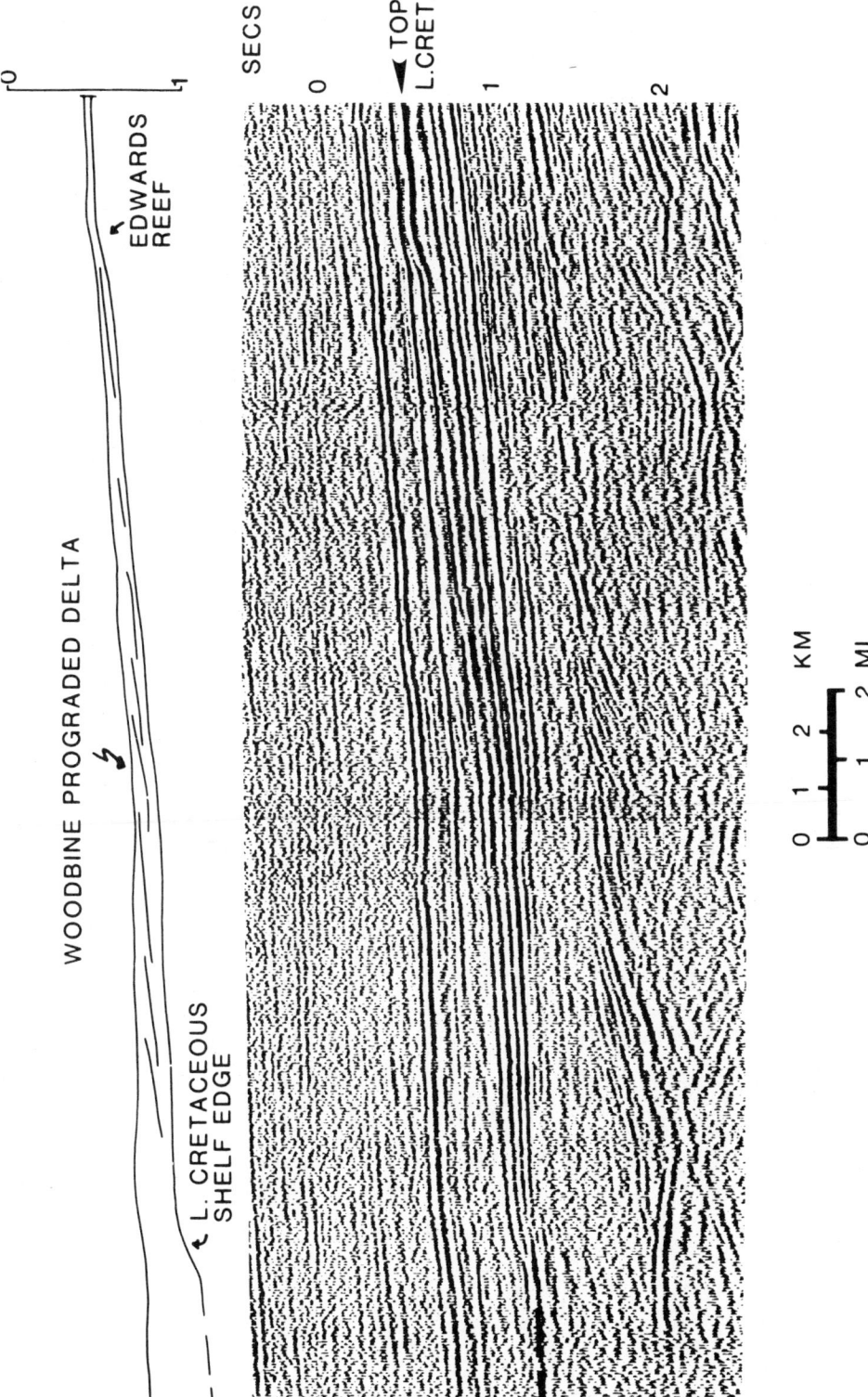

**Figure 5.16:** *Shingled seismic reflections in the Woodbine sandstone indicates a wave-dominated deltaic system (Seismic section courtesy Seiscom Delta Inc., Berg, 1982).*

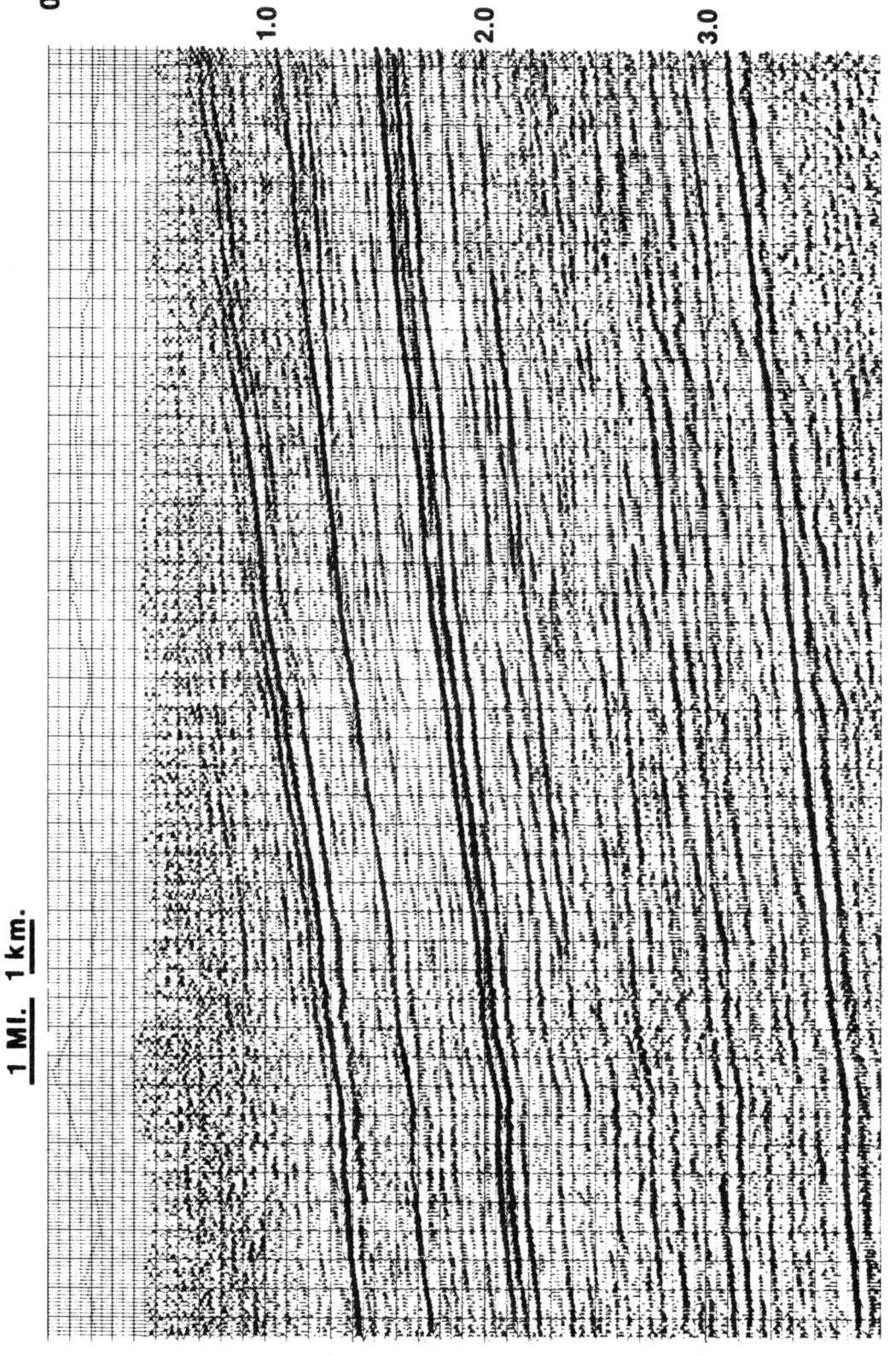

**Figure 5.17:** *West to east, 5-fold CDP seismic line located in the Greater Green River Basin, Wyoming. On the east side of the line reflections at 3.1 represent the Mississippian; 2.7, Cretaceous Dakota; 2.5, Cretaceous Mesaverde; 1.0, Cretaceous Lewis; 0.7, Tertiary Fort Union (Laing, 1972).*

nated delta. It occurs in the lower Wilcox (Tertiary) of southwest Texas. Production from this deltaic system occurs at the Washburn Ranch Field, La Salle County, Texas.

The significance of fan delta systems to exploration is just now being realized by explorationists. Re-study of the Quirquire Field, Venezuela, may reveal that it is all or in part of a fan deltaic system rather than an alluvial fan deposit (Chapter 4). Harms *(1980)* considers the conglomerates and coarse-grained sandstones of upper Jurassic age in the Brae Field area in the North Sea to be fan deltaic deposits. Dutton *(1982)* reports that the main reservoir at Mobeetie Field, Texas, is part of a fan deltaic system. It is probable that the Fahler member of the Spirit River formation at Elmworth Field, Alberta, was deposited in a fan deltaic complex *(Masters, 1979 and McLean, 1979)*.

## PRUDHOE BAY FIELD, ALASKA

Prudhoe Bay Field is located on the Barrow Arch, south of Point Barrow, Alaska. The arch is a major feature of the Arctic Slope Province. Several attempts to develop oil seeps were made by private companies in the early 1900's but exploration ceased in 1923 with the establishment of Naval Petroleum Reserve No. 4. The Navy conducted two exploration plays in the area between 1944 and 1953. These programs did discover oil and gas but not in commercial quantities. Also, the Navy did not adequately evaluate the pre-Cretaceous section. Private companies drilled nine wildcats during 1963-1967 but commercial production was not established. The Arco-Humble 1-Prudhoe Bay discovered 400 feet of gas column in the Sadlerochit Formation (Permian-Lower Triassic). A conirmation well, drilled 7 miles southeast and more than 400 feet down structure from the discovery well, penetrated a 400-foot oil column. The field contains a 900-foot oil column that covers 125,000 acres *(Morgridge and Smith, 1972)*.

A combination trap accounts for the ultimate recovery of over 9 BBO and 26 TCFG. The Prudhoe Bay structure is a northwesterly plunging anticline that was severely altered by

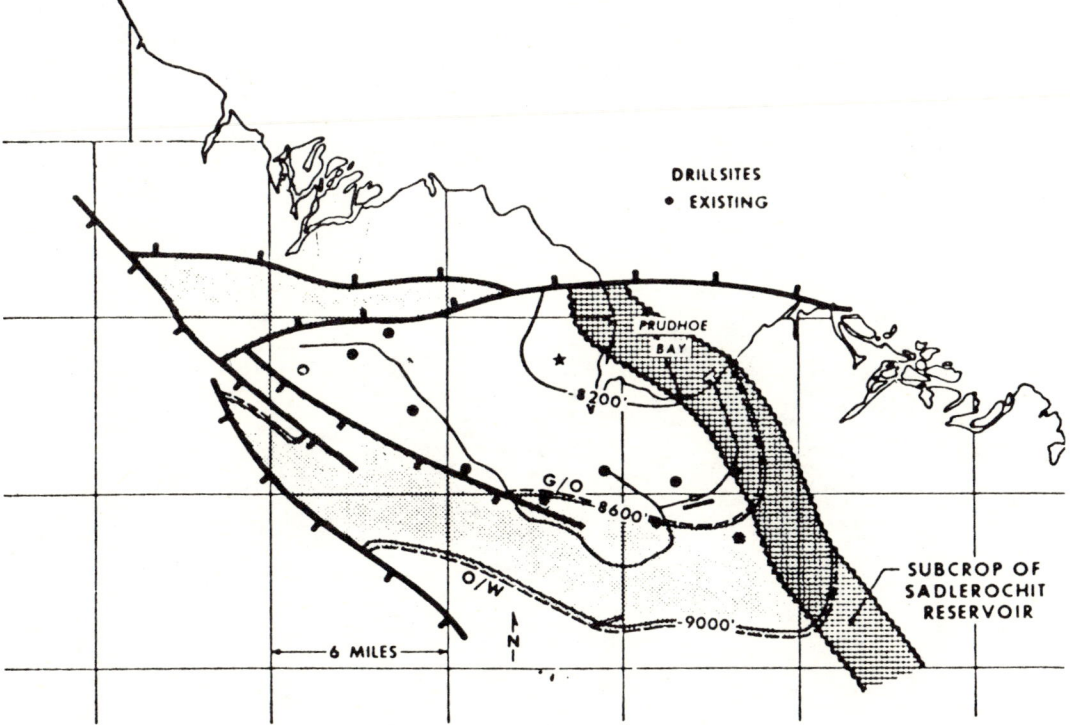

**Figure 5.18:** *Structural contours and fluid contacts at the top of the Sadlerochit Reservoir, Prudhoe Bay Field, Alaska (Morgridge and Smith, Jr., 1972).*

early Cretaceous erosion and Permian-Jurassic faulting (Fig. 5.18). Normal faults along the northern flank of the structure are so numerous and have such large vertical movement that they obscure the original dip. In contrast, normal faulting along the southwestern margin are less intense and original dip is less than 2° *(Morgridge and Smith, 1972).* Critical to the trap is basal Cretaceous shales that provide a seal along the eastern margin of the field (Fig. 5.19).

The vast majority of reserves at Prudhoe Bay Field are contained in the upper part of the Sadlerochit Formation (Permian-Triassic). The formation constitutes two major episodes of deposition (Fig. 5.20). The lower part of the Sadlerochit Formation consists of deltaic and shallow marine clastics. Basal strata of the Sadlerochit are represented by irregularly bedded, parallel laminated, silty shales that contain shallow water marine forams. The laminations are slightly inclined and the section is interpreted to represent prodeltaic deposits. Thickness is variable because it infills lows produced by pre-Sadlerochit erosion. This prodeltaic marine sequence grades upward into a predominantly sand sequence that is interbedded with shales. Typically, the lower sands are laminated, burrowed, and fine-grained. Upper sands are coarser-grained and cross-bedded. They have been interpreted as channel-mouth-bars *(Morgridge and Smith, 1972).* Porosity and permeability are low and it is doubtful that any significant recovery of oil will be made from this part of the Sadlerochit.

The upper part of the Sadlerochit Formation represents various alluvial processes of deposition. The lower part of the unit was constitute sand and shale that was deposited

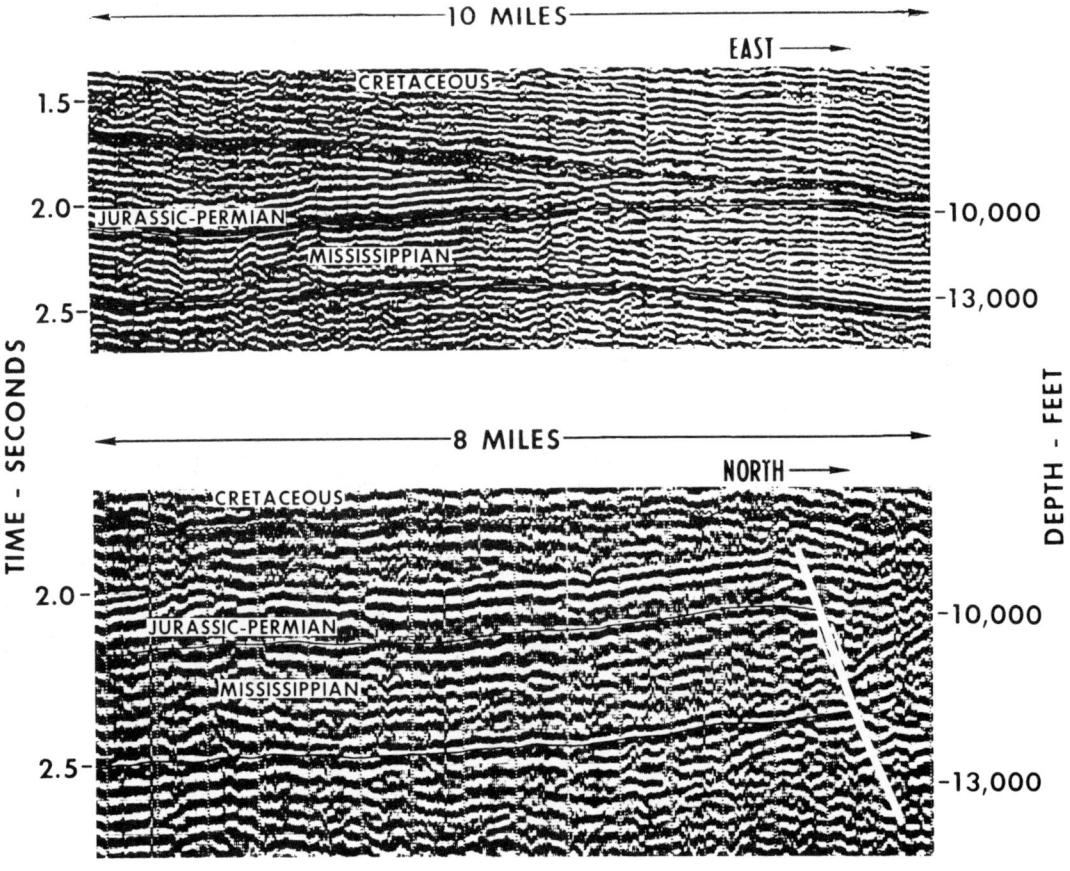

**Figure 5.19:** *Seismic profiles showing unconformity and faults that are responsible for trapping oil and gas at Prudhoe Bay Field, Alaska (Morgridge and Smith, Jr., 1972).*

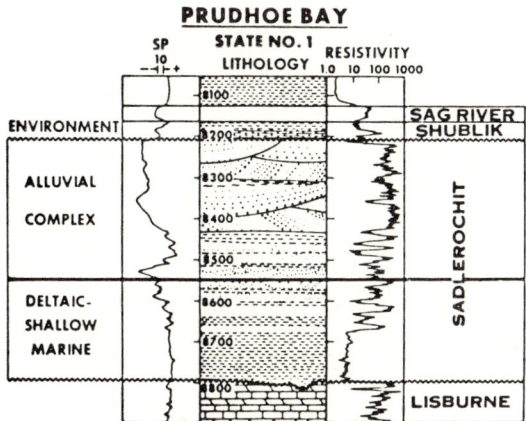

**Figure 5.20:** *Lithologies and sedimentary structures are based on core descriptions from the Prudhoe Bay State No. 1 (Morgridge and Smith, Jr., 1972).*

on top of the older Sadlerochit deltaic plain strata. The sands filled in the older distributary channels of the once active deltaic complex and the shales spread across to topographically featureless plain of the delta. The upper part of the alluvial complex is a braided stream deposit. The repetitious sands grade upward from pebble conglomerates to fine-grained sands. Typical of braided stream deposits, the sands contain large scale cross beds at their base, in contrast to laminated and rippled upper surfaces. Typically this lower flow regime sequence is absent due to truncation. Porosity is excellent, it ranges from 20 to 25%. Permeabilities range from 300md to several darcies. Consequently, this part of the Sadlerochit Formation contains the most recoverable reserves in the Prudhoe Bay Field.

## WEST TUSCOLA FIELD, TEXAS

West Tuscola Field is located in west central Texas, north of the Concho Arch and west of the Bend Arch. The field is small, less than 7 million barrels of oil in place, but it is an excellent example of a stratigraphic trap produced by deltaic sedimentation *(Shannon and Dahl, 1971)*. Regional dip is to the northwest and there is no structural closure within the limits of the field (Fig. 5.21).

Producing sandstones in the field are Strawn in age (Pennsylvanian) and they occur in an interval that ranges from 10 to 100 feet thick. Individual productive sands rarely exceed 25 feet thick with an average porosity of 12% and permeabilities of 44md.

The sandstones are interpreted to reflect distributary channel, channel-mouth-bar and deltaic front sand deposition *(Shannon and Dahl, 1971)*. The distributary channel sands fine upward; they contain large scale cross beds and basal contacts are sharp. In contrast, the channel-mouth-bar sands coarsen upward; basal contacts are gradational with inclined, interlaminated silt and shale that are burrowed, reflecting the transition from bar-front to bar crest deposition (Fig. 5.22).

Shannon and Dahl *(1971)* infer that the delta prograded into the Tuscola Field area from the east. Evidence to support their interpretation includes the geometry of the sandstone bodies and the character and distribution of the sands (Fig. 5.23). The productive limits of the field is "T" shaped. The shank of the "T" trends east-west and typically the sands along this trend are characterized by composite funnel and bell-shaped SP curves. Cores from several of the wells depict an upward coarsening sand that reverses to an upward finning sand. The authors interpret the sands as distributary-channels that prograded over narrow channel-mouth-bar sands. Sands along the top of the "T" are typically upward coarsening sands, although several composite sands occur within the trend. Shannon and Dahl *(1971)* infer that these sands represent well developed channel-mouth-bar and bar-slope deposits. Progradation of the deltaic system stalled near the western limits of the field.

## MOBEETIE FIELD, ANADARKO BASIN, TEXAS

Mobeetie Field is located in the panhandle of Texas along the southern margin of the Anadarko Basin, Texas. The field is principally a structural closure that was caused by the draping of strata over an underlying horst block (Fig. 5.24). Ultimate recovery is estimated to be 5 millions barrels of oil and 60 billion cubic feet of gas *(Sahl, 1970)*.

Missourian reservoirs at Mobeetie Field consist of three cycles of alternating sandstone and limestone units (Fig. 5.25). The limestone deposition occurred during limited influx of terrigenous clastics and a relative high stand of sea level. Progradation of coarse-grained terri-

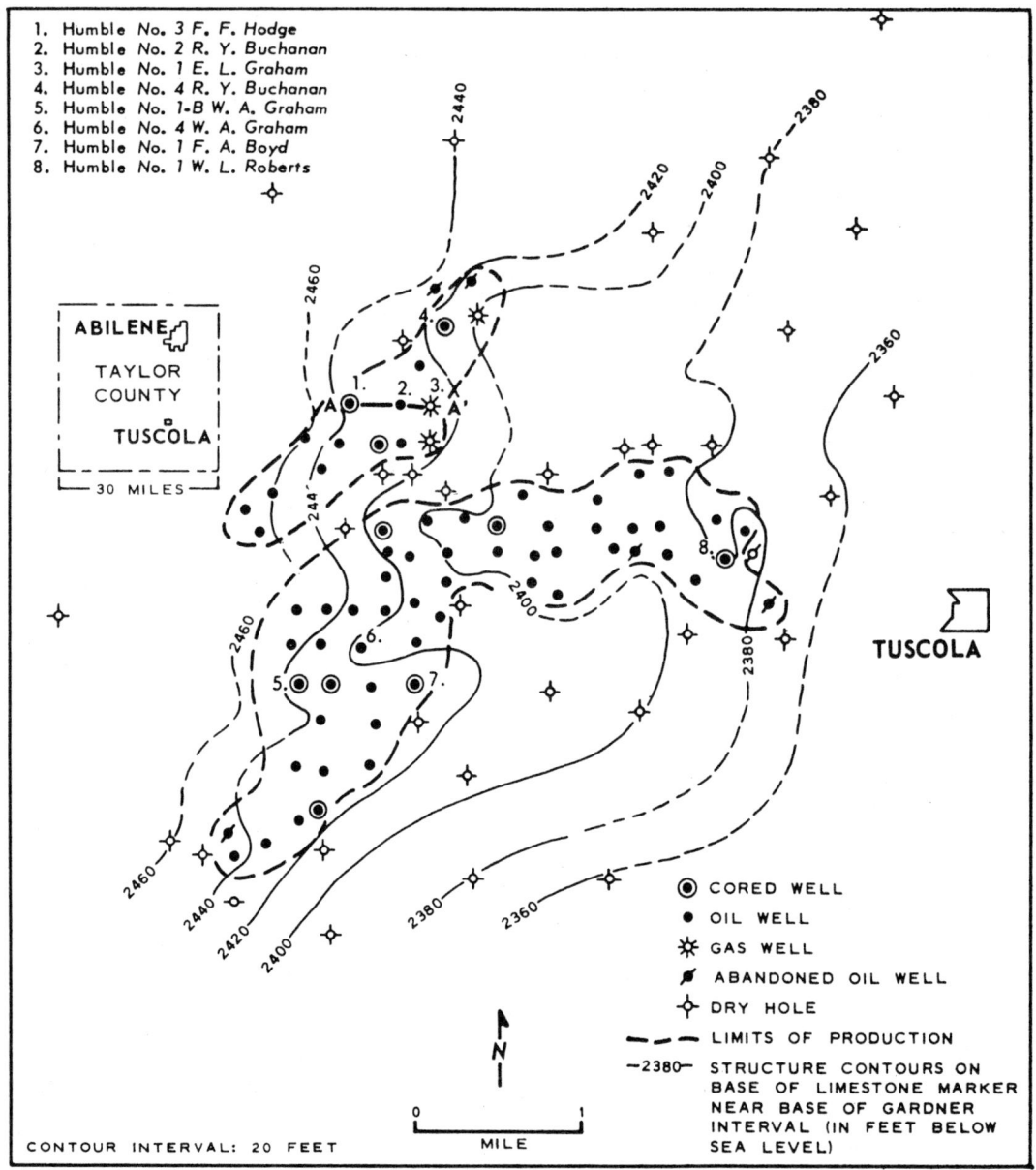

**Figure 5.21:** *Structure map of West Tuscola Field, Texas (Shannon and Dahl, 1971).*

# EXPLORATION FOR DELTA AND FAN DELTA RESERVOIRS

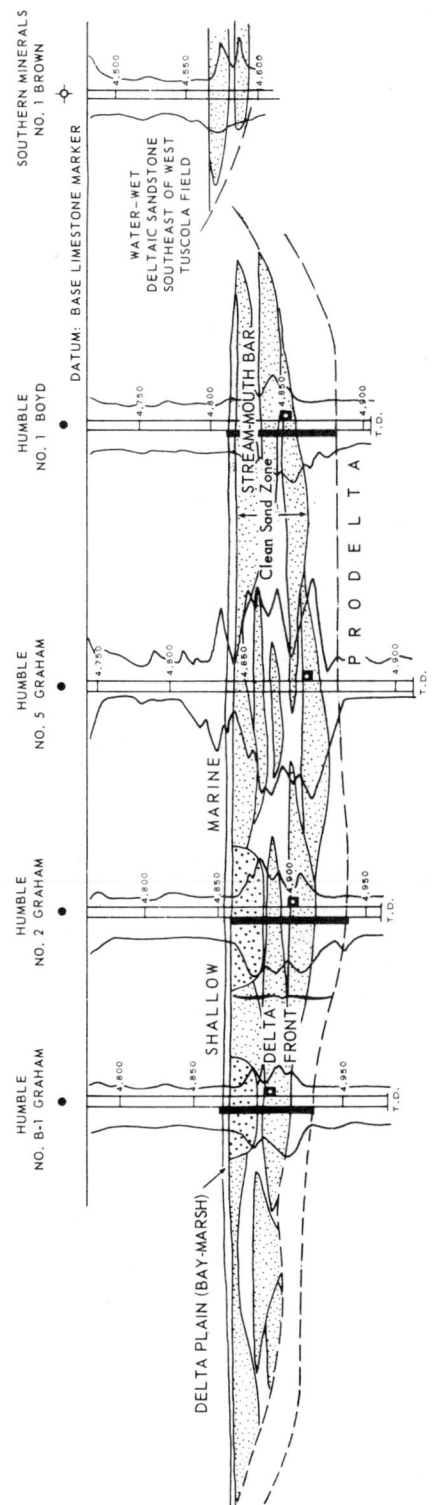

**Figure 5.22:** *West to east stratigraphic cross section located on the southern flank of West Tuscola Field, Texas (Shannon and Dahl, 1971).*

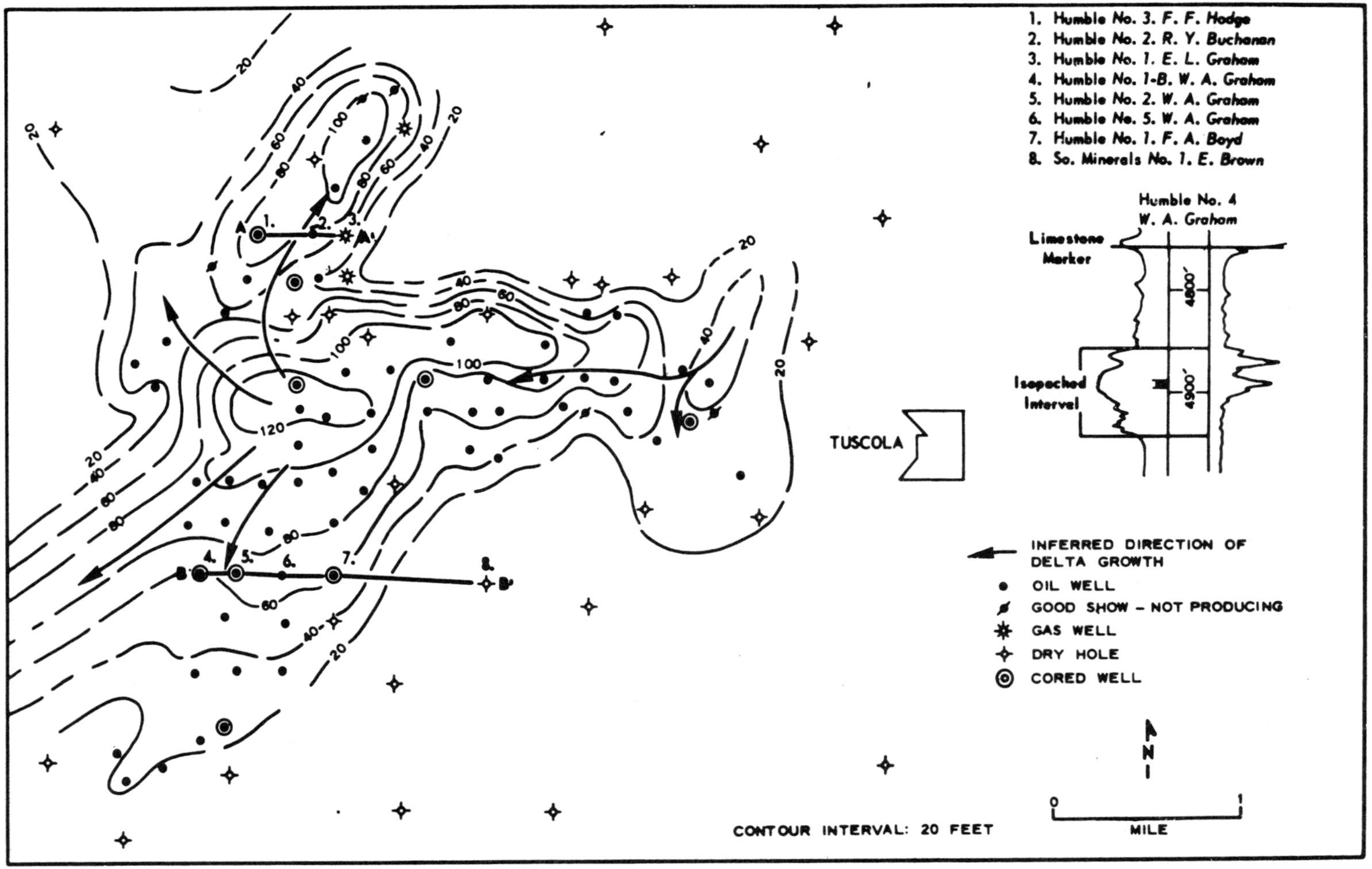

**Figure 5.23:** *Isopach map of deltaic front facies, West Tuscola Field, Texas (Shannon and Dahl, 1971).*

# EXPLORATION FOR DELTA AND FAN DELTA RESERVOIRS

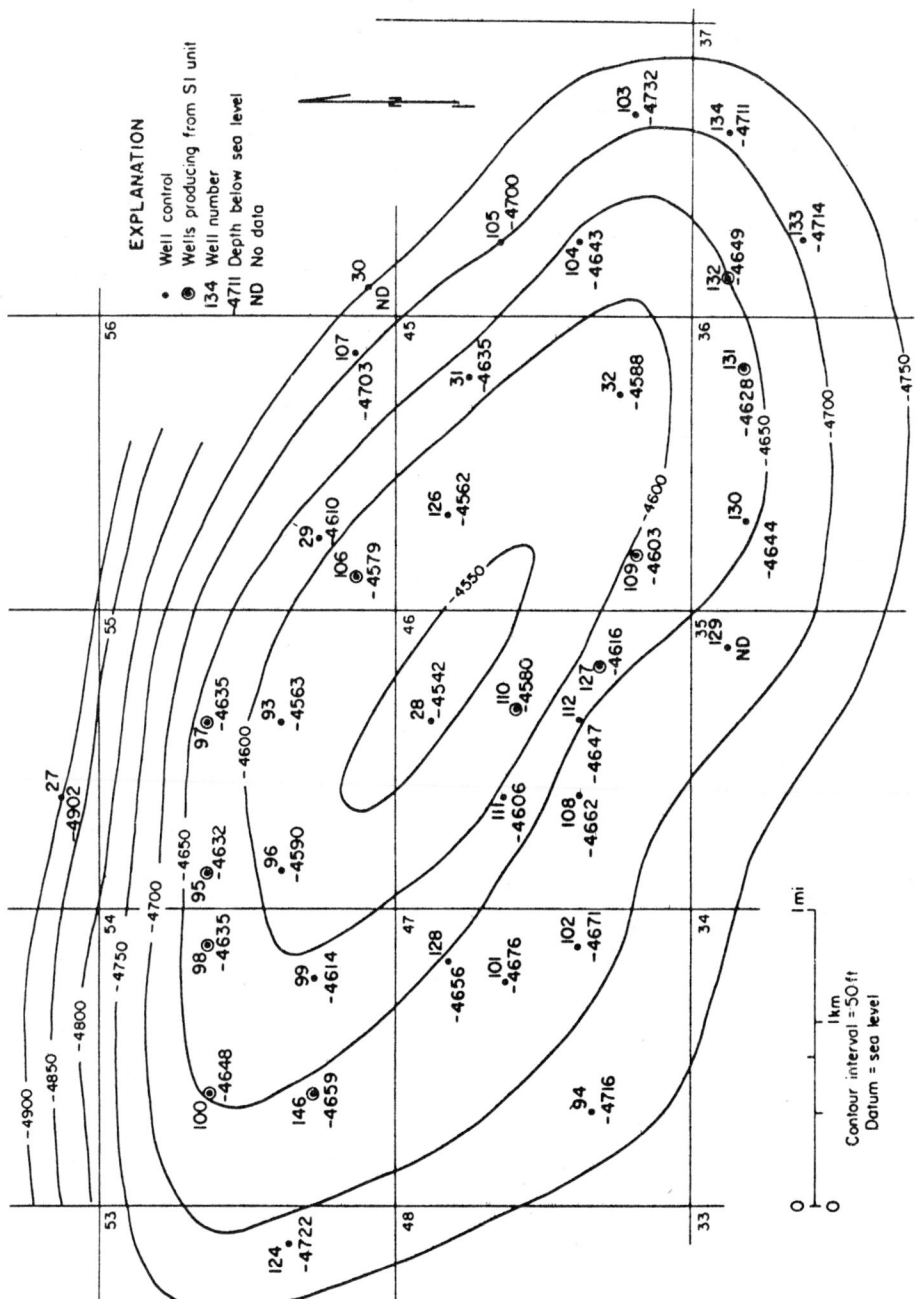

**Figure 5.24:** *Structural contour map on top of S1 sandstone, Mobeetie Field, Texas (Dutton, 1982).*

genous clastics terminated carbonate production by algae, foraminifera and corals. Braided streams breached the low areas of the carbonate environment and provided the distributary system for fan deltaic sedimentation. The upper two sandstone units were reworked along sedimentary strike into spits and bars. These reworked fan deltaic sands were cemented prior to migration of hydrocarbons and therefore, they are not productive in the field area *(Dutton, 1982)*.

The presence of fan-shaped deposits of coarse-grained clastics that were transported a short distance and that interfinger with marine sediments is the key to delineating fan deltaic deposits *(Handford, 1980)*. The terrigenous clastics at Mobeetie Field meet these criteria *(Dutton, 1982)*. The entire clastic interval at Mobeetie Field probably represents the distal part of a fan delta system.

The S1 sandstone is the principal reservoir at Mobeetie Field (Fig. 5.26). Normally fan deltas that prograde onto a shelf depict a coarsening upward sequence *(Wescott and Ethridge, 1980)*. The S1 sandstone depicts a variation of this ideal sequence. Prodeltaic muds are normally incorporated in the upper part of each underlying carbonate unit. In addition, Dutton *(1982)* reports that at many localities, the S1 sandstone is characterized by a sharp change from carbonates to coarse, braided-channel clastics. Progradation of S1 clastics into shallow water resulted in thin upward-coarsening sequences that are capped by coarse-grained channel deposits. It is also probable that the prograding braided channels scoured and removed the finer grained, distal-fan deposits.

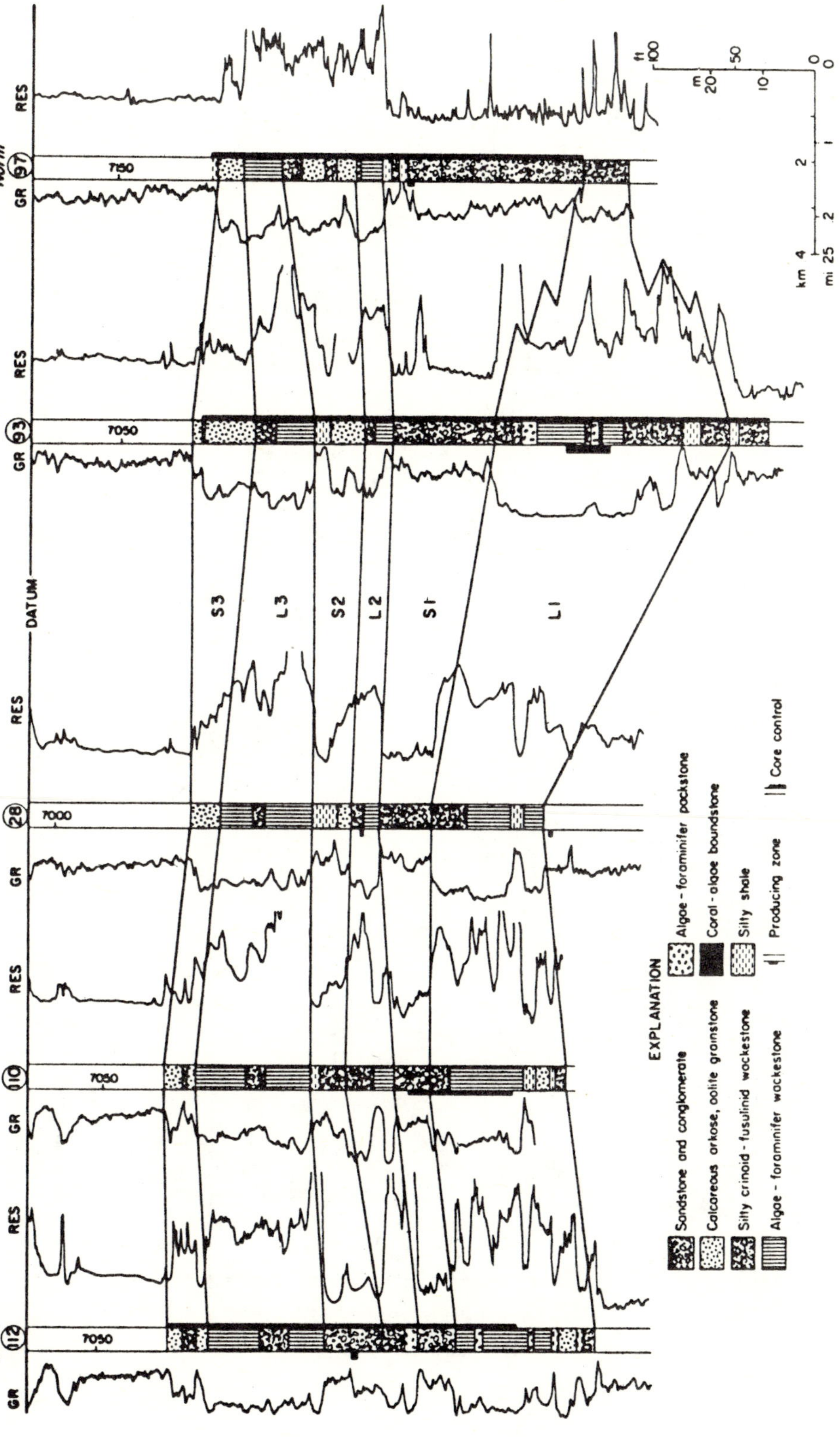

EXPLORATION FOR DELTA AND FAN DELTA RESERVOIRS 121

**Figure 5.25:** *North-south cross section, Mobeetie Field, Texas (Dutton, 1982).*

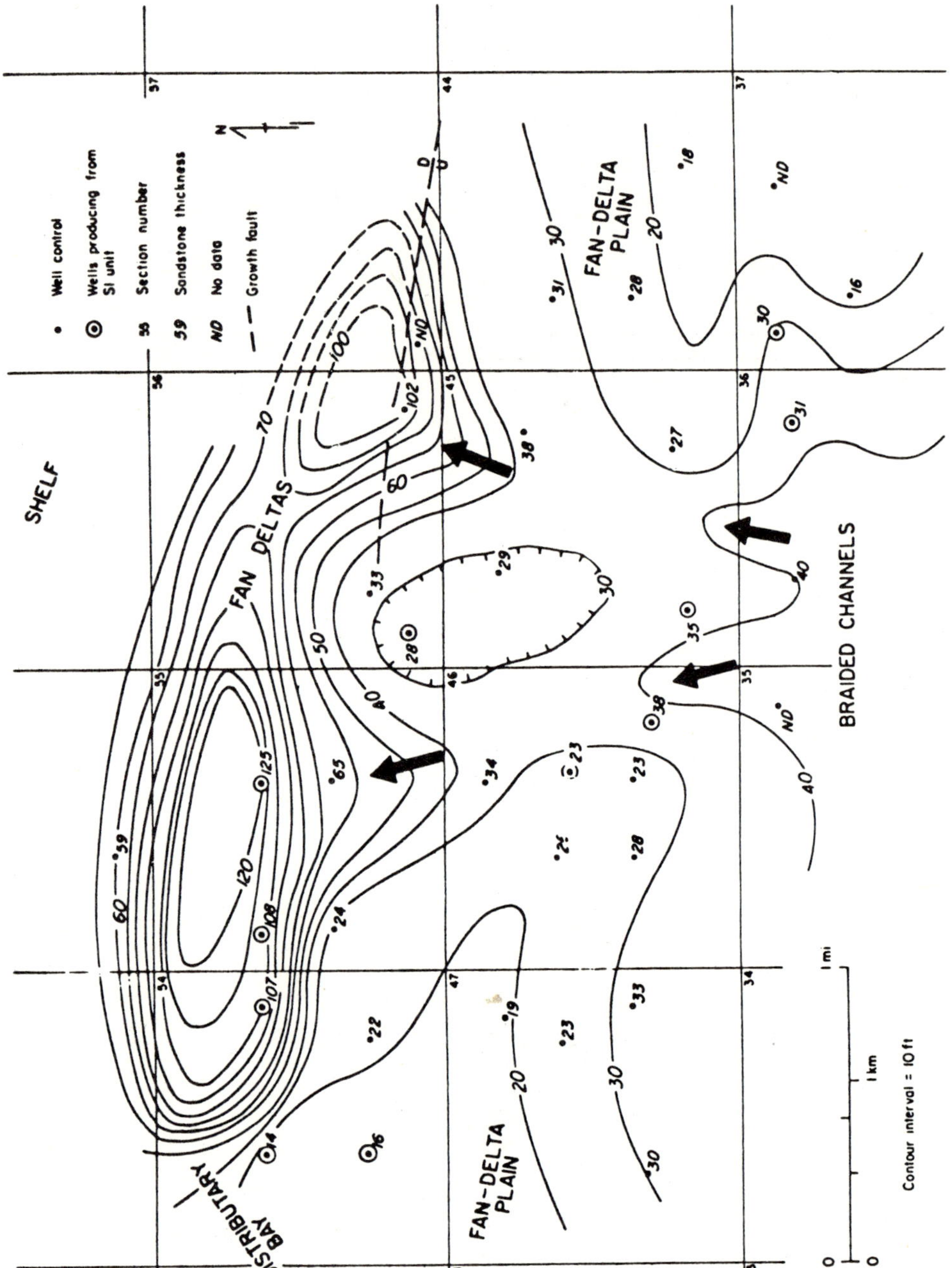

**Figure 5.26:** Isopach of S1 fan deltaic sandstone, Mobeetie Field, Texas (Dutton, 1982).

# 6
# Exploration for Interdeltaic Sandstone Reservoirs

## INTRODUCTION

Microtidal coasts (wave dominated), such as the Holocene U.S.A. Gulf Coast, are characterized by well developed, river dominated deltas that are separated by long, continuous barrier island deposits. Much of the production from the Cenozoic reservoirs of the U.S.A Gulf Coast is obtained from microtidal coastlines. Mesotidal (mixed energy) coasts, including the Holocene coast of North Carolina and Georgia, are typified by short barrier islands that are cut by numerous tidal inlets and their associated tidal deltas. Deltaic sedimentation is not nearly as common along mesotidal coasts as it is in microtidal coasts. Macrotidal (tide-dominated) coasts, such as the northern end of the Gulf of California, represent the other end of the hydrologic spectrum from wave dominated coasts. Typically, the mouths of major rivers are occupied by estuaries and extensive inter-estuary tidal flats and salt marshes. In general, these coastlines contain few sandstone facies and therefore, interdeltaic reservoirs are rare *(Hayes, 1977)*.

Interdeltaic reservoirs are especially sensitive to relative changes in sea level. For example, Figures 6.1 and 6.2 depict sea level changes for the U.S.A. Gulf Coast and the mid-Atlantic coast. Although sea level stand is a global process, local changes may occur as a result of local adjustments to tectonics. Data for the Gulf Coast area shows a stillstand near the present level that dates back to about 3,000 years. Barrier islands along this coast offlap towards the Gulf. In contrast, data from the mid-Atlantic coast depicts a sea level rise of about 6 inches per century for the last 3,000 years. Barrier islands along this coast onlap the Holocene surface *(Kraft and Chacko, 1979)*.

## SHORELINE RESERVOIRS

### BEACH RESERVOIRS

**Site and Depositional Processes** Beach sands accumulate along interdeltaic coastal areas (Fig. 4.3). Tidal and longshore currents range from weak to very strong. Rate of influx of clastics to the depositional site, rate of subsidence, rate of global sea level change, and character of marine conditions highly influence the thickness and geometry of this progradational deposit.

**Internal Features** The beach sequence is typically composed of four units; they are, from base to top, offshore, shoreface, beach, and dune (Fig. 6.3). The offshore facies is composed of interbedded fine-grained sand and silt. Skeletal debris is common. Few sedimentary structures other than laminations occur in this facies. The shoreface facies is typically highly burrowed, and it contains abundant shell debris. Silt and clay beds are rare. In gradational contact with shoreface deposits, the beach facies is typically well sorted, tabular cross bedded sand. The angle of cross beds is highly influenced by the tidal range; the higher the tides, the steeper the cross beds. The upper part of the beach sand sequence may contain cross beds that dip landward. These, of course, represent the backshore of the beach. Grain size and sorting increase upward in the beach sequence. In arid to semiarid climates the beach sequence may be capped by dunes.

**External Features** Basal contacts of the beach sands are gradational with shoreface deposits. Upper contacts are gradational with dune sands, high flat, or low flat shales. Beach deposits are typically long, narrow, bifurcating

124                                          EXPLORATION FOR SANDSTONE RESERVOIRS

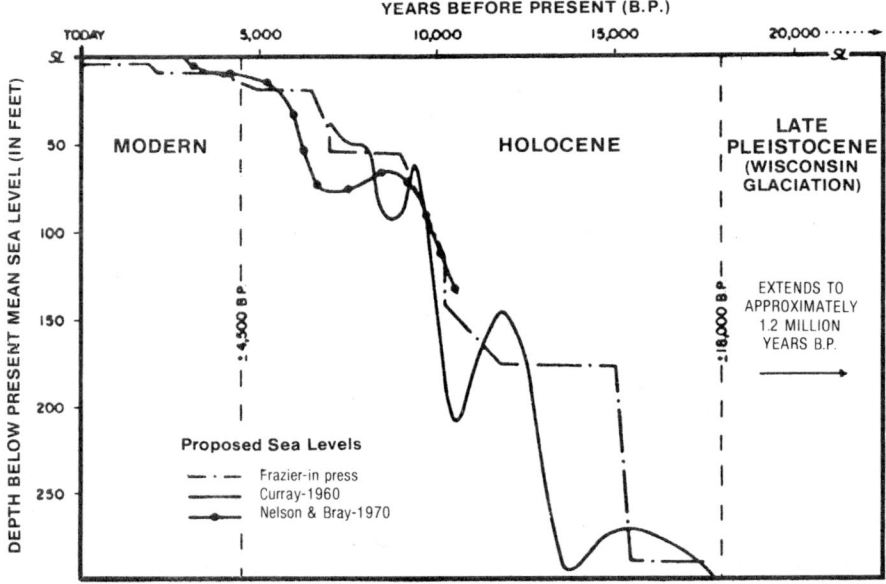

**Figure 6.1:** *Proposed sea-level changes during the last 20,000 years for the Gulf Coast area (Fisher, et al., 1973).*

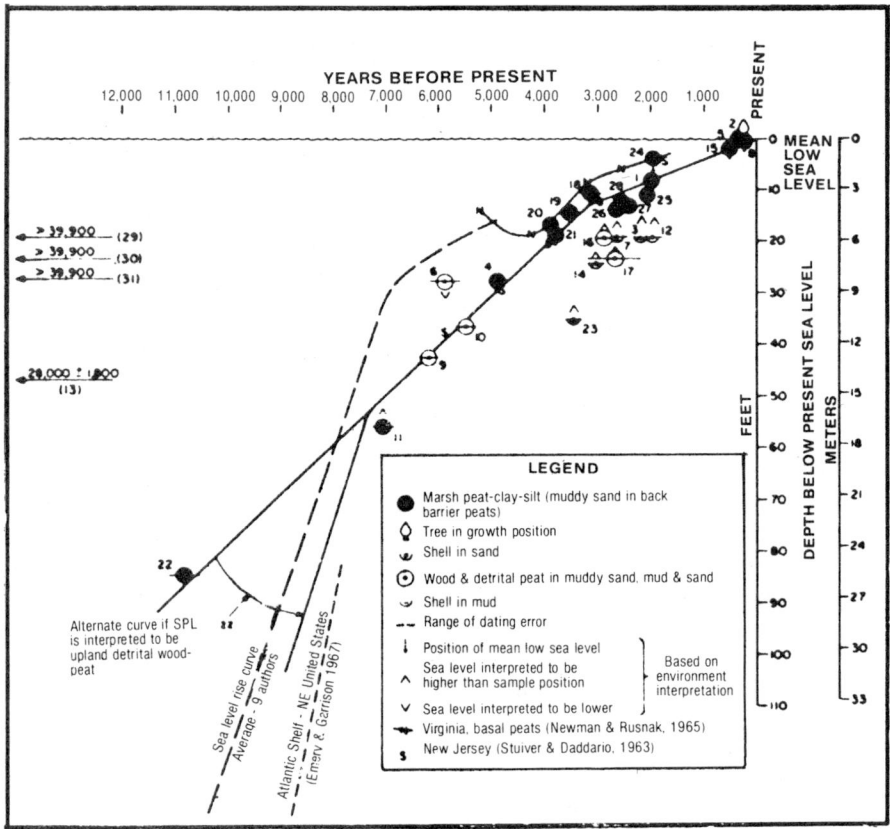

**Figure 6.2:** *Relative rise in local sea level for the Holocene Epoch in the mid-Atlantic area (Kraft, 1971).*

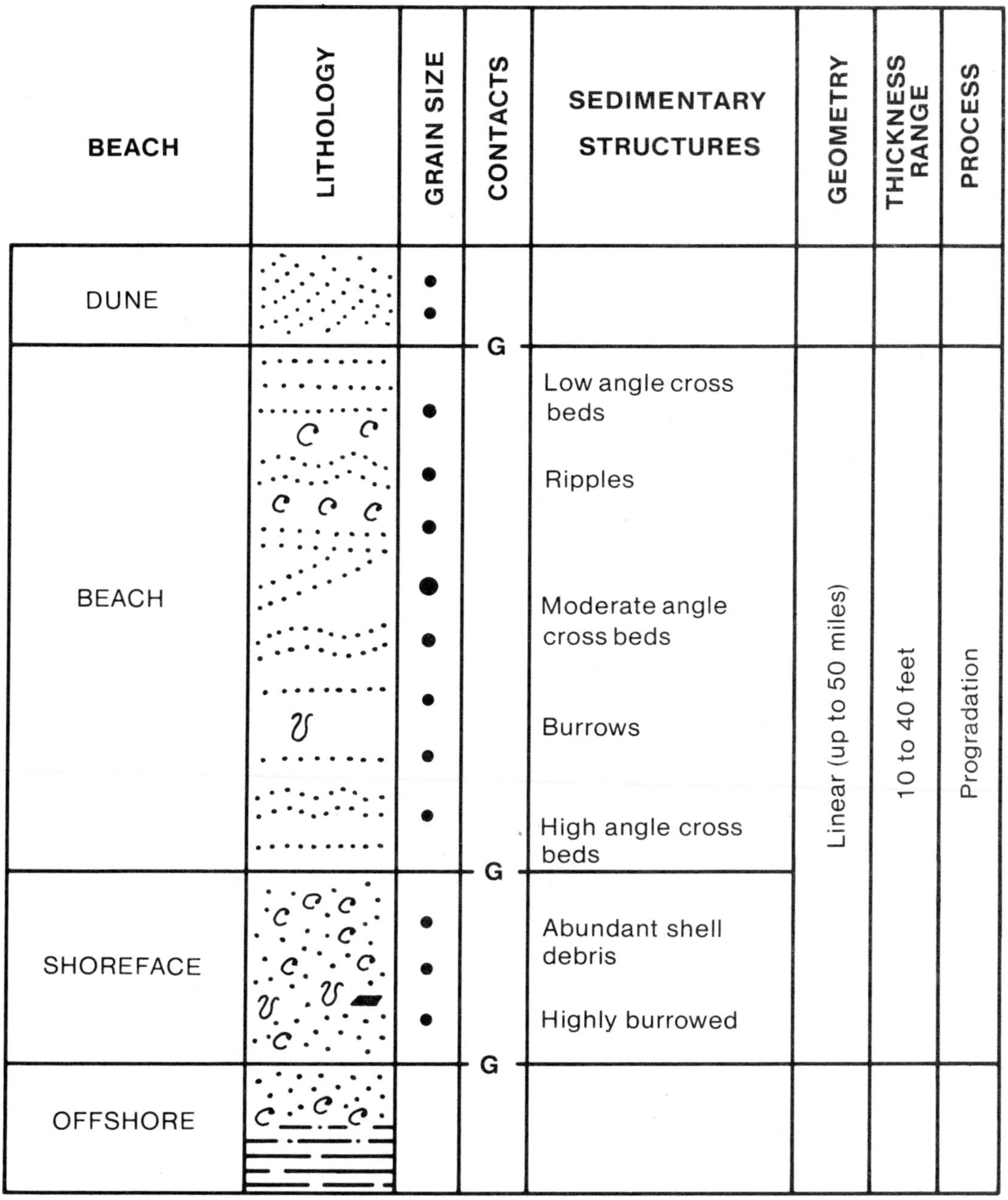

# EXPLORATION FOR INTERDELTAIC SANDSTONE RESERVOIRS

| BEACH | LITHOLOGY | GRAIN SIZE | CONTACTS | SEDIMENTARY STRUCTURES | GEOMETRY | THICKNESS RANGE | PROCESS |
|---|---|---|---|---|---|---|---|
| DUNE | | • | | | | | |
| | | | G | | | | |
| BEACH | | • | | Low angle cross beds | Linear (up to 50 miles) | 10 to 40 feet | Progradation |
| | | • | | Ripples | | | |
| | | • | | | | | |
| | | • | | Moderate angle cross beds | | | |
| | | • | | | | | |
| | | • | | Burrows | | | |
| | | • | | | | | |
| | | • | | High angle cross beds | | | |
| | | | G | | | | |
| SHOREFACE | | • | | Abundant shell debris | | | |
| | | • | | | | | |
| | | • | | Highly burrowed | | | |
| | | | G | | | | |
| OFFSHORE | | | | | | | |

**Figure 6.3:** *Depositional characteristics of beach deposits.*

ridges. Individual ridges range in width from several hundred feet to about a mile. However, onlap of beach ridges can produce very widely-distributed, almost tabular, sand deposits. Length varies up to forty miles or more. Thickness is dependent upon tidal ranges and normally varies from fifteen to thirty feet.

**Reservoir and Stratigraphic Trap Potential** Vertical and lateral continuity of porosity and permeability in beach deposits range from good to excellent; therefore, these sands represent some of the best reservoirs discovered to date. The best opportunity for the development of a stratigraphic trap occurs along the landward margin of the beach, particularly where the beach forms a lobate pattern into the tidal flat facies. In some cases, where tidal channels dissect the beach, the tidal channels may be filled with clay-sized material that may form a trap.

## STRANDED BEACH RESERVOIRS

Stranded beach reservoirs share most of the same depositional and reservoir characteristics as the beach reservoirs (Fig. 6.4). However, three significant differences should be mentioned. They are:
1. The upper contact of a stranded beach is sharp, and the facies is almost always tidal flat deposits.
2. The upper part of the stranded beach facies is subaerially exposed, and internal sedimentation of silt and clay-sized material will partially occlude pore space.
3. Stranded beach deposits are not as likely to become a part of the rock record as beach deposits that accumulated during onlap.

## TIDAL CHANNEL RESERVOIRS

**Site and Depositional Processes** Tidal channel deposits occur marginal to the coastline, trending nearly perpendicular to beach, barrier island, marsh-swamp, or mud flat environments. Lateral accretion is the dominant process of accumulation (Fig. 6.5).

**Internal Features** Grain size of the tidal channel deposits is controlled by the source of sand. Typically, these deposits decrease in grain size upward; sorting is poorest near the top of the sequence, and rounding is dependent upon the source of sand. Thick cross bedded sand occurs in the lower part of the tidal channel deposits. Laminations and ripples are common. Burrows are especially abundant in the finer-grained tidal channel deposits.

**External Features** Lower contacts are sharp with foreshore, beach, or tidal flat deposits. Upper contacts are gradational with mudflat, swamp-marsh, beach, or offshore deposits. The tidal channels form irregular sand bodies that are typically narrow and rarely more than several hundred feet long. Where they are entrenched in the mudflat or tidal flat, they are typically sinuous. Normally the sands are less than ten feet thick. Tidal channel sands in the Niger River delta are, of course, an important exception. Here the sands are thick, extend for as much as 25 to 30 miles into the deltaic plain, tidal swamps, and marshes, and they are typically anastomosing sand bodies.

**Reservoir and Stratigraphic Trap Potential** Normally the reservoir potential of a tidal channel deposit is so limited in aerial extent that explorationists cannot afford to prospect for this group of sands. However, they can cause havoc with reservoir continuity in beach or barrier island reservoirs. In tidal-dominated deltas, for example, the Niger River delta, tidal channel sands can be important reservoir rocks, and their reservoir continuity resembles that of a point bar. Maximum porosity and permeability occur near the base of the tidal channel deposit. Vertical continuity is subordinate to lateral continuity.

## OFFSHORE BARRIER ISLAND RESERVOIRS

### SITE AND DEPOSITIONAL PROCESSES

Interdeltaic barrier island sands are separated from the mainland by a lagoon of varying width and depth. The sand may be derived from a point source, *i.e.,* a delta or a multi-source, such as alluvial fans, sea cliffs, or a mainland beach (Fig. 6.6). The barrier island complex typically trends parallel to the margin of the basin. Progradation and, to a lesser extent, lateral accretion are the dominant processes of deposition.

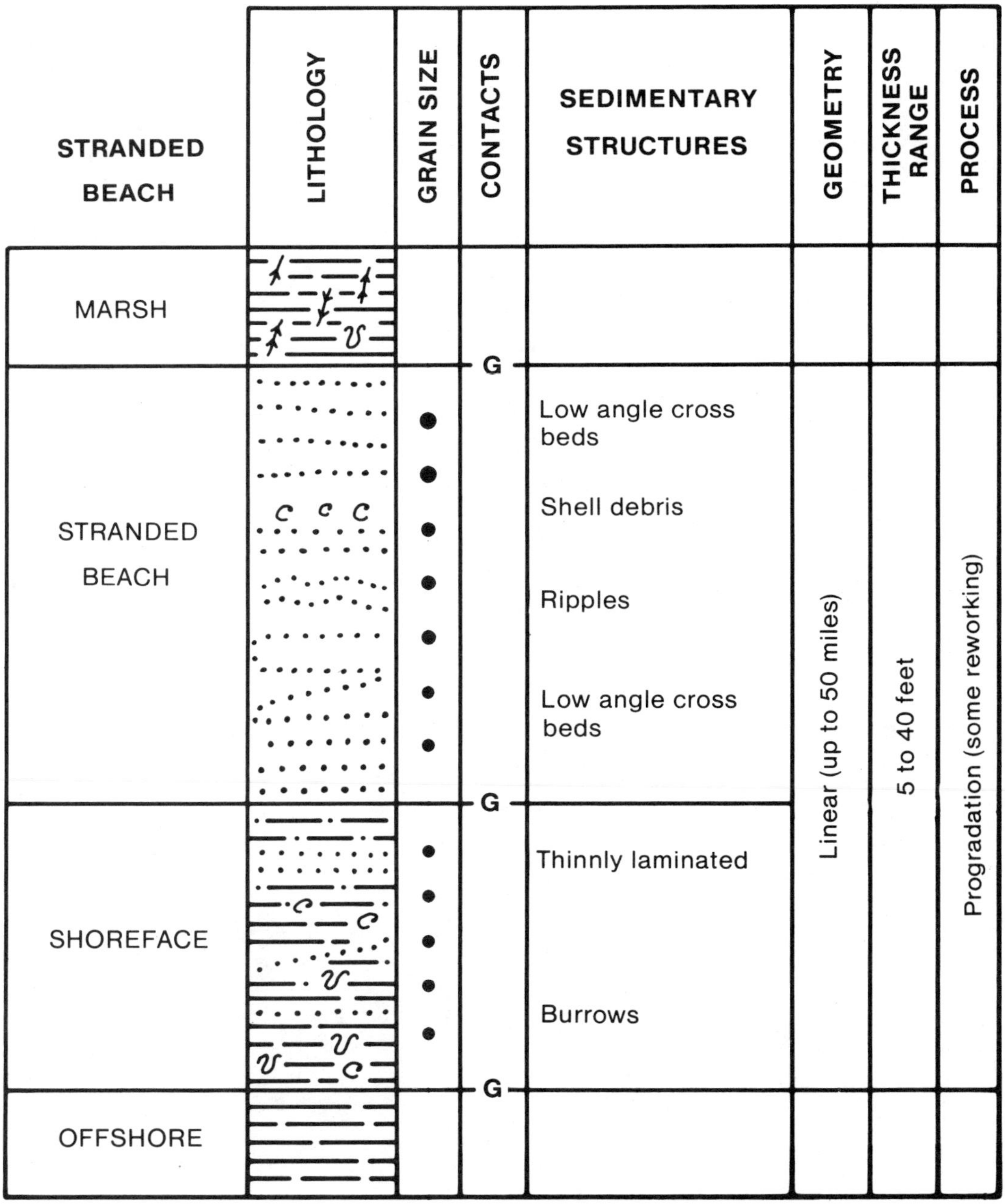

# EXPLORATION FOR INTERDELTAIC SANDSTONE RESERVOIRS

| STRANDED BEACH | LITHOLOGY | GRAIN SIZE | CONTACTS | SEDIMENTARY STRUCTURES | GEOMETRY | THICKNESS RANGE | PROCESS |
|---|---|---|---|---|---|---|---|
| MARSH | | | G | | | | |
| STRANDED BEACH | | | • • • • • • • | Low angle cross beds<br><br>Shell debris<br><br>Ripples<br><br>Low angle cross beds | Linear (up to 50 miles) | 5 to 40 feet | Progradation (some reworking) |
| SHOREFACE | | | G<br>•<br>•<br>•<br>•<br>•<br>G | Thinnly laminated<br><br>Burrows | | | |
| OFFSHORE | | | | | | | |

**Figure 6.4:** *Depositional characteristics of stranded-beach deposits.*

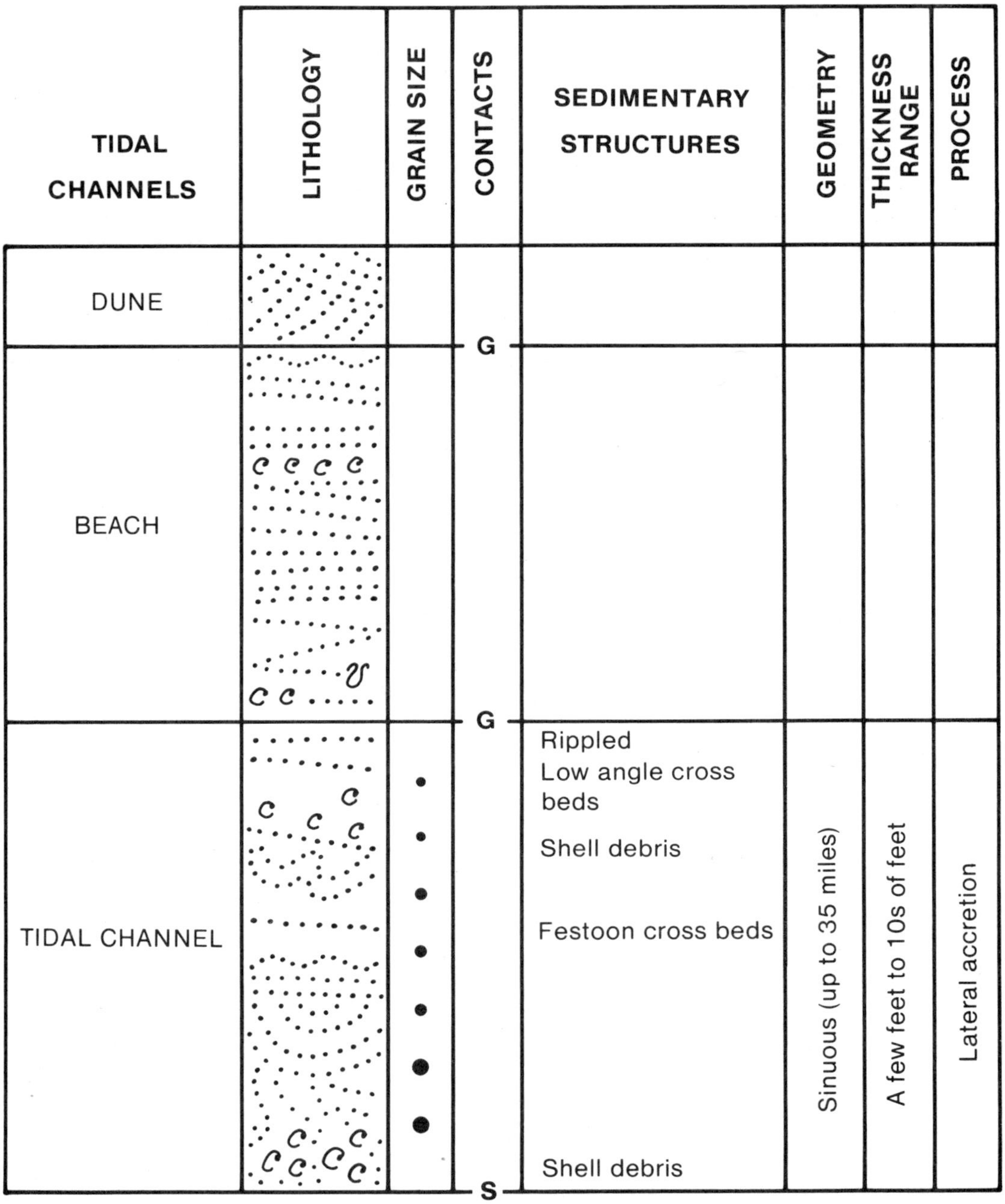

| TIDAL CHANNELS | LITHOLOGY | GRAIN SIZE | CONTACTS | SEDIMENTARY STRUCTURES | GEOMETRY | THICKNESS RANGE | PROCESS |
|---|---|---|---|---|---|---|---|
| DUNE | | | G | | | | |
| BEACH | | | G | | | | |
| TIDAL CHANNEL | | | S | Rippled<br>Low angle cross beds<br>Shell debris<br>Festoon cross beds<br>Shell debris | Sinuous (up to 35 miles) | A few feet to 10s of feet | Lateral accretion |

**Figure 6.5:** *Depositional characteristics of tidal channel deposits.*

# EXPLORATION FOR INTERDELTAIC SANDSTONE RESERVOIRS 129

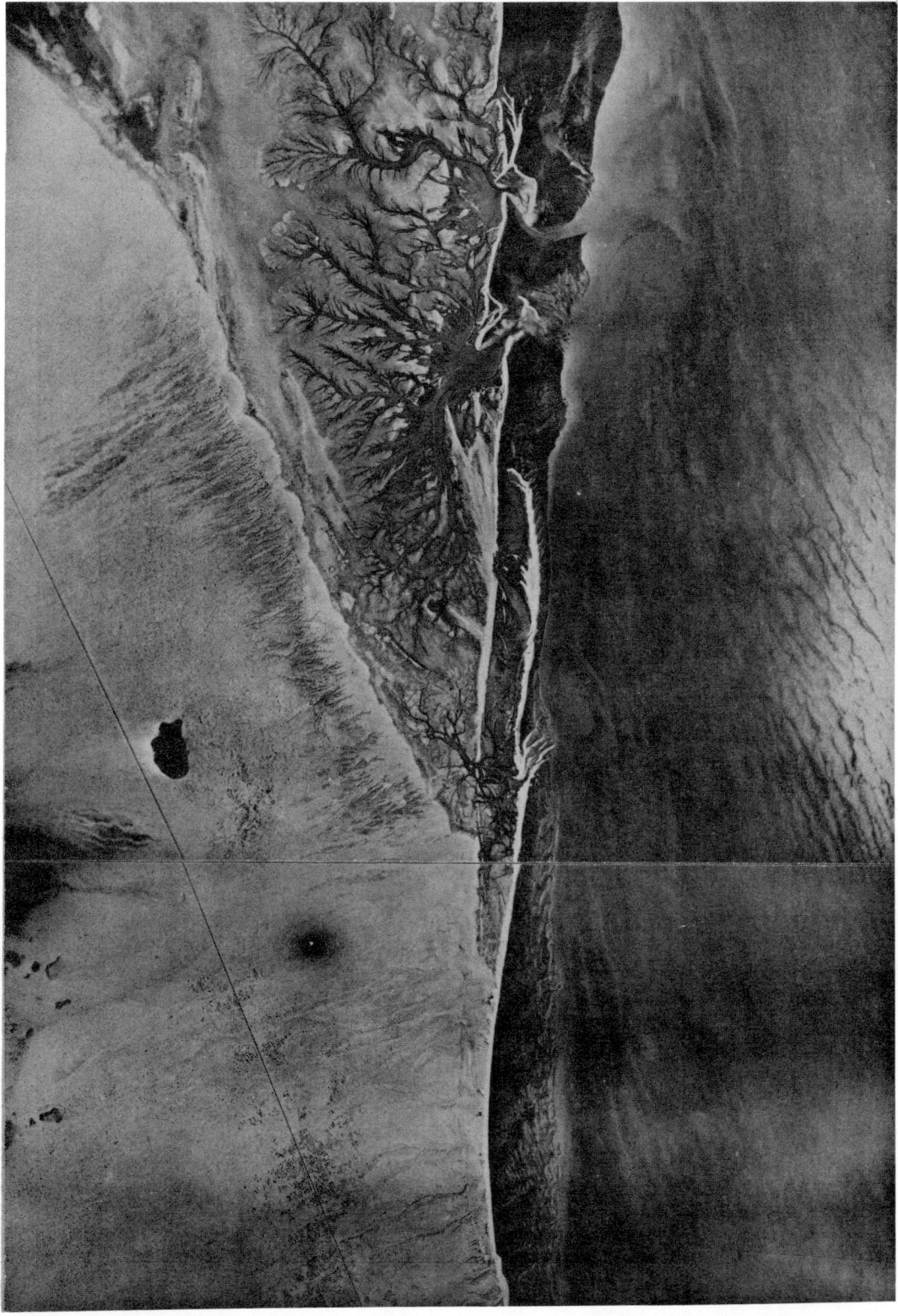

**Figure 6.6:** *Aerial photograph of beach-barrier island complex located just north of San Felipe, Baja California. Pleistocene sea cliffs and alluvial fans are the sources of sand.*

## DEVELOPMENT AND PRESERVATION OF BARRIER ISLAND SANDS

The depositional evolution of a barrier island complex favors its preservation. Barrier islands can originate on an offshore high or as a mainland beach (Stage 1, Fig. 6.7). Longshore currents that are generated by winds transport river, sea-cliff, or offshore sourced sands along the mainland or offshore high. The resultant sand body develops a typical shoreface-beach-dune profile and progrades seaward. Subsidence along the basin margin and progradation of the beach-ridge produces a narrow lagoon (Stage 2). The barrier island contiues to prograde seaward by accretion of the beach ridges. Inlets, produced by storms or differential progradation of the beach ridges, are formed and in some cases may become tidal channels.

Subsidence via differential compaction of shallow marine pro-island silts and clays causes the lagoon to widen (Stage 3). Tidal channels become incised, and tidal deltas form on both the lagoon and seaward sides of the barrier island. Compaction of the underlying pre-barrier-island sediments and older offshore deposits gives the barrier-island sand-body a convex-downward base. A beach ridge may close off a tidal inlet that in turn permits transport of the seaward tidal delta deposits. Preservation of the lagoonal tidal delta is enhanced by the fact that it differentially subsides into the lagoonal silts and clays.

Progradation of the barrier-island is terminated by either the cut-off of a source of sand or the progradation of the barrier-island too far seaward which permits longshore currents to flow behind the complex rather than in front of it (Stage 4). Wave action begins to attack the barrier-island, which in turn, forms a transgressive beach on the island complex. Subsidence of the sand body by differential compaction permits preservation by "transgressive" shallow marine silts and clays (Stage 5). Davies and others *(1971)* present an excellent review on the recognition of barrier island environments.

## INTERNAL FEATURES

Textural changes within a barrier-island sequence are the same as the offshore, beach, and dune sequence described for the beach in Figure 6.3. Cross bedding, laminations, ripples, and burrows are distributed throughout the barrier-island complex (Fig. 6.8). Their distribution is the same as those previously described for the beach sequence.

## EXTERNAL FEATURES

Basal contacts of the barrier-island sequence normally are gradational with offshore marine deposits. Upper contacts are gradational to sharp with marsh, lagoon, or tidal flat deposits. Lateral contacts are gradational landward with lagoon or marsh deposits, gradational seaward with shallow offshore marine deposits, and sharp with tidal channel or continental deposits.

Barrier-islands normally are long and linear with a smooth seaward section, and the top may be flat due to erosion prior to preservation. Barrier-island deposits vary in thickness from a few tens of feet to more than eighty feet. Normally, they are wider than beach deposits, with their width dependent upon the duration of progradation; and they are one to five miles wide, ranging in length from a few miles to more than a hundred miles.

## RESERVOIR AND STRATIGRAPHIC TRAP POTENTIAL

Barrier-islands, like beach sands, have good to excellent lateral and vertical continuity of porosity and permeability. Maximum stratigraphic trap potential occurs along the landward margin of the complex where porous and permeable sands interfinger with silts and clays of the tidal flat, lagoon, and/or marsh deposits. Onlapping shallow marine or offlapping lagoonal silts and clays provide the upper seal.

## SEISMIC CHARACTERISTICS OF SANDSTONE INTERDELTAIC RESERVOIRS

Seismic reflection characteristics of interdeltaic facies are dependent upon the amount and thickness of the sand facies. If sand is abundant and it is deposited principally by fluvial grading to wave processes, the seismic reflections depict a wedge to sheet geometry. They are continuous, and they have a high amplitude (Fig. 6.9, 2.1-3.2 sec. interval). Basal seismic reflections generally depict gentle onlap and/or

# EXPLORATION FOR INTERDELTAIC SANDSTONE RESERVOIRS 131

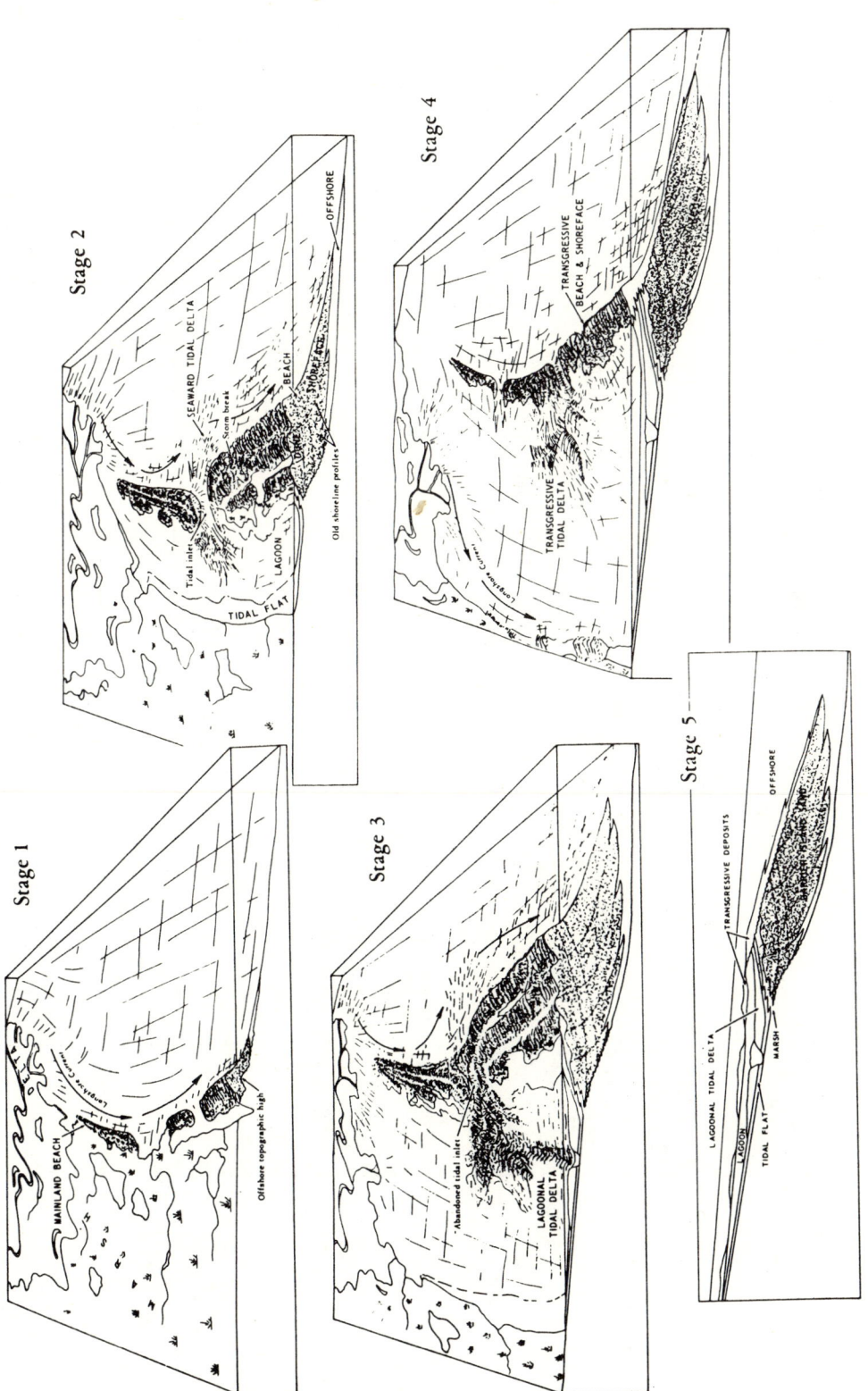

**Figure 6.7:** *Development and preservation of barrier island sands.*

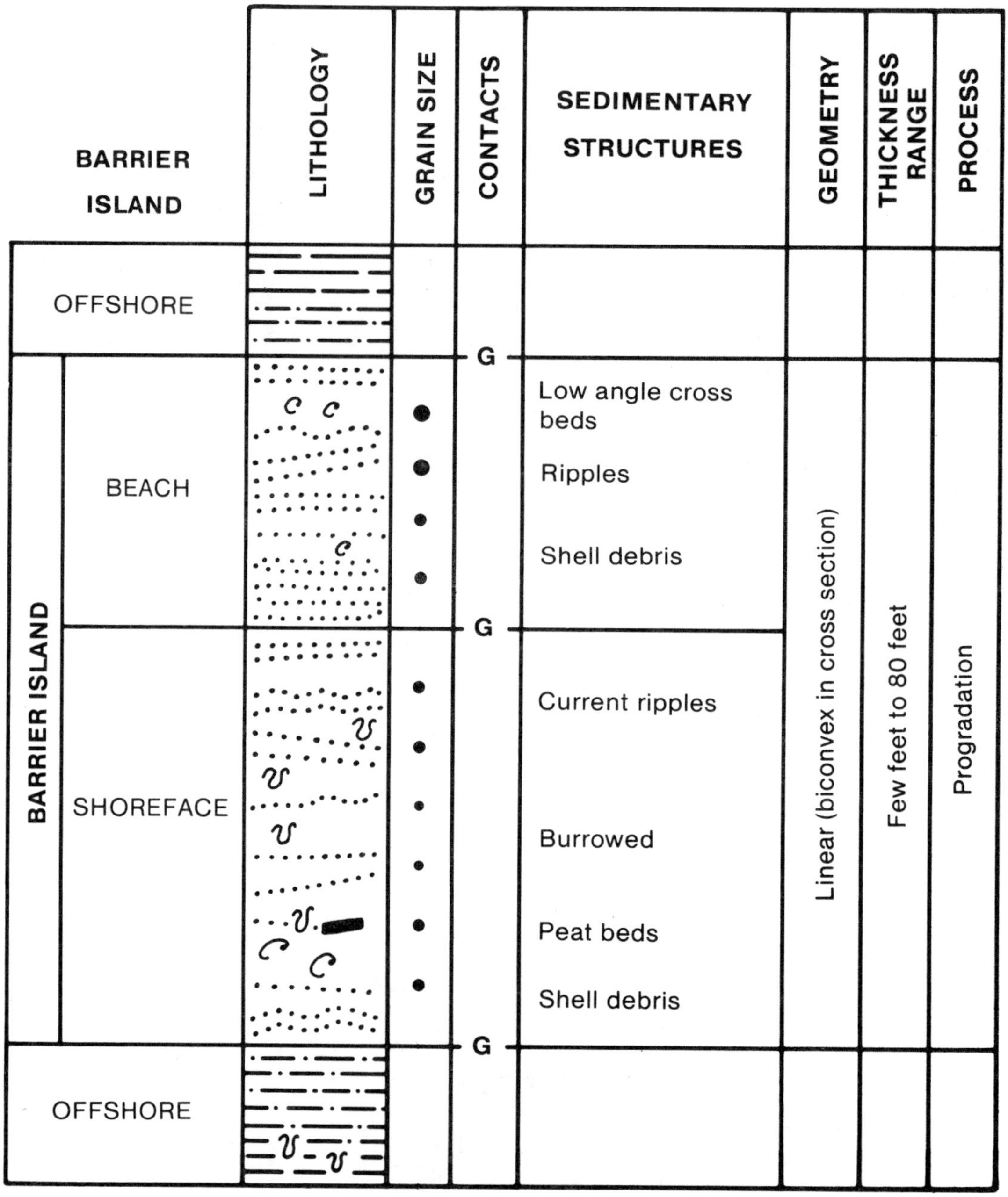

| BARRIER ISLAND | | LITHOLOGY | GRAIN SIZE | CONTACTS | SEDIMENTARY STRUCTURES | GEOMETRY | THICKNESS RANGE | PROCESS |
|---|---|---|---|---|---|---|---|---|
| | OFFSHORE | | | | | | | |
| BARRIER ISLAND | BEACH | | | G •  •  •  • | Low angle cross beds  Ripples  Shell debris | Linear (biconvex in cross section) | Few feet to 80 feet | Progradation |
| | SHOREFACE | | | G •  •  •  •  •  • | Current ripples  Burrowed  Peat beds  Shell debris | | | |
| | OFFSHORE | | | G | | | | |

**Figure 6.8:** *Depositional characteristics of barrier island deposits.*

# EXPLORATION FOR INTERDELTAIC SANDSTONE RESERVOIRS

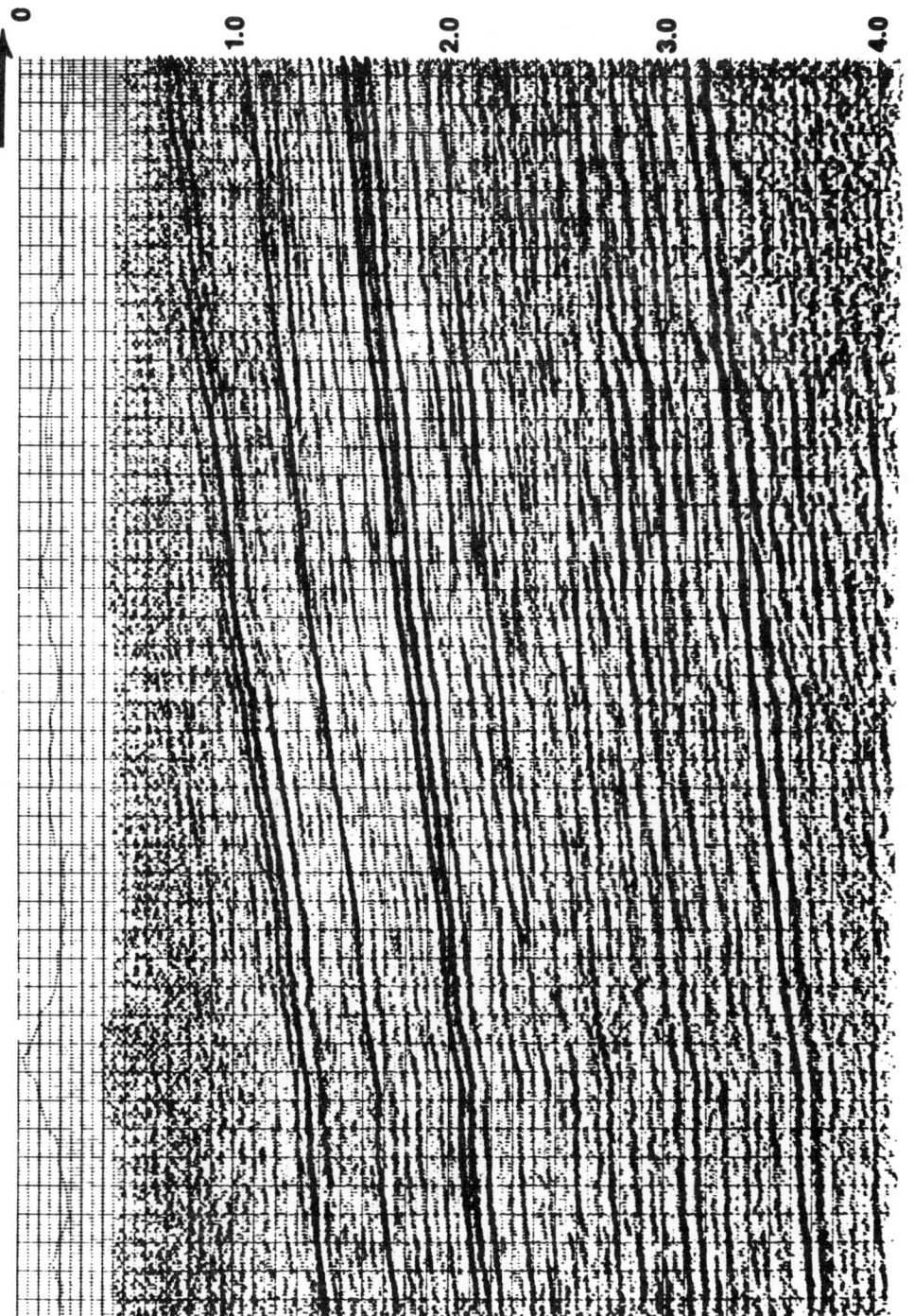

**Figure 6.9:** *Seismic line located in the Green River Basin, Wyoming, illustrates deltaic and interdeltaic sedimentary patterns (after Laing, 1972).*

downlap patterns, and upper reflections are generally concordant. Internal reflections are commonly parallel within the shale intervals and divergent in the sand facies. On the other hand, if sand is not abundant, a thin wedge (or in some cases, sheet) of sand accumulates in the interdeltaic environment. Reflections are low in amplitude and generally variable in continuity. Basal reflections depict gradual onlap and/or downlap sedimentary patterns. Upper patterns are typically concordant. Like in a sand-rich interdeltaic environment, internal reflections are parallel to divergent.

## EXAMPLES OF INTERDELTAIC RESERVOIRS

### INTRODUCTION

Numerous examples of reservoirs deposited in interdeltaic environments have been reported in the literature. Some of these include the Bisti Field, upper Cretaceous, San Juan Basin, New Mexico *(Sabins, 1963)*; Louden Field, Illinois Basin, Illinois *(Cluff and Lasemi, 1980)*; Hilight Field, Powder River Basin, Wyoming *(Berg, 1976)*; Horseshoe Field, San Juan Basin, New Mexico *(McCubbin, 1969)*; Borden Island Gas Field, western Canadian Arctic Archipelago *(Douglas and Oliver, 1979)*; Patrick Draw Field *(Weimer, 1966)*; and Bell Creek Field, Montana *(McGregor and Biggs, 1968)*.

### BARRIER ISLAND RESERVOIR - BELL CREEK FIELD, MONTANA

The Bell Creek Field is located in southeastern Montana on the northeastern flank of the Powder River Basin. The field was discovered in 1967 and total recoverable reserves are estimated to exceed 200 million barrels of 30° API all from over 15,000 productive acres. Shallow depths (about 4,500 feet) and low drilling costs make this field very profitable. The field is a classic example of a stratigraphic trap that contains significant recoverable reserves *(McGregor and Biggs, 1968)*.

Exploration drilling prior to the discovery of Bell Creek Field was based on the surface and subsurface interpreted closures. However, the structural interpretations were incorrect and the wells were dry. Subsequent drilling in the Bell Creek Field area depicts a remarkably uniform homocline that dips northwest at about 100 ft/mile (Fig. 6.10). There is no structural control on the updip productive limits of the field.

The major reservoir in Bell Creek Field is the Muddy Sandstone (Cretaceous). The sandstone reaches a maximum thickness of more than 40 feet along the northwest central flank of the field, but in the productive limits of the field, the Muddy ranges in thickness from 20 to 30 feet (Fig. 6.11). Thickness trends parallel structural strike, and isopach contours depict a northeast trending elongated sand body that pinches out in all directions. The Muddy Sandstone in the Bell Creek Field has excellent reservoir qualities (Fig. 6.12). Porosities range from 25 to 35% and permeabilities as high as 2.5 darcies occur in the field. Vertical communication is excellent but lateral communication is disrupted by apparent pinchouts of individual reservoirs (Fig. 6.13).

The Muddy Sandstone in the Bell Creek Field area is interpreted to have been deposited as barrier island complex *(McGregor and Biggs, 1968)*. The paleogeography for the area, as interpreted by McGregor and Biggs, is depicted in Figure 6.14. At least four distinct environments are represented in the area, namely, meandering channel, marsh, deltaic front, and barrier island. Each of the environments depict the classic grain size changes, sedimentary structures, and contacts as previously described. At least two, and possibly three discrete barrier island bars produce in the field. This is indicated by the occurrence of different oil-water contacts within the field (Fig. 6.13). Locally the field is cut by sands that fine upward. These sands may represent tidal channel or fluvial channel environments. The updip seal is formed by the facies change of the barrier island sands into the fine-grained lagoonal facies.

### BARRIER ISLAND RESERVOIR - PATRICK DRAW FIELD, WYOMING

Patrick Draw Field, discovered in 1959, is one of the most significant discoveries in the Rocky Mountains during the 1950's. The field is located on the east flank of the Rock Springs uplift and on the south flank of the Wamsutter Arch (Fig. 6.15). Production is from the upper sands of the Almond Formation (Cretaceous) at depths ranging from 3,500 to 5,400 feet. It

# EXPLORATION FOR INTERDELTAIC SANDSTONE RESERVOIRS

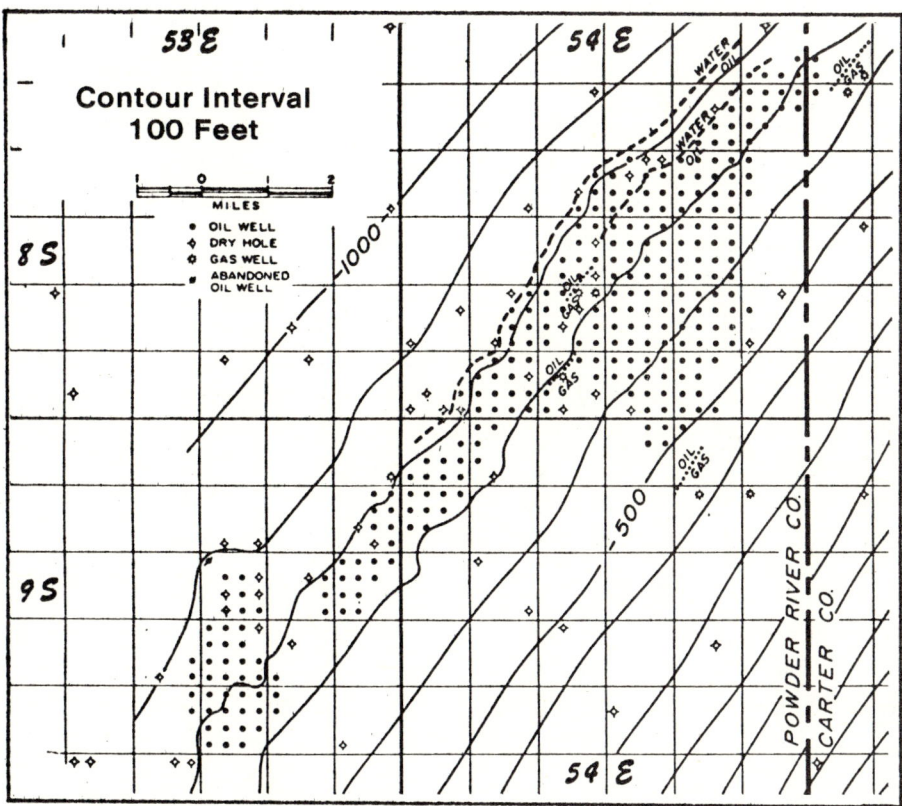

**Figure 6.10:** *Structure contour map on top of Muddy Sandstone, Bell Creek Field area, Montana (McGregor and Biggs, 1968).*

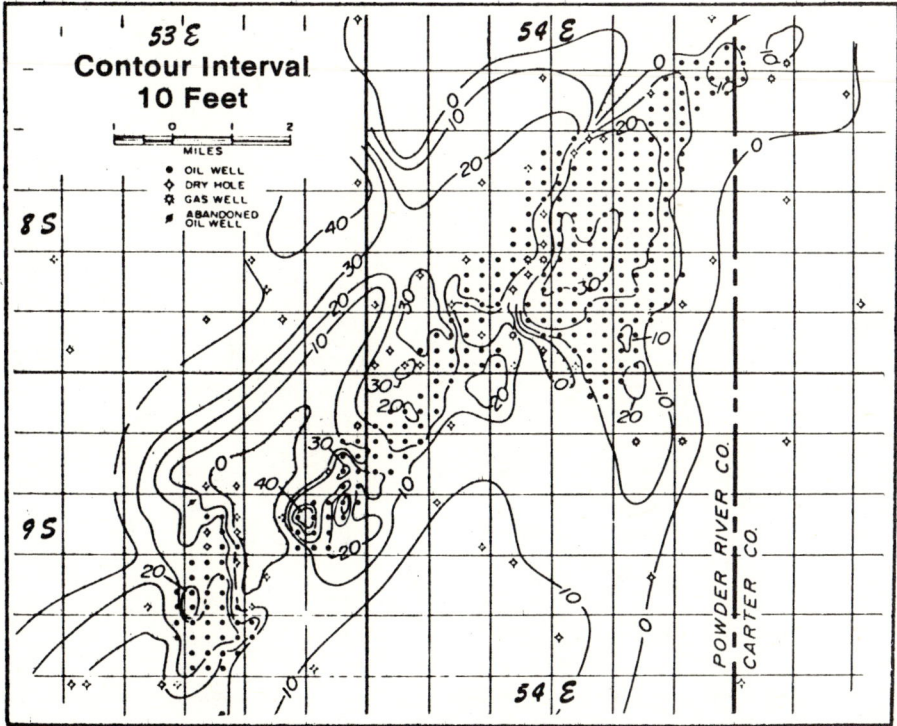

**Figure 6.11:** *Isopach map of Muddy Sandstone, Bell Creek Field area, Montana (McGregor and Biggs, 1968).*

136                               EXPLORATION FOR SANDSTONE RESERVOIRS

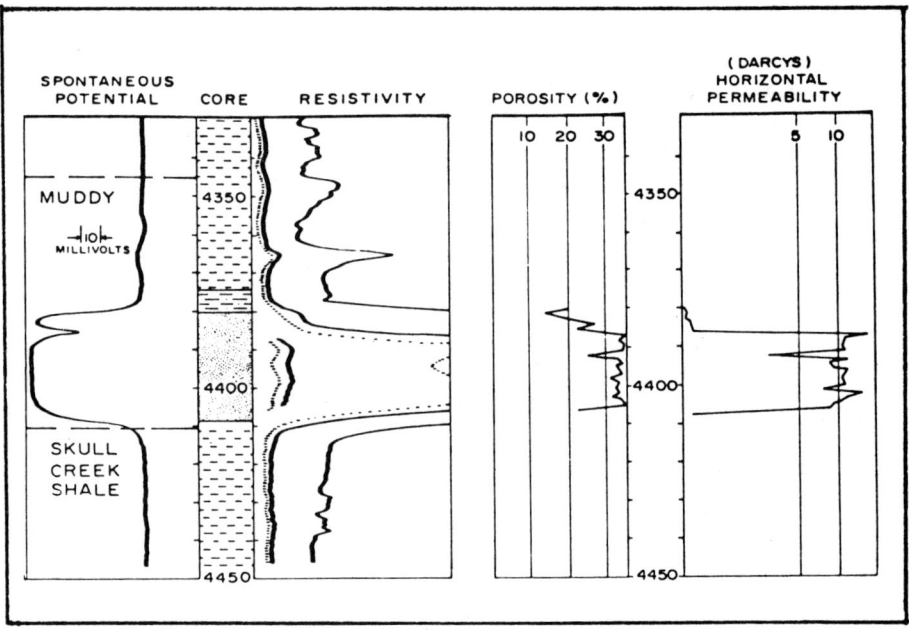

**Figure 6.12:** *Electric log and core analysis of typical Muddy Sandstone section at Bell Creek Field, Montana (McGregor and Biggs, 1968).*

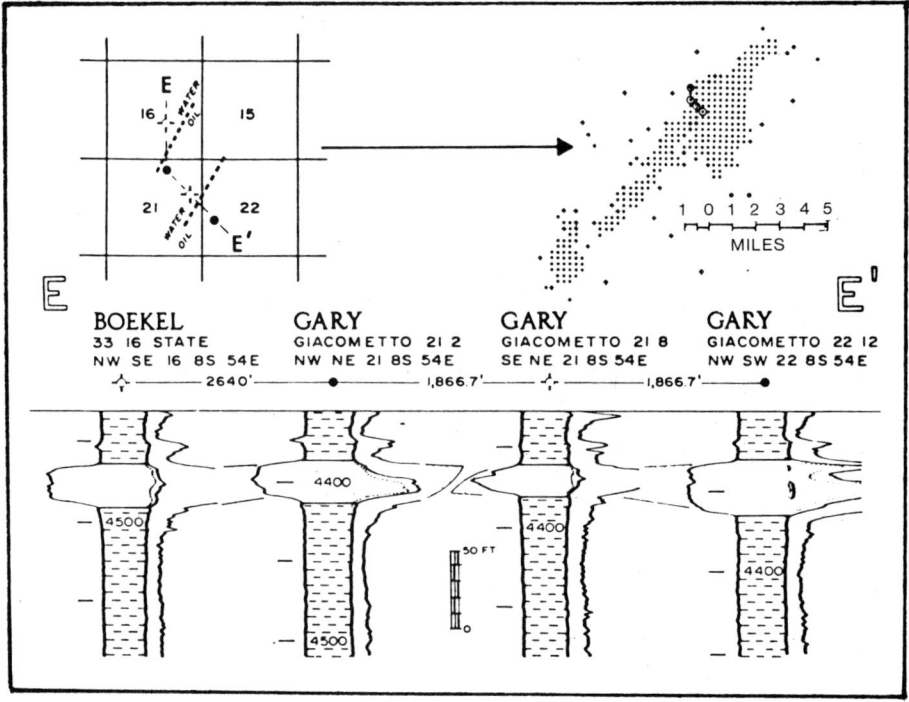

**Figure 6.13:** *Fluid content in the Muddy Sandstone at Bell Creek Field (Montana) demonstrates that at least two barrier island deposits occur in the field (McGregor and Biggs, 1968).*

# EXPLORATION FOR INTERDELTAIC SANDSTONE RESERVOIRS

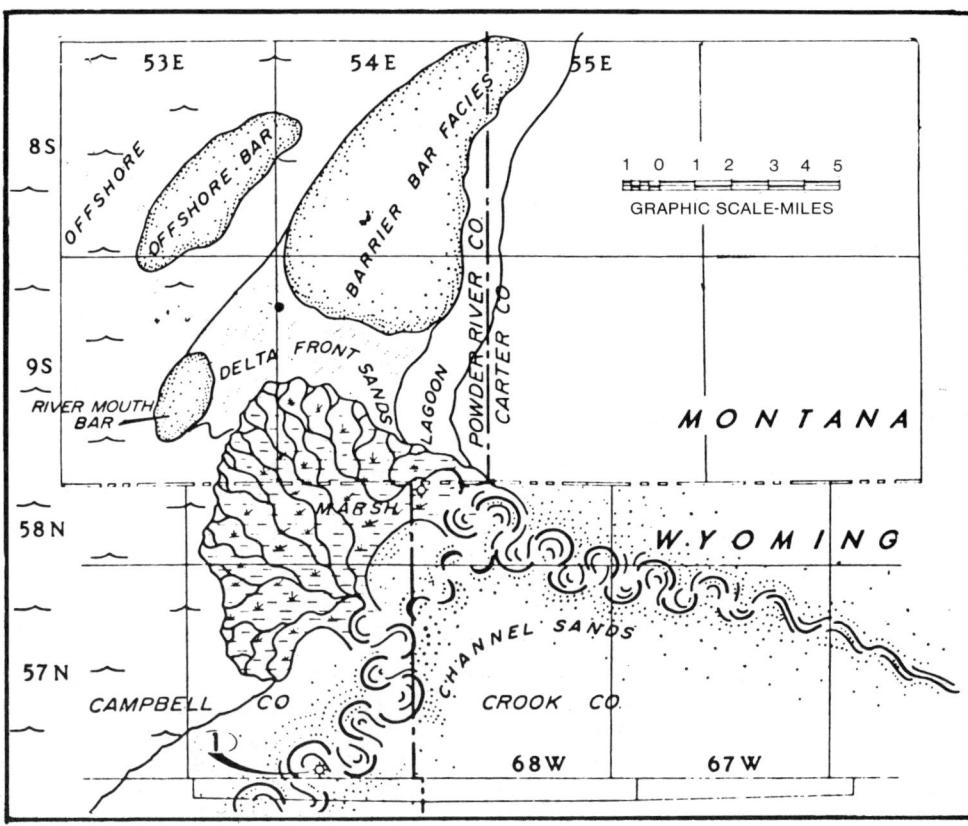

**Figure 6.14:** *Paleogeomorphic map of the Bell Creek Field area (Montana) during Muddy Sandstone deposition (McGregor and Biggs, 1968).*

is estimated that 200 to 250 millions barrels of 40°API gravity oil are in place *(Weimer, 1966)*.

The field is unique for a couple of reasons. A structure map on top of the Almond Formation depicts a homocline that initiates at the surface about 8 miles east of the field, and dips southeast at 4° (Fig. 6.15). The updip limits of the field are defined by the facies change from sandstone to shale. A second unique characteristic of Patrick Draw is that this large oil field is positioned stratigraphically in a part of the section that is typically gas productive rather than oil in the Rocky Mountains.

The stratigraphic section in the Patrick Draw area is represented by complexly interfingering continental, interdeltaic, and marine clastics. Sandstones are abundant and lenticular. Therefore, accurate reconstruction of the Almond depositional history is impractical prior to the development of a valid stratigraphic framework. Upper Cretaceous formations in the Patrick Draw area were deposited along the western margin of the Cretaceous seaway that occupied much of the Western Great Plains of North America. Consequently, the coastline transgressed or regressed in response to sediment supply, rates of subsidence, and/or global sea level change. Because of the shifts in environments, the vertical sequence of environments are varied and complex which in turn make lateral correlations difficult. Coeval relations can only be determined by detailed pattern correlation techniques that are described in Chapter 2. Fortunately, the Lewis Formation in the Patrick Draw area contains several bentonite beds that provide excellent time surfaces to which pattern correlation of resistivity markers can be related (Fig. 6.16).

The main reservoir is interpreted to have been deposited as a barrier island sequence *(Weimer, 1966)*. The sandstone is composed of quartz (64%), chert (32%), feldspar (4%),

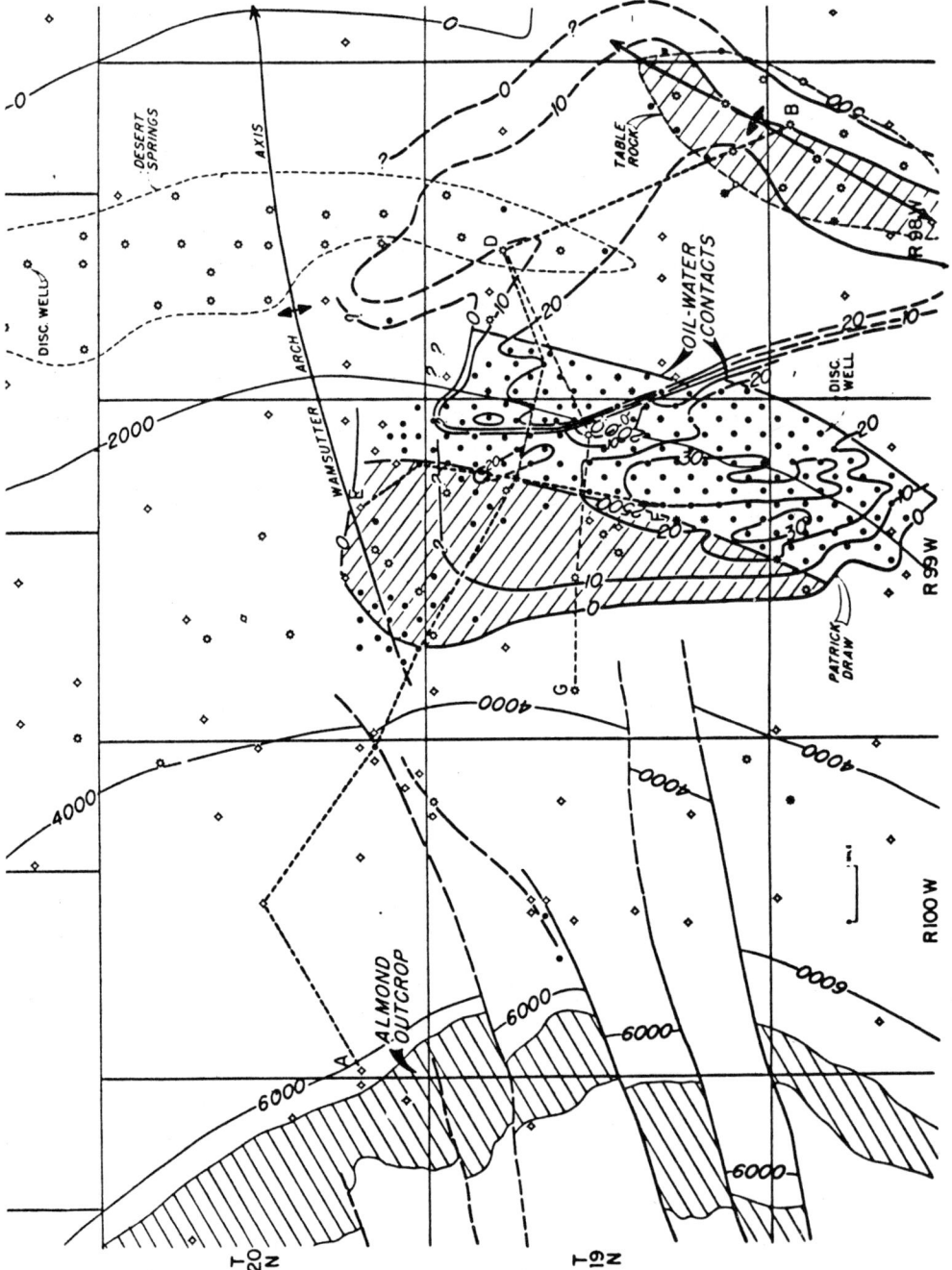

**Figure 6.15:** *Isopach of principal reservoir at Patrick Draw Field. Structural contours on top of Almond Formation. Isopach interval is 10 feet, structural contour interval is 2,000 feet. Diagonal pattern depicts gas cap (Weimer, 1966).*

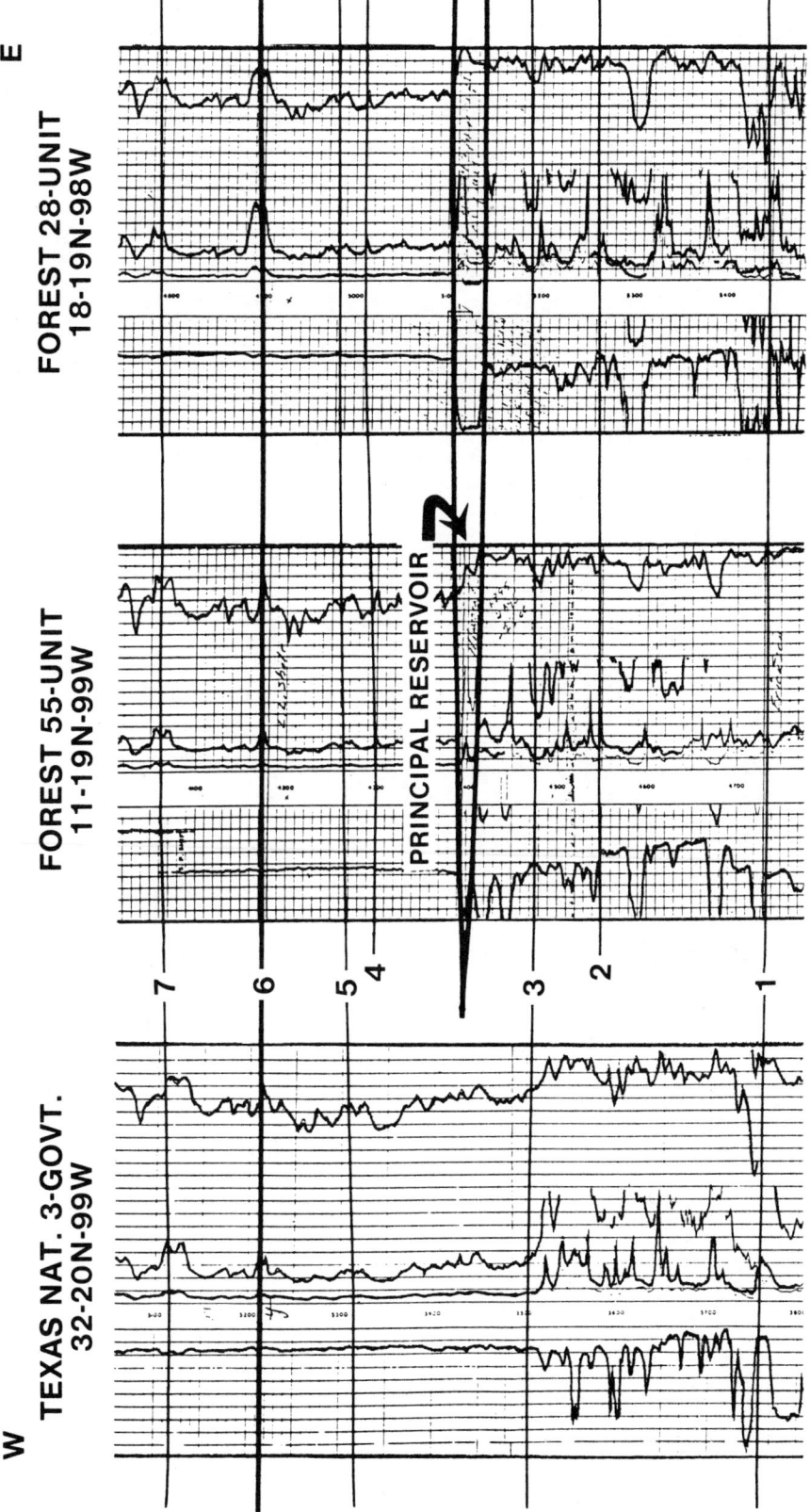

**Figure 6.16:** *Stratigraphic cross section across Patrick Draw Field, Wyoming. Pattern correlation lines 1–7 support the interpretation that the principal reservoir at Patrick Draw pinches out to the west. The facies change from barrier island sands to lagoonal shales provides the updip seal for the field.*

and trace amounts of biotite, musconite, chlorite, and zircon *(Lawson and Crowson, 1961)*. Grain size increases upward slightly from fine to medium. Cross stratification at the base of the sand ranges from 4° to 5°. The sand is biconvex with 5 to 10 feet of relief in relation to marker beds above the sand. Basal contacts are sharp whereas upper contacts are gradational (Fig. 6.17). At least two and possibly three discrete bars occur in the same stratigraphic position as indicated by different oil/water contacts within the field *(Weimer, 1966)*.

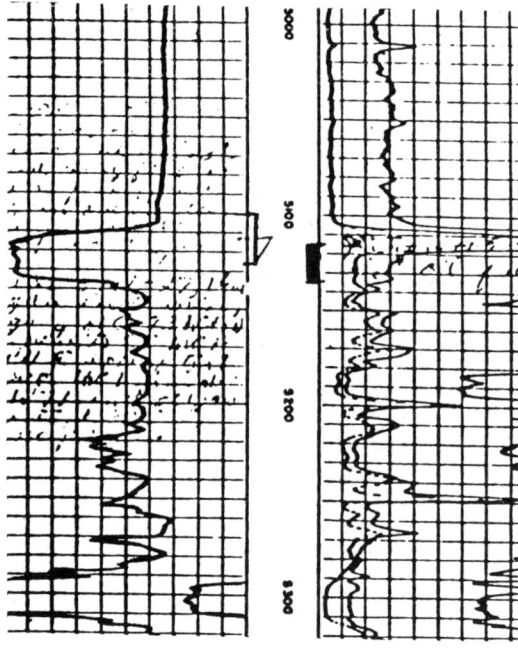

**Figure 6.17:** *Mechanical log of Forest Unit 28, 18-19N-98W, Wyoming. Sand at 5110' is the main reservoir at Patrick Draw Field.*

## BEACH-SHOREFACE RESERVOIR - HELCA & DRAKE POINT FIELDS, CANADA

Helca and Drake Point Gas Fields are located adjacent to the Sabine Peninsula in northeastern Melville Island in the western Canadian Arctic Archipelago. Helca contains 11.4 TCF of gas and Drake Point has 9.1 TCF of marketable gas. The fields are the largest discovered to date in the Sverdrup Basin of northwest Canada. Both fields are structural traps; Helca contains a gas column of at least 800 feet whereas a 250-foot or greater column occurs at Drake Point *(Douglas and Oliver, 1979)*.

The Borden Island Formation (Jurassic) constitutes the main reservoir rock at Helca and Drake Point fields *(Douglas and Oliver, 1979)*. However, Douglas and Oliver were unable to document correlations from the fields to the type area of the Borden Island Formation and therefore, they used the informal term *Borden Island.* The productive zone is represented by two or three coarsening-upward cycles. Each cycle is represented by basal siltstones and claystones that contain soft sediment deformation and burrowed structures. Typically, the sandstone cycles are crossbedded and contain massive intervals near the top of each cycle. Basal contacts are gradational and upper contacts are sharp (Fig. 6.18). The zone thickens in a basinal direction (Fig. 6.19).

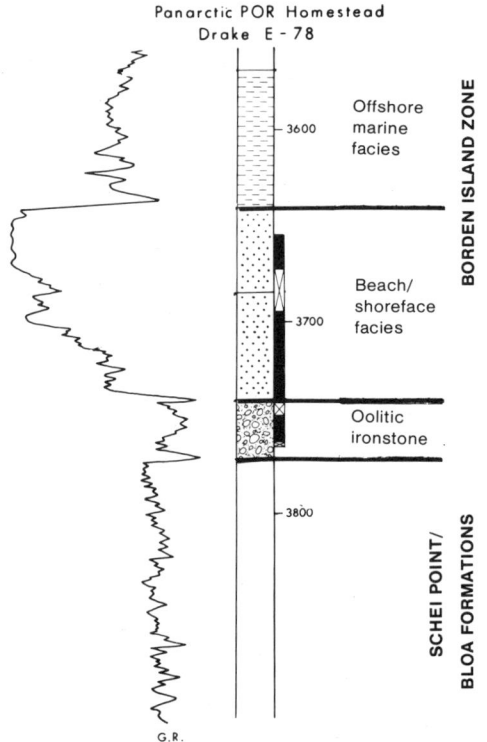

**Figure 6.18:** *Gamma ray curve through Borden Island gas zone, Drake Point Field, Canada (Douglas and Oliver, 1979).*

Douglas and Oliver *(1979)* interpret the productive Borden Island gas zone to have been deposited in an interdeltaic beach/offshore environment. Sparse subsurface control reduced the sophistication of their interpretations and therefore, they proposed the most simple coastline geomorphology, that is—a land-attached beach. A barrier island system was ruled out mainly because the authors did not see any evidence in the available core data to support the existence of a lagoonal facies. However, it is possible that the "Borden Island" coastline was mesotidal (North Carolina model) rather than the microtidal (Galveston Island model) which would not require the preservation of a lagoonal environment.

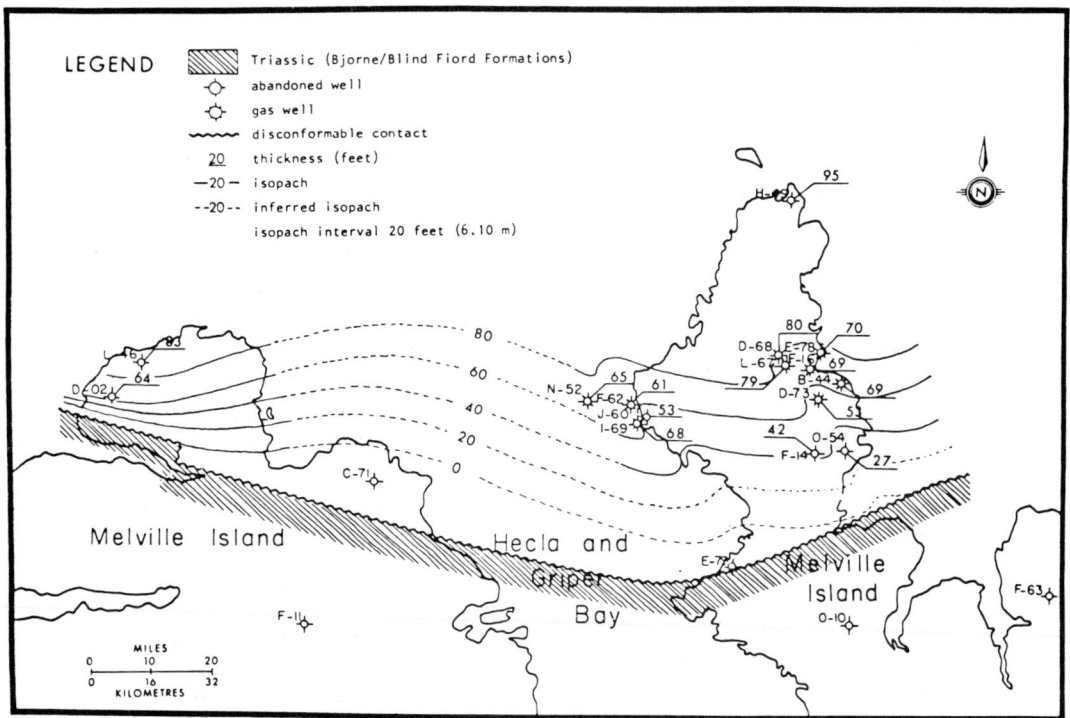

**Figure 6.19:** *Isopach map of the Helca and Drake Point gas fields, northwestern Canada (Douglas and Oliver, 1979).*

# 7

# Exploration for Offshore Marine Sandstone Reservoirs

## INTRODUCTION

In the not too distant past, explorationists thought that base level, the boundary between erosion and deposition, was sea level. Consequently, they assumed that deposition of sand was restricted to the continental, interdeltaic and deltaic environments. Beginning around 1955, research by the oceanographic institutes demonstrated that not only sands can be deposited in offshore shallow marine waters, but sands may accumulate in deep water tens of miles from a coastline. It is now realized that most of the processes that operate in the alluvial and coastal models of sedimentation also operate in deep marine environments.

## SHALLOW MARINE RESERVOIRS

### SITE AND DEPOSITIONAL PROCESSES

Shallow marine shelf sands are those that accumulated in water depths that range from 30 to 600 feet *(Walker, 1979)*. Although excellent studies of modern sands of offshore Great Britian and offshore east coast United States have been made, the origin of these sands are not totally understood. Numerous studies have documented the distribution, thickness, and internal features of the sands but because sea level has risen over 300 feet during the Holocene (Fig. 6.1), the primary origin of the sands has not been documented. Consequently, it is difficult to interpret ancient shallow water marine sand tones based on what is known about modern shallow marine sands.

Modern shallow marine sands that occur offshore of the United States east coast are grouped into three types *(Swift, 1975; Swift and Sears, 1974; and Duane and others, 1972)*. *Shoal retreat massifs* are sand bodies that are oriented perpendicular to the present coastline (Fig. 7.1). They are tens of miles

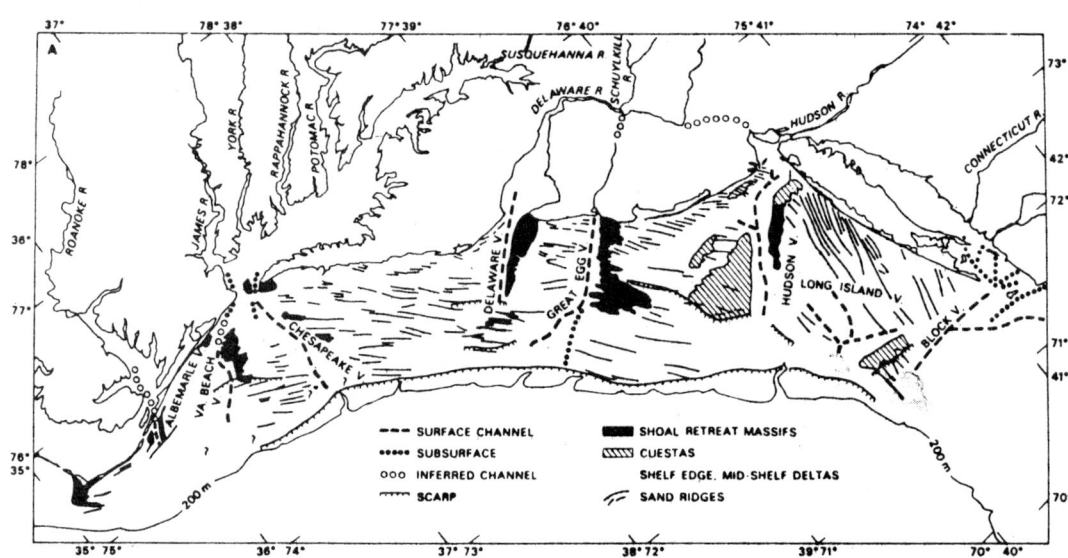

**Figure 7.1:** *Major morphological features of the middle Atlantic Shelf, United States (Swift and others, 1973).*

long, a few miles wide, and they range in thickness from 30 to 90 feet. Normally these sands can be traced landward into present-day areas of preferred sand accumulations such as estuaries. *Linear sand ridges* are tens of miles long but only hundreds of feet wide and tens of feet thick. Typically they form on retreat ridge massifs with an average spacing of 1.5 miles. *Sand waves* are the smallest sand bodies on the continental shelf. They represent features that resemble periodically spaced bedforms that occur on the ridges and massifs. Normally they are only a few feet thick.

Processes that are responsible for transporting and depositing sand on modern continental shelves include tidal currents, storm currents, intruding ocean currents, and density currents. Of these, tidal and storm currents are the most important *(Walker, 1979)*. Shoal retreat massifs represent local coastline depocenters that may correspond to topographic lows that were produced during the Pleistocene low stand of sea level. As such, they are not directly related to modern mechanisms of transport.

Linear sand ridges and the smaller superimposed sand waves are storm-dominated features, both in terms of their initiation and their subsequent transgression *(Walker, 1979)*. After detachment as shoreline features, these ridges were modified by storm-generated currents (Fig. 7.2). Burrowing is active during fair weather conditions but during minor storms, the crests of the ridges are winnowed. During major storms, coarse sand is moved to the crests and fine sand remains on the flanks which in turn develops the typical coarsening, upward grainsize observed in ancient counterparts *(Stubblefield and others, 1975)*. Field *(1980)* supported this observation through his studies of the sand bodies that occur on the inner shelf off Maryland. These sands are typically gray to brown, fine to coarse-grained, well sorted, and highly bioturbated. He interpreted the sands as reworked remnants of Holocene back barrier and lagoonal deposits.

## INTERNAL FEATURES

Grain size and rounding is dependent upon the relic environment. Typically, there is no significant change in grain size upward in the so-called shoestring sand sequence (Fig. 7.3). It is generally conceded that little additional rounding is produced in the secondary stage of reworking. Sorting is generally poor near the top of the sequence, however, the middle sand interval contains the best sorting. The bars are cross bedded, horizontally laminated, and normally highly burrowed. Clay clasts that are marine in origin may occur near the top of the sand. LaFon *(1981)* describes in detail from

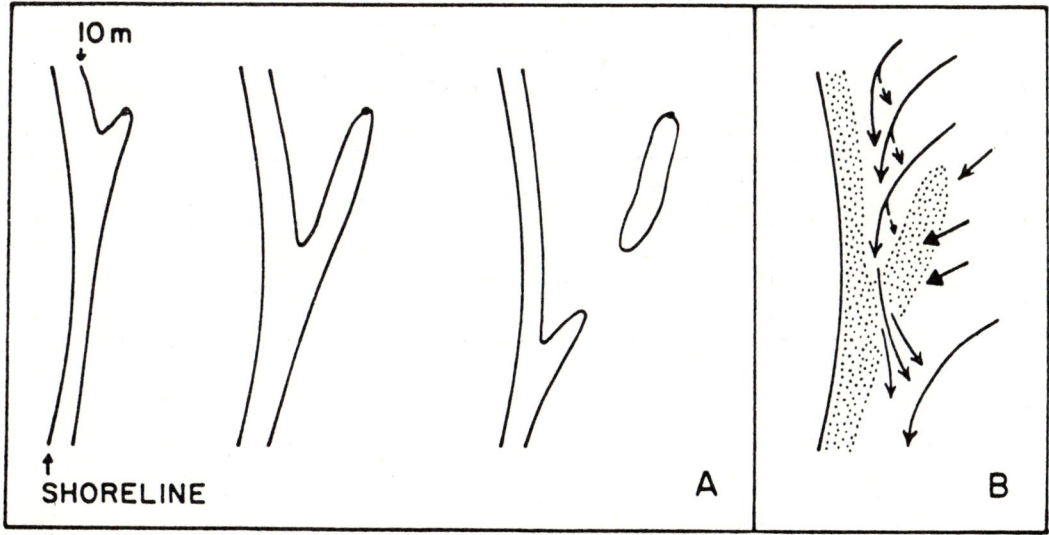

**Figure 7.2:** *Evolution of storm-dominated linear sand ridges, mid-Atlantic Shelf, United States (Walker, 1979).*

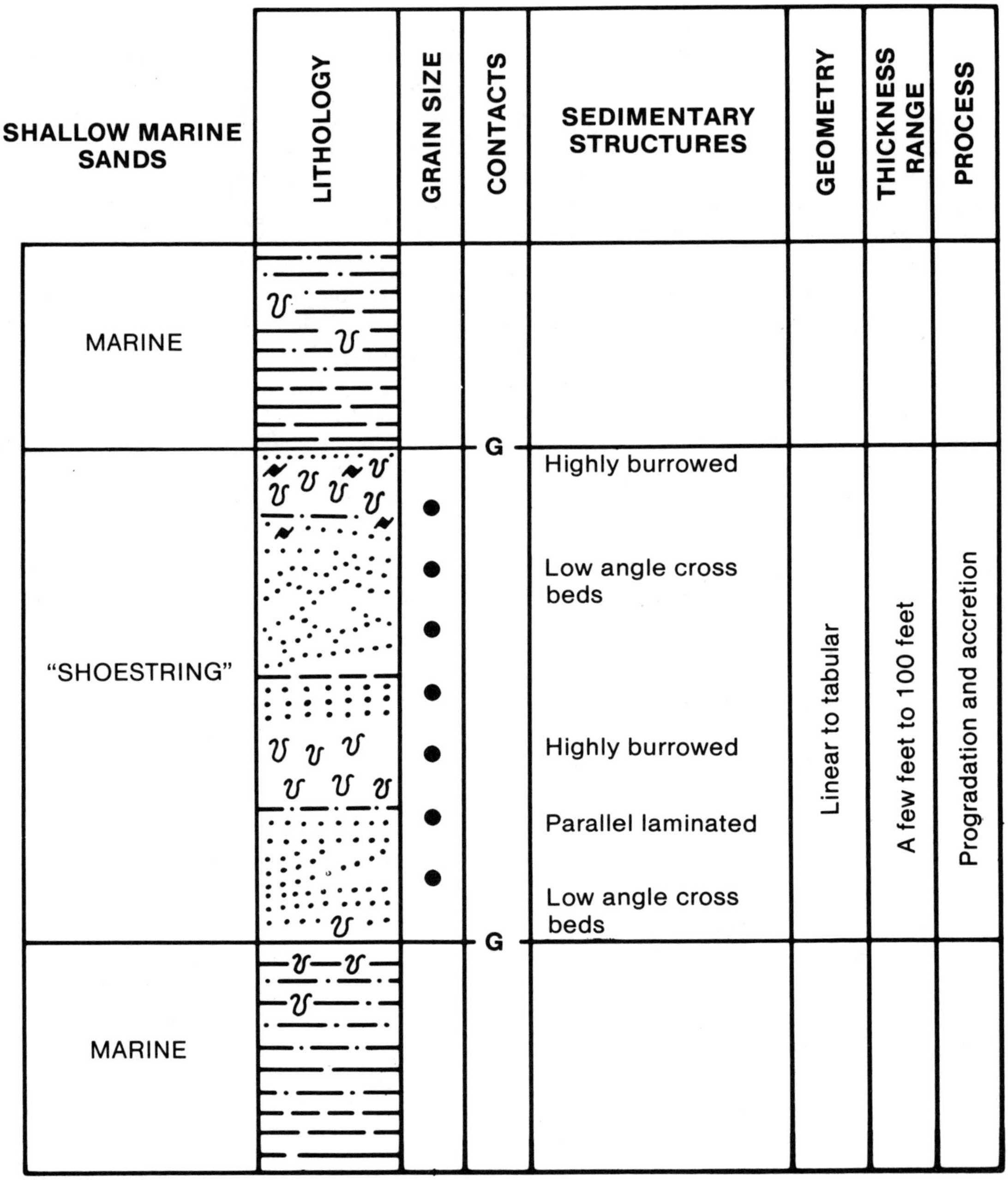

| SHALLOW MARINE SANDS | LITHOLOGY | GRAIN SIZE | CONTACTS | SEDIMENTARY STRUCTURES | GEOMETRY | THICKNESS RANGE | PROCESS |
|---|---|---|---|---|---|---|---|
| MARINE | | | | | | | |
| | | | G | | | | |
| "SHOESTRING" | | • | | Highly burrowed | Linear to tabular | A few feet to 100 feet | Progradation and accretion |
| | | • | | Low angle cross beds | | | |
| | | • | | | | | |
| | | • | | | | | |
| | | • | | Highly burrowed | | | |
| | | • | | Parallel laminated | | | |
| | | • | | Low angle cross beds | | | |
| | | | G | | | | |
| MARINE | | | | | | | |

**Figure 7.3:** *Depositional characteristics of shallow marine sands.*

# EXPLORATION FOR OFFSHORE MARINE SANDSTONE RESERVOIRS

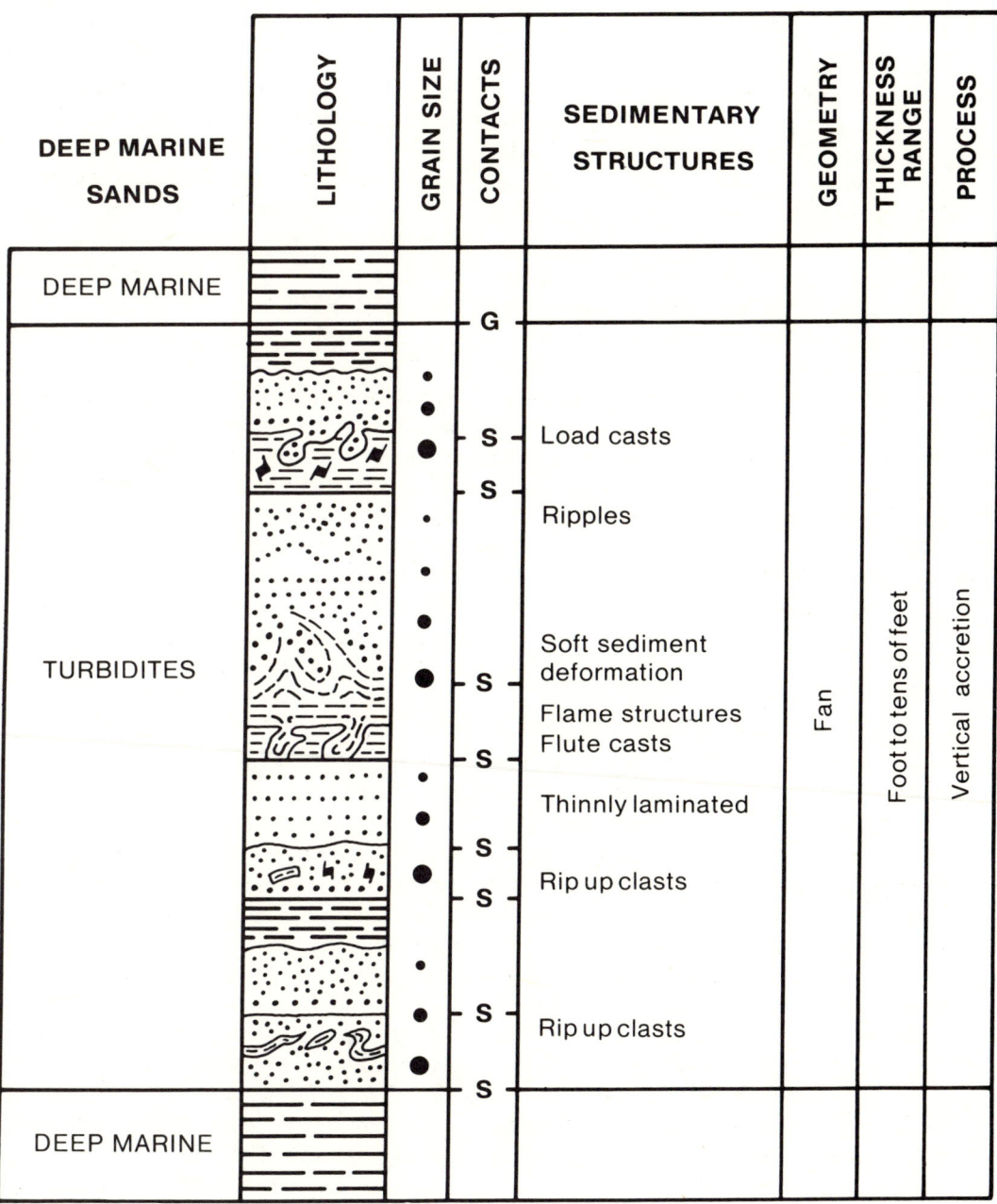

**Figure 7.4:** *Depositional characteristics of deep marine deposits.*

surface exposures, the internal features of an offshore bar in the Mancos shale (Cretaceous) in New Mexico.

## EXTERNAL FEATURES

Lower contacts of the sand body are normally gradational with shallow marine shales. If the basal contact is sharp, this frequently implies that the sand is a relic that was produced in the course of reworking during rapid sea level rise. Upper contacts are normally gradational with shallow marine shales.

The bars range in thickness from a few feet to more than 1,000 feet. Several geometries are possible. Widespread sheet sand (Vaqueros, Miocene), broad "pods" (Cardium at Pembina Field, Cretaceous), and "shoestring" sands (Cardium at Cross Field, Cretaceous) have been considered to be reworked relic sands. The geometry of the sand bodies is influenced by the strength, consistency, and orientation of the marine currents, the bathymetry of the sea floor, and the amount of sand that can be reworked.

## RESERVOIR AND STRATIGRAPHIC TRAP POTENTIAL

Vertical continuity of reservoirs is not as good as many other bar deposits because they contain abundant interbeds of shale and/or silt. Lateral continuity is normally good. Permeability is reduced by abundant burrows. Individual sand units may be imbricated with respect to one another; therefore, discrete sand units can occur more frequently than mapped, using only mechanical logs. Frequently, marine currents tend to drop their traction load around growing structures, and these sands will tend to onlap contemporaneous growing structures. The use of paleogeographic maps for exploration of these is questionable because marine currents can trend at any angle to the depositional strike.

## SUBMARINE FAN AND TURBIDITE RESERVOIRS

### INTRODUCTION

Two models, the rapid transport by turbidity flows (Rabbit Model), or mass transport by slow creep (Turtle Model), are in vogue. Turbidite flow is attributed to currents in which the density difference between sea water and the current is due to dispersed sediment. Evidence for turbidity currents includes cutting of transatlantic cables, direct observation, and destruction of offshore rigs. Slow creep of sediment across a break in slope or down a submarine canyon can also produce a subaqueous fan deposit. Evidence for this model of deposition includes leveed channels that decrease in size towards the toe of the fan. The only depth connotation is that the sedimentation occurs below storm wave base so the sands cannot be reworked by marine wave generated currents.

### SITE

Submarine fan and turbidite deposits typically accumulate downslope from a river mouth or submarine canyon. Tectonically active basins are excellent sites for the accumulation of fan deposits. They are normally characterized by rapid rates of sedimentation that can result in over-steepened sedimentary dips, and can be "cut loose" by earthquakes. Storms can also produce subaqueous density flows both in tectonically active and "quiet" basins.

### DEPOSITIONAL PROCESSES

Submarine fans and turbidity reservoirs are classified by the nature of the dominant sediment-supporting mechanisms of transport (Fig. 7.5). *Debris flows* are characterized by a matrix of interstitial fluids and/or fine sediment. They are usually massive, poorly sorted, mud-supported boulder conglomerates. *Grain flows* are grain-supported and during transport the individual grains undergo collisions. Usually these sands have erosional bases, massive bedding and/or reverse grading, and a flat top. *Fluidized flows* occur when the individual grains are supported by water that escapes during deposition of the grains by gravity. These deposits are poorly graded; dish structures and convolute laminations are common. *Turbidity flows* are produced by upward movement of water due to turbulence. The resultant deposits are grain-supported; graded and upper flow regime structures are common. It is probable that a combination of two or all of the above processes occur during the deposition of a discrete sand body *(Mutti and Ricci Lucchi, 1972).*

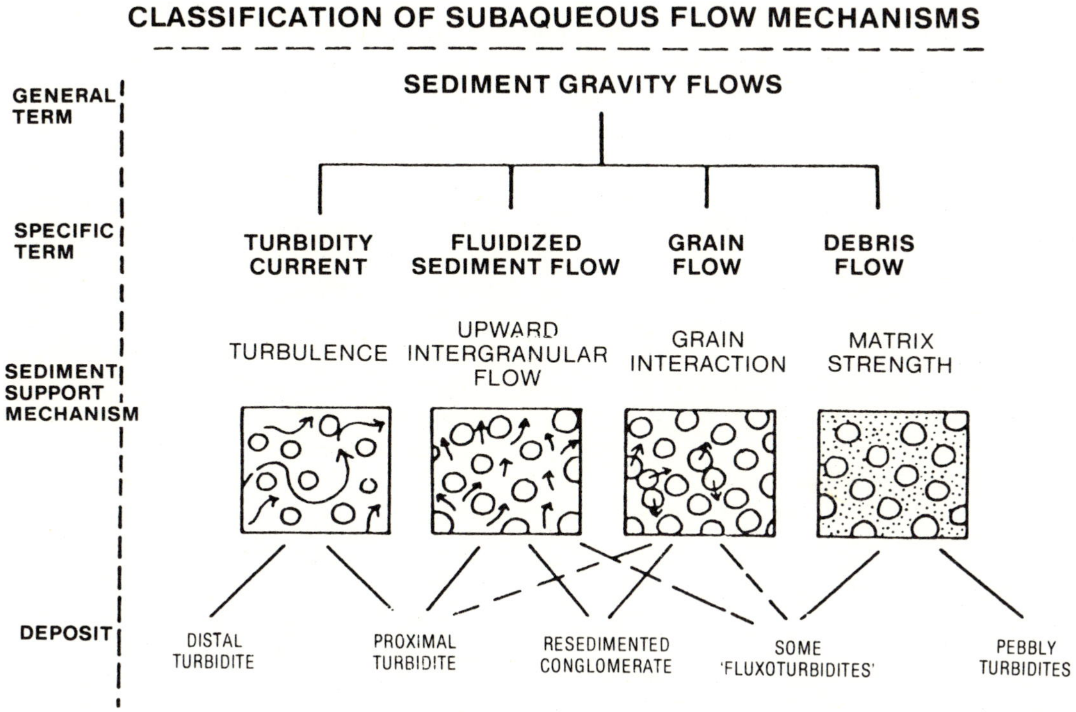

# EXPLORATION FOR OFFSHORE MARINE SANDSTONE RESERVOIRS

## CLASSIFICATION OF SUBAQUEOUS FLOW MECHANISMS

| | SEDIMENT GRAVITY FLOWS | | | |
|---|---|---|---|---|
| **GENERAL TERM** | | | | |
| **SPECIFIC TERM** | TURBIDITY CURRENT | FLUIDIZED SEDIMENT FLOW | GRAIN FLOW | DEBRIS FLOW |
| **SEDIMENT SUPPORT MECHANISM** | TURBULENCE | UPWARD INTERGRANULAR FLOW | GRAIN INTERACTION | MATRIX STRENGTH |
| **DEPOSIT** | DISTAL TURBIDITE / PROXIMAL TURBIDITE | RESEDIMENTED CONGLOMERATE | SOME 'FLUXOTURBIDITES' | PEBBLY TURBIDITES |

**Figure 7.5:** *Classification of subaqueous sediment gravity flows (Middleton and Hampton, 1973).*

## SOURCE CONCEPTS

Submarine fan deposits may be sourced from a single point, *i.e.,* a submarine canyon, or from an area, *i.e.,* a deltaic complex. Rapid sedimentation of deltaic front sands into deep water may result in over-steepening of the channel-mouth-bar at the seaward edge. Sand is put into turbid flow by earthquakes, severe storms, or excessive discharge of the river during flood stage. Velocities from a few miles per hour to fifty miles per hour have been recorded.

In contrast to a multi-source, the source of sand may be a submarine canyon. River or beach sands may be swept into a submarine canyon by longshore currents. Typically, this model is characterized by significant sea floor topography. Fans develop at the mouth of the canyon, and individual fans may coalesce at later stages which form an extensive almost submarine deltaic complex.

## INTERNAL FEATURES

**Hydrodynamic Structures** Submarine fan deposits contain a host of unique hydrodynamic sedimentary structures. They include laminations, cross beds, current ripples, and flute clasts. Flute clasts are common in the base of a flow. They are produced by loading of the overlying deposit and may be formed contemporaneous to or after deposition (Fig. 7.4). During the waning stages of a flow, current ripples are formed at the top of the sand. They are commonly destroyed by the next flow. Large scale tabular cross beds are formed during turbid deposition. They provide an excellent method to determine the direction of flow. Alternating layers of organic rich silts and "clean" sands are formed during low flow regime in the upper part of a flow. Like current ripples, they normally are scoured by the next flow.

**Soft Sediment Deformation Structures**
Rapid sedimentation produces a unique set of penecontemporaneous and early post depositional sedimentary structures. They include load casts, convolute bedding, crinkled bedding, and pull-apart structures. Tensional features that are produced by slumping sediments along bedding planes are called pull-aparts. Contortion of pelagic shale by the turbid flow form "tails" of pelagic shale that flare up into

the overlying graded bed. Fold-like structures that are composed of individual laminae in juxtaposition with parallel laminae may be formed during deposition and are typical of the upper flow regime of a turbidite. Load casts are formed by plastic deformation of soft pelagic clay under the load of a turbidite. They are most common where the sand is coarse-grained.

**Textures** Grain roundness is a function of source, and it is not significantly altered during turbid transport. Graded bedding, both upward and lateral, is one of the most diagnostic features of submarine fan deposits. Turbidites are much poorer sorted than most other sandstone reservoirs.

## EXTERNAL FEATURES

Basal contacts of turbidite deposits are sharp and commonly scoured. Upper contacts are normally gradational with deep marine pelagic shales. Typically individual sand beds vary in thickness from a few inches to tens of feet. Composite flows may obtain thicknesses of greater than 1000 feet.

These wedge-shaped deposits are in many respects the submarine equivalents of alluvial fans. The submarine fans can be divided into inner, mid, and outer fan deposits (Fig. 7.6). The inner fan is normally characterized by a concave-up profile with an irregular base that reflects rugged topography. Basal contacts are sharp. Inner fan channels, including the submarine canyon, may be filled with grain-supported and mud-supported conglomerates (Fig. 7.6I). The middle fan area is characterized by a convex-up profile. This portion of the fan may contain channel-fill deposits that coarsen upward (Fig. 7.6II). Interchannel deposits typically are composed of shale and/or siltstone. The outer fan area has a concave-up profile and a basal contact that dips uniformly into the basin. It may contain small channels (Fig. 7.6III). Lateral progradation of the fan system would produce an upward succession of outer fan, mud fan, and inner fan deposits (Fig. 7.7).

## RESERVOIR AND STRATIGRAPHIC TRAP POTENTIAL

Normally, reservoir quality of turbidites is poor to moderate. Vertical and lateral continuity normally is poor; consequently, reservoir pressures usually drop rapidly during the first year or two of production. Reservoir quality is highly influenced by grain size.

Turbidite flows are influenced by sea floor topography; therefore, "buttress" unconformities where the flow "runs up hill" against a topographic high produces a sand wedgeout and a possible trap. Truncation by shale-filled channels and canyons also produce excellent possibilities for stratigraphic traps. Where faulting intersects the sea floor, the downthrown side of the fault may localize the mass-transported sand.

## SEISMIC CHARACTERISTICS OF MARINE SANDSTONE RESERVOIRS

Basin slope and floor facies may be represented by several different sand-rich environments. Submarine fan deposits vary from fan-shaped to elongate mounds. Basal reflections generally depict downlap patterns (Fig. 7.8). Upper patterns may be concordant or truncated. Amplitude and continuity of reflections are variable. Turbidite deposits vary in form in a similar manner as fan deposits. However, turbidite deposits are more likely to parallel the axis of a basin than submarine fans (Fig. 7.8). Internal reflections normally are chaotic. Amplitude is variable. Reflections generally are very discontinuous. It is frequently difficult to document the geometric pattern of basal and upper reflections because continuity of reflections is so poor (Fig. 7.8).

Berg *(1982)* concluded that turbidite deposits may be more widely distributed than previously recognized and therefore, they may represent one of the major exploration objectives of the future. He cited numerous lithogenetic relationships between deltaic and turbidite deposits. One of these included the Cretaceous Tuscaloosa delta and turbidites in central Louisiana (Fig. 1.9). The Tuscaloosa delta prograded southward where it was stalled at the edge of the steep slope produced by the barrier reef margin of the Edwards Limestone (Cretaceous). Continued influx of terrigenous clastics produced oversteepened deltaic front sands which in turn were transported into the deep water by mass transport processes. Reflections generated by the turbidites are variable amplitude and discontinuous.

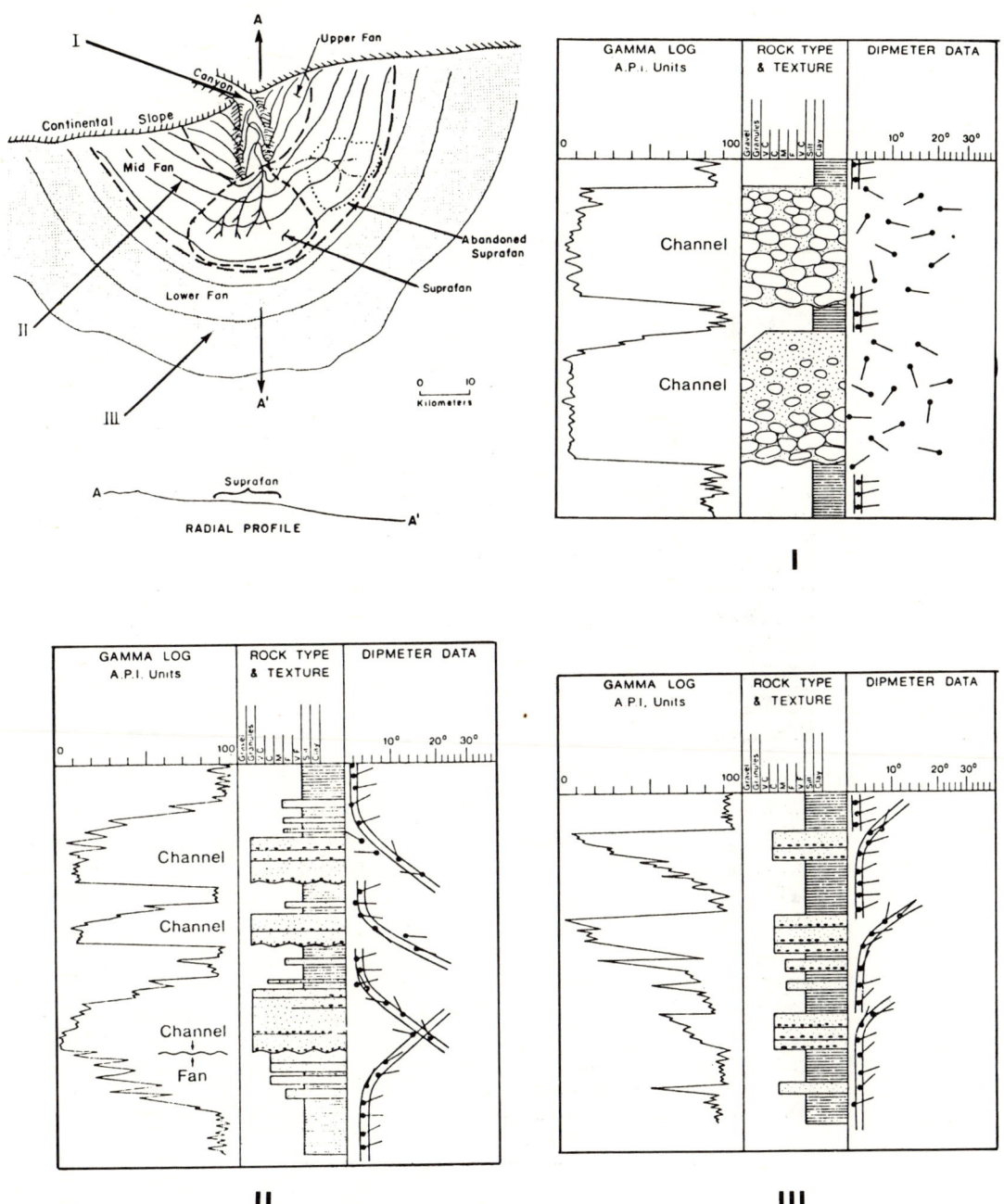

**Figure 7.6:** *Morphology and nomenclature of submarine fans (Selley, 1979).*

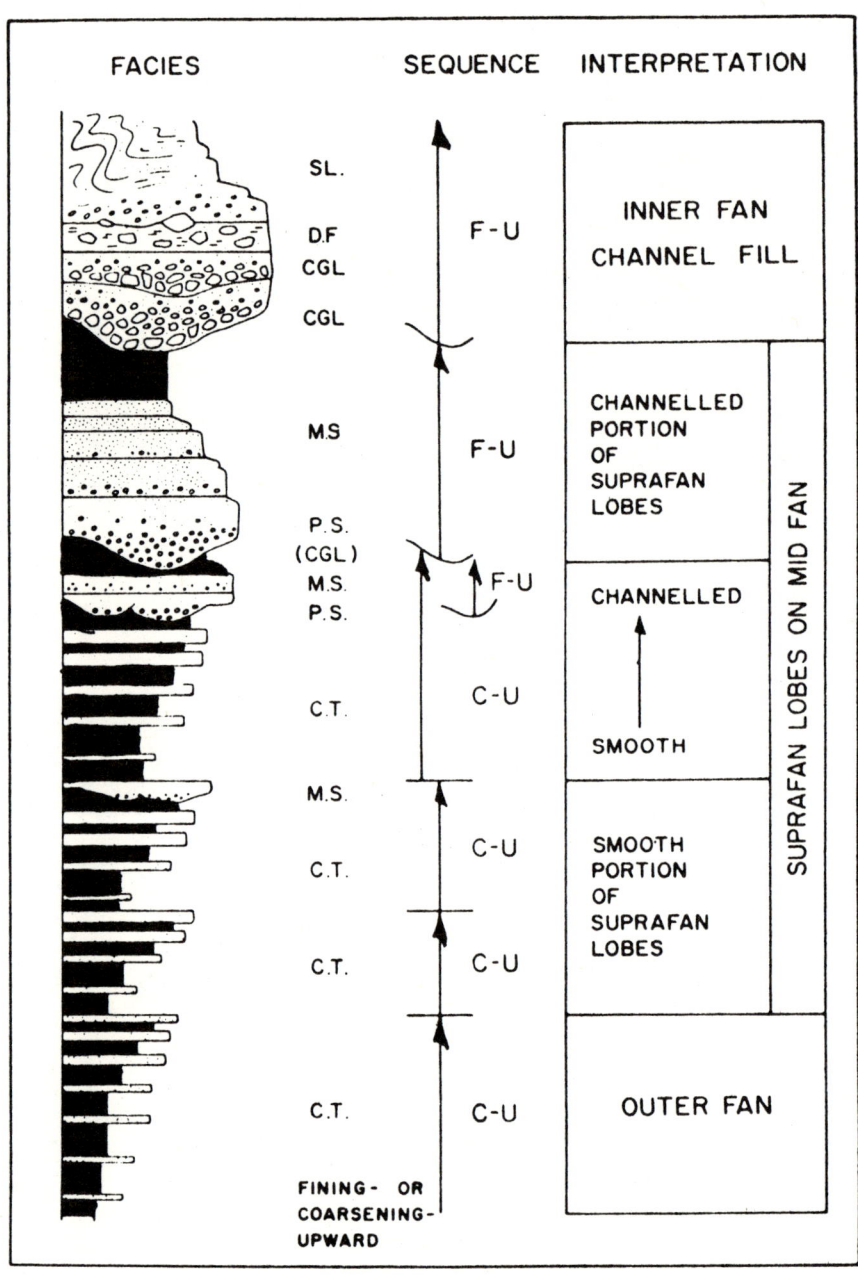

**Figure 7.7:** *Hypothetical upward succession of facies produced by a prograding submarine fan system (Walker, 1979).*

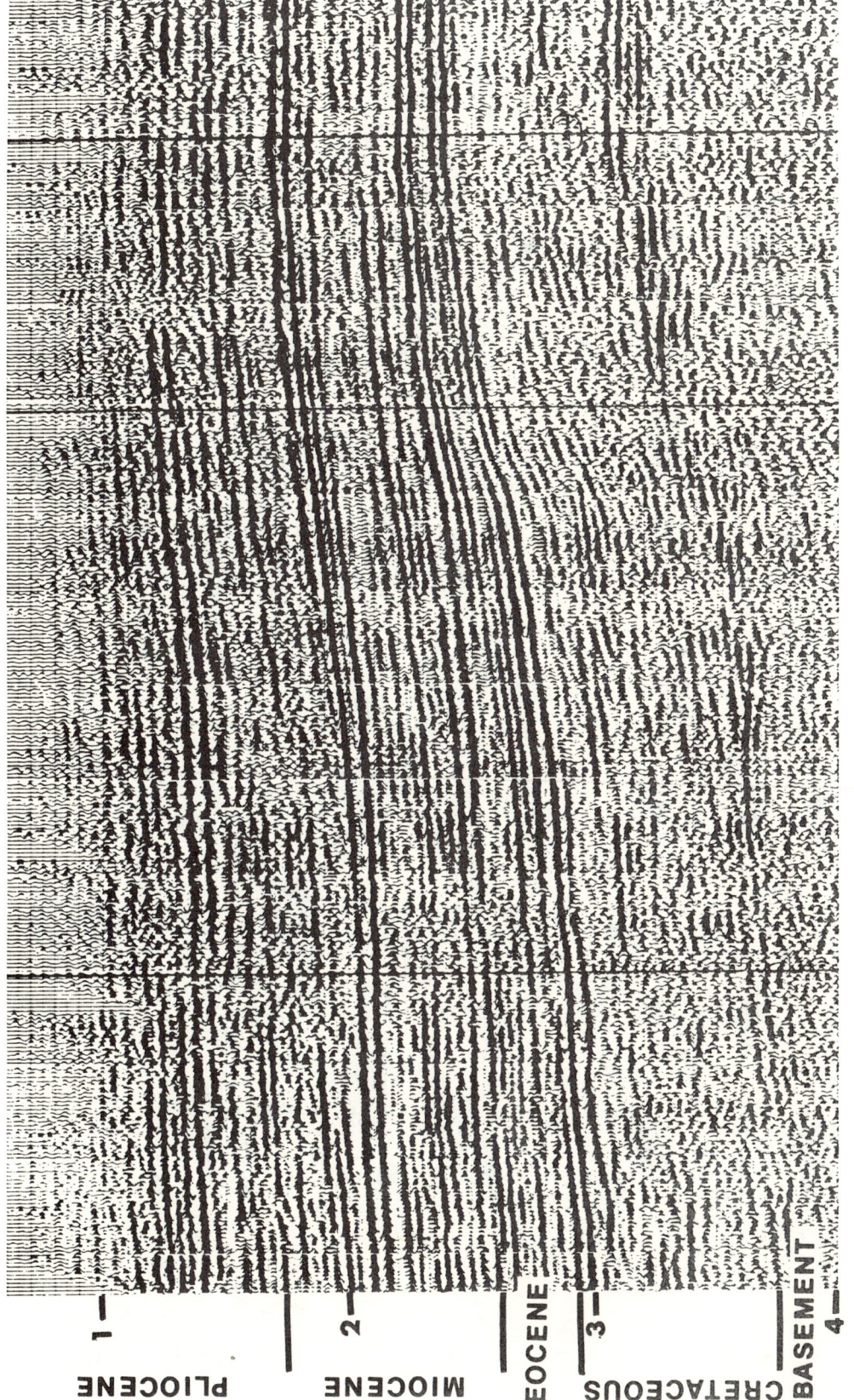

Figure 7.8: *West to east seismic section showing discontinuous variable amplitude reflectors that represent turbidite and submarine fan deposits in the San Joaquin Basin, California (Laing, 1972).*

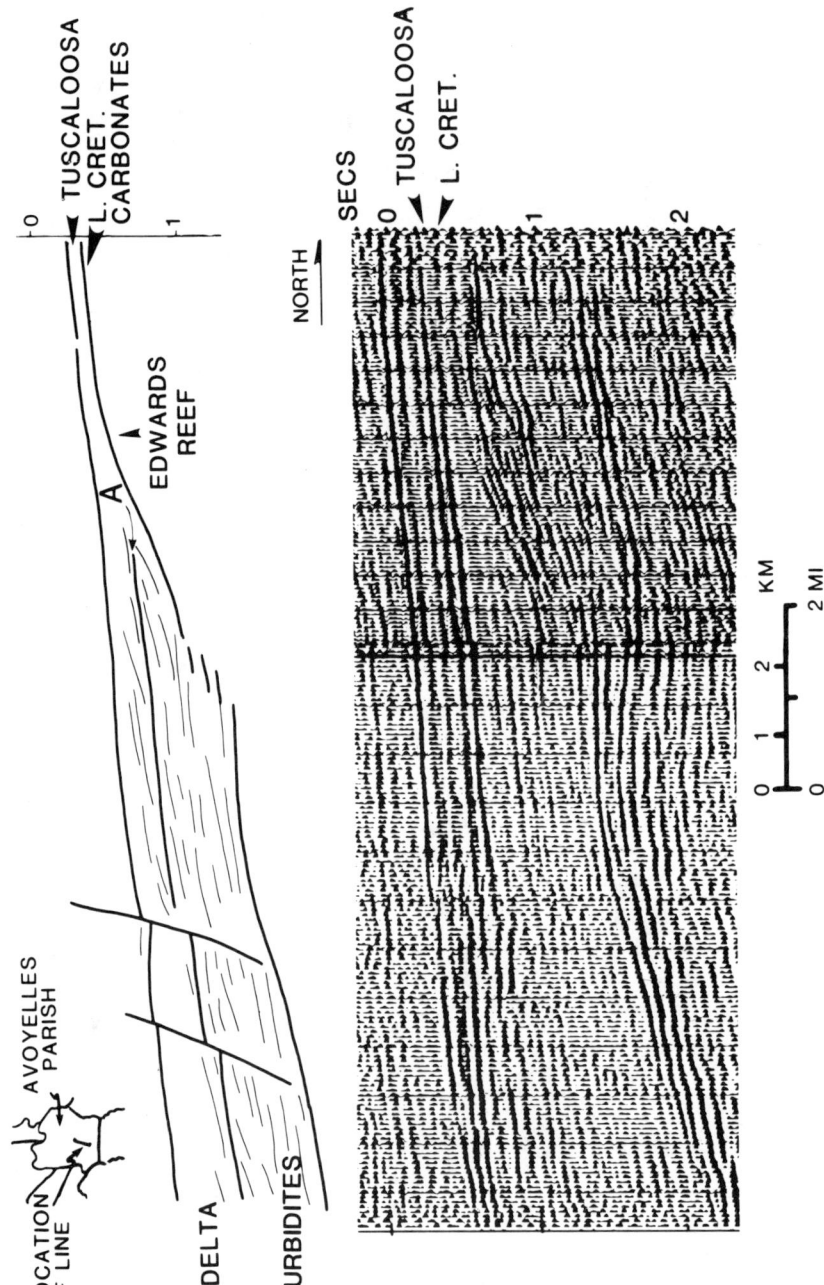

**Figure 7.9:** Seismic section in central Louisiana showing the lithogenic relationship of Tuscaloosa deltaic and turbidite systems (Berg, 1982).

# EXAMPLES OF SHALLOW MARINE SANDSTONE RESERVOIRS

## INTRODUCTION

Shallow marine shelf sandstones depict a wide range in size and shape. Some appear to be sheetlike; for example, the Viking Sandstone, Alberta *(Evans, 1970).* In other cases, the reservoirs occur in irregular-shaped "pods" as in the case of the Shannon Sandstone at Heldt Draw Field, Wyoming *(Seeling, 1978)* and the Cardium Formation at Pembina Field, Alberta *(Michaelis and Dixon, 1969).* Shallow marine sandstones can also occur as elongated belts or "shoestrings" as is the case for the Sussex Sandstone at House Creek Field, Wyoming *(Berg, 1975).* In most cases, these shallow marine deposited reservoirs produce stratigraphic traps. However, large structural-stratigraphic traps do occur; for example, the Viking Sandstone in the Suffield area, Alberta *(Tizzard and Lerbekmo, 1975).*

## HOUSE CREEK FIELD, WYOMING

House Creek Field lies on the east flank of the Powder River Basin, Wyoming. The field is about one mile wide and it is at least 28 miles long. It is a simple stratigraphic trap that was formed by the updip loss of permeability to oil in the Sussex Sandstone member of the Pierre Shale (Fig. 7.10). Average net pay is about 15 feet. More than 150 wells have been drilled to develop the more than 100 million barrels of oil that are in place *(Hobson and others, 1982).*

Hobson and others *(1982)* recognized six, generally gradational facies within the Sussex Sandstone (Upper Cretaceous). Typically the basal part of the Sussex (Facies 1) contains silty claystone that is burrowed and poorly fissile. In transitional to abrupt contact, interlaminated sandstone and claystone (Facies 2) overlie the silty claystone facies 1. This is in turn overlain by rippled, fine to very fine-grained sandstone with laminations of claystone spaced throughout the facies (Facies 3). Upper and lower contacts are transitional. Facies 4 is composed of cross-bedded sandstone that is fine to medium-grained and contains some clay clips. Intransitional con-

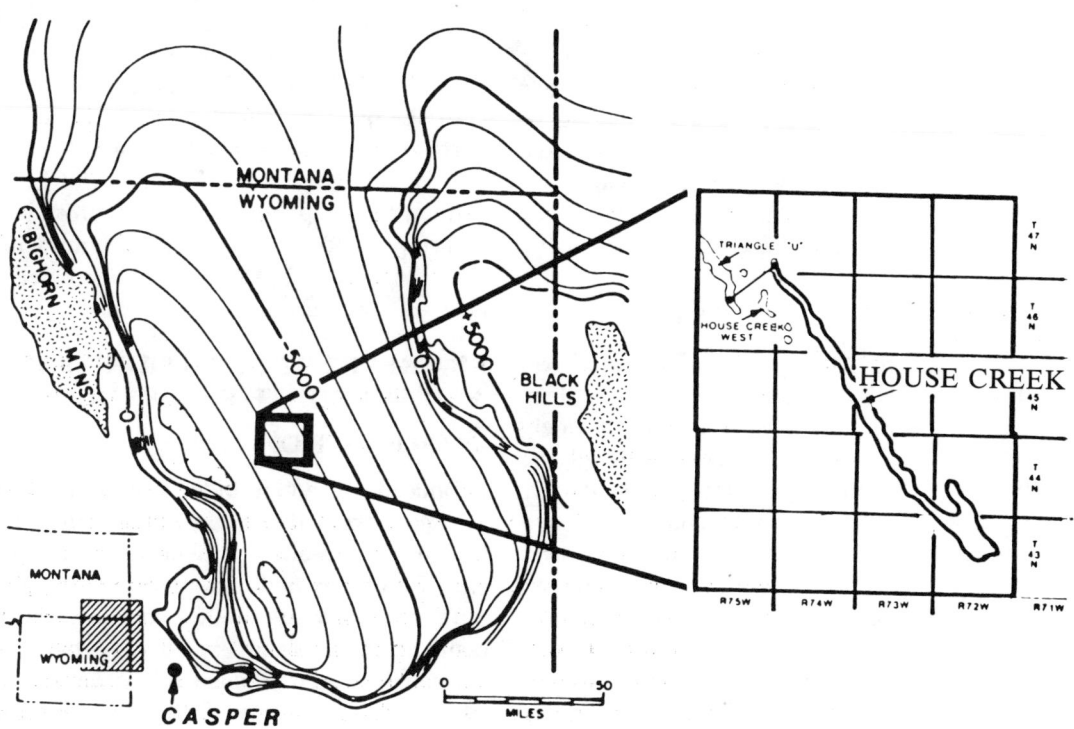

**Figure 7.10:** *House Creek Field is a stratigraphic trap that is located on the northeast flank of the Powder River Basin, Wyoming (Hobson and others, 1982).*

tact, facies 5 comprises the coarsest grain facies of the Sussex Sandstone. It is composed of pebbly, medium to coarse-grained sandstone with low angle crossbeds. Clay clips, reworked sideritic nodules, and chert grains are common. In conformable but abrupt contact, facies 6 is a highly bioturbated sandstone. It is very argillaceous and silty, and grainsize ranges from pebble to fine.

The Sussex Sandstone was deposited and re-structured during three significant stages *(Hobson and others, 1982)*. Stage one (Fig. 7.11A) was characterized by a system of wave-dominated, prograding deltas that were located along the western margin of the narrow Champanian seaway that extended from the Arctic to the Gulf of Mexico *(Asquith, 1970 and Williams and Stelck, 1975)*. Silty claystones (Facies 1) were transported into the House Creek area by river mouth bypassing, density flows, turbulent suspension, and shelf currents. During stage two, a rapid sea level rise drowned the low relief coastlines, produced sand-choked estuaries and transgression of the shoreface environments. The relic shelf sands were above wave base and were reworked into migrating sand-rich sheets. During continued transgression (stage 3) the sheets of sands were sculpted into offshore bar complexes. Although the productive limits of the field depict a shoestring sand geometry (Fig. 7.10), the sand body is much wider and downdip production is attributed to an oil/water contact within the reservoir.

## PEMBINA FIELD, ALBERTA

The principal objective of the Pembina Field discovery well, drilled in 1953, was the deep Devonian carbonate section that is extremely productive in Alberta, Canada. Although the principle objective was dry, the shallower Cardium sandstone (Cretaceous) was tested and resulted in a major discovery. Total estimated reserves are 2.2 billion barrels of recoverable oil from thin but extensive sandstones and conglomerates *(Michaelis and Dixon, 1969)*. Net pay ranges from 1 to over 40 feet. Regional dip is about 0.5° to the northeast (Fig. 7.12). The trap consists of updip (southwest) pinchout of porous sandstone-conglomerate into marine shale.

In the foothills of the Canadian Rocky Mountains west of Pembina Field, the Cardium Sandstone outcrops. On the basis of surface mapping, the Cardium was divided into three cycles of interdeltaic coastal deposits. Consequently, when the discovery well was announced, little excitement was generated by the industry because the thin sandstone and conglomerates were considered to be stray turbidite deposits encased in marine shales *(Michaelis and Dixon, 1969)*. Data obtained from subsequent drilling and core descriptions demonstrate that the geometry and internal features of the Cardium Sandstone do not support a turbidite origin for the reservoir.

The Cardium Sandstone is interpreted to have been a reworked shallow shelf, offshore marine bar system. Geometry of the sandbody is perhaps the most damaging evidence for a turbidite origin for the Cardium. The Cardium Formation contains two narrow, long, echelon sandstone bodies that extend northwesterly from Calgary for more than 120 miles *(Berven, 1966)*. In the Pembina area, three almost sheetlike sand bodies have superimposed northwest trending ridges (Fig. 7.13). Each sand body is in transitional contact with a marine shale. The basal part of each sand body is represented by thinly interlaminated, burrowed sandstone and claystones. Contacts are gradational. Each of the sand-rich bodies contain fine to pebble grainsizes, cross-beds, and ripples. Reservoir communication within each sandstone is excellent as indicated by reservoir pressure data.

## EXAMPLES OF SUBMARINE FAN AND TURBIDITE RESERVOIRS

### INTRODUCTION

Submarine fan and turbidite systems that have not been involved in intense organic processes frequently contain excellent reservoir qualities. Their proximal relationship with pelagic muds, that may be excellent source rocks, has resulted in the discovery of numerous commercial fields that produce for this depositional system. Consequently, the geologic literature contains a wealth of excellent publications that describe submarine fan and turbidite reservoirs.

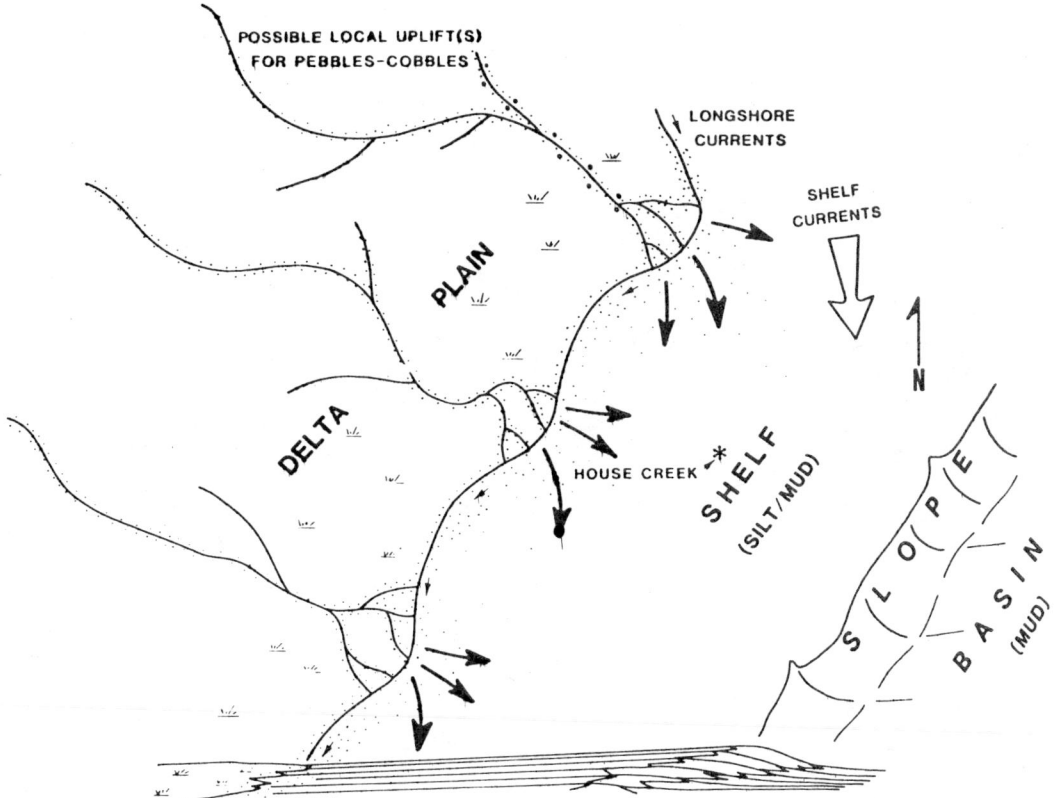

**Figure 7.11**A: *Stage I of the Sussex sandstone depositional history (Hobson and others, 1982).*

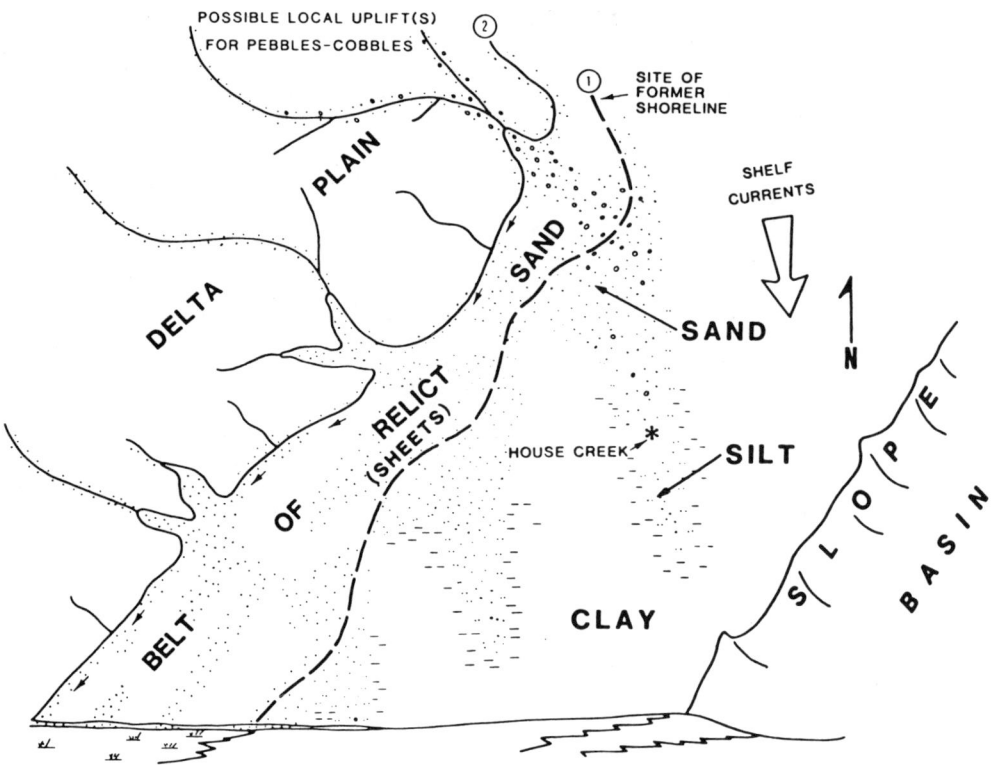

**Figure 7.11B:** *Stage II of the Sussex sandstone depositional history (Hobson and others, 1982).*

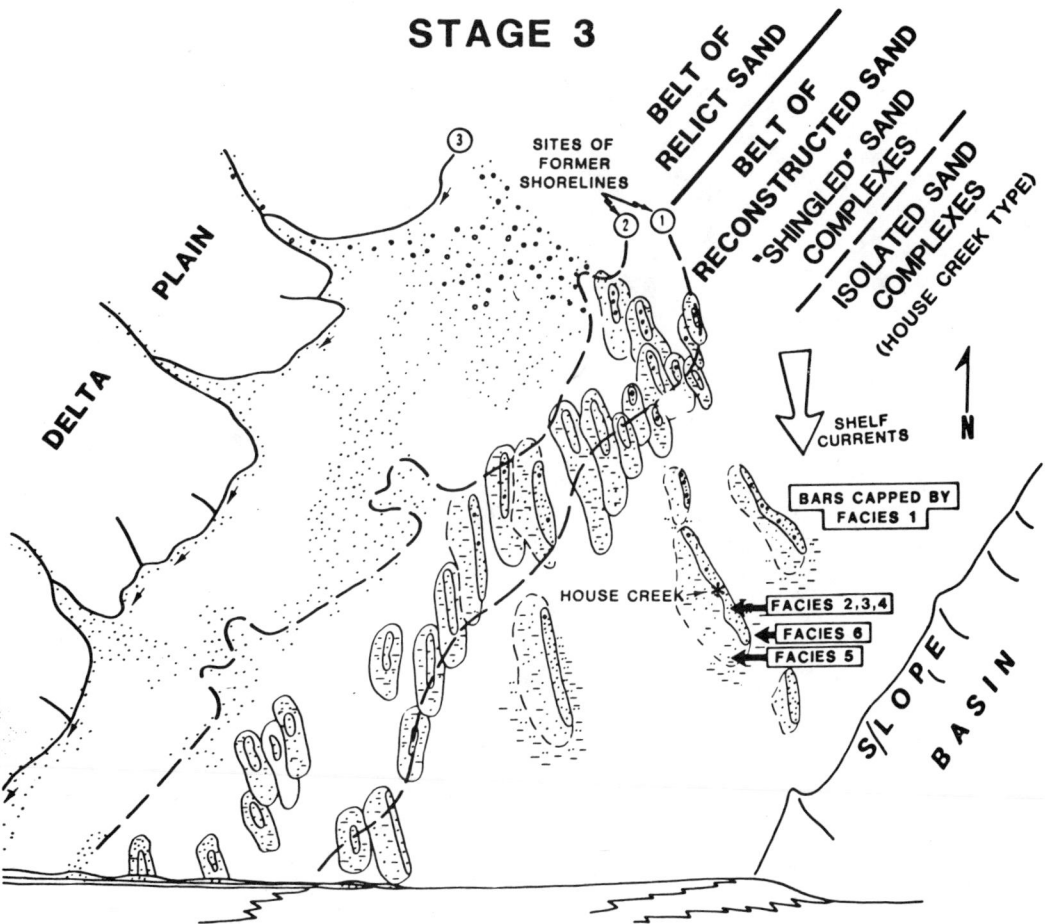

**Figure 7.11C:** *Stage III of the Sussex sandstone depositional history (Hobson and others, 1982).*

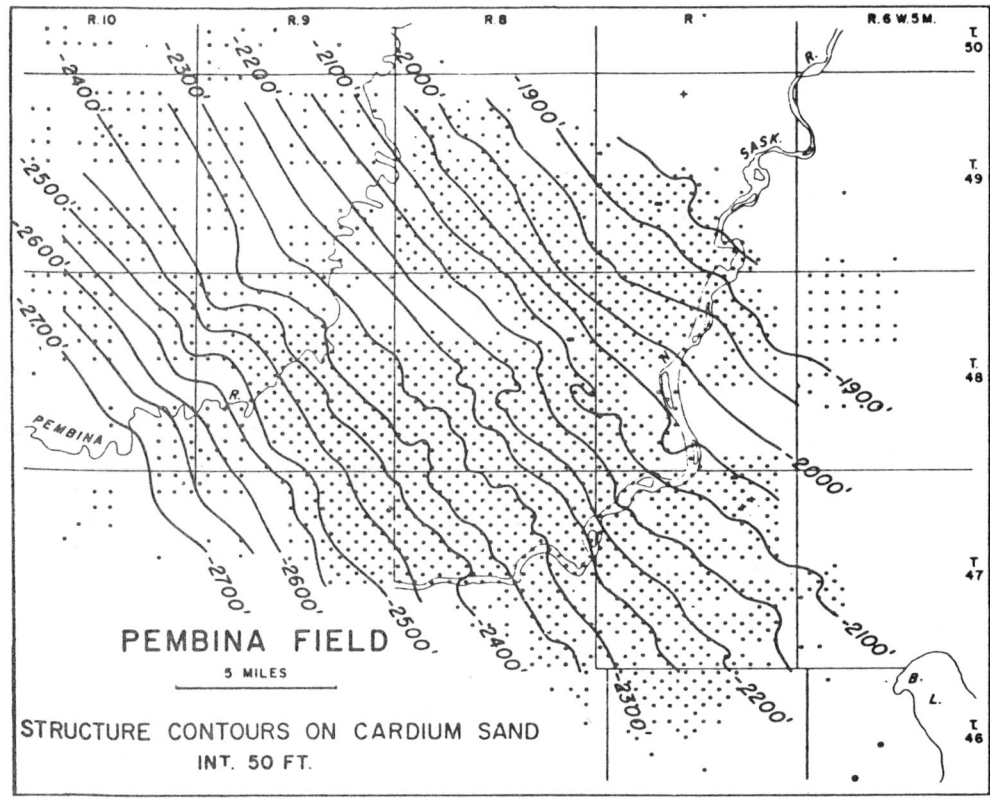

**Figure 7.12:** *Structure contour map on top of the Cardium Sandstone (Cretaceous), Pembina Field, area, Alberta (Neilsen, 1957).*

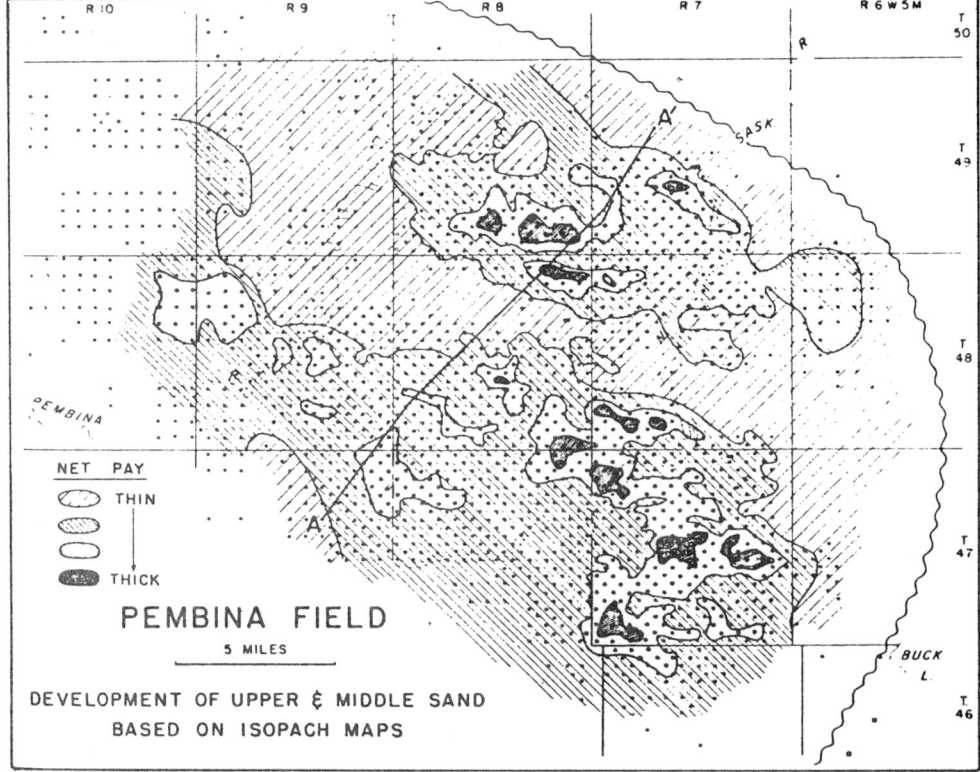

**Figure 7.13:** *Net pay isopach map of the Cardium Sandstone (Cretaceous), Pembina Field, Alberta (Neilsen, 1957).*

The downfaulted Tertiary basins of California contain nearly 30,000 feet of submarine fan and turbidite sandstones. The principal reservoirs at Rosedale Ranch and Fruitale Fields are submarine-canyon channel fill deposits *(MacPherson, 1978)*. The bulk of the reservoirs of the fields located on the Bakersfield arch are interpreted as submarine fan deposits by Sullwald *(1961)*. Webb *(1981)* reports that the best reservoirs in the Stevens sands (Miocene) are braided-channel sands that occur in deep marine fans. Walker *(1978)* interprets the lower Pliocene Repetto Formation in the Ventura fields as submarine fan deposits. Silver and Todd *(1969)* concluded that the Sprayberry and Bone Springs (Permian) sands of the Midland and Delaware Basins were deposited by submarine fan and turbidite processes. Their interpretations were confirmed by Mazzullo *(1982)*. Details of individual turbidite sequences within the mid-Pennsylvanian submarine fan at the Canyon Field, Valverde Basin, Texas, were made by Berg *(1978)*.

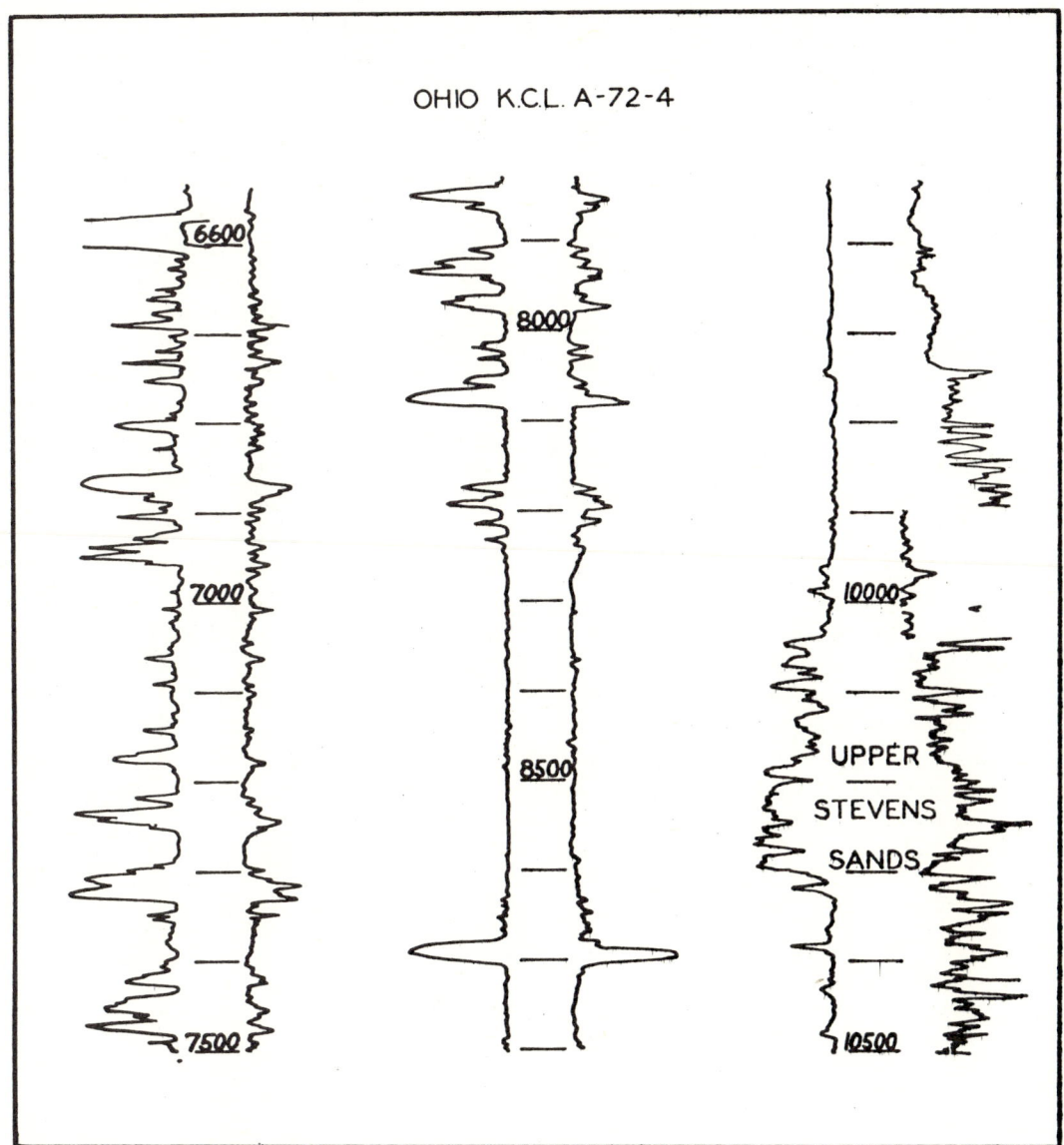

**Figure 7.14:** *Parts of the deep Paloma test in the Greely area, San Joaquin Basin, California. The left and center columns depict the log character of the shallow marine Pliocene sands and clays; the right—deep marine sands of the Stevens (Webb, 1981).*

## GREELEY FIELD, SAN JOAQUIN BASIN

Discovery of commercial quantities of oil in the Stevens sandstones date back to the 1930's. Initially these sandstones were considered to have been deposited in same spectrum of environments (deltaic, barrier-island, and tidal channels) as the shallower Pliocene sand in the San Joaquin Basin (Fig. 7.14). Perhaps Sullwald *(1961)*, based on ideas of Natland *(1957)*, was the first to propose that the Stevens sandstones were deposited in various submarine fan-turbidite environments.

The Greeley Field area represents an excellent example of the influence of depositional topography and sand distribution as a trapping mechanism. Figure 7.15 depicts a northwest plunging anticline in the Greeley Field area *(Webb, 1981)*. The Stevens thins and wedges out down-plunge thus causing a reversal of plunge from northwest to southeast on the much shallower Verdder sand. The lensing of the individual Stevens sands serve to trap oil at Greeley Field despite the lack of southeast closure (Fig. 7.16).

## VENTURA FIELD, CALIFORNIA

One of the most complete published studies of a turbidite deposited reservoir was made by Hsu *(1977)*. More than 10,000 feet of deep-water Pliocene sediments occur in the Ventura Field, Ventura Basin, California. The field is a large east-west trending anticlinal structural trap that will produce in excess of 860 million barrels of oil. Average porosities are not high, but permeabilities range up to 256md in the better sands of the Repetto Formation.

The Repetto Formation comprises more than 5,000 feet of conformably deposited sands, silts, and shales *(Hsu, 1977)*. The formation was divided into microfaunal zones that could be correlated to electric logs and three main producing zones were recognized (Fig. 7.17). Individual sands within each zone typically have sharp basal contacts. Grainsize ranges from fine to coarse. Texture varies considerably from sand to sand as well as within a single sand unit. Most sands have grain to grain contacts. Sorting varies from poor to excellent. Degree of sorting typically increases upward within each sand unit. Graded beds are normally only a few feet thick and rarely exceed more than 15 feet in thickness. Cross laminations are common, particularly in the finer-grained sands near the top of each sand body. Dip of individual cross laminae range from 15° to 20°. Geometry of individual sands is difficult to map because correlations within the field are a problem because of the number of sands and lack of shales for reference of pattern correlations. Perhaps the best mappable sand is located in the upper producing zone (AO-1 sand, Fig. 7.18). This sand is an elongate, lenticular deposit with an east-west trend that nearly parallels the anticlinal crest of the Ventura Field. Within the limits of the field, the AO-1 sand is about 7 miles long and more than 1 mile wide. It has a maximum thickness of about 80 feet and apparently thins in all directions. Geometry of the sands and their sedimentary structures lead Hsu and others *(1980)* to conclude that east to west longitudinal transport by turbidity currents was responsible for sand deposition.

EXPLORATION FOR OFFSHORE MARINE SANDSTONE RESERVOIRS 161

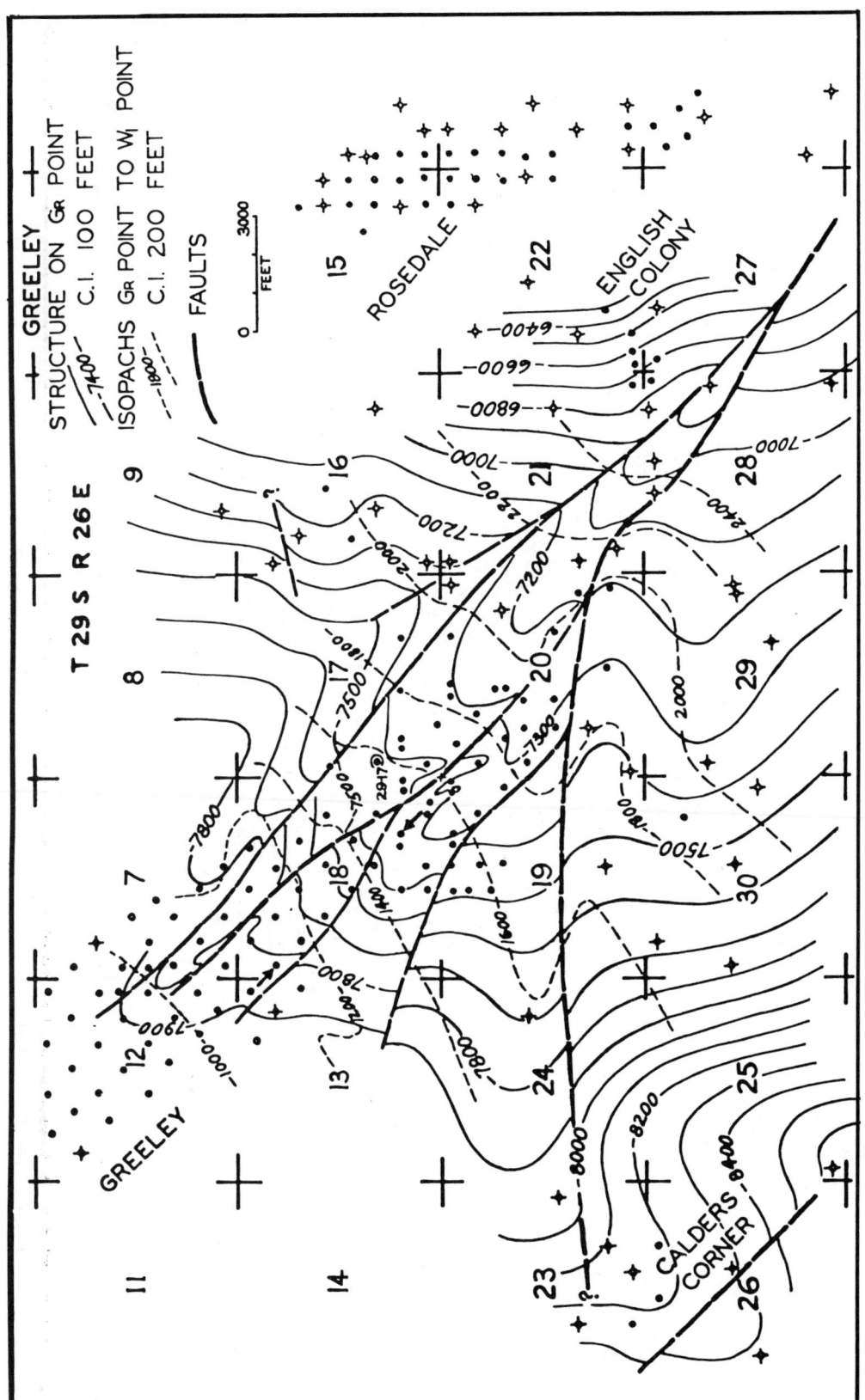

**Figure 7.15:** *Structure contours on the "Gr" point in the uppermost Stevens Sandstones, Greeley area, California. The dashed isopach contours represent thickness change in the "Gr" to W1 point. (Webb, 1981).*

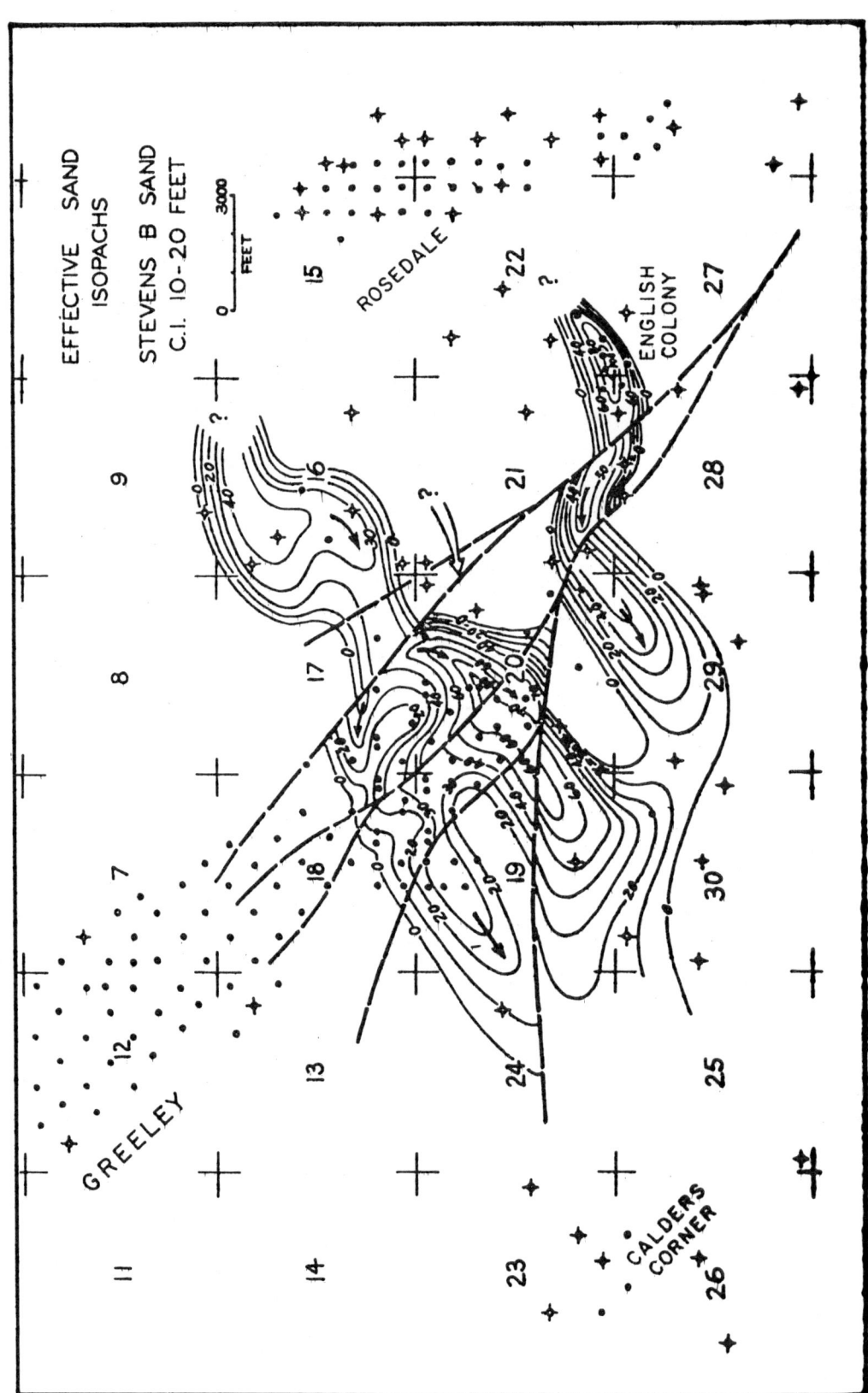

**Figure 7.16:** *B sand effective sand isopach map, Greeley area, California. The B sand produces at Greeley and English Colony fields because of lateral shale-outs of the channel sands (Webb, 1981).*

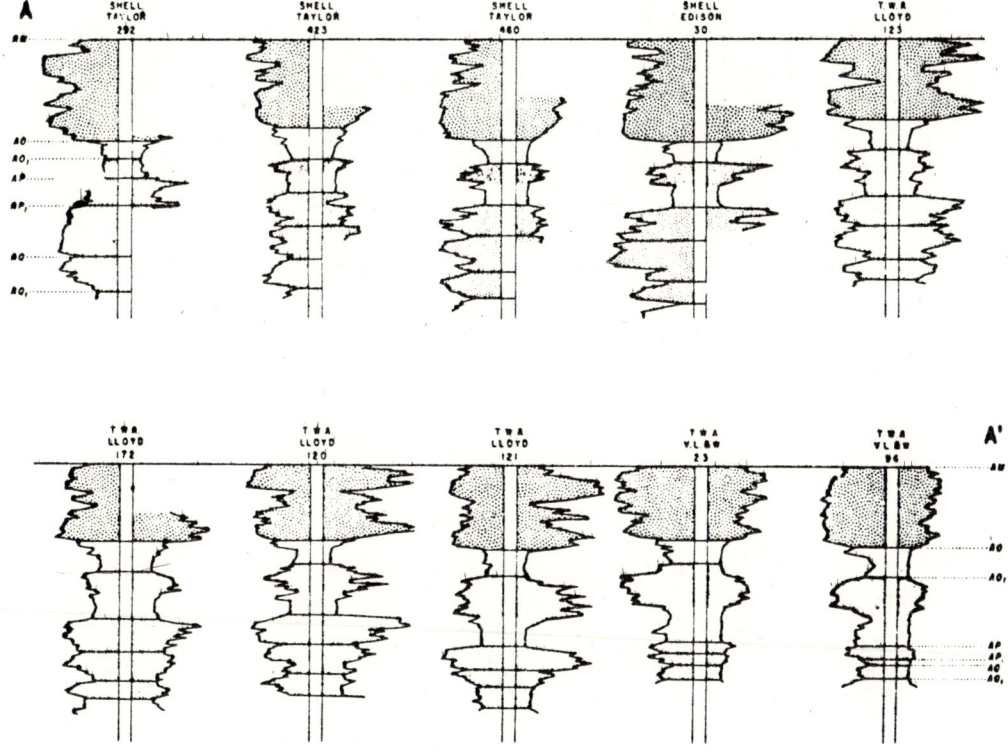

**Figure 7.17:** *Electric log correlation of upper Repetto Formation along the Ventura Anticline, Ventura Field, California (Hsu, 1977).*

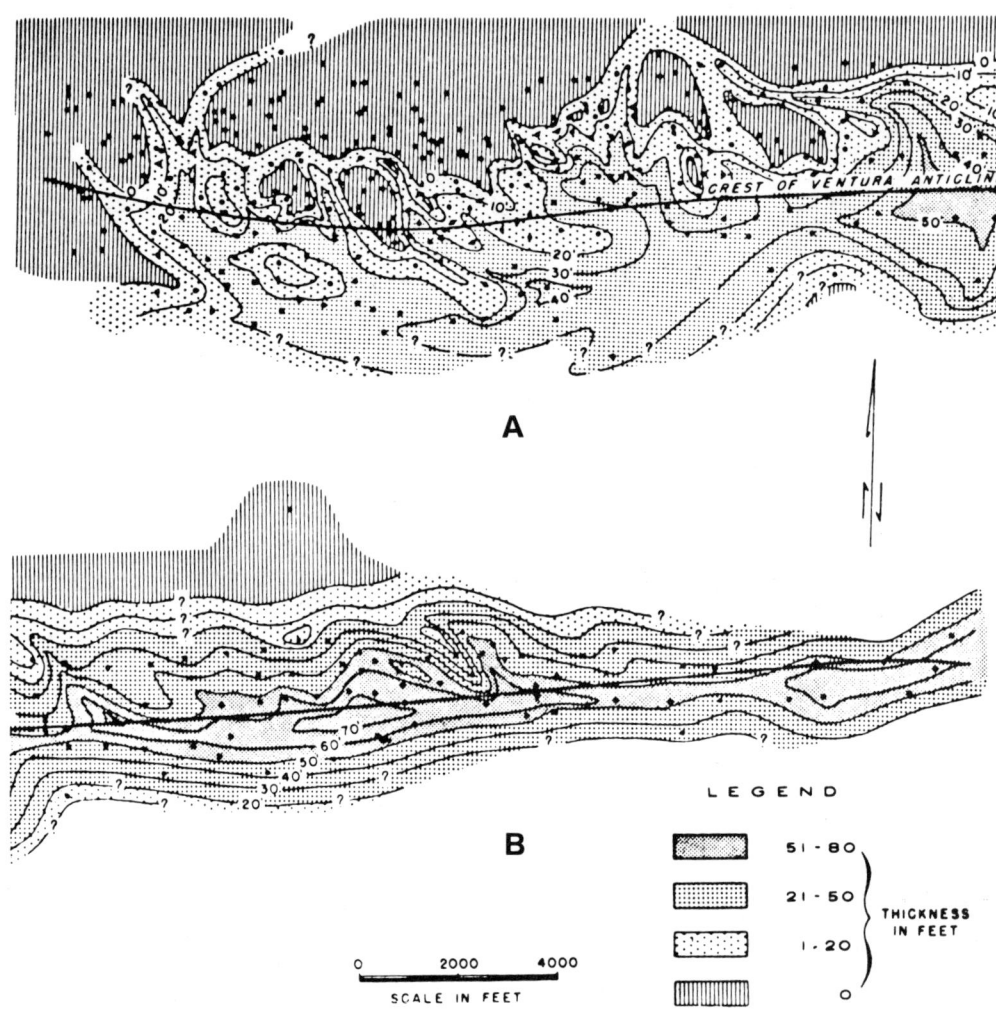

**Figure 7.18:** *Isopach map of the DA Sandstone, Repetto Formation, Ventura Field, California. Map B is the eastern continuation of map A (Hsu, 1977).*

# 8
# Diagenesis of Potential Sandstone Reservoirs

## INTRODUCTION
### NATURE OF PROBLEM

Petroleum explorationists have long since realized the need for interpreting the environment of deposition of potential reservoir strata. Consequently, they have made exhaustive studies of modern depositional models — for example, the Mississippi River delta — with the knowledge that the sediment reflects its terminal environment. Therefore, by using the principle of uniformitarianism and geology by analogy, specialists in sedimentology can apply their observations of modern sedimentary processes to ancient rocks. As a result, geologists have become reasonably confident in the interpretation of ancient sandstone sedimentary environments.

The delineation of depositional environments has not proved sufficient for the efficient exploration and exploitation of sedimentary reservoirs. For example, sandstones deposited by wind or water typically have porosities of 35% to 40% and several darcies of permeability. Sandstone reservoirs normally range in porosity from 15% to 25% with permeabilities of a few millidarcies to a few 100 millidarcies. The primary intergranular porosity and permeability are reduced by post depositional processes such as compaction, cementation, recrystallization, and replacement.

The diagenetic history — the post-depositional history of a rock — may be characterized by a number of diagenetic environments that may act on the sediment over a continuum of geologic time. Diagenetic environments pass through the sediment rather than the sediment passing through the environment. Therefore, geology by process is the best approach for interpreting the diagenetic history of sediments rather than geology by analogy.

## DEFINITION

Diagenesis includes all processes, excluding metamorphism, that act on sediments after their initial deposition. Generally, the processes operate at temperatures less than 300°C (Fig. 8.1). However, the boundary between diagenetic alteration of a sediment and mineralogical change of a rock (metamorphism) is gradational and, in many cases, difficult to differentiate.

## DIAGENETIC ENVIRONMENTS

**Marine** There are four major diagenetic environments: marine, brackish, phreatic, and vadose (Fig. 8.2). The marine diagenetic environment can be subdivided into the marine surface and marine lens environments. Accretion, burrowing, boring, corrosion, and cementation are the dominant diagenetic processes which occur in the marine lithosphere milieu. Leaching, solution, inversion of carbonate minerals, dolomitization, and cementation are the major processes which operate in this connate water environment.

**Fresh Water** The phreatic fresh water diagenetic environment is characterized by pore water that contains less than 1,000 milligrams per liter dissolved solids (a specific conductance of less than 2,000 micromhos/cm). Leaching, solution, corrosion, cementation, and fracturing may occur in this environment. This environment is associated generally with subaerially exposed strata, but phreatic lenses may extend out into a marine lens or a thin Ghyben-Herzberg lens may occur on a mid-oceanic island. Depth of the phreatic environment is variable and is dependent upon climate and communication of the potential reservoir with meteoric water.

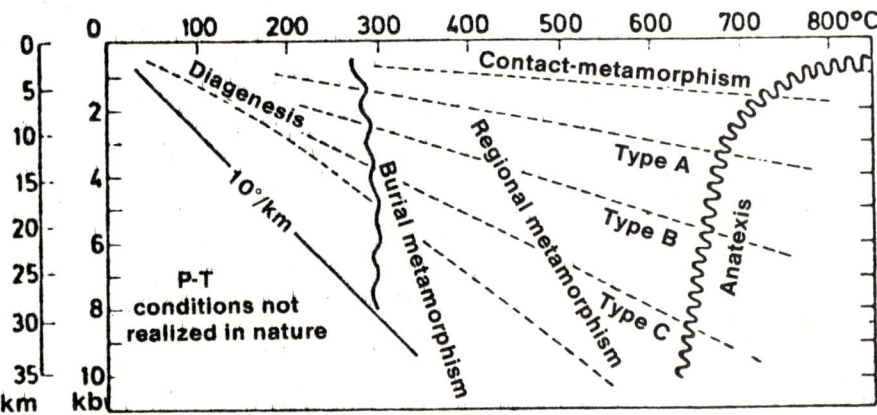

**Figure 8.1:** *Pressure-temperature ranges of diagenesis, metamorphism, and anatexis (Pettijohn and others, 1972).*

**Brackish Water** In gradational contact with the phreatic fresh water and/or marine lens diagenetic environments, the brackish environment contains pore water in excess of 1,000 milligrams per liter, dissolved solids, and a specific conductance greater than 2,000 micromhos/cm. Solution, fracturing, pressure solution, and cementation dominate the diagenetic processes of this environment. The milieu is perhaps the most aerially extensive of the diagenetic environments, but it does not necessarily alter the potential reservoir rock as much as the marine or phreatic environments. If petroleum has not migrated into the strata prior to the sediment entering the environment, the pores commonly have been occluded, which inhibits further alteration.

**Vadose** The vadose environment is defined as the zone of aeration above the ground-water table. This zone may be above fresh, brackish, or marine lens (Fig. 8.2). If the vadose zone is above the marine lens, it generally coincides with the tidal flat, patch reef, or other frequently subaerially exposed depositional environment. Such processes as leaching, solution, desiccation, internal sedimentation, and cementation characterize this early stage of vadose diagenesis. On the other hand, if the vadose zone overlies the fresh and/or brackish environments, it probably represents a late stage of diagenesis prior to weathering. Solution, crumbly fracturing, internal sedimentation, and cementation are the characteristic processes of this stage of vadose diagenesis. Exceptions to the above include point-bar and other alluvial deposits which are characterized by an early vadose environment in vertical contact with the phreatic zone. In general, the vadose environment is dominated by destructive diagenetic processes — namely, solution and crumbly fracturing. Cementation and, to a lesser extent, internal sedimentation, have been reported, but these constructive processes are rare.

## DIAGENETIC ROUTES

In an undisturbed area, the normal diagenetic route (or sequence) is characterized by the sediment passing from the marine through the phreatic, brackish, and vadose environments (Fig. 8.3). This route requires continuous subsidence until the brackish environment is reached, followed by epeirogenic uplift until the strata enters the vadose environment. Numerous interruptions to the normal route are possible. Perhaps the most common alternate route is associated with the various alluvial and tidal flat depositional models where the vadose and phreatic, or vadose and marine lens environments frequently exchange positions. Other alternate routes include marine to fresh and marine to brackish.

# DIAGENESIS OF POTENTIAL SANDSTONE RESERVOIRS

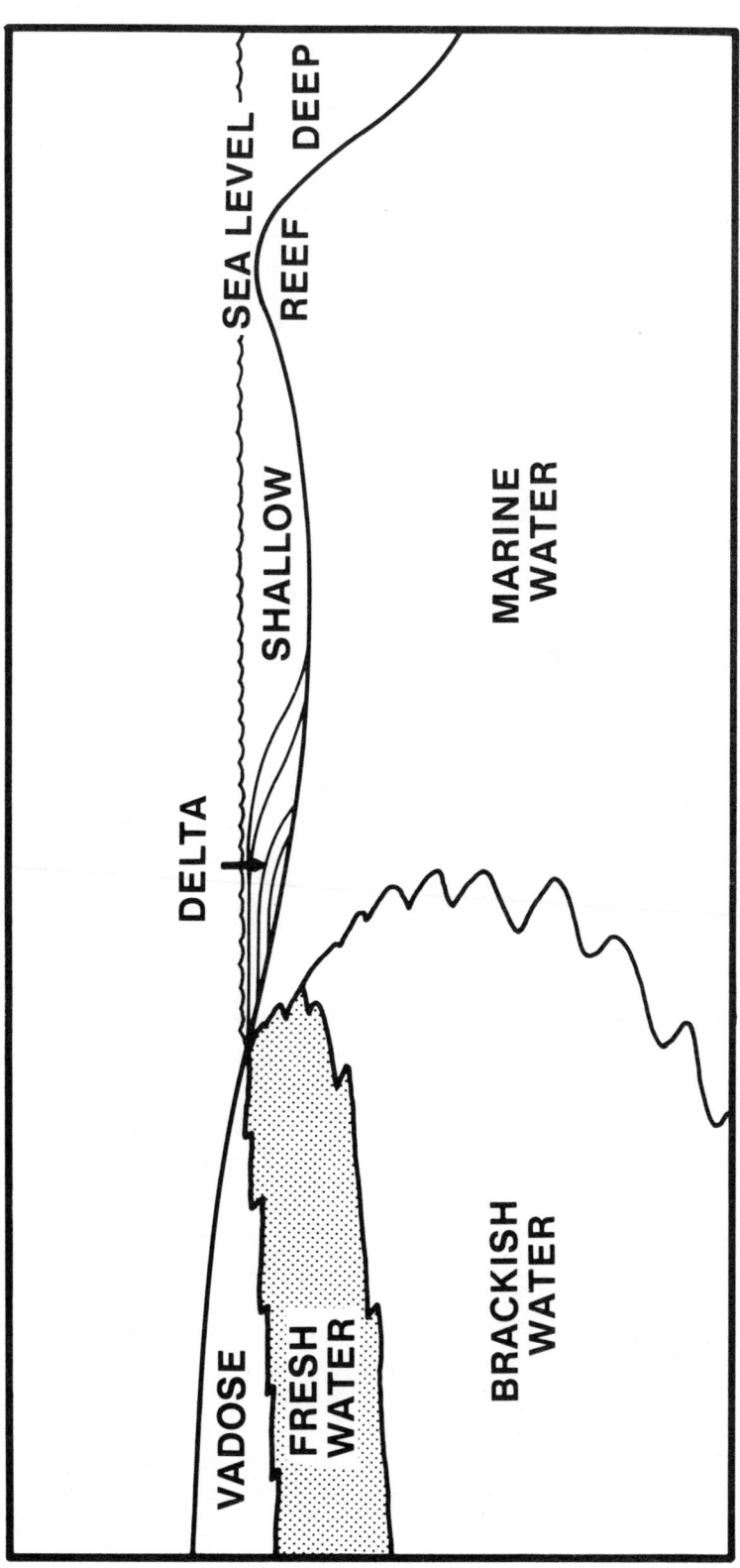

**Figure 8.2:** *Diagenetic environments (Silver and Wermiel, 1976).*

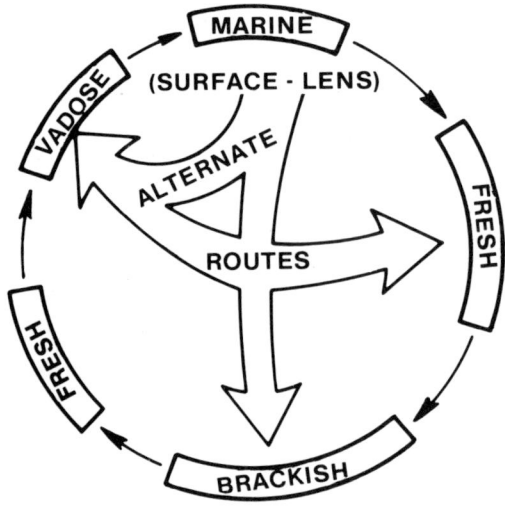

**Figure 8.3:** *Diagenetic environments pass through sedimentary rocks (Silver and Wermiel, 1976).*

## PROCESSES OF PORE OCCLUSION

### PROGRAMMED PORE HISTORY

The pore history of a sandstone is programmed in part by prediagenetic history of the provenance, depositional environment, and tectonic setting. Figure 8.4 depicts the pathway to porosity destruction as a function of detrital

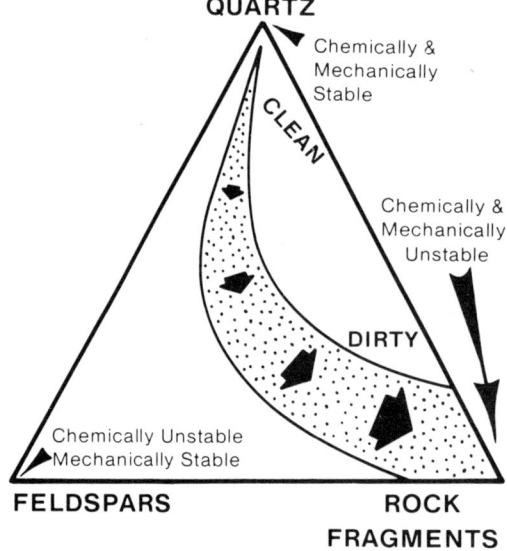

**Figure 8.4:** *The pathway of porosity destruction as a function of mineralogy (Schmidt and McDonald, 1979).*

sand composition. Dirty and feldspar rich sands, as dictated by the provenance, rarely contain reservoir quality primary porosity and permeability in the subsurface. Rapid depositional rates reduce the winnowing effect of water; therefore, clay content may be high. Once again, it is the rare exception that a clay-rich sandstone contains sufficient primary porosity and permeability in the subsurface to be a reservoir rock (Fig. 8.5). Tectonics highly influence the rate of burial. Figure 8.6 illustrates the effect of depth of burial on porosity.

### CHEMICAL PROCESS

The processes of pore occlusion that effectively destroy primary porosity and permeability of sandstones can be classified as chemical, mechanical, and combination. Chemical processes include cementation, replacement, and recrystallization.

**Cementation**   Silica and calcite are the two most important pore-filling cements in sandstones. Dolomite, siderite, various zeolites, sulfates, and halides constitute the other cements. Silica is typically a late stage cement, whereas the other cements normally are early stage cements (Fig. 8.7). Ordinary quartz cement forms as an overgrowth on framework quartz grains, the cement always develops crystal facies if pore space permits, and it is in crystallographic continuity with the quartz framework grains (Fig. 8.8). The initial quartz cement commonly nucleates at several positions. The mini-crystals enlarge and merge to form straight edge prisms and rhombohedral crystal facies. If space permits, the rhombohedral facies grow faster, overtake the prism facies, and a hexagonal bipyramid results *(Waugh, 1970)*. Table 8.1 summarizes the possible sources for silica to form the quartz cement.

**Dissolution of Quartz Grains in Sandstone**
Calcite is the second most common cement in sandstones. There are important differences in the mineralogy and texture between Quaternary and ancient calcite. Some differences reflect the diagenetic changes where unstable phases of $CaCO_3$ invert to or are replaced by calcite, whereas the reasons for other differences are as yet unknown. Calcite, Mg-calcite, and aragonite occur in Holocene sands, but

# DIAGENESIS OF POTENTIAL SANDSTONE RESERVOIRS

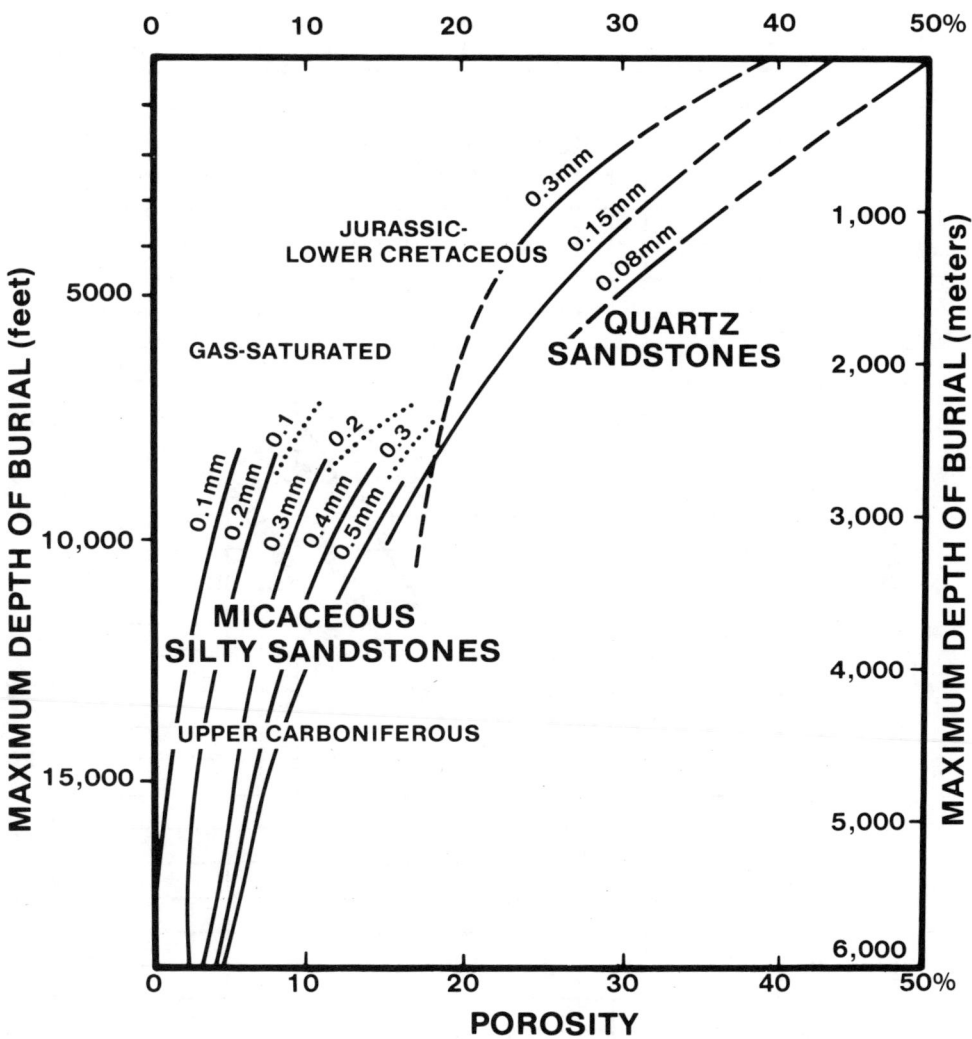

**Figure 8.5:** *Decrease of sandstone porosity with increasing depth (Fulchtbauer, 1974).*

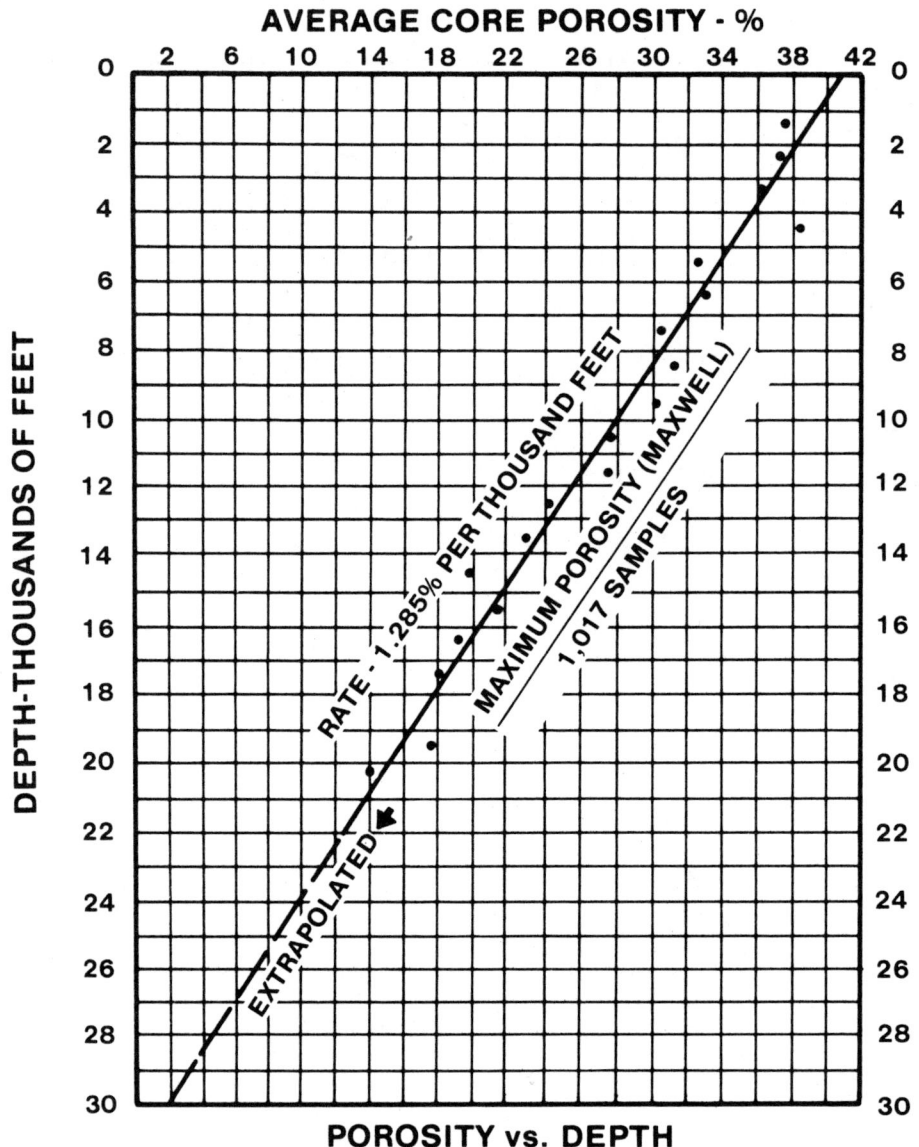

**Figure 8.6:** *Porosity vs. depth plot of 17,367 samples from south Louisiana cores (Atwater and Miller, 1965).*

# DIAGENESIS OF POTENTIAL SANDSTONE RESERVOIRS

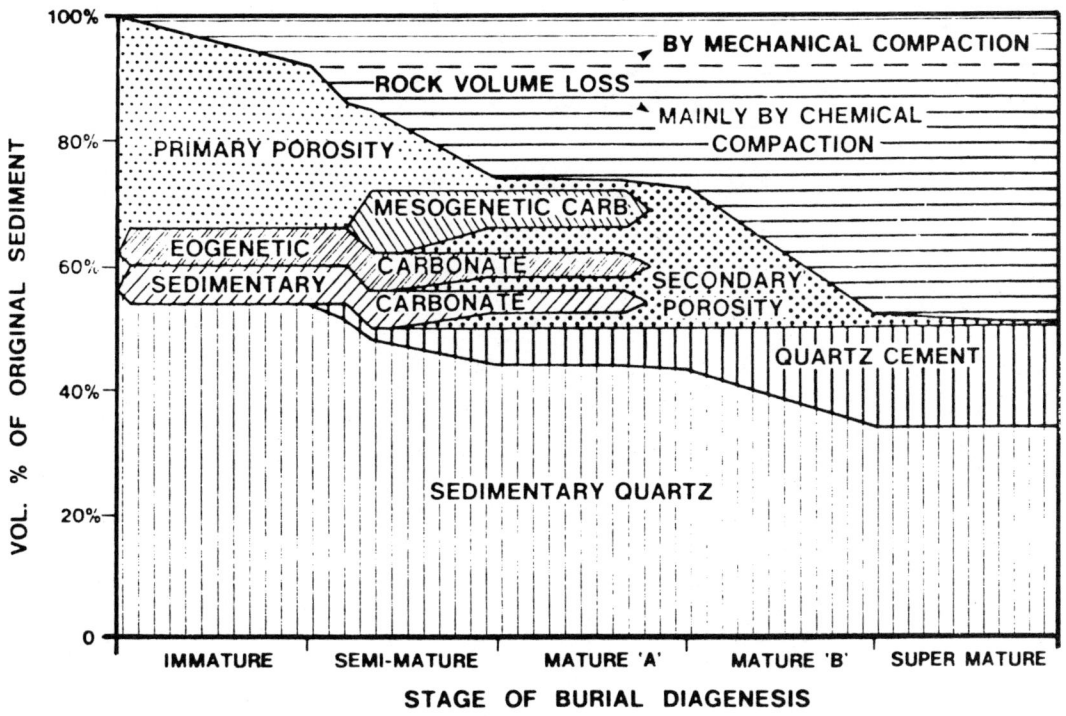

**Figure 8.7:** *Burial diagenesis of a quartz arenite (Schmidt and McDonald, 1979).*

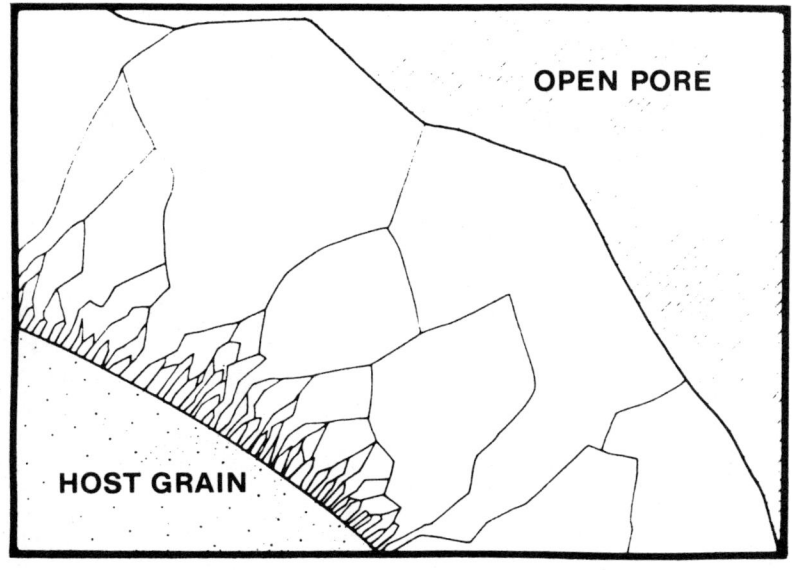

**Figure 8.8:** *Progressive growth in crystal size as silica precipitate grows outward from host grain into pore (Pettijohn and others, 1972).*

## TABLE 8.1

### Sources of Silica for Quartz Cements

Pressure solution of quartz framework grains

Clay mineral alterations during burial

Replacement of quartz and silicate grains by carbonate

Dissolution of opaline skeletal grains

Kaolinization of feldspar

Dissolution of quartz silt grains in shale

Ground water

Expulsion of shale pore water

Sea water

Hydrothermal water

---

only calcite is present in ancient sands. Ancient sands are cemented chiefly with spar crystals in mosaic fabric such that from one to six crystals fill individual pores. Skeletal grains, sea water, connate water, mixing of ground water, dewatering of shales, and by-products of redox reactions are possible sources of calcite as cementing agents in sandstones.

**Replacement** Replacement is the solution of one mineral and precipitation of another mineral in its place on a volume-for-volume or molecule-for-molecule basis. Replacement occurs when pore fluids that are not in equilibrium with the solids in the sandstone promote solution-precipitation reactions. Replacement of zeolites, feldspars, and clay minerals can result in an increase in volume which, in turn, occludes pore space.

**Recrystallization** Recrystallization is a "catch-all" term used to describe the formation of new, crystalline mineral grains in a rock. Originally the term was limited to essentially a solid state process, but has been expanded to describe wet processes. Recrystallization of feldspar, clay, halide, and carbonate grains frequently results in an increase in volume and a consequent decrease in porosity and permeability.

## MECHANICAL PROCESSES

Compaction is an important mechanical process that results in porosity and permeability reduction. Rotation of platy grains, deformation of ductile grains, and breakage of brittle grains can alter the texture and fabric of a rock such that reservoir quality is reduced (Fig. 8.9).

## COMBINATION PROCESSES

**Biotite** Early in the diagenetic history of a sandstone, biotite alters to fine-grained carbonates and hydrous clay minerals *(Hayes, 1979)*. The increased volume results in "squeezing" the clay globs between the rigid grains.

**Pressure Solution** Pressure solution refers to the dissolution of minerals under lithostatic pressure. Solution of framework grains takes place at grain-to-grain contacts (Fig. 8.10). The loss of quartz at the points of contact of the subspherical grains causes their centers to move closer during pressure solution which results in a loss of pore space. Rittenhouse *(1971)* computed the loss of porosity with several different packing patterns (Fig. 8.11).

The material dissolved during pressure solution may be removed by ground water or precipitated in the pores where pressure is

# DIAGENESIS OF POTENTIAL SANDSTONE RESERVOIRS

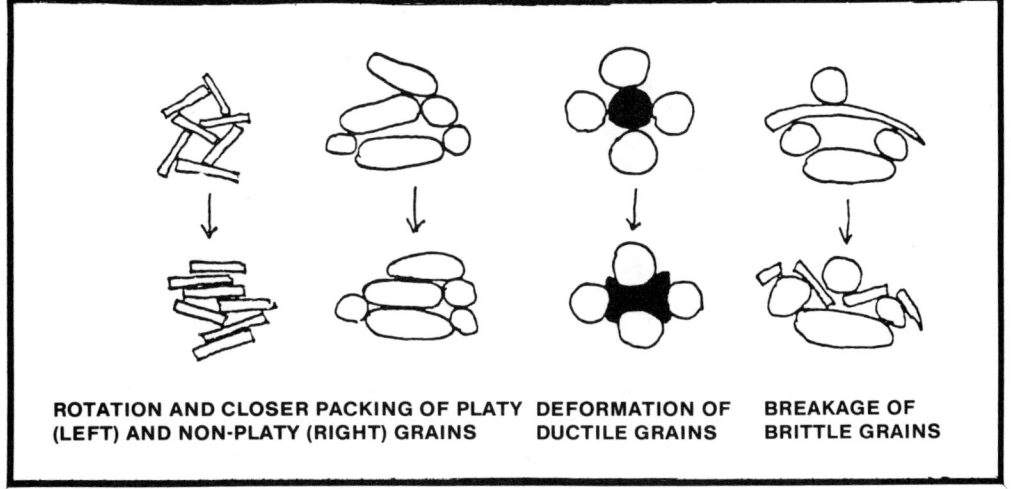

**Figure 8.9:** *Mechanics of compaction (Jonas and McBride, 1977).*

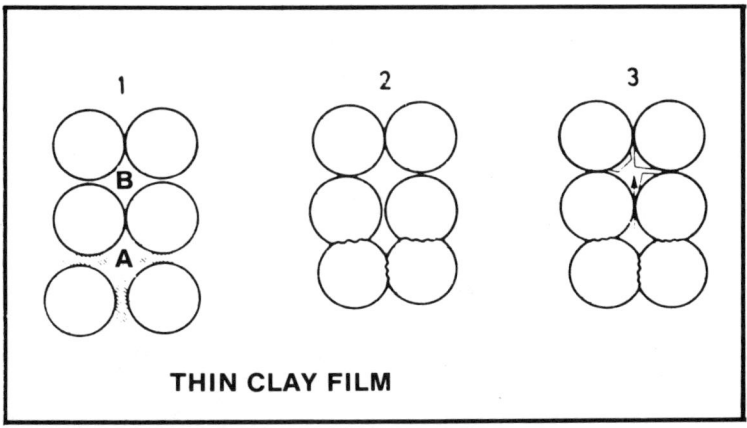

**Figure 8.10:** *Pressure solution occurs at grain-to-grain contacts (Pettijohn and others, 1972).*

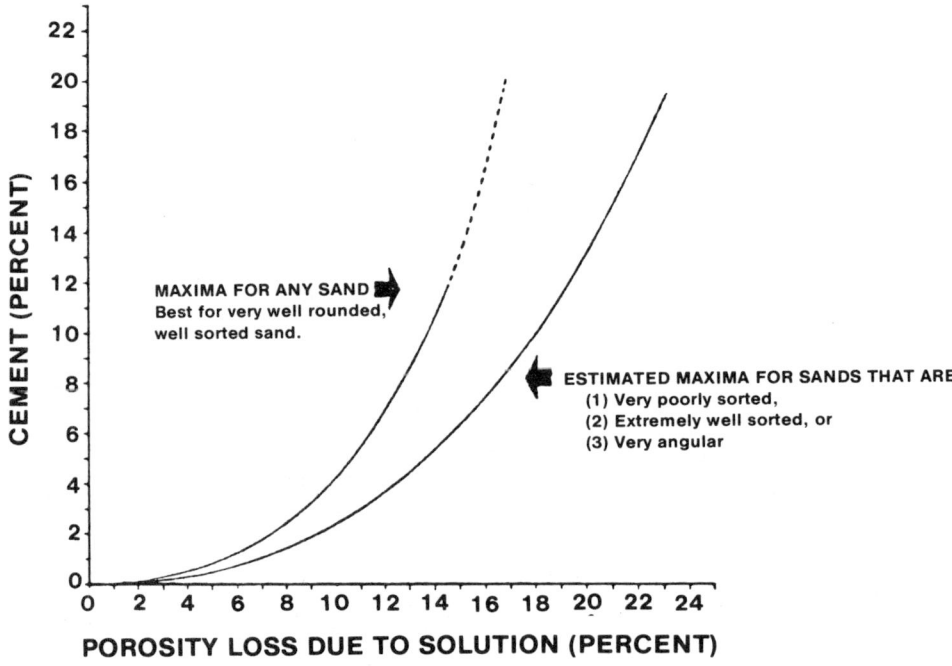

**Figure 8.11:** *Maxima cement for different amounts of porosity loss due to solution for sands (Rittenhouse, 1971).*

less. In the latter case, pore space reduction is not only a result of packing, but also of cementation. Calcite, dolomite, feldspars, and quartz are subject to pressure solution.

## PROCESSES OF SECONDARY POROSITY FORMATION

### INTRODUCTION

Prior to 1960, geologists considered that the vast majority of porosity in sandstone reservoirs was primary in origin (Fig. 8.12). The late seventies saw nearly a 100% turnaround. It is now recognized that secondary porosity in sandstones is real, important, and perhaps in the future, it will be considered to account for the bulk of the porosity in sandstone reservoirs. Porosity enhancement occurs by fracturing, shrinkage, and dissolution (Fig. 8.13).

Destruction of primary porosity initiates during early burial of the sandstone (Fig. 8.14). The depth at which complete destruction occurs is dependent upon the mineralogy of the sandstone, migration of pore waters, rate of pore water recharge, chemistry of the pore waters, and rate of burial. Formation of secondary porosity may occur prior to or after the primary porosity has been destroyed. Continued burial results in occlusion of secondary porosity by cementation, particularly by quartz cements.

### BASIC TYPES OF SECONDARY POROSITY

**Fractures** Fracturing of rocks, grains, and intergranular constituents is not an important process in creating secondary porosity *(Friedman and Stearns, 1971)*. In some cases, particularly in matrix rich sandstones, fracturing can be a major porosity type. Also, if the fractures are held open by later precipitation of quartz cement, pores can exist in the fracture *(Pittman, 1979)*. Fracturing does, however, enhance permeability *(Friedman and Stearns, 1971)*.

**Shrinkage Voids** Like fracturing, shrinkage of clay minerals in the matrix is not generally an important secondary porosity mechanism. Shrinkage, due to dehydration of various clay minerals, probably occurs around 2500 to 3000 feet *(Pittman, 1979)*.

# DIAGENESIS OF POTENTIAL SANDSTONE RESERVOIRS

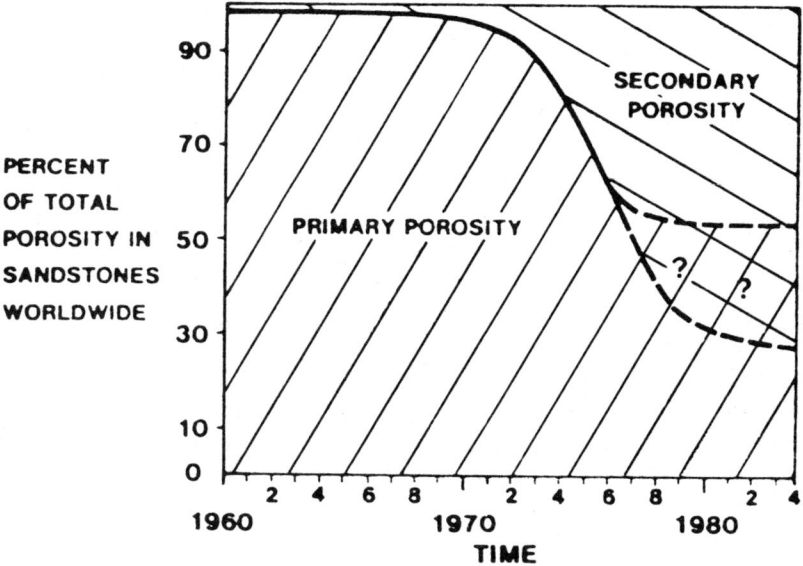

Figure 8.12: *Change of interpretation of nature of sandstone porosity (Schmidt and McDonald, 1979).*

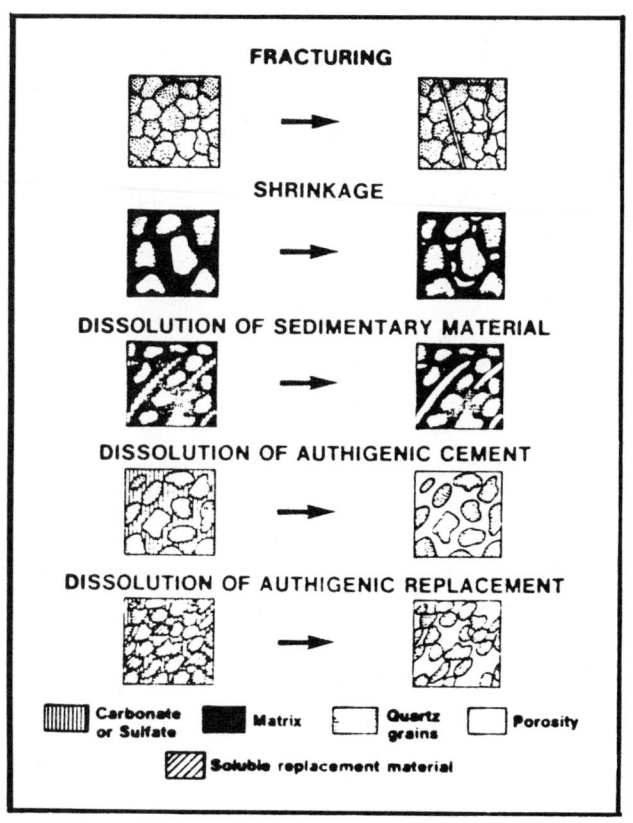

Figure 8.13: *Textural origin of secondary sandstone porosity (Schmidt and McDonald, 1979).*

**Dissolution of Authigenic Cements** Dissolution of authigenic cements is perhaps the most important process of secondary porosity development. Of particularly importance is the dissolution of carbonate cements such as calcite, dolomite, and to a lesser extent, siderite. To date, dissolution of silica cements is not considered to be an important process *(Schmidt and McDonald, 1979)*.

*Chemical Processes:* Dilution of interstitial water or changes in the ratios of ion species may cause dissolution. Dissolution can also be caused by the formation of carbonic acid in the interstitial water. It produces a lower pH and dissolution of carbonate minerals occurs. Interlayer water that is released from montmorillonite during its transformation to illite, and $CO_2$ originating from oxidation of organic matter, can cause subsurface brines to become so undersaturated in respect to carbonate minerals that dissolution occurs *(Savkevic, 1971)*.

*Physiochemical Processes:* Dissolution can be caused by changes in temperature and pressure if the salinity and ionic ratio of the interstitial waters remain constant. Pore waters that move from zones of hydrostatic pressures into zones of geopressures may become undersaturated with respect to carbonates because of the pressure increase. This can produce dissolution *(Schmidt and McDonald, 1979)*.

**Dissolution of Authigenic Replacements**
Where calcite, dolomite, or siderite have replaced feldspars or lithic rock fragments, dissolution of the carbonate minerals can produce secondary porosity. The causes of dissolution are the same as those described for dissolution of authigenic cements.

## CONCLUSIONS

The timing of generation-expulsion and migration of hydrocarbons relative to destruction of primary porosity in sandstones is of ultimate importance. It is probable that future research will demonstrate that only in abandoned rift zones, such as the Sirte Basin, or other thermally hot areas, can generation occur early enough to preserve primary porosity. It is further probable that the future research will demonstrate that normal burial diagenesis of source rocks will result in the generation of hydrocarbons too late for the preservation of primary porosity, and therefore the timing of generation-expulsion and migration relative to the formation of secondary porosity is critical. Regional isopach and isocron maps will become increasingly important in order to delineate trends that relate generation-expulsion and migration to secondary porosity development.

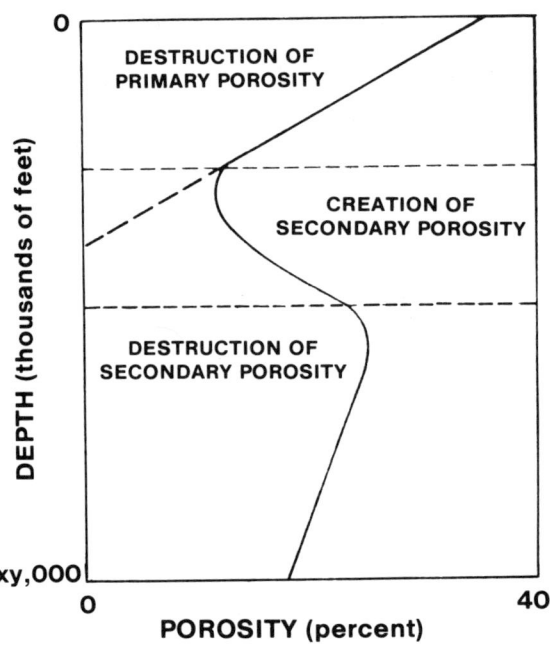

**Figure 8.14:** *Depth distribution and evolutionary sequence of primary and secondary porosity in sandstones (Hayes, 1979).*

# PART III:
# EXPLORATION FOR CARBONATE RESERVOIRS

- *9   Overview of Carbonate Fundamentals*
- *10  Carbonate Depositional Environments and Stratigraphy*
- *11  Diagenesis of Potential Carbonate Reservoirs*
- *12  Reservoirs and Carbonate Traps*

# 9
# Overview of Carbonate Fundamentals

## NATURE OF PROBLEM

Geoscientists who specialize in carbonate geology are at least a decade behind in their understanding of their speciality than those who specialize in sandstone geology. This is in part attributed to the fact that major petroleum reserves were discovered in carbonate reservoirs 20 to 30 years after giant oil fields had been put on stream from sandstone reservoirs. Also, early carbonate geologists were principally interested in describing the rocks with particular attention focused on their biologic content. Over the last decade, carbonate researchers have begun to realize that carbonates are more of a chemical problem that a biologic one.

Not only are potential carbonate reservoir rocks a product of both deposition and diagenesis, but the combined effect of these control the reservoir, seal, and source potential of the rock. Moreover, these processes can produce a trapping mechanism that may be independent or in conjunction with tectonics. Thus, prediction of potential carbonate reservoirs is frequently more difficult than sandstone reservoirs because not many carbonate facies are systematically distributed, and diagenesis is a major process that preserves, occludes, or creates secondary porosity and permeability. Consequently, a more fundamental approach is required to reconstruct the depositional environment of potential carbonate reservoirs.

## FUNDAMENTALS OF SHALLOW WATER SEDIMENTARY CARBONATES

### PRINCIPALLY ORGANIC IN ORIGIN

Most carbonate particles originate directly or indirectly by biochemical processes. The carbonate particles are derived from skeletons of organisms, or they are indirectly precipitated by organisms. Consequently, the rate of carbonate production is controlled by those factors that influence life. They include salinity, nutrient supply, light, temperature, and turbidity. Because the particles are directly or indirectly sourced by organisms, particle size, shape, and sorting are not always an indication of the energy level of the depositional environment.

### LITHOFACIES PARALLEL BIOFACIES

Because carbonate sediments are directly or indirectly sourced by organisms, a particular lithofacies represents one or more biofacies (Fig. 9.1). For example, the back reef facies of a typical Silurian Thornton reef complex is composed of two biofacies, the Gastropod and the Brachiopod.

Two other important spin-offs of this concept are illustrated by Figure. 9.1. Stromatoporoids are extinct, therefore, it is impossible to directly apply geology by analogy to interpret the depositional environment of this biofacies. Furthermore, by association with other biofacies, Paleozoic Crinoid biofacies have been considered to represent shallow water conditions, but today Crinoids occur in water depths exceeding 500 feet. Care must be taken to not only correctly interpret the significance of extinct biofacies, but also to be aware that a modern biofacies may be a misleading counterpart to its ancient analog.

### COMPETITIVE SEDIMENTATION

Since lithofacies parallel biofacies, a lithofacies is a function of the ability of an organism(s) to occupy a given environment. Moreover, recall that one of the pitfalls of using the modern to interpret the ancient, was that not all organisms contain hard parts (*i.e.,* grasses, algae, etc.) and, therefore, their influence on competitive sedimentation in ancient strata may not be recognized by the explorationists.

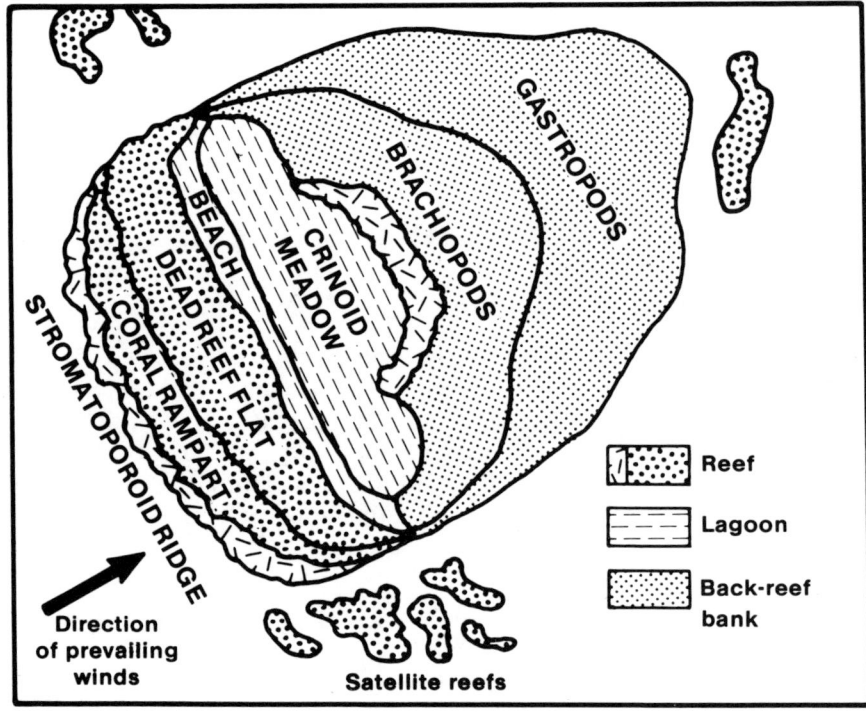

**Figure 9.1:** *Middle Silurian Thornton Reef complex located in Illinois is a good example where biofacies and lithofacies are parallel (Ingels, 1963).*

## INTRABASINAL ORIGIN

Carbonate sediments are derived near or at the site of deposition. Therefore, dispersal patterns and sediment transport are not important concepts in predicting carbonate facies. This further reduces the importance of size, rounding, and sorting to aid the explorationist to reconstruct carbonate depositional environments.

## TROPICAL OR SUBTROPICAL ORIGIN

Excellent nutrients, light, and temperature conditions occur in the shallow waters of tropical and subtropical climatic belts. It is, therefore, these belts in which most carbonate deposits occur. Reef development is even more restricted; all modern reefs occur in tropical climates (Fig. 9.2).

## TOPOGRAPHIC CONTROLS

Carbonate sediment production is most rapid on subsiding, topographically high continental shelves where conditions are most favorable for active organic processes. Rates of carbonate production in deep water are restricted and, therefore, accumulation rates are minimal. Carbonate turbidites are an exception because, like their clastic counterparts, they obtain maximum thickness in topographic lows. However, as will be discussed later, early cementation of carbonate sediment reduces the importance of turbidite deposition relative to their quartz sand counterparts.

## RESERVOIR - SEAL - SOURCE ROCK RELATIONSHIPS

Reservoirs, seals, and source rocks can be deposited in close vertical and/or lateral proximity. In fact, current research suggests that in specific situations, the biodegradation of organic matter in a mudstone will generate $CO_2$ which, in turn, will facilitate solution and development of secondary porosity. In these cases, it is possible for the source rock and reservoir rock to be the same. In other cases, due to the unstable mineralogy of carbonate sediments, the grainstone facies which contains high primary porosity can become the seal, and the mudstone facies which contains low primary porosity can become the reservoir.

# OVERVIEW OF CARBONATE FUNDAMENTALS

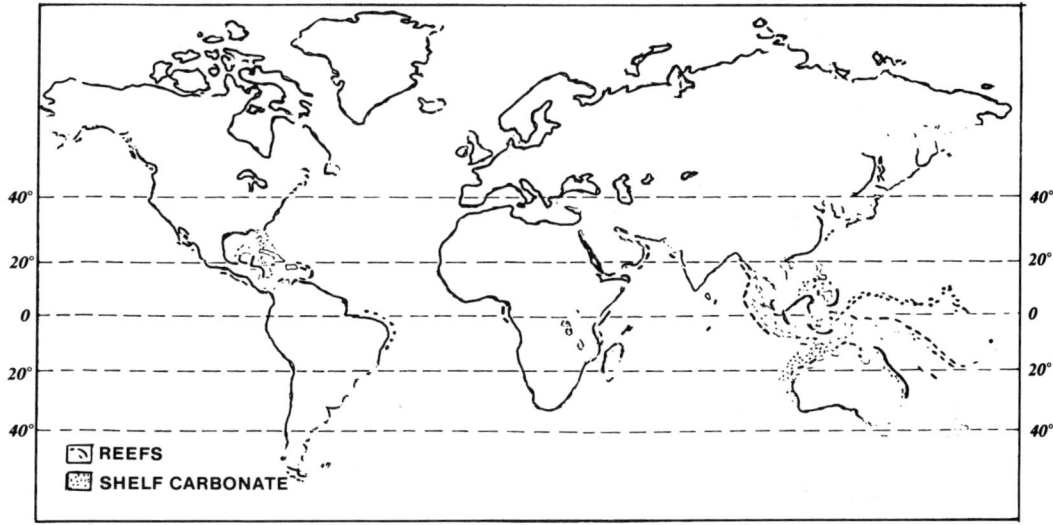

**Figure 9.2:** *Distribution of modern shallow water carbonate sediments (Wilson, 1974).*

## DIAGENESIS - THE RULE

Most carbonate particles are composed of an unstable form of calcite at the time of deposition. Therefore, they readily react with naturally occuring weak acids and bases. Diagenetic processes may reverse reservoir-seal relationships and vice versa. Moreover, diagenetic processes can rearrange porosity to improve reservoir quality even during pore occlusion. Fortunately, diagenetic overprints generally occur throughout the burial history of a carbonate rock, which permits the explorationist to reconstruct the diagenetic history in order to relate the relative timing of petroleum migration to pore occlusion.

## MINERALOGY OF MODERN CARBONATE SEDIMENTS AND YOUNG ROCKS

### MODERN CARBONATE SEDIMENTS

Modern carbonate sediments are composed of four different forms of calcite: aragonite, Mg-calcite, calcite, and disordered dolomite. Aragonite is orthorhombic calcium carbonate with a relatively high specific gravity (2.9). It is the most common skeletal mineral in shallow water carbonates where it occurs in corals, many mollusks, green algae, etc. Aragonite is also a common constituent of early cements in carbonate sediments.

Mg-calcite (sometimes referred to as high Mg-calcite) is a rhombohedral calcium carbonate. This crystalline structure has a lower specific gravity than aragonite (2.7). Most red algae, some mollusks, and many worms are composed of Mg-calcite. Like aragonite, Mg-calcite can occur as a cement. It typically contains 4 to 18% $MgCO_3$.

Echinoderms, some foraminifera, and some pelagic algae precipitate the more stable low Mg-calcite. Specific gravity of low Mg-calcite is 2.7. Normally, low Mg-calcite is referred to as calcite. Calcite contains less than 4% $MgCO_3$.

Several modern carbonate environments are favorable for dolomitization that produces a disordered dolomite that contains more *Ca* than *Mg*. This ratio varies from 56/44 to 50/50. The specific gravity of dolomite is 2.85. Dolomite is very common in ancient rocks and relatively rare in modern carbonate sediments.

### YOUNG CARBONATE ROCKS

Under appropriate diagenetic conditions aragonite and high Mg-calcite will rapidly alter to the more stable calcite. Disordered dolomite will alter to more ordered dolomite. The process of alteration is usually by solution, reprecipitation, or by neomorphism. Associated with this change is an isotopic enrichment toward more positive $\delta C$ values.

## CARBONATE TERMINOLOGY

### INTRODUCTION

Most explorationists have some knowledge of carbonates that range from elementary to highly sophisticated, and attached to that knowledge is a set of terms with which they are comfortable. There is no universally accepted terminology for describing carbonate rocks. Therefore, communication between explorationists is, at best, difficult. Do not let this become a barrier to your understanding. If corals and associated organisms construct a porous carbonate mass that is encased in carbonate mud, its reservoir properties, shape, size, and location do not change regardless if we choose to describe it as a pinnacle reef, patch reef, or organic buildup. The important thing is that the explorationist develop a mental picture of the feature and its characteristics in order to determine his or her exploration strategy.

### CLASSIFICATION OF CARBONATE ROCKS

**Basis for Classification**

*Grains:* The most important carbonate grains include detrital, skeletal, pellets, lumps, and coated grains.

1. **Detrital (Intraclasts) grains** are derived from pre-existing rock. Generally they are intraformational and are characterized by a wide range in shape, size, and age.
2. **Skeletal grains** are part of whole fragments derived from organisms that secrete hard parts. They are composed of aragonite, high Mg-calcite, or low Mg-calcite. They are highly variable in shape, size, and texture.
3. **Pellets** are grains that are composed of micritic material (fine-grained carbonate mud). They lack internal structure. They are generally ovoid in shape, commonly opaque, and are fine to silt-sized.
4. **Lumps** are composite grains formed by aggregation. They commonly contain superficial re-entrants and have a lobate outline. They are texturally similar to the micrite matrix.
5. **Coated grains** include oolites, pisolites, and various algal encrusted grains.
    a. **Oolites** are small spherical or subspherical accretionary grains less than 2mm in diameter. They have a concentric structure and may be formed by physical, chemical, and/or biochemical processes.
    b. **Pisolites** are similar to oolites, but are longer than 2.0mm in diameter. They are less regular in shape than oolites. Some pisolites are concretionary rather than accretionary in origin.
    c. **Algal encrusted grains** are highly variable in shape and size. They are accretionary in origin.

*Carbonate Mud:* Carbonate mud (micrite) is consolidated or unconsolidated ooze or mud less than 0.03mm in size. Its origin may be (1) biochemical precipitation, (2) grain size reduction by biologic or physical processes, and (3) diagenetic.

*Cement:* Crystalline aragonite, high Mg-calcite, low Mg-calcite, or calcite material that occurs in the interstices or in between the grains or matrix of a carbonate rock is termed cement. Grain size varies from micritic to coarse, and texture is variable. Origins of cement include (1) biochemical, (2) physiochemical, and (3) neomorphic.

*In-Place Organic Structures:* Corals, stromatoporoids, algal, or any other organism that produces a wave resistant framework is incorporated into classifications of carbonate rocks.

*Fabric:* Fabric, as used by carbonate geologists, describes the grain-to-grain relationships within rock.

**Folk Classification** The classification includes *particles* (skeletal and nonskeletal), matrix, and cement (Fig. 9.3). Pigeonholes include grain size, particle content, and whether the latter are encased in a matrix or cement together. The particles are assigned a prefix:

$$\begin{aligned}\text{skeletal} &= \text{bio} \\ \text{oolites} &= \text{oo} \\ \text{pellets} &= \text{pel} \\ \text{intraclastics} &= \text{intra}\end{aligned}$$

The classification contains "pigeonholes" for carbonate mud without particles, framework material, replacement dolomite, and neomorphic processes.

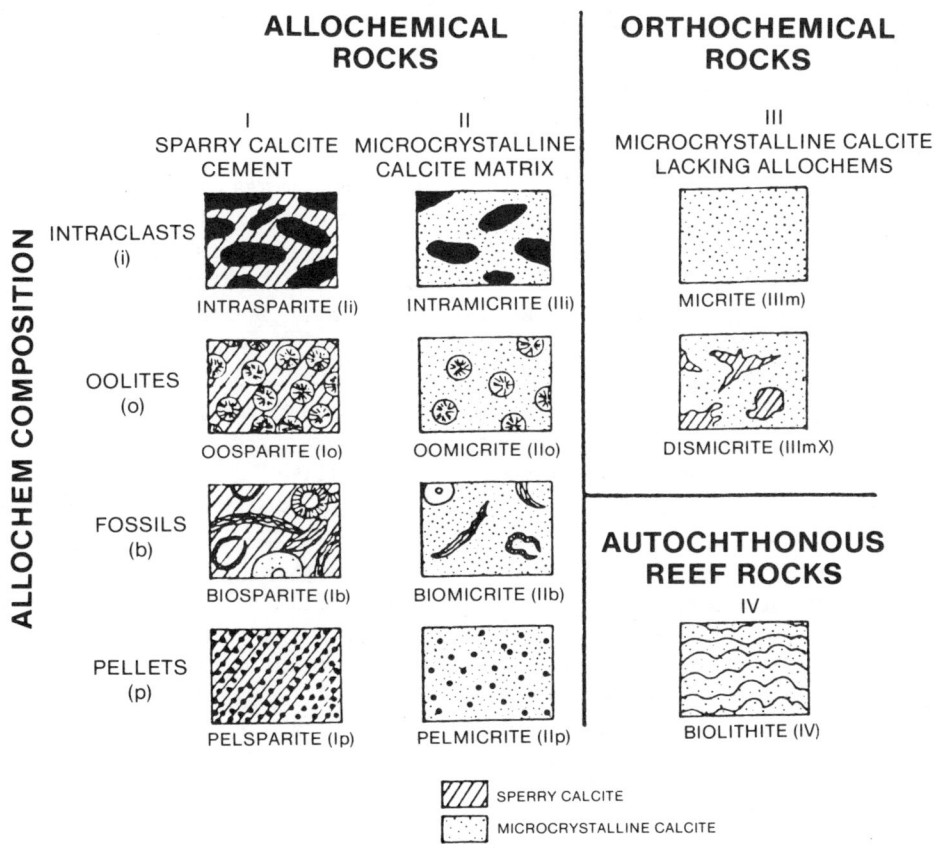

**Figure 9.3:** *Classification of carbonate rocks proposed by Folk, 1962.*

The Folk classification has many strong points. It describes most rock types, therefore, it is considered flexible and adaptable to most ancient carbonate rocks. It also describes the entire rock including particle, matrix, and cement. The classification provides descriptive code names for logging samples and cores that are easily plotted on strip logs.

No carbonate classification proposed to date is perfect. For example, the classification mixes diagenetic and depositional aspects of a rock. It is frequently difficult to determine the origin of a cement. Additional modifiers are required to describe textural variations. The classification is frequently misused by inexperienced explorationists as one that implies the environment of deposition. It is also difficult to apply to a new suite of rocks without the aid of thin sections or acetate peels. This is an important disadvantage because the pace of an exploration program is too rapid to delay it for a week or so while waiting on a thin section to be prepared. Once the explorationists become familiar with the suite of rocks, thin sections are not normally required.

**Dunham Classification** The Dunham classification is based primarily on depositional texture (Fig. 9.4). The "pigeonholes" are divided solely on the degree and nature of support of constituent particles. Grain-supported rocks have particles that are in contact, and mud-supported rocks are those where the constituents are "floating" in carbonate mud. Sub-divisions in the classification are based on the ratio of mud to particles and the occurrence of organically bound and crystalline rocks.

The Dunham classification has some distinct advantages over the Folk classification. The Dunham classification is easily grasped and can be used without the aid of thin sections or acetate peels. Only etched slabs are necessary. The textural categories have genetic significance for reconstruction of depositional environments.

Like the Folk classification, the Dunham classification has some disadvantages. The classification essentially ignores cements and requires modifiers for describing particle type. It is also probable that compaction of carbonate sediments occurs and that this process could significantly alter the genetic implications of the classification.

Figure 9.5 is a shorthand description of carbonate rocks that is useful for the rapid description of carbonate rocks either in the field or in the core or sample laboratory.

**Pore Space Classification** The simple intergranular porosity and rare moldic porosity common in sandstone reservoirs are normally

| DEPOSITIONAL TEXTURE RECOGNIZABLE ||||| DEPOSITIONAL TEXTURE NOT RECOGNIZABLE |
|---|---|---|---|---|---|
| ORIGINAL COMPONENTS NOT BOUND TOGETHER DURING DEPOSITION |||| Original components were bound together during deposition.... as shown by intergrown skeletal matter, lamination contrary to gravity or sediment-floored cavities that are roofed over by organic or questionable organic matter and are too large to be interstices. | Crystalline Carbonate (Subdivide according to classifications designed to bear on physical texture or diagenesis) |
| Contains mud (particles of clay and fine silt size) || Lacks mud and is Grain-supported | |||
| Mud-supported || Grain-supported | ||
| less than 10% grains | more than 10% grains | | ||
| Mudstone | Wackestone | Packstone | Grainstone | Boundstone | |

**Figure 9.4:** *Dunham (1962) classification of carbonate rocks.*

# OVERVIEW OF CARBONATE FUNDAMENTALS

**Figure 9.5:** *Shorthand technique for describing carbonate rocks.*

exceptions in carbonate rocks. Furthermore, predictability of porosity in sandstones is much higher than for carbonate rocks because compaction, cementation, and in some cases, clay diagenesis are the only important diagenetic processes that reduce primary porosity in sandstones. Stability and solubility of carbonate minerals are such that porosity reduction in carbonate rocks is the rule rather than the exception. Two questions must be asked by the explorationist who is attempting to predict carbonate porosity: what diagenetic processes preserve primary porosity, and if it is destroyed, what diagenetic processes "create" porosity?

While gross porosity controls the storage capacity of a carbonate reservoir, it is the degree of inter-connection (permeability) that controls the flow of hydrocarbons. In certain carbonate rocks, it has been demonstrated that pore occlusion can occur at the same time as permeability enhancement *(Pittman, 1974)*. It may be that preservation of primary porosity is related to early generation, expulsion, and migration of hydrocarbons into the reservoir.

Choquette and Pray *(1970)* recognize fifteen basic pore types (Fig. 9.6A), that when coupled with modifiers (Fig. 9.6B), characterize the genesis, size, and abundance of most of the pore types occurring in carbonate reservoirs. Figure 9.6B illustrates the technique of combining the pore types and modifiers.

The pore system type can be compared to log response and reservoir performance in known fields to provide a calibration set for prediction of reservoir performance in prospective areas. Most pore types are readily recognized by beginning explorationists. Therefore, the classification is strongly recommended. However, the modifying terms are difficult to use without a fairly sophisticated petrographic study. Although the classification is not designed to describe pore networks (permeability), this shortcoming can produce some erroneous predictions of carbonate reservoir quality.

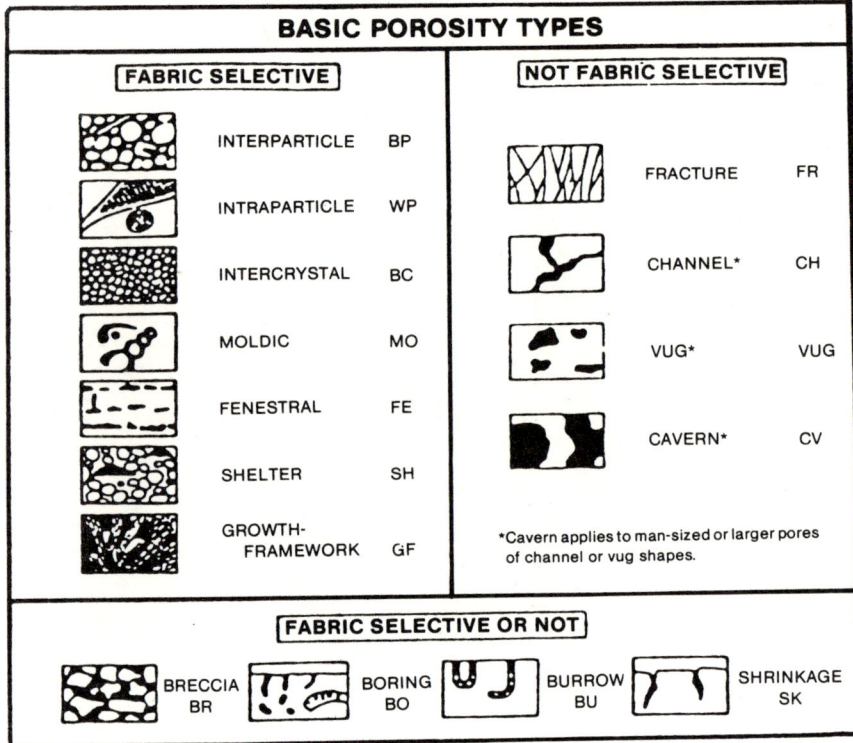

**Figure 9.6:** *Classification of pores in carbonate rocks (Choquette and Pray, 1970).*

# CLASSIFICATION OF CARBONATE ROCK BODIES

**General Terms** Sedimentary carbonates are highly influenced by the initial sea-floor topography, but they generally construct geometric patterns that, in turn, control further aspects of carbonate accumulation. Some general terms include a *Carbonate Buildup* and *Geologic Reef*. A carbonate buildup is a carbonate body of rock that possesses topographic relief. The term carries no inference about internal composition, origin, or scale of the rock body. A geologic reef is a carbonate body that possesses a topographic relief. It includes both local mound-like and regional curvilinear mounds. The term is frequently interchanged with *carbonate buildup* and carries no genetic implication.

## Terms Base on Configuration of Regional Topography

*Carbonate Ramps:* Large scale (hundreds of kilometers) carbonate bodies that accumulate on subsiding topographically high areas are termed *ramps* (Fig. 9.7A). The ramp may have slight variations in relief but, in general, it slopes seaward with no significant break in gradient. The ramp is characterized by three broad facies belts. The coastal zone is represented by extensive tidal flat and salt flats. The middle zone corresponds to the area that the sea floor is agitated by waves. Random development of high energy shoals and carbonate buildup are typical of this environment. Packstones, grainstones, and boundstones are the dominant lithologies that represent the environment. The outer zone conforms to that area which is below wave base. Pelagic muds are the dominant lithology.

*Carbonate Steps:* Large scale (hundreds of kilometers) carbonate bodies that accumulate on a subsiding topographically high area characterized by a break-in-slope is termed a *step* (Fig. 9.7B). The step may be two-sided to form a platform. The step or platform is divisible into three major environments; shelf, shelf margin, and basin. The *shelf* (undaform) is nearly horizontal and the facies belts are similar to those in the coastal and middle zones of the ramp. The shelf may contain salt flats, tidal flats, shoals, and randomly developed carbonate buildups that may include framework organisms. Facies belts on the shelf margin are curvilinear and predictable. A leeward shelf margin may contain back-bank, bank, and fore–bank deposits whereas the windward shelf margin will be characterized by back-reef, reef, and fore-reef deposits. Like the outer zone of the ramp, the basinal environment in the step is characterized by deep water and slow carbonate production rates, but unlike the ramp, basinal step deposits may be interbedded with turbidite deposits.

Because carbonate production rates are highest along the middle zone of a ramp, the ramp may locally or regionally develop into a step model of carbonate deposition. In order for this to occur, subsidence and/or sea level rise, interlaced with stillstands, must occur over a significant interval of geologic time.

## Terms Based on Configuration Without Genetic Implications:

A circular to longate carbonate buildup that covers a relatively limited area is descirbed as a *mound* (Fig. 9.8). Mounds can occur on a ramp or the shelf of a step. A carbonate *pinnacle* is conical in shape, generally steepsided, and normally it is tapered. Pinnacles are rare on a ramp. They can form in a deep shelf lagoon or on the shelf margin of a step. *Atolls* are ring-like carbonate buildups that develop in an offshore, oceanic environment or in a deep shelf lagoon. They are rare in the ramp model of deposition. They consist of a lagoon of variable depths that is protected from the offshore or oceanic waves by a boundstone lithofacies. Mud, skeletal, or oolitic material may form a carbonate *bank* of relatively large areal extent to its thickness. They may form on the shelf of a ramp or step, or along the shelf margin of a step.

## Terms Based on Configuration with Genetic Implications

*Patch reefs* are isolated, more or less circular carbonate buildups that are constructed by framework organisms. They may occur on a shelf of either a ramp or step, in the lagoon of an atoll, or in the basin. A curvilinear belt of organic framework constructed carbonate buildup is described as a *barrier reef*. Barrier reefs may occur on a shelf or along the break in slope of a step. They are frequently massive and contain an extensive back and fore reef facies. A curvilinear belt of organic framework constructed near the coastline is termed a *fringing reef*. They are restricted to the break-in-slope of a step. *Bioherms* are carbonate buildups that represent in situ production of lime-secreting organisms or framework or encrusting growth. They normally occur on the shelf of ramps and steps.

**Recommended Usage of Carbonate Terminology** Explorationists are concerned with mapping and predicting the geometry of carbonate rock bodies. Although lithofacies and biofacies can be described in cores and samples, it is generally very difficult to delineate precisely the origin of a carbonate body. Therefore, it is recommended that emphasis should be placed on the geometry of the rock body and its occurrence within a facies model, rather than spending a great deal of effort worrying about its genesis.

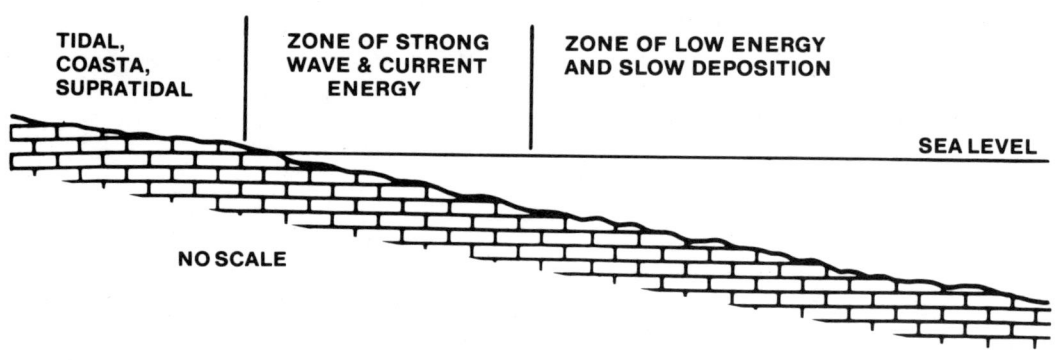

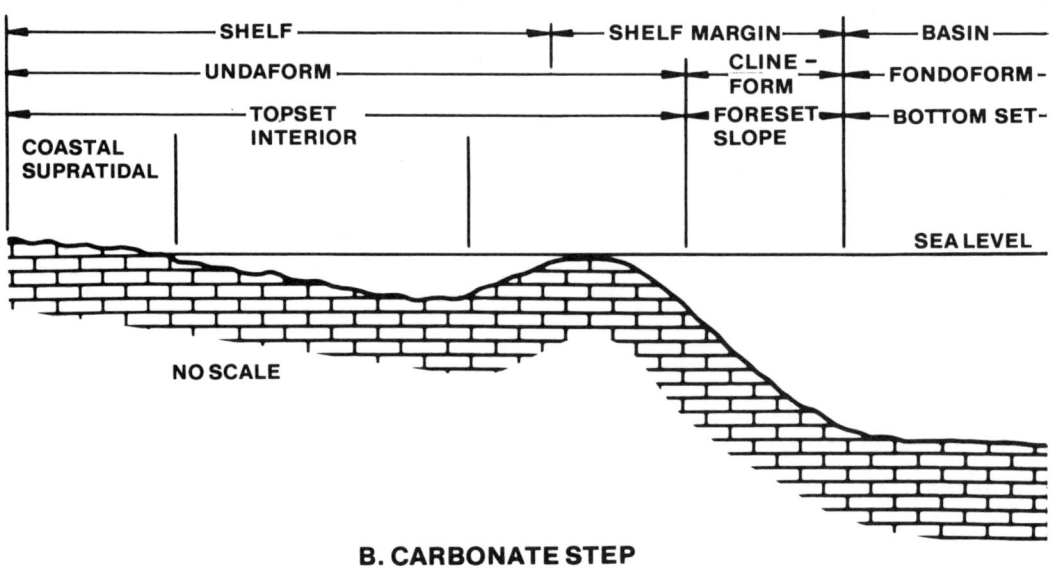

**Figure 9.7:** *Carbonate sediments may accumulate on (A) ramps, (B) steps.*

# OVERVIEW OF CARBONATE FUNDAMENTALS

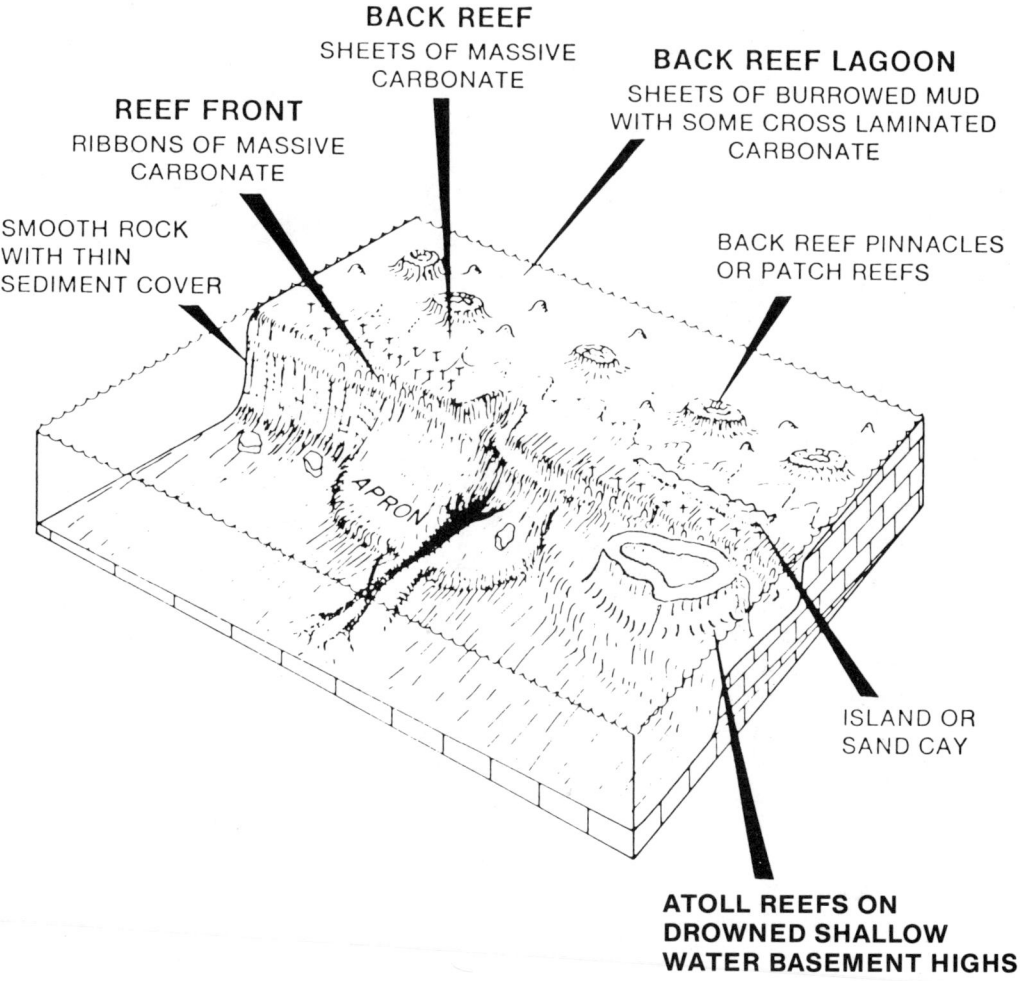

**Figure 9.8:** *Typical geometry of potential carbonate buildups (adapted from SEPM Core Workshop No. 4, 1983).*

# 10
# Carbonate Depositional Environments and Stratigraphy

## INTRODUCTION

Major discoveries in carbonate reservoirs located in the Eastern Hemisphere, West Texas, U.S.A., and central Alberta, Canada resulted in demonstrating the ignorance of petroleum explorationists and exploitationists regarding development of strategies for exploration and exploitation of carbonate reservoirs. Therefore, those major oil companies who owned interest in these major discoveries initiated a worldwide study to describe modern carbonate environments of deposition. The intent of these studies was to understand the factors that controlled the production of carbonate facies and porosity in order to integrate this data with information obtained from studies of ancient carbonate rocks. The approach was to: (1) relate the types of carbonate deposition to the physical, chemical, and biological features of the environment, (2) catalog those physical and biological features of the deposits that are best preserved and which are independent of local tectonics, but related to global sea level changes, (3) place the depositional features into the regional geographic framework of a basin, and (4) interpret how the three-dimensional accumulations would respond to variations in global sea level, subsidence rates, production rates, and climate.

Geology by analogy resulted in the interpretation of recent and ancient models that permitted the formulation of a limited number of carbonate models of deposition. These models permit explorationists to extrapolate between points of control on maps and cross sections, and to make predictions on the distribution of potential carbonate reservoirs, seals, and source rocks. However, explorationists must keep in perspective that direct application of the modern models to ancient rocks may produce erroneous predictions because these models only represent (1) a snapshot in time, (2) present ecologic niches of organisms which may have changed through geologic time, (3) present sea level conditions and the environments may have not completed equilibration with the last sea level rise and/or tectonics, (4) the total organic community of which not all will be preserved in the ancient record, and (5) only marine diagenetic processes (which represent a few of the processes of pore occlusion and/or secondary porosity development) have altered the depositional fabric of the sediment.

Perhaps even more basic is the fact that application of modern carbonate rocks has been based on the assumption that sea level has been more or less static over the past 3,000 years. As pointed out by an excellent editorial by Wanless *(1982)*, most shallow water carbonate environments were inundated about 3,000 years ago when sea level was within 1m of the present level in south Florida. Between 8,000 to 3,000 years ago, sea level rose at the rate of 2.6mm per year. During most of the last 3,000 years, sea level rise slowed to only 0.4mm per year in south Florida. Hence, this minor rate was not considered to be important by early workers in modern carbonates. However, tidal data over the last 48 years shows that the rate of sea level rise has recently taken a dramatic increase to 2.3mm per year in south Florida *(Wanless, 1982)*.

Most of our data regarding modern carbonate sedimentation was collected over this time span. Consider the erroneous conclusions that we may have made from this data. For example, perhaps we need to re-evaluate our present understanding of tidal and supratidal flats and the protodolomite that frequently occurs within these environments. It is also probable that much of our knowledge of shallow lagoonal and bay environments needs to be re-examined. And finally, our zonation of shallow water benthonic organisms and their ecologic niches may be in error.

## STEP CARBONATE RESERVOIRS— BELIZE

### PHYSICAL SETTING

**Structural Evolution** Belize is situated on the Yucatan continental block just south of Mexico (Fig. 10.1). The continental margin of Belize was evolved in two distinctly different sets of plate motions *(Silver et al, 1981)*. The first episode occurred during Paleocene time. Left-lateral movement of a transform fault which trends parallel to the Montagua-Cayman Trench, produced the Gulf of Honduras. The second episode caused an eastward shift in the Caribbean plate during Early Oligocene time. This change in plate motion formed several north-south trending extensional faults that are subparallel with the Cayman trough. Basement structure of the continental margin of Belize is characterized by a series of aligned block-faulted ridges that trend parallel to the coast and they, in part, control the trend of the shelf margin.

**Bathymetry** Modern carbonate deposition in Belize occurs on a rimmed continental margin that can be subdivided into three general bathymetric zones — shelf, shelf margin, and basin (Fig. 10.2). The shelf varies in width from just a few kilometers at its northern limits to over 40 kilometers to the south where it opens into the Caribbean. Corresponding to the increase in width of the shelf to the south is an increase in water depth. The northern shelf is typically less than a few meters in depth, but the southern shelf contains water depth of greater than 50 meters.

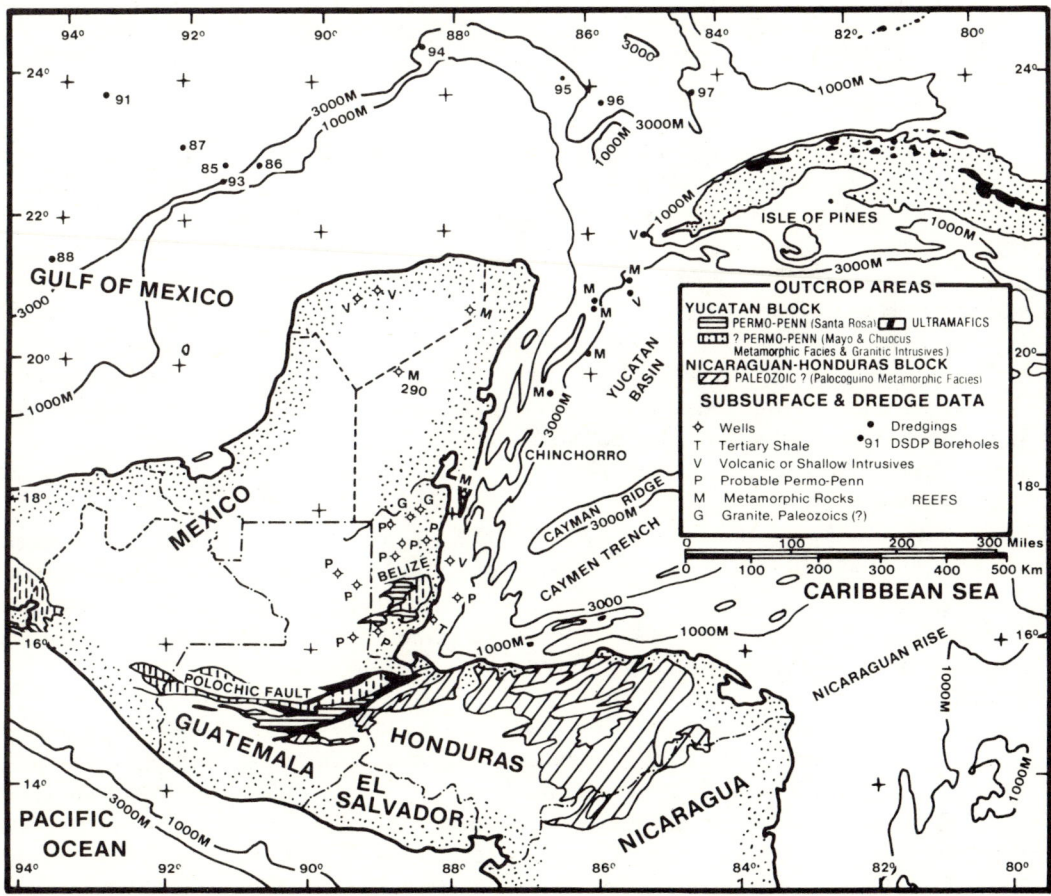

**Figure 10.1:** *Belize comprises the southeastern margin of the Yucatan Peninsula (Dillon and Vedder, 1973).*

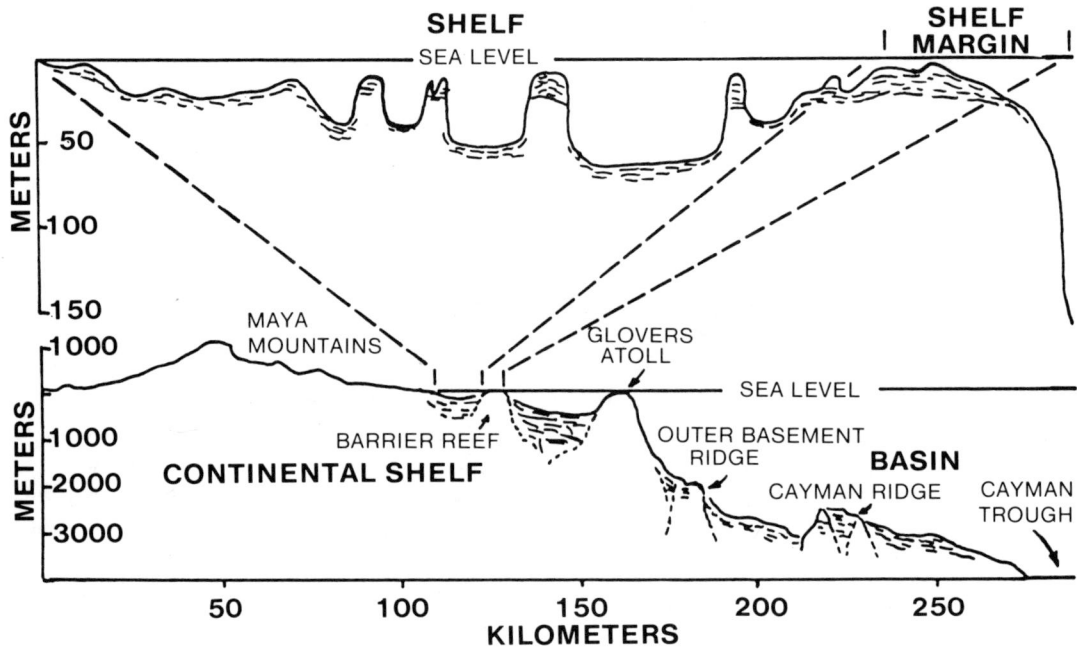

**Figure 10.2:** *Geologic subdivisions of the Belize Complex (partially adapted from Dillon and Vedder, 1973).*

Bathymetry along the shelf margin can be subdivided into two distinct profiles. The most common profile is a slope that extends, without major interruption, into water depths of 1,000 meters or more. This profile is typical of the southernmost part of the shelf margin where it is juxtapositional with the Cayman trough. It also occurs on the eastern side of Glovers Atoll and Turneffe Island and on both sides of Lighthouse Reef. The second distinct shelf-margin profile flattens out at a depth of 300 to 450 meters and grades into a gently sloping basin. This profile is best illustrated by the shelf margin north of Turneffe Atoll.

**Climate** Belize and its continental shelf are located within the tropical climate belt. An in-depth discussion of the climate has been done by Wright *et al,* 1950. Annual rainfall ranges from 125 centimeters a year near Corozal at the Mexican border to 175 centimeters at Belize City to 450 centimeters at Punta Gorda to the south. Air temperatures range from an average of 27°C for the May to September period to 24°C for the November to January period. Dominant wind direction is from east to west.

**Water Characteristics** The general patterns of salinity in the surface waters of the shelf were reported by Purdy *(1974).* He noted that salinities decrease northward into Chetumal Bay in addition to southward into the Gulf of Honduras. He attributed these lower salinities to freshwater runoff from land. Slightly higher salinities occur in the northern lagoon, because of the combination of lower rainfall and of a reduced drainage area.

## SHELF LITHOFACIES

The shelf lithofacies can be broadly subdivided into terrigenous, mixed-terrigenous-carbonate mud, carbonate mud, island, carbonate grain, and reef (Figs. 10.3 and 10.4).

**Terrigenous** Terrigenous influx of clastics is significantly higher in the southern part of Belize than in the norther part, because the former contains more relief and drainage area than the latter *(Krueger, 1963).* The northern shelf contains one small area of coarse quartz sand that is located in the Cabbage Ridge area, whereas, the southern continental coastal zone contains a linear belt of coarse quartz sand that is parallel to the coastline. Most of the

# CARBONATE DEPOSITIONAL ENVIRONMENTS AND STRATIGRAPHY

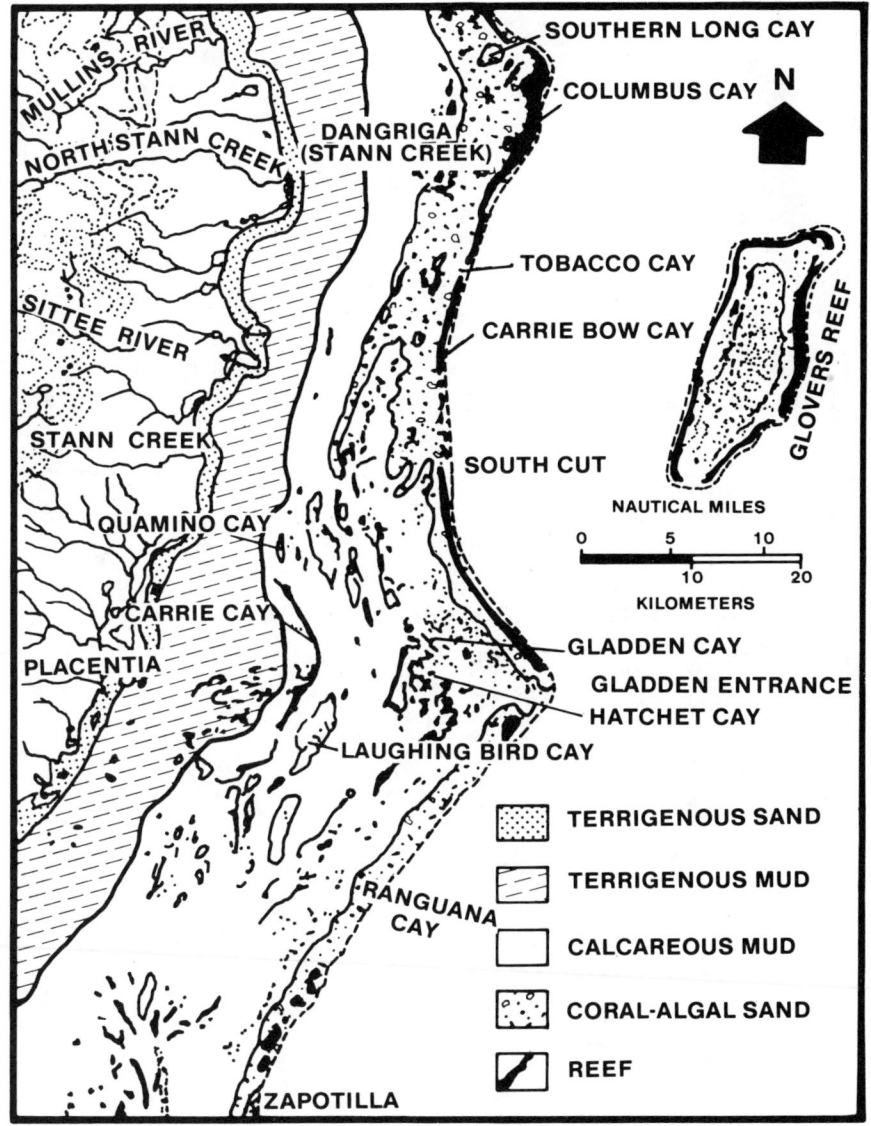

**Figure 10.3:** *Distribution of surficial sediment in the southern Belize shelf lagoon (in part adapted from Purdy, et al., 1975 and Silver, et al., 1981).*

coarse terrigenous materials are deposited very near shore as small lobate deltas at the mouth of rivers (Fig. 10.3). It is reworked toward the shore by heavy surf as well as transported southward by longshore currents to form cuspate headlands and prominent beach ridges. Clays remain in suspension and they are transported to all parts of the deep southern shelf lagoon. Clay content decreases toward the barrier reef such that, near the eastern edge of the deep lagoon, the sediments are pre-dominantly carbonate mud *(Scott,* 1966). A linear belt of clay borders the terrigenous sand and silt facies. It floors much of the Bay of Honduras.

**Carbonate Mud and Montmorillonite Mud**
Carbonate mud is the dominant lithology of the northern shelf, whereas, montmorillonite mud covers much of the shelf floor south of Belize City. The southern mud facies averages 46% montmorillonite, 30% kaolinite, and 24% illite. The carbonate mud contains two major skeletal components — pelletoids and

*Miliolids*. The pelletoid muds are concentrated in the Chetumal Bay area, and the *Miliolid* muds dominate a north-south belt just seaward of the terrigenous sand and silt lithofacies. A mixed calcareous montmorillonite mud occurs on the southern shelf. It is bordered on the east by the montmorillonite mud and on the west by *Halimeda* sands. This lithofacies surrounds most of the major shelf reefs in the southern part of the Belize complex.

**Carbonate Sand** The northern shelf contains an almost rectangular belt of *Peneroplid* sand that occurs just behind the shelf-margin barrier reef (Fig. 10.4). The sands are fine to medium in grain size, moderately sorted, and contain some mud-sized carbonate material. South of Belize City, a medium-to-coarse grained *Halimeda* sand comprises a north-south linear lithofacies belt behind the barrier reef. Mud content is very low.

**Shelf Reefs** The shelf reefs can be grouped into four principal forms: (1) pinnacle, (2) loaf, (3) cellular, and (4) ring *(Ginsburg and James, 1979)*. The ring reefs are subcircular

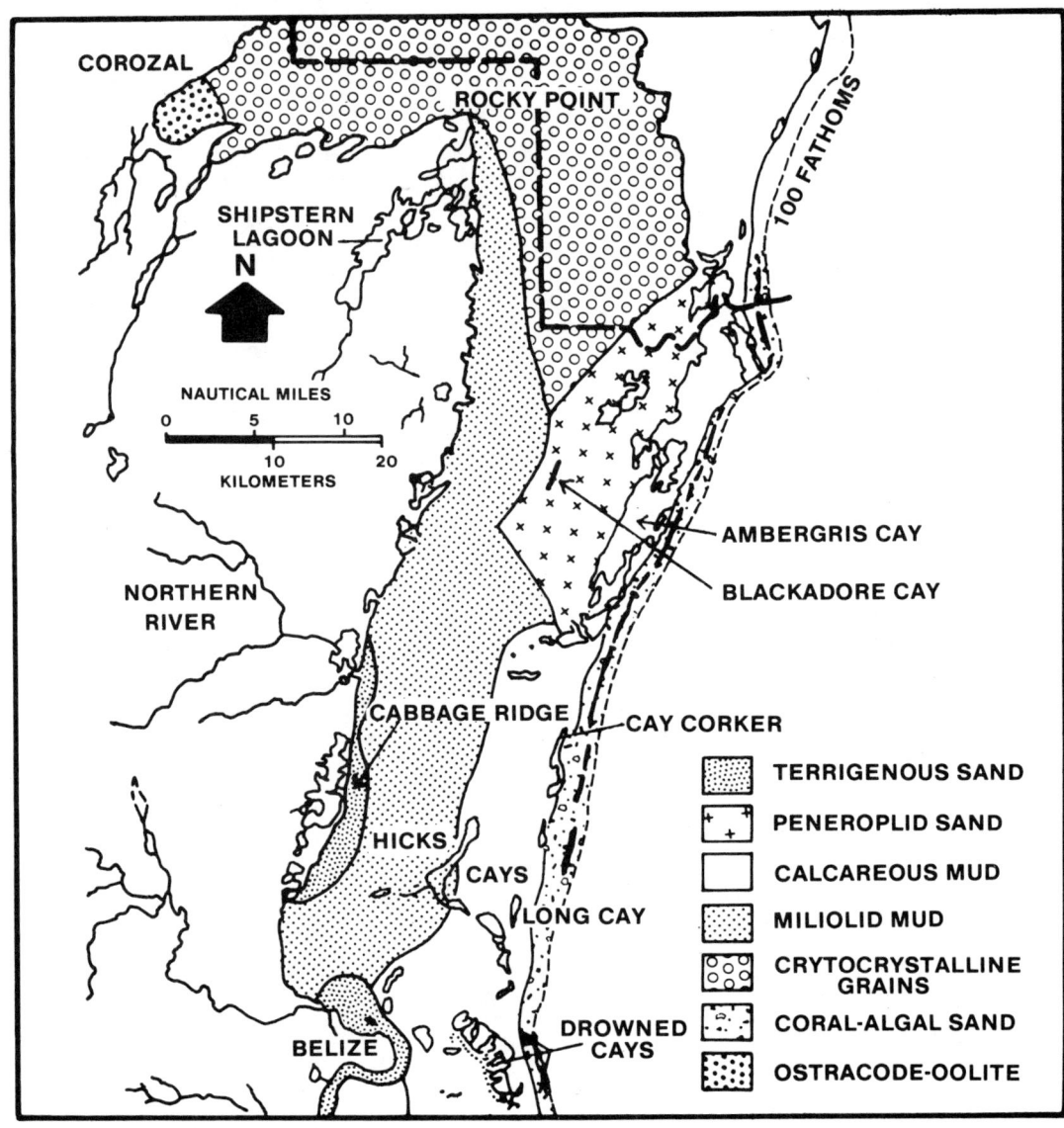

**Figure 10.4:** *Distribution of surficial sediment in northern Belize shelf lagoon (in part adapted from Purdy, et al., 1975 and Silver, et al., 1981).*

# CARBONATE DEPOSITIONAL ENVIRONMENTS AND STRATIGRAPHY

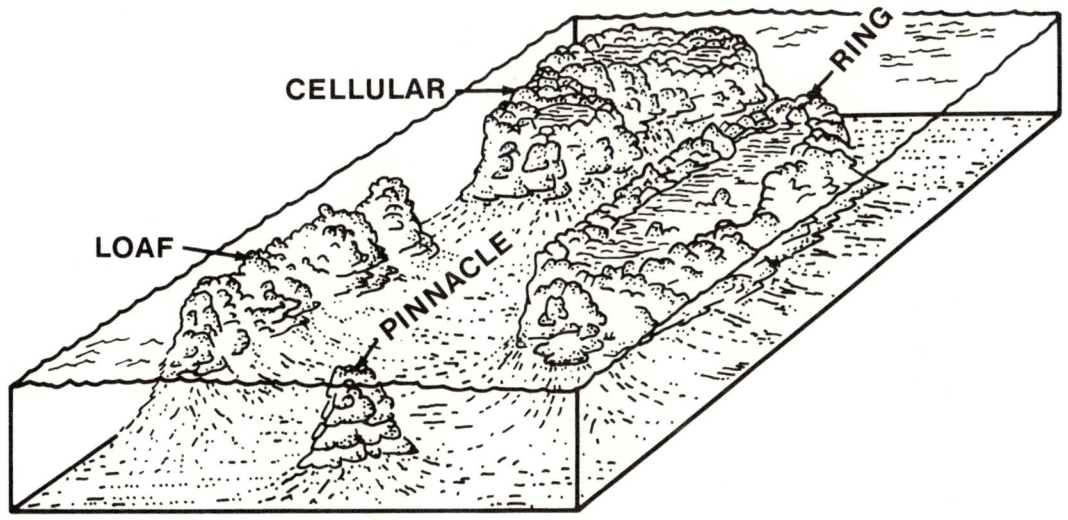

**Figure 10.5:** *Typical geometric of shelf reefs, Belize (Ginsburg and James, 1979).*

to polygonal. They are typically 2 to 8 kilometers long and 15 to 30 meters thick (Fig. 10.5). Normally, the cellular reefs are irregular networks and branching arrays of varying dimensions, but generally, they are less than 10 meters high. The loaf reefs are up to 50 meters wide, are 200 to 300 meters long, and have 4 to 8 meters of relief. The pinnacle reefs generally are 10 to 100 meters wide with 2 to 20 meters of relief. Coral growth typically is concentrated on the margins surrounding the reefs, but, in the case of the ring reefs, coral growth also occurs adjacent to the sandy center of the reefs.

## SHELF MARGIN LITHOFACIES

**Barrier Reef** The shelf margin lithofacies may be subdivided into two major environments of deposition — fore reef and barrier reef. The Belize shelf margin consists of a relatively continuous series of barrier reefs interrupted in places by channels that connect the basin with the large shelf lagoon. Corals and encrusting red algae have colonized depths from 0 to 20 meters and have constructed a barrier reef that is 275 to 365 meters wide and 270 kilometers long *(Stoddart, 1960)*.

**Fore Reef** Two types of slope margins occur along the Belize barrier reef. They are a depositional slope margin and a by-pass margin (Fig. 10.6). The traverse from the reef scarp into the basin between the barrier reef and Glovers Atoll is a classic example of a depositional slope shelf margin. The lithofacies is composed of intraclast bearing wackestones and packstones. The by-pass slope lies along the western border of the Yucatan basin. Sediments include lithified rock with a thin cover of perennial (pelagic and hemipelagic) sediments on the ridges. It is dissected by channels that contain coarse-grained turbidite layers of shallow-water material that were redeposited in water depth of as much as 3,100 meters.

## BASIN LITHOFACIES

The basin deposits are composed of quartz sand and terrigenous clay that is being transported through the Bay of Honduras. *Halimeda*-rich lime muds and sands make up much of the sediment immediately east of the toe of the slope. Planktonic foram-rich lime mud and *Halimeda* plates comprise the sediment accumulating on the fault block ridges *(James and Ginsburg, 1979)*.

## ATOLL LITHOFACIES

Three large, elongated atolls are located on horst blocks seaward of the Belize shelf margin — Turneffe, Lighthouse, and Glovers. Turneffe reef is approximately 16 by 43 kilometers. It is surrounded as are the others, by fringing reefs. The interior lagoon is very shallow and contains many palm-covered and mangrove-

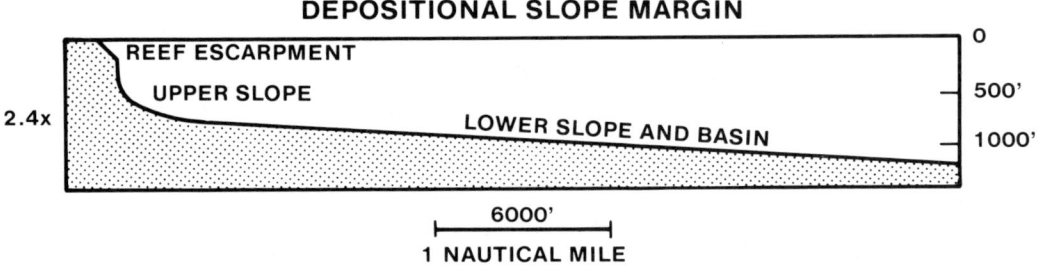

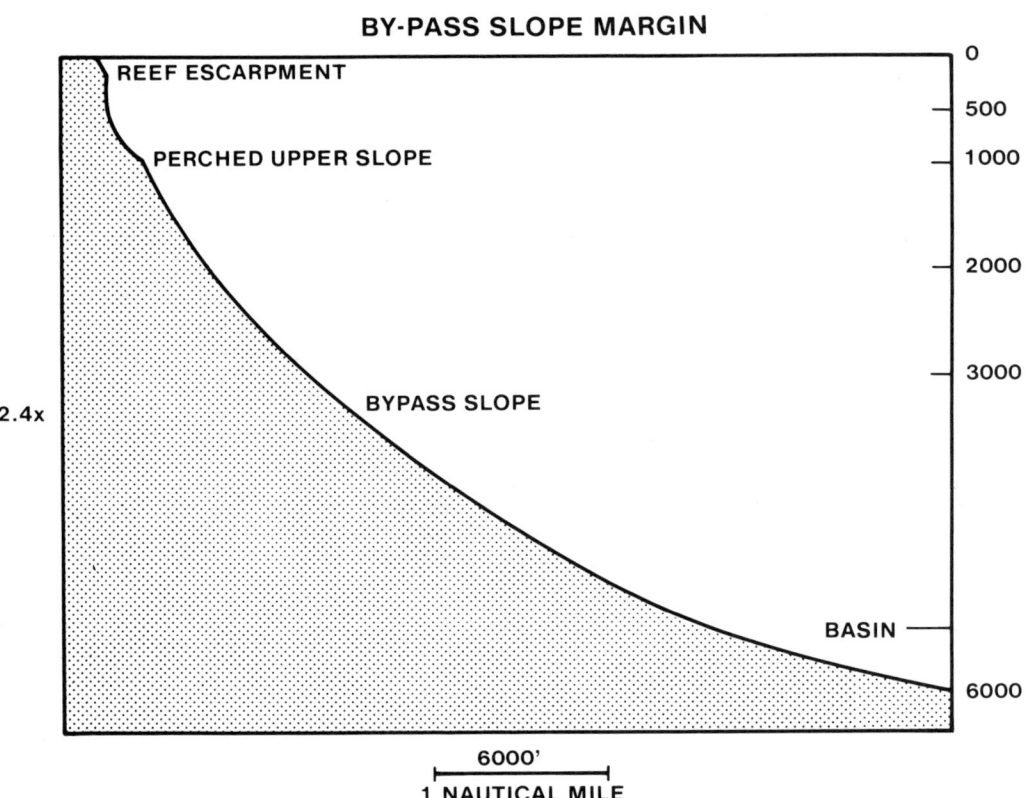

**Figure 10.6:** *Schematic topographic profiles of margin types and related pysiographic elements (Koch, 1973).*

covered islands. Lighthouse reef is approximately 8 by 39 kilometers, and the lagoon is shallow.

Glovers reef has been studied more than the other two atolls and is about 10 by 28 kilometers (Fig. 10.7). This fringing reef contains five small islands that surround a lagoon which is generally 7 to 18 meters deep. Over 700 pinnacle reefs are present within the lagoon. It is probable that explorationists would not recognize Glovers as an atoll in 20,000,000 years, because the pinnacle reefs will probably converge sufficiently to produce a large shoaling platform. A windward and leeward margin would develop. Facies patterns, similiar to the Bahamian Platform, might develop.

# CARBONATE DEPOSITIONAL ENVIRONMENTS AND STRATIGRAPHY 197

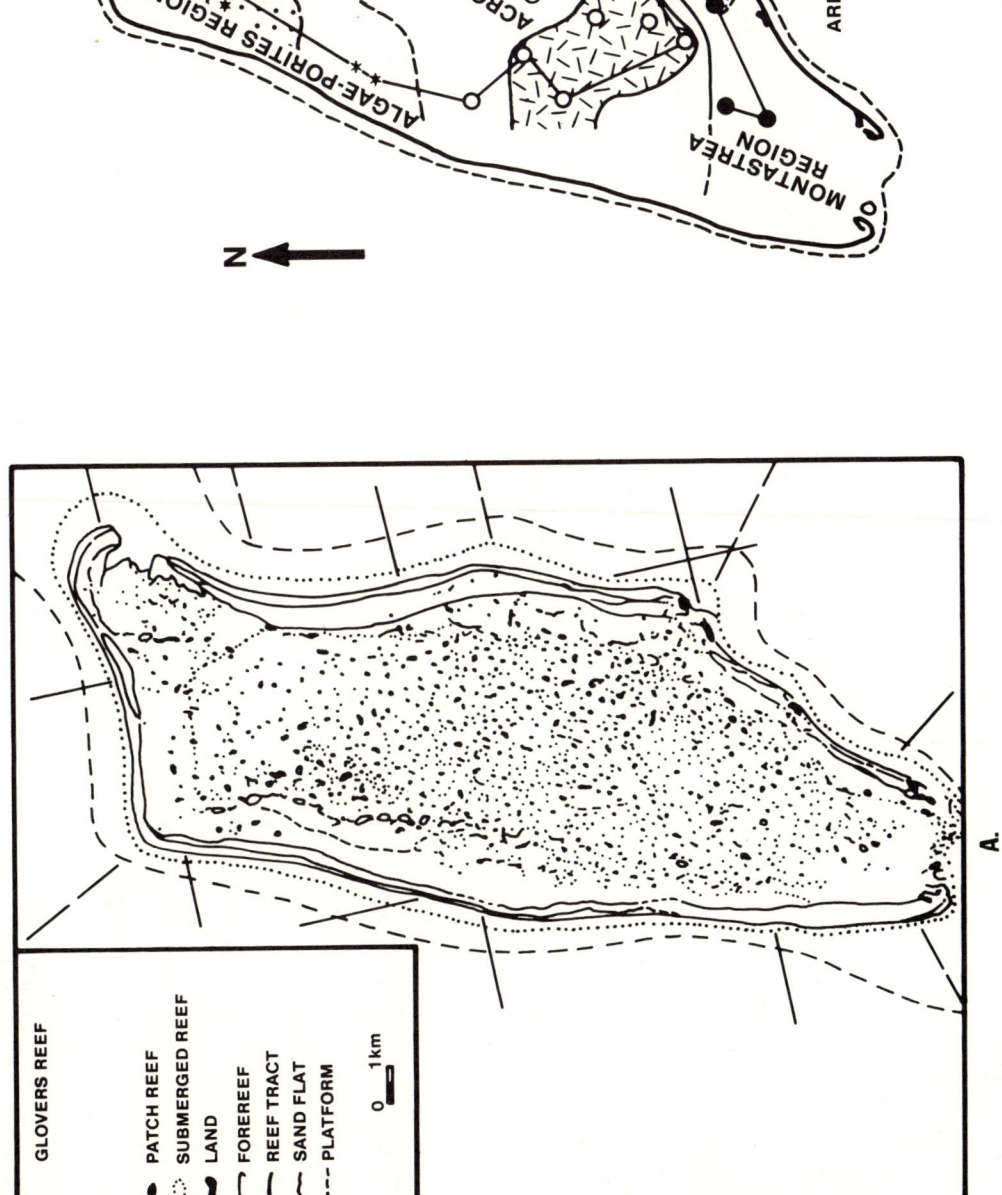

**Figure 10.7:** *A. Map of Glovers Reef. The major tidal channels are located at the eastern corner of north reef. B. Map of patch-reef assemblage distribution in the lagoon of Glovers Reef (Wallace and Schafersman, 1977).*

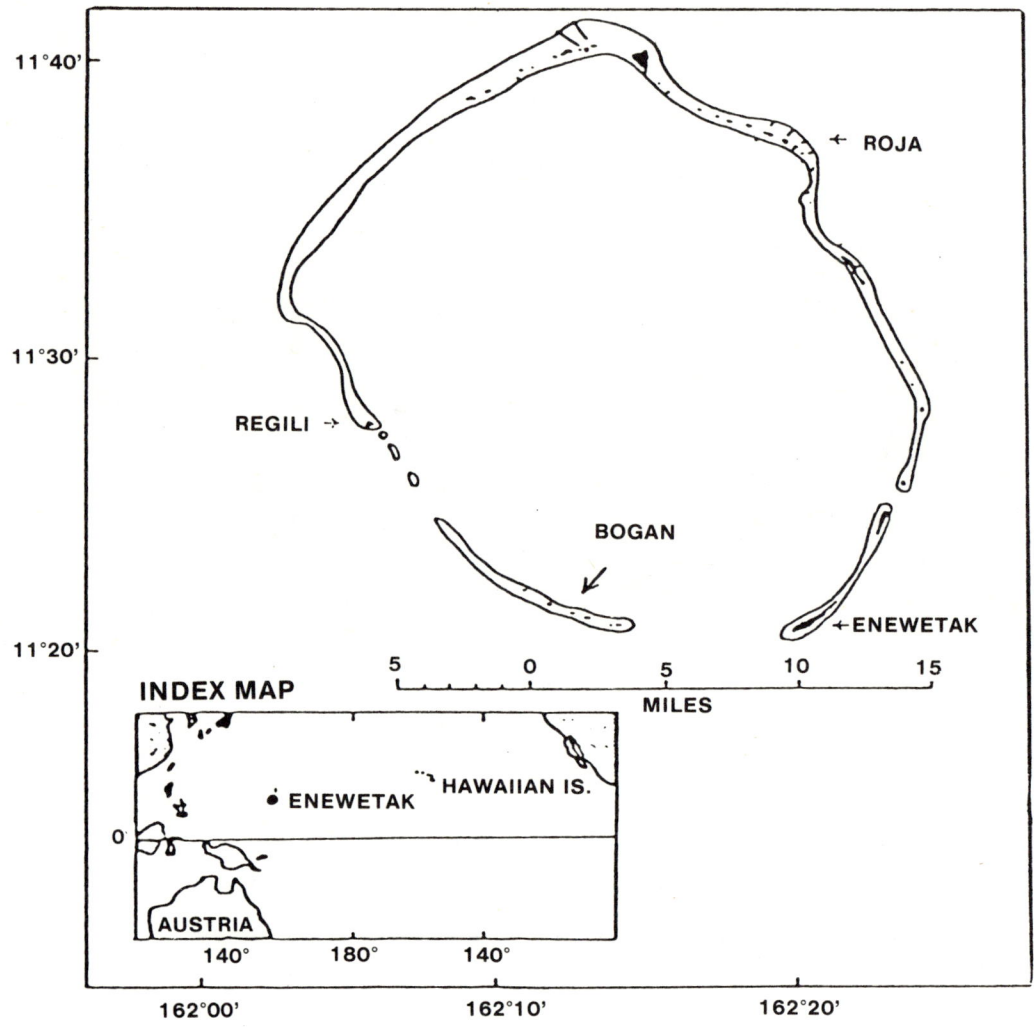

**Figure 10.8:** *Location map, Enewetak Atoll, Marshall Islands.*

## ATOLL CARBONATE RESERVOIRS - ENEWETAK ATOLL

### PHYSICAL SETTING

Perhaps the most geologically studied atoll is Enewetak. The atoll is located about 9,200 km west-southwest of Los Angeles and 4,890 km north-northwest of Sydney (Fig. 10.8). Its shape is broadly elliptical, 41 km long and 31 km wide. The atoll contains about 36 islands with a total area of 5.9km². Maximum elevation is less than 3.5m above mean sea level. Prevailing winds are from the northeast, and tidal range is -0.1m to 1.8m. Oceanic water depth adjacent to the atoll is in excess of 4,000m, maximum depth of the lagoon is about 65m.

## LITHOFACIES

**Reef Lithofacies** The reef facies can be subdivided into backreef (reef flat), reef crest, and forereef. The reef flat and reef crest are similar in organic content and texture. Both facies consist of coral and coralline algal boundstones (Fig. 10.9). Accessory components include mollusks, algae, and echinoderms. Maximum development of these facies occurs on the windward side of the atoll. Transitional and lee reef crest and reef flat facies are narrow and poorly developed.

Forereef facies are composed of cobbles up to 60cm in size. Slopes up to 60° occur on the margin of the atoll.

# CARBONATE DEPOSITIONAL ENVIRONMENTS AND STRATIGRAPHY

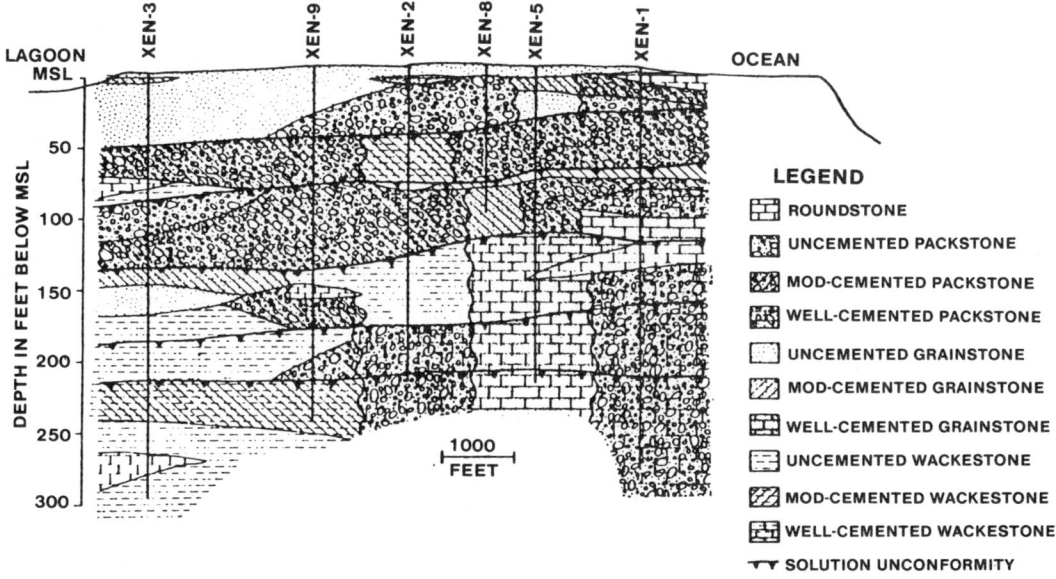

**Figure 10.9:** *Cross section illustrating facies and cementation zones, Enewetak Atoll (Couch and others, 1977).*

**Beach Facies** Not volumetrically important, but almost spectacular beach facies composed of coral and coralline rubble comprise the beach facies. A poorly sorted, partially cemented, planar cross-bedded grainstone is the dominant lithic type. The beach rock is genetically related to island development, but locally, especially on the transitional flank of the atoll, beach rock occurs between islands on the reef flat.

**Lagoon Lithofacies** The lagoon lithofacies is by far the most voluminous facies on the atoll. It is composed of a heterogeneous sequence of muddy carbonates, grain-rich sands, and coral framework lithologies. The coral boundstones represent patch reefs in the lagoon and are an important contributor of carbonate grains. The patch reefs are most abundant and best developed on the southwest side of the lagoon. This is a result of the concentration of nutrients that are collected and transported across the lagoon from northeast to southwest by wind generated waves. Interpatch reef deposits are sourced from the patch and outer reef facies, as well as tests from organisms living in the lagoon. The lagoonal interreef facies is not cemented.

## PLATFORM CARBONATE RESERVOIRS - GREAT BAHAMA BANK

### PHYSICAL SETTING

A platform is a two-sided step model of carbonate deposition. The Great Bahama Bank is a vast, nearly flat area that is characterized by shelf and shelf margin environments that occur on three sides of the platform (Fig. 10.10). The platform is located about 80 km east of Florida and covers about 60,000 square miles of which only about 4,000 square miles are subaerially exposed. The platform is located in the path of the Easterly Trade Winds. Tides are about 1 meter, and tidal currents range from one to three knots. Longshore currents are dominantly northeast through the Florida Straits and counterclockwise in the Tongue of the Ocean. Water depths along the margins of the platform vary from 600 to 1,000 meters in the Florida Straits to more than 2,000 meters in the Tongue of the Ocean. The platform is depressed toward the center where water depths normally are less than 20 feet.

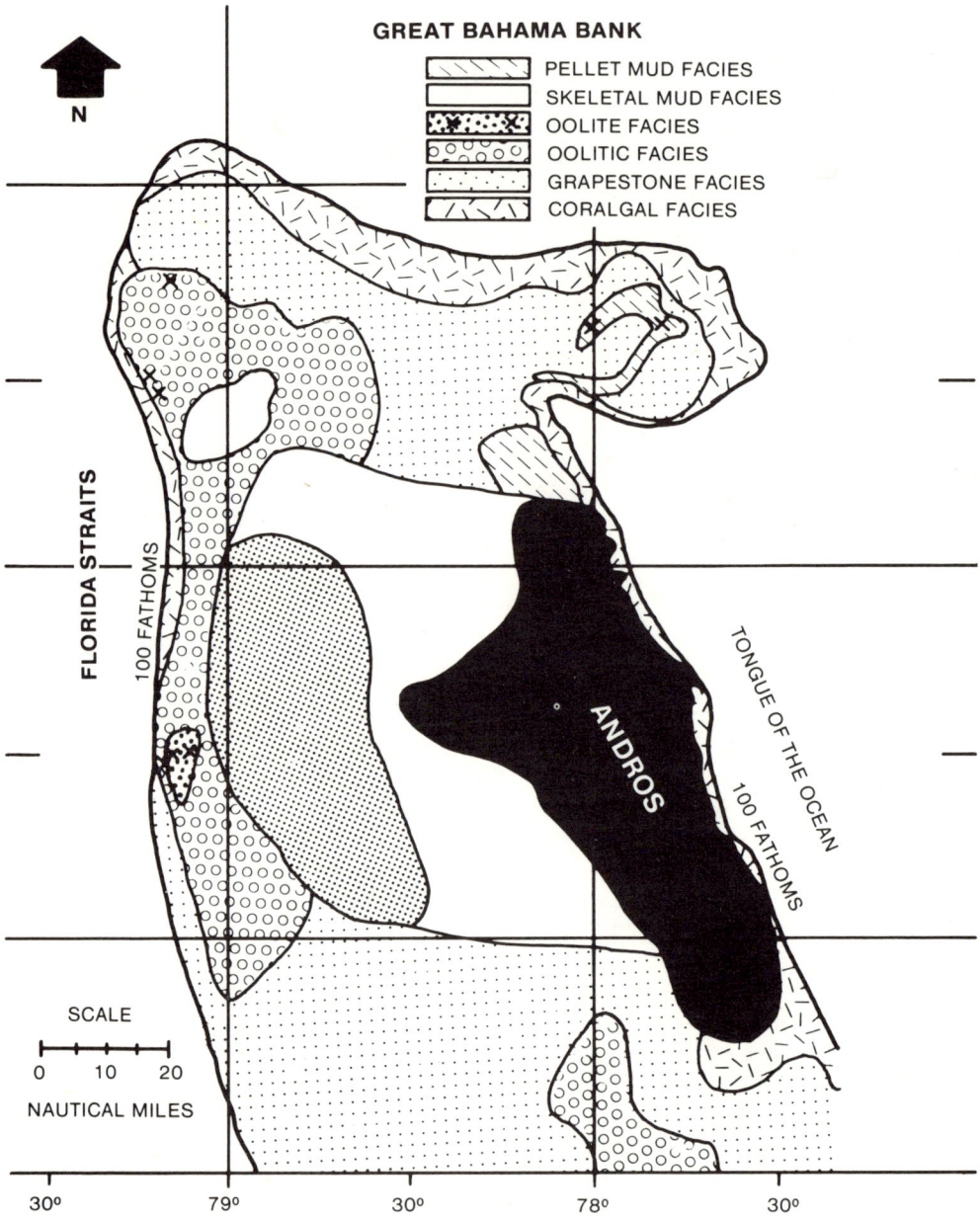

**Figure 10.10:** *Lithofacies of the Great Bahama Bank (Purdy, 1963).*

## SHELF LITHOFACIES

The shelf may be subdivided into four facies: pelletal mud, skeletal mud, grapestone, and oolitic.

**Pelletal Mud Facies** The pelletal mud facies is best developed in the lagoon on the lee side of Andros Island (Fig. 10.10). This area is characterized by quiet water containing a high content of carbonate. The ellipsoidal pellets of clay and silt-size carbonate particles are probably fecal in origin. Algae and *Thallasea* (turtle grass) are common. Few primary sedimentary structures are preserved because the sediment is highly burrowed by various clams and worms. Locally, in the lagoon, pellets are not abundant, and the facies gives way to a predominantly mud facies.

**Skeletal Mud Facies** The pelletal mud facies is bordered on the west and east by the skeletal mud facies on the lee of Andros Island (Fig. 10.10). Mollusks are the dominant skeletal component. However, coral, foraminifera, and algae are also present. Pellets are locally abundant, and the boundary of the facies is gradational with the pelletal mud facies.

**Grapestone Facies** The grapestone facies is best developed on the north and south ends of the Andros lobe where it covers several hundred square miles (Fig. 10.10). Grapestones range in size from 2.5 to 0.5mm. They consist of randomly oriented aggregates of various nonskeletal grains. Although skeletal grains of corals and mollusks are present, they generally are not abundant.

The grains are cemented at their contact with clay and silt-sized, irregularly-shaped aragonite crystals. Generally, the cement is restricted to the periphery of the grapestone grain. Thus, primary porosity is high.

Grapestones are thought to form in shallow, quiet water, although some gentle agitation is required to move the grains in order to permit peripheral cementation. The lagoon bounded by the Berry Islands (Fig. 10.11) contains a relatively thick section of grapestones that have common facies boundaries with relatively high energy coralgal and oolitic grains.

**Oolitic Facies** This facies is the least common of the shelf facies. It normally occurs in patches. It is commonly associated with the coral-algal and/or oolitic facies of the platform margin (Figs. 10.10 and 10.12). This facies is best developed in the normally low energy lagoons such as Bimini (Fig. 10.12). The oolites are 90 to $15\mu$ in diameter, and individual rhines are composed of aragonitic layers that are about $3\mu$ thick.

## SHELF-MARGIN LITHOFACIES

**Reef Facies** The shelf-margin facies are subdivided into reef, coralgal, and oolite deposits. The reef facies occurs on the windward side of Andros (Fig. 10.10). The reef track is about 250 km long, but only 50 to 60 meters wide. It is broken by numerous tidal channels. The reef is bordered on the west by a windward lagoon and on the east by the Tongue of the Ocean. The facies can be subdivided into backreef, reef crest, and forereef environments, each reflecting nutrient supply, wave energy, and depth of water. *Millapora, Acropora, Monastria,* and *Porities* are the major organisms of this boundstone facies (Fig. 10.13).

**Coralgal Facies** The coralgal facies occurs along the outer margin of the platform. Coral fragments, algae (including *Halimeda* and *Penesoplidae,* foraminifera, and mollusks) are the principal organisms of this facies. It is best developed on the northern margin of the Andros lobe, especially in the region of the Berry Islands (Fig. 10.11).

**Oolitic Facies** This facies is best developed on the lee margin of the platform (Fig. 10.10). It is composed of well-sorted oolites that are normally greater than 2mm in diameter. The nuclei of the oolites include relatively large recrystallized mud aggregates, *Halimeda,* mollusk, coraline algae, or echinoid fragments. The environment is highly agitated. The facies is bordered on the west by the relatively cool, deep waters of the Florida Strait and on the east by the shallow, warm, carbonate-saturated waters of the platform. Mixing of the two waters results in the precipitation of $CaCO_3$. This mechanism produces the $3-10\mu$ rhines.

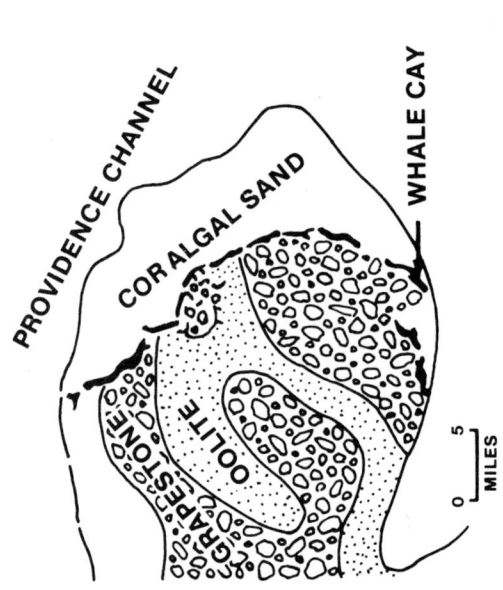
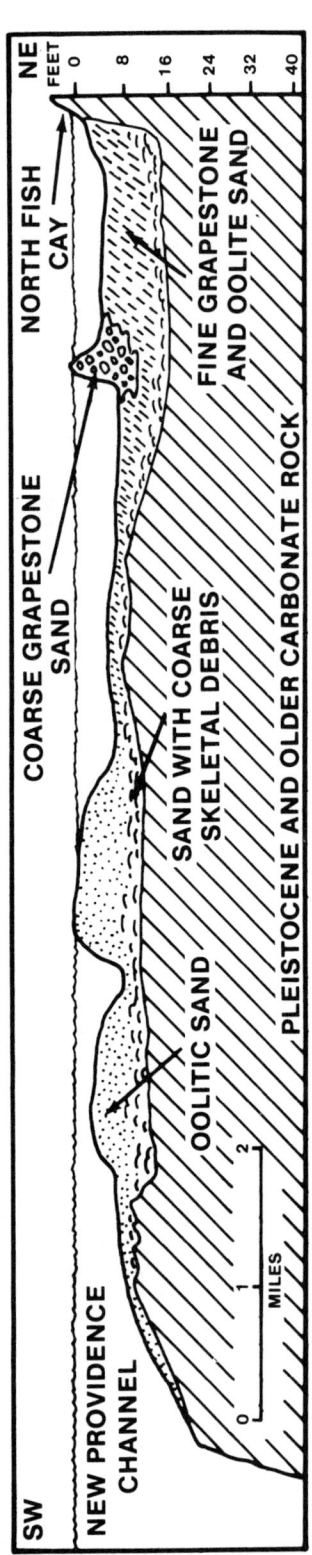

**Figure 10.11:** *Berry Islands are located on the northeast flank of the Great Bahama Bank. Grapestones, oolites, and coralgal sands are the typical carbonate particles.*

# CARBONATE DEPOSITIONAL ENVIRONMENTS AND STRATIGRAPHY

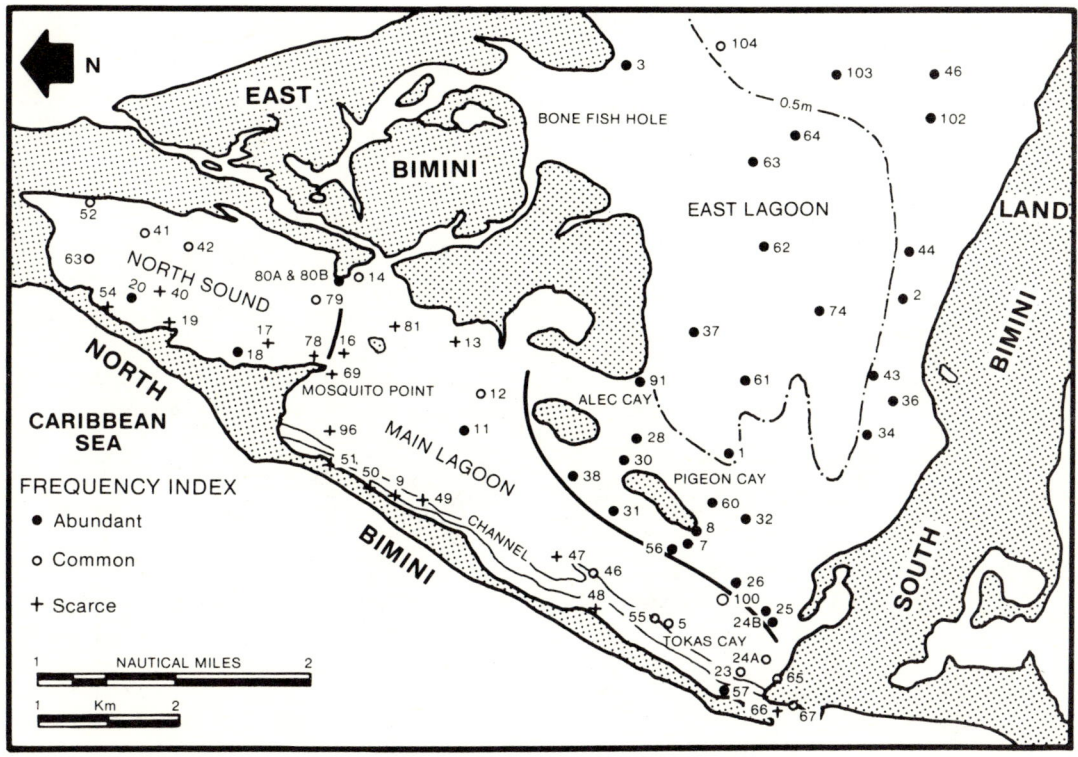

**Figure 10.12:** *Distribution of coated rounded grains, Bimini lagoon (Bathurst, 1967).*

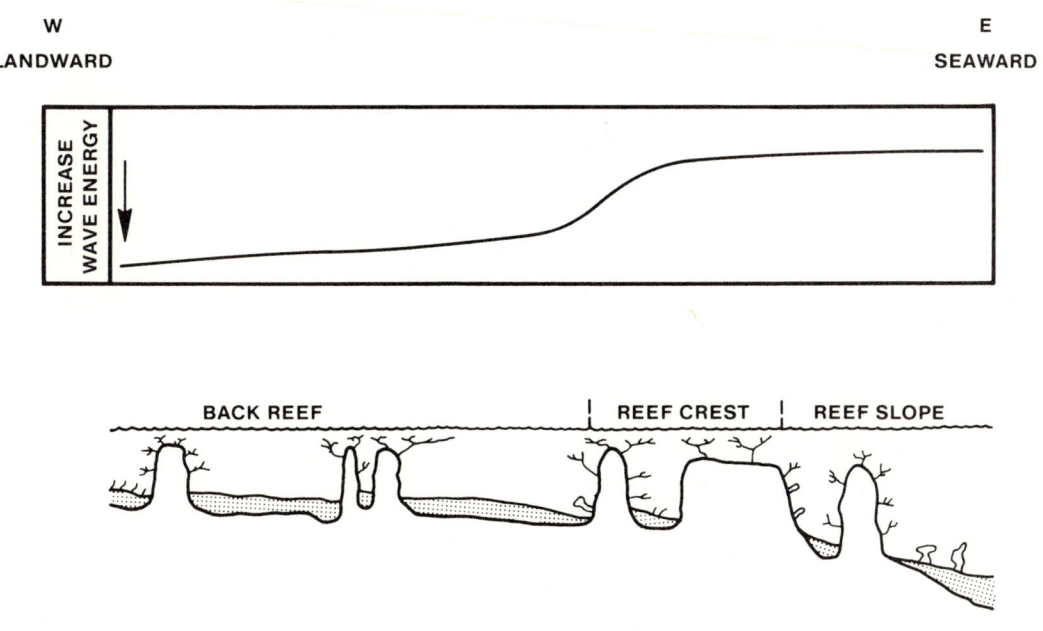

**Figure 10.13:** *Cross section of Andros reef tract showing sea-floor topography distribution of corals and wave energy.*

## BASIN ENVIRONMENTS - EXUMA SOUND

**Introduction**  Exuma Sound is one of three intraplatform basins of the Bahama Bank (Fig. 10.14). The elongated NW-SE trending trough is almost completely bordered by open marine platform and/or eolianite clays except for the southeastern border, where it is open to the Atlantic Ocean. Figure 10.15 illustrates the bathymetry of Exuma Sound. Table 10.1 summarizes the physiographic features of the bathymetric survey.

### TABLE 10.1

Submarine Physiographic Features Defined from Exuma Sound Survey

| | |
|---|---|
| Gullied slope: | Topography is irregular and hummocky, composed of gullies and spurs, dips greater than 3° and in excess of 20°. |
| Toe-of-slope: | Smooth, gently dipping sea floor of 0.5 to 2.2°, transitional area between gullied slope and basin. |
| Basin floor: | Flat, generally featureless, dips less than 0.5°. |

*Crevello, 1978*

**Peri-platform Muds**  Perennial rain of fine suspended sediment produces a carbonate mud composed of planktonic ooze.

**Graded Sand**  The graded sands form many of the thin and very thin layers interbedded with the basinal peri-platform muds. These sands are turbidite deposits as indicated by (1) grain size — it is much coarser than the intervening perennial sediment, (2) graded bedding, (3) abundant shallow-water grains, (4) sharp basinal contact and a gradational upper contact, (5) ordered sequence of internal sedimentary structures described by Bouma *(1962),* and (6) the presence of burrows in the upper few centimeters of the sand layers.

**Rubble Sands**  Rubble occurs in over 80% of the deposits in Exuma Sound. The rubble is composed of angular to sub-rounded chalk clasts in a muddy matrix. Some clasts are composed of micrite. It is poorly sorted, frequently graded, and grades upward into the graded sand facies.

**Pebbly Muds**  The pebbly mud facies is the least common of the facies in Exuma Sound. It comprises less than 3% of the stratigraphic record and about 12% of the debris deposited facies. The facies is characterized by dispersed textures and structures. It generally grades upward into the graded sand facies.

## SHELF-BUILDUPS - SOUTHWEST PUERTO RICO

### INTRODUCTION

Shelf margin barrier reefs and/or banks are not always the best reservoirs in a step model of deposition. Frequently, the shelf margin facies are cemented, as in the Edwards (Cretaceous) shelf margin of South Texas, or the facies lacks a seal, as in the Clear Fork (Permian) of West Texas. In these cases, the most common reservoir rock is typically a carbonate build-up located in the shelf lagoon behind the barrier reef (or bank). The buildups are typically small mounds, banks, or reefs, and they may be distributed randomly across the shelf. These buildups are difficult to explore for without some knowledge of their possible origin and seismic character.

The insular shelf of southwest Puerto Rico extends southward 3 to 7 kilometers from the island. The margin of the shelf consists of an inactive barrier reef that is submerged in 20 meters of water. Water depths on the shelf vary from a few centimeters to more than 30 meters (Fig. 2.4). The shelf contains three different trends of carbonate buildups that are separated by barrier inter-reef troughs *(Stewart and Silver, in preparation).*

### GENERAL DESCRIPTION OF BUILDUPS

Three types of carbonate buildups are recognized. Type I buildups range in thickness from 14 to 18 meters. Coarse coralgal sand with numerous thin interbeds of lime mud are the dominant lithologies. Basal sands are slightly coarser-grained than upper sands. *Acropora palmata* is the dominant coral, but *Diploria*

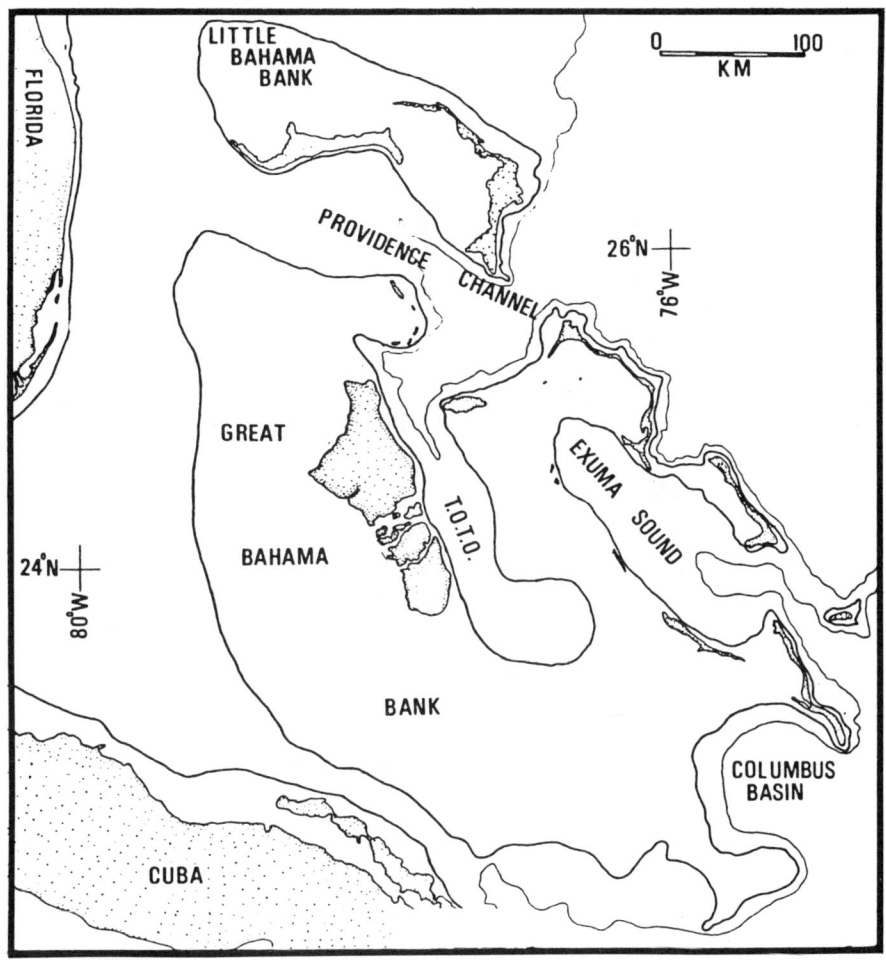

**Figure 10.14:** *Index map of northwestern Bahama Platform (Crevello, 1978).*

*sp, Monastrea annularis,* and *A. cervicornis* are common. *Halimeda* is the dominant algae. Thin beds (0.2m) of lime mud are interbedded with the coralgal sand (Fig. 10.16).

Type II buildups range in thickness from 14 to 18 meters. Like Type I buildups, Type II buildups are composed of coralgal sand and thin beds of lime muds. Unlike Type I, Type II are comprised of three zones, the lower and upper thirds of the buildups are composed of coarser grained sands than the middle zone. *M. annularis* and *A. cervicornis* are the dominant corals. *Milleporia sp.* is present but not common.

Type III buildups range in thickness from 6 to 10 meters. They contain a poorly diversified biota. *A. cervicornis,* and *Porites porities* are the dominant corals but the algae *Halimeda* represents the dominant biota.

## ORIGIN AND PREDICTABILITY

The geometry and 130° orientation to prevailing wind and current directions of the Holocene buildups suggest that pre-Holocene events and global sea level changes over the past 6,000 years have controlled their development. Type I buildups are thought to be formed on post-Miocene *en* echelon anticlines that were formed by left lateral northeast striking strike slip faults. These faults are associated with the Caribbean margin located south of the insular shelf. Distribution of the buildups are predictable, once the structural framework is documented. These buildups formed in deeper water than in the other two types of carbonate buildups. Sparker reflections depict vertical accretion and the buildups form oval sedimentary bodies (Fig. 10.17). Maximum primary

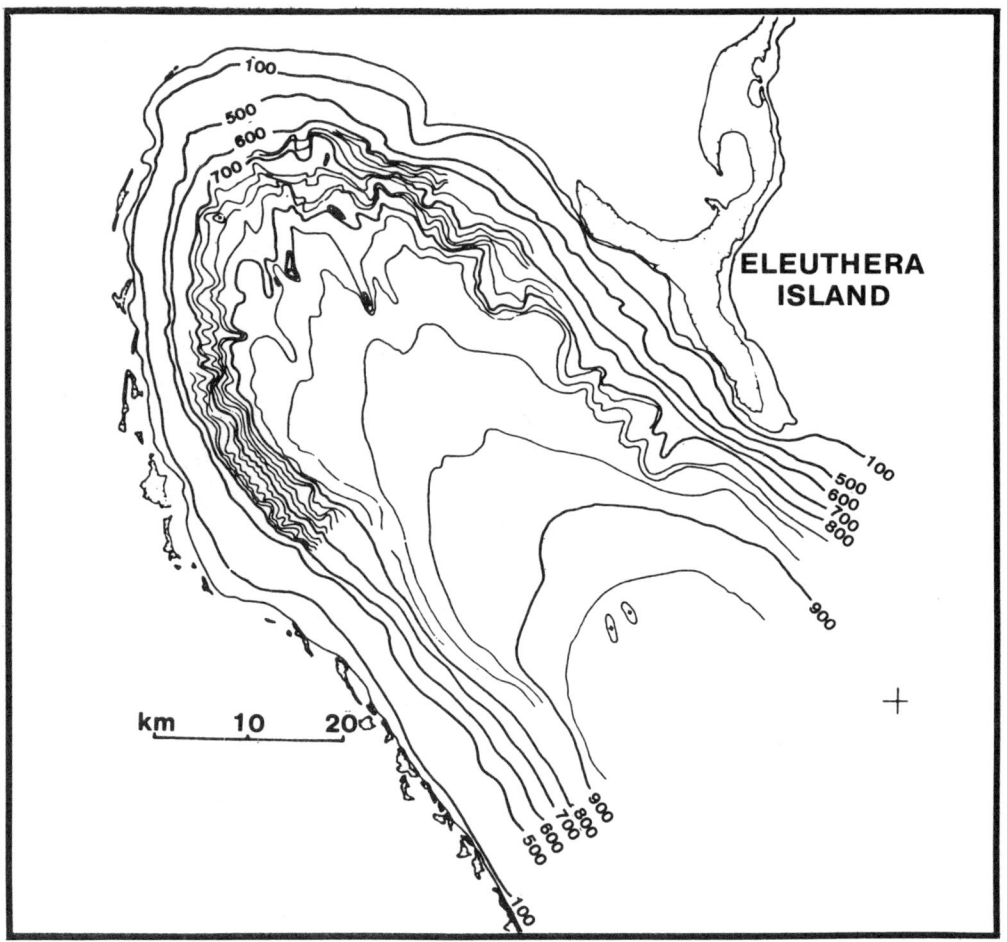

**Figure 10.15:** *Precision bathymetric map of northern Exuma Sound. Contours in 20 and 100 fathom intervals (Crevello, 1978).*

reservoir quality occurs in the upper, coarse grained coralgal sands *(Stewart and Silver, in preparation).*

Type II buildups form landward of Type I buildups. The buildups occur along a curvilinear break-in-slope that was produced by the post Cretaceous drape folds which also controlled the position of the Pleistocene shoreline. Predictability is excellent. Sparker reflections depict oblique progradational patterns that are in response to carbonate production during stillstands (Fig. 10.18) *(Stewart and Silver, in preparation).*

Type III buildups occur landward of Type II buildups. The former are low relief mounds that initiated in topographic lows. It is probable that the lows were formed by solution during the Pleistocene. Distribution of these buildups is random and predictability is poor. Sparker reflection data indicate that the buildups accreted in a landward direction during sea level stillstands (Fig. 10.19). Maximum primary reservoir quality occurs in the upper coarse grained coralgal sands *(Stewart and Silver, in preparation).*

## RAMP CARBONATE RESERVOIRS - PERSIAN GULF

### INTRODUCTION

Epeiric seas are epicontinental, generally shallow, and may be subdivided into three major zones — nearshore (coastal), middle (transitional), and outer. The latter is characterized by a sea floor that is below wave base; carbonate textures are generally fine-grained,

# CARBONATE DEPOSITIONAL ENVIRONMENTS AND STRATIGRAPHY

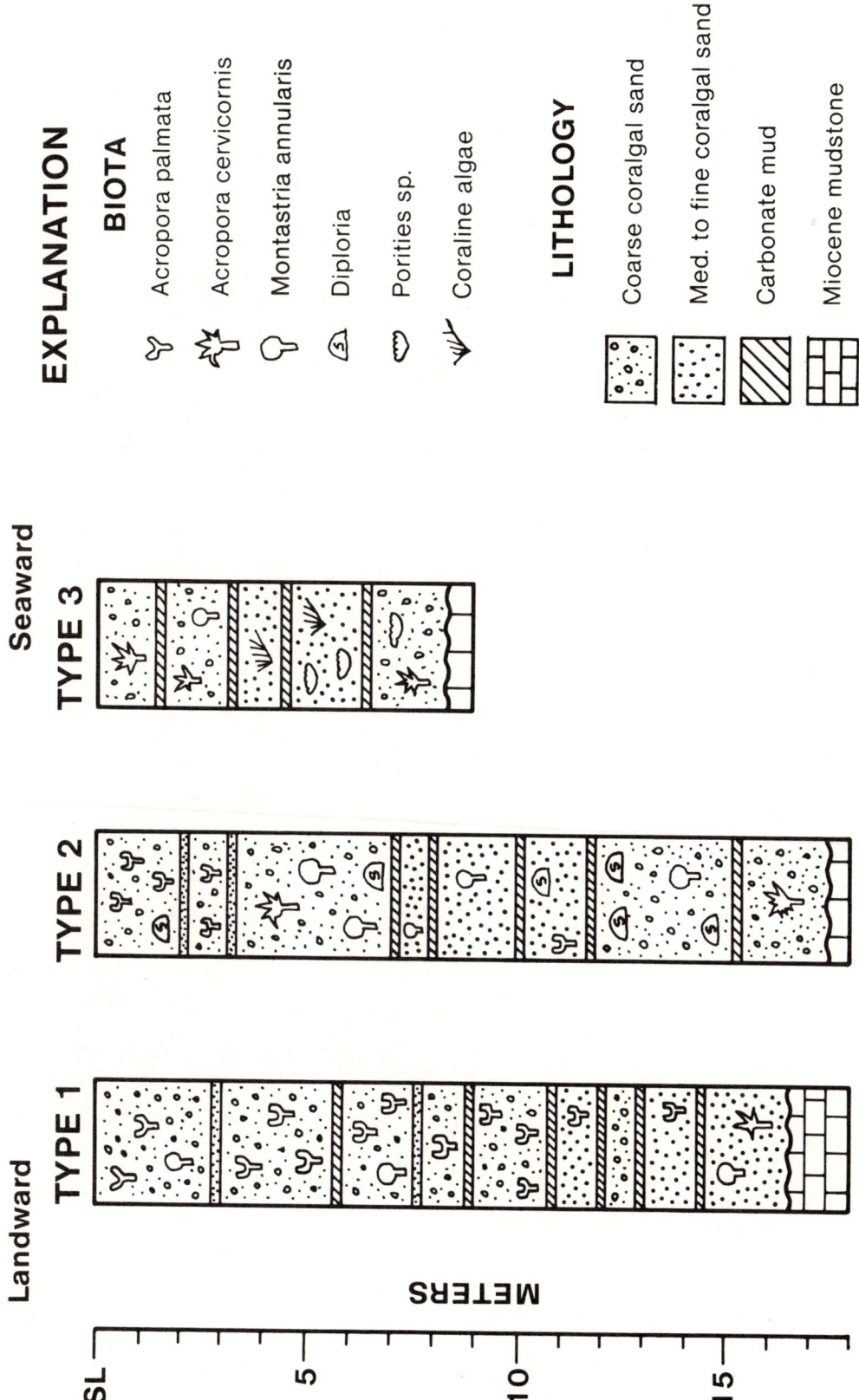

Figure 10.16: *Lithology and biologic content of carbonate buildups, southwestern Puerto Rico (Stewart and Silver, in preparation).*

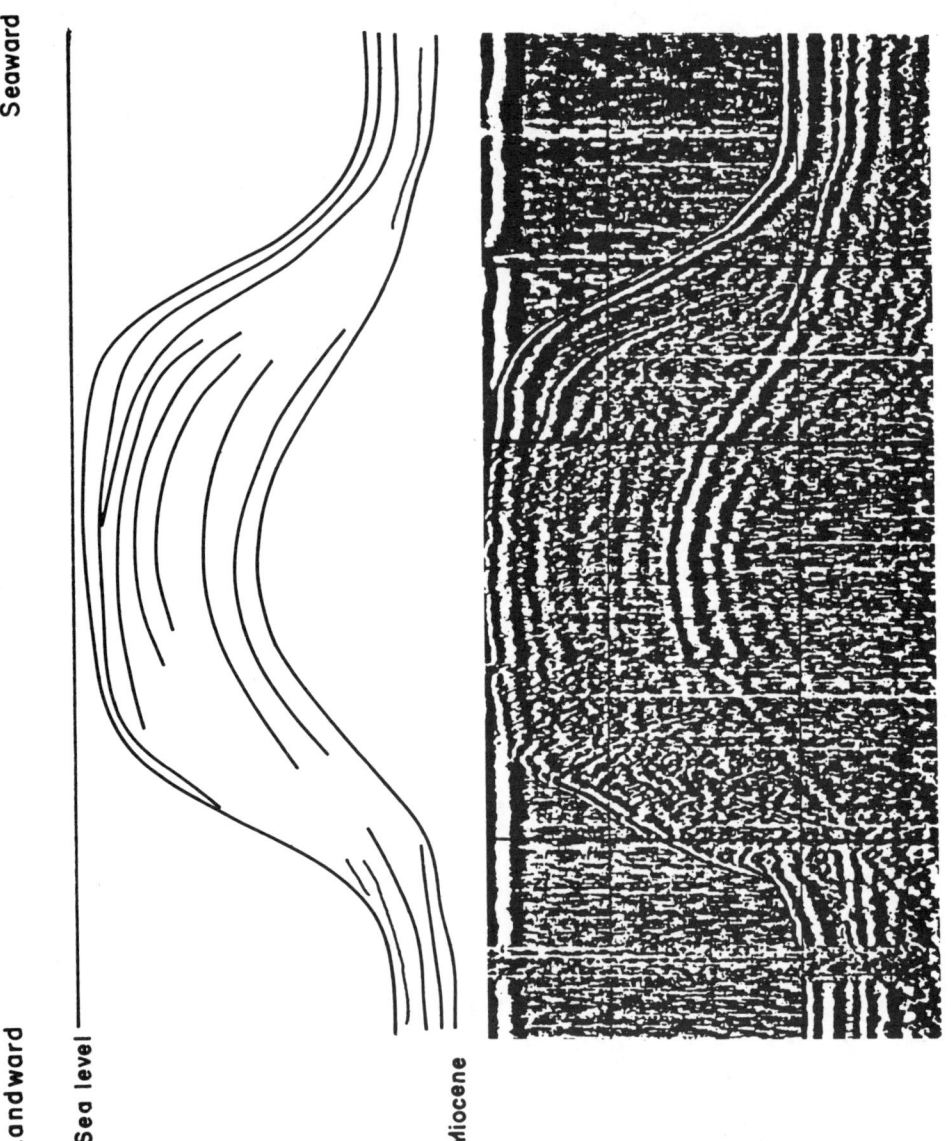

**Figure 10.17:** *Seismic signature of Type I carbonate buildups, southwestern Puerto Rico (Stewart and Silver, in preparation).*

# CARBONATE DEPOSITIONAL ENVIRONMENTS AND STRATIGRAPHY 209

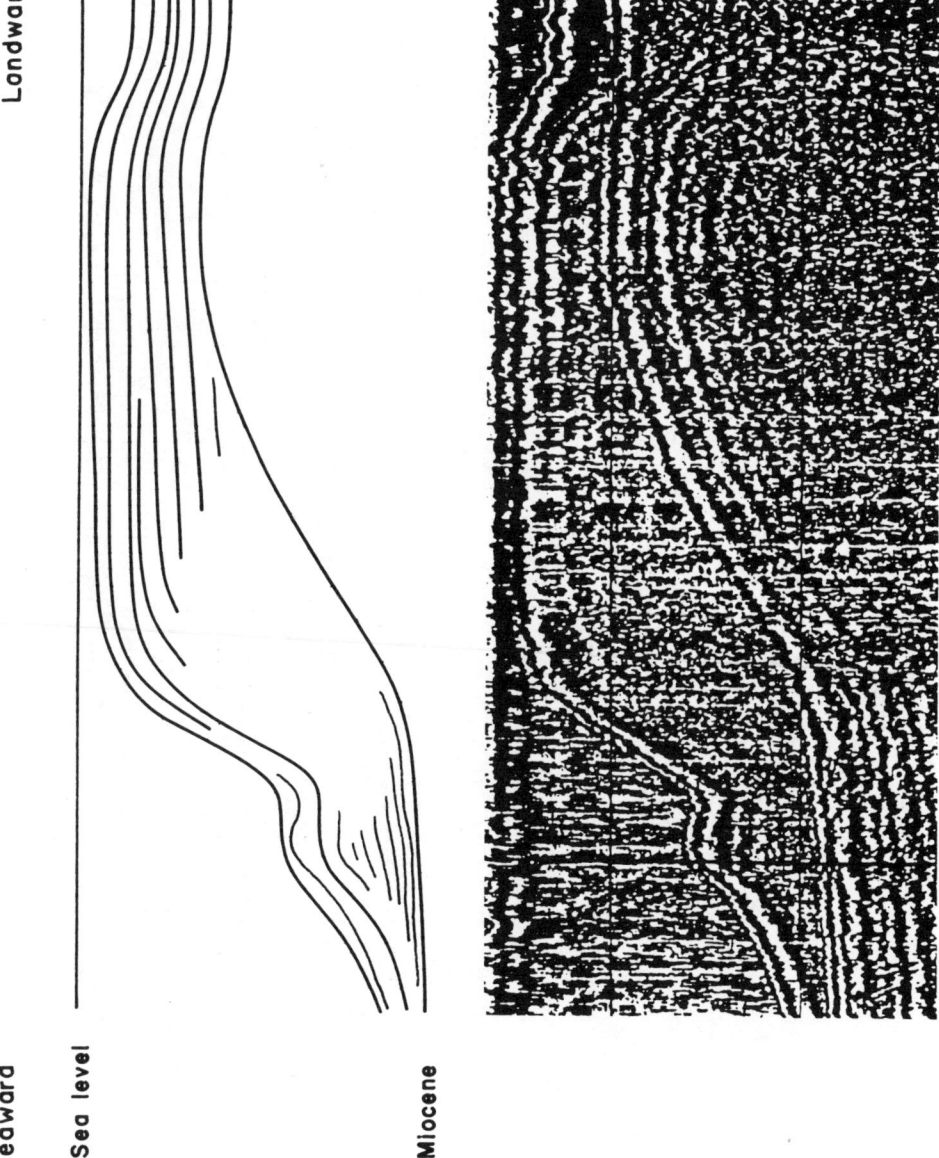

**Figure 10.18:** *Seismic signature of Type II carbonate buildups, southwestern Puerto Rico (Stewart and Silver, in preparation).*

210 EXPLORATION FOR CARBONATE RESERVOIRS

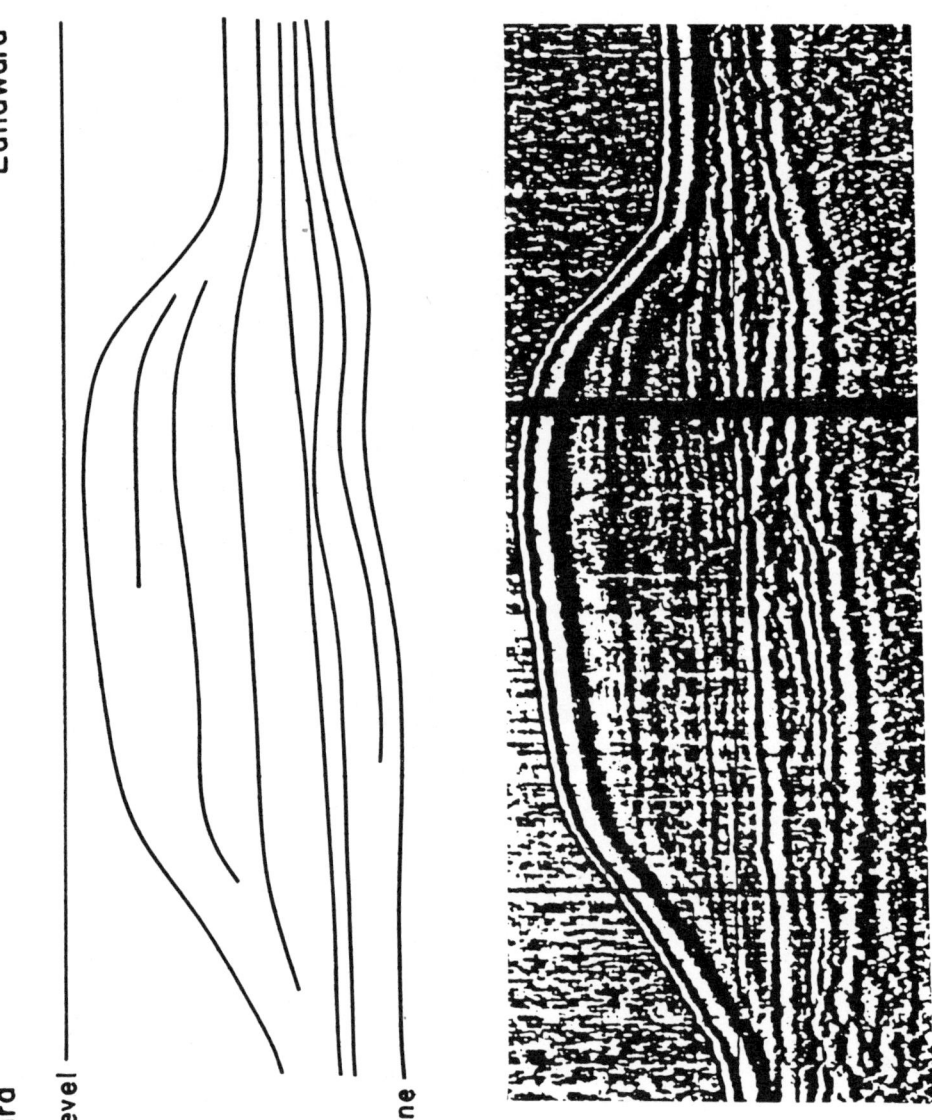

**Figure 10.19:** *Seismic signature of Type III carbonate buildups, southwestern Puerto Rico (Stewart and Silver, in preparation).*

and silt and clay are minor constituents. It is generally hundreds of miles wide and very low energy. In contrast, the sea floor of the middle zone is at, or near, wave base; therefore, it is highly agitated. Nutrients are abundant, and the resultant carbonate sediments are rich in skeletal grains. The zone is normallly tens of miles wide. It is possible for the zone to extend from the coastline to deeper water. Epeiric coastal deposits generally accumulate in quiet, extremely shallow water, and evaporites and dolomites are the dominant lithology. Like the outer zone, the coastal zone may be hundreds of miles in width.

## PHYSICAL SETTING OF THE PERSIAN GULF

A brief review of geologic history suggests that present-day topography and land surface may be greater than at any other time during the last 570 million years. Consequently, there is not an excellent modern day analog to the extensive epeiric seas of 300 to 570 million years ago. The Trucial Coast, however, is a scaled-down version that, in part, illustrates some of the carbonate lithologies and textures that accumulated in ancient epeiric seas. The area is located in the Persian Gulf (Arabian Gulf). It is bounded by the Qatar Peninsula, Iran, Straits of Hormuz, and Arabia (Fig. 10.20). The asymmetrical gulf dips gently into an east-west trending trough that contains about 40 fathoms of water. The "shamal winds" are from the northeast and are generated in an arid climate.

## COASTAL LITHOFACIES

**Evaporites-Dolomites** Laminated, dolomitized carbonate muds and bedded evaporite deposits represent the tidal and supratidal environments in the Trucial Coast (Fig. 10.21). The evaporites give way to nodular dolomitized carbonate muds which, in turn, are facies of the seaward algal laminated dolomitized muds. Pelecypods, gastropods, and foraminifera are the dominant contributors of skeletal grains.

## TRANSITIONAL LITHOFACIES

**Carbonate Sands** Coralgal, oolite, gastropods, and pellets are the major carbonate particles in this facies. The sands reflect wind direction and are most abundant on the lee side of Qatar Peninsula. The oolites, however, are concentrated in the tidal channels adjacent to coral reefs. The sands, in general, reflect shoaling conditions proximal to the reefs as well as sub-tidal wave base conditions.

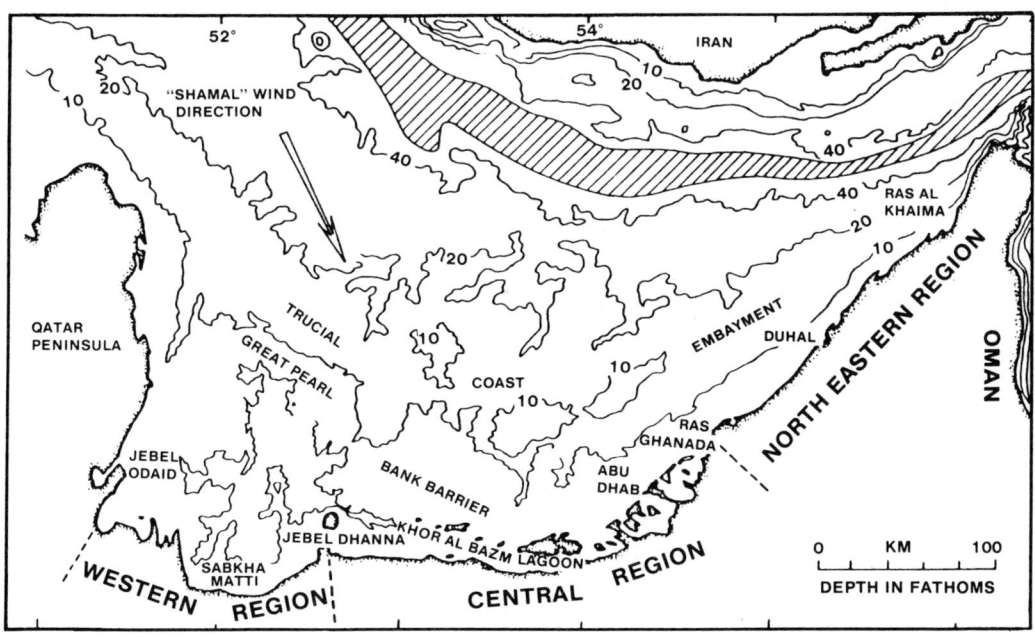

**Figure 10.20:** *Location and bathymetric map of the Persian Gulf.*

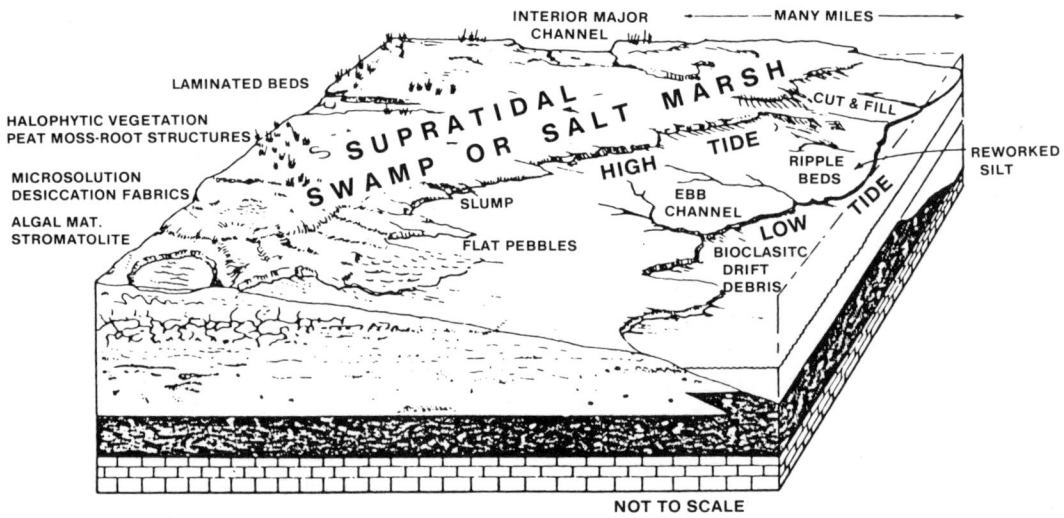

**Figure 10.21:** *Conceptual scheme of epeiric tidal flat-shoreline environment of carbonate sedimentation.*

**Reefs** Coral reefs are located on "wrinkles" along the transitional zone. They are generally seaward of the islands and are thoroughly dissected by tidal channels. Oolites are common in the tidal channels.

## OFFSHORE FACIES

Fine-grained skeletal muddy carbonates are the dominant offshore facies. The facies is laminated and burrowed. Hardgrounds formed by cement are commonly underlaid by non-cemented, disordered dolomitic skeletal muds.

## "SPREAD OUTS" - CHALKS

### INTRODUCTION

Major oil and/or gas production in the United States, Middle East, Mexico, and the North Sea is from deep-water, aerially extensive pelagic limestone reservoirs termed *chalk*. A number of connotations of *chalk* occur in the literature, and each of these need to be defined. *Chalk* is a carbonate rock composed of micro-sized particles. It has a homogeneous microscopic fabric and tends to stick to the tongue when licked. *Pelagic chalk* is formed by the accumulation of calcitic micro-organisms. Coccolith algae are the major skeletal contributors (Fig. 10.22). Chalkify is a diagenetic process that results in grain-size reduction of coarse carbonate particles, such as ooliths, pelletoids, and skeletal grains so that the resultant diagenetic fabric has a chalk-like appearance. Most Paleozoic chalks are probably diagenetic in origin.

## FUNDAMENTALS OF CHALK DEPOSITION

Chalks are dominantly composed of coccoliths and foraminifera. Both consist of low Mg-calcite. Subordinate constituents include benthonic foraminifera, pelecypods, and in some older chalks, ammonites. Clay is a minor constituent. Chalks accumulate in water depths that range from 100 to greater than 4,500m. The limiting factors for accumulation are rates of accumulation and rates of chemical and/or physical removal. Most chalk accumulates at rates of 1 to 30 m/my although some Cretaceous shelf chalks are thought to accumulate at as high as 54 m/my.

There are no climatic controls on chalk deposition; modern chalk is accumulating in nearly all climatic belts. Furthermore, there are few biochemical controls in chalk accumulation because coccoliths in particular have extremely broad tolerances ranging from tropical to boreal waters and from normal marine to extremely reduced salinities. Therefore, chalks can be deposited in a wide range of salinities and temperatures. However, in deep water areas, the dissolution of $CaCO_3$ affects the rate of accumulation. The calcite compen-

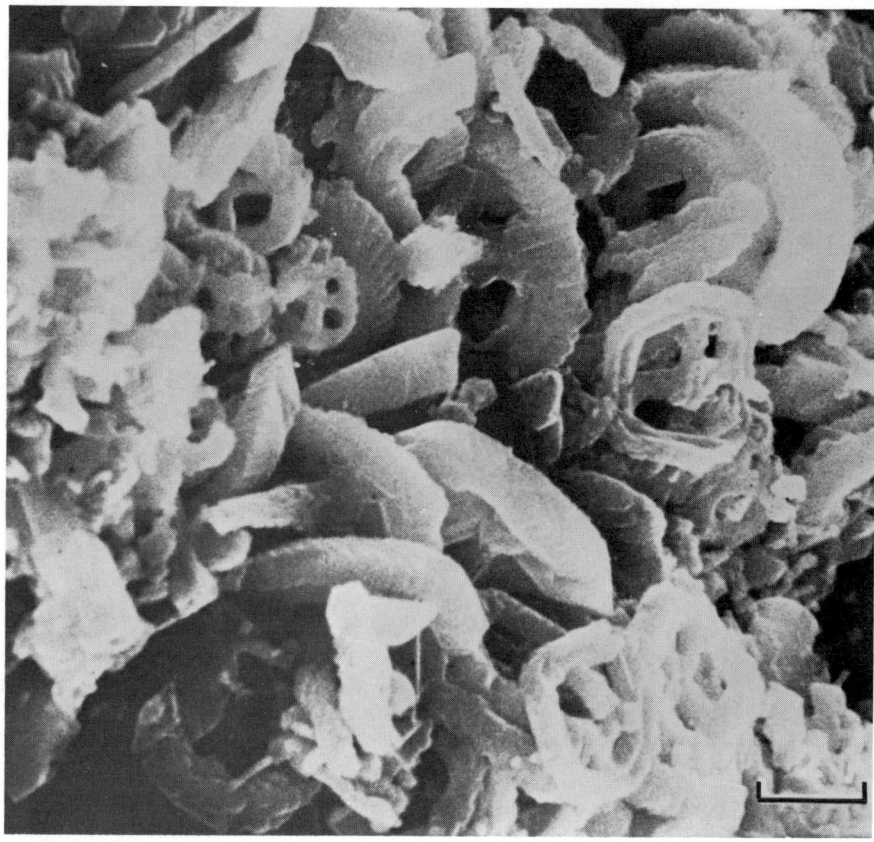

**Figure 10.22:** *Coccoliths are the major constituent of pelagic chalks (SEM photograph).*

sation depth (CCD) is normally at 4,000 to 5,000m. The removal of calcite minerals is very complex. It is a function of water chemistry, sedimentation rates, water depth, and biological productivity.

Chalks are rare in Paleozoic strata and become increasingly more abundant during and after the Jurassic Period. During post-Jurassic time, pelagic carbonates have tremendously increased in importance so that in the last 100 my, chalk comprises about 67% of the total carbonate rock record. This, of course, is due to the rapid evolution and expansion of coccoliths and planktonic foraminifera.

Because chalks accumulate in deep water, far from the influx of terrigenous derived clastics, they form aerially extensive (sheet-like) tabular rock bodies. Large scale sedimentary structures are rare. Smaller scale structures include rhythmic bedding, horizontal laminations, and extensive burrows.

## STRATIGRAPHY OF CARBONATE DEPOSITS

### BASIC FACIES PATTERNS

Modern carbonate depositional systems are characterized by as many as nine distinct "facies belts" *(Wilson, 1975)*. Explorationists working with mechanical logs, limited sample and/or core data, and seismic data rarely can delineate each of the nine facies belts, particularly in areas of limited subsurface control. Frequently, explorationists lump these facies into three zones. In step-and-platform depositional systems we can delineate shelf, shelf margin, and basin environments; and in ramp depositional models—coastal, transitional, and outer zones (Fig. 10.23). In mature basins (areas of abundant subsurface control) explorationists frequently can subdivide the three main environments

| RAMP | COASTAL | TRANSITIONAL | ABSENT | | | OUTER | |
|---|---|---|---|---|---|---|---|
| STEP & PLATFORM | SHELF | | SHELF MARGIN | | | BASIN | |
| | DISTAL | PROXIMAL | BACK | CREST | SLOPE | PROXIMAL | DISTAL |
| LITHOLOGY & TEXTURE | (1.) Nodular anhydrite and dolomite on salt flats<br>(2.) Laminate evaporites | (1.) Carbonate grainstones, packstones, and wackestones. Boundstones in carbonate buildups. | Grainstones & packstones | Boundstones, grainstones, & packstones | Grainstones, packstones | (1.) Non Fan: Laminated & burrowed wackestone & mudstone<br>(2.) Fan: Bedded packstone, grainstone, & wackestone | (1.) Non Turbidite: Laminated mudstones & evaporites<br>(2.) Turbidite: Packstones |
| COLOR | Red, yellow, brown | Dark to Light | Light | Light | Light to Dark | Dark to Light | Dark |
| BIOTA | Few indigenous fauna except stromatolictic algae | Abundant brachiopods, mollusca, sponges, cephalopods, echinoderms, forams, & algae. Various framework organisms in buildups | Few indigenous organisms | Major framebuilding organisms, *i.e.*, stroms, corals, specialized Brachiopods, algae, etc. | Local concentration of framebuilding organisms. Open marine fauna, *i.e.*, forams | Diverse fauna including both infauna and epifauna except in the fan and atolls | Exclusively pelagic fauna except in the turbidite deposits |
| POTENTIAL PRIMARY RESERVOIRS: ENVIRONMENT | Supratidal and Tidal | Lagoon & Bay shoals and local buildups | Shoal | Reef (windward) Bank (leeward) | Fore reef or fore bank | Fan and Atolls (1.) | Turbidites (1.) |
| POTENTIAL PRIMARY RESERVOIRS: GEOMETRY | Tabular | Local Buildups: Linear, pinnacle, ring, cellular<br>Buildups Abs: Tabular | Curvilinear | Curvilinear | Curvilinear | Single Fan: Tear-shaped<br>Multi Fan: Biconvex wedge | Single: Tear-shaped<br>Multi: Tabular |
| POTENTIAL PRIMARY RESERVOIRS: DOLOMITIZATION | Abundant | Abundant | Common | Common | Rare | Rare except atoll | Very Rare |
| MECHANICAL LOG CORRELATIONS | Straightforward, particular attention should be placed on shale. Disrupted locally in area of local buildups. | | Difficult, particularly in areas characterized by cycles. Look for shale breaks that correspond to coeval breaks on shelf. Probably reflect progradation during sea level stillstands. | | | Generally straightforward within basins only. Locally disrupted in areas of fan, atoll, or turbidite deposits. Difficult to correlate basin to shelf however. | |

(1.) Absent in ramp depositional models

(2.) Generally platform facies belts will be more restricted than step facies belts and the platform will contain both a leeward and windward environment.

**Figure 10.23:** *Generalized characteristics of ramp, step, and platform carbonate deposits.*

into seven facies belts. A brief description of each of these follows.

**Shelf** After a network of sedimentary strike and dip stratigraphic cross sections have been constructed, explorationists can subdivide the shelf facies into distal and proximal. Distal facies correspond to the coastal zone of the ramp; proximal facies are similar to the transitional zone of the ramp. Typically, the distal (coastal) zone is characterized by nodular anhydrite and dolomite. Laminated evaporites are common, particularly in tidal pools. Deposits are commonly red in color. They contain few indigenous fauna; however, algal and stromatolictic structures may be locally common. These strata are tabular to linear in geometry, and traps may develop where the dolomites "wedge out" into the evaporites. Mechanical log response to this facies is depicted on Figure 10.24A, and correlations are normally straightforward. Progradation is the dominant sedimentary process.

Characteristics of the proximal shelf of the step and transitional zones of the ramp are similar. Both zones are composed of extensive grainstones and packstones with local buildups of boundstones. Shoal deposits of oolites and/or pellets are common. They are typically light in color, but deep shelf deposits can contain abundant organic material that gives them a dark color. Biota is abundant and includes brachiopods, mollusca, sponges, cephalopods, echinoderms, forams, and algae. Local buildups are composed of various growth from building organisms such as corals, rudestids and specialized algae, bryozoans, and brachiopods. Potential reservoir rocks in stratigraphic and combination traps include oolitic and/or pellet shoals as well as carbonate buildups. The almost tabular grainstone and packstone facies can produce on structure. Mechanical log response typically depicts progradation, but upward accretion is also an important process (Fig. 10.24B). Shale breaks occur, but they normally are not as abundant or as thick as those in the distal shelf (coastal zone).

**Shelf Margin** Windard shelf margins are characterized by barrier and/or fringing reefs, whereas leeward shelf margins are composed of barrier and/or fringing banks. Typically, explorationists can recognize reef (bank) core and slope deposits, but precise definition of the backreef (backbank) deposits is sometimes difficult. Backreef and backbank deposits are normally composed of grainstones and packstones that are made up of coated grains, pellets, and skeletal fragments. The light-colored deposits contain few indigenous organisms. The facies are typically curvilinear and, of course, their distribution is parallel to trend of the reef or bank. Typically this facies is less porous than the reef or bank facies, and production from the former facies may be marginal.

The reef is composed of frame-building organisms (stromatoporoids, corals, and/or various specialized algae, bryozoans, and brachiopods). Boundstones, grainstones, and packstones are common. In contrast, bank deposits are dominated by grainstones that are composed of oolites and/or skeletal fragments. Both deposits are light in color. Perhaps shelf margin deposits, more than any other carbonate deposit, are the easiest to map and predict because they typically form curvilinear rock bodies. Numerous successful "lease plays" have been conducted on the basis of "straight edge" geology. Mechanical log response through shelf-margin crest deposits are usually characterized by continuous, well-developed SP and/or GR curves (Fig. 10.24C). Correlations within the deposits are normally difficult, and arbitrary slice maps are frequently the best method to map reservoir characteristics within the deposit.

Shelf-margin slope deposits are typically composed of grainstones and packstones. Individual grains may have been derived from both the shelf and shelf margin. These light-to-dark colored deposits contain local frame building organisms and a diverse, open marine fauna. Like the reef and bank, slope deposits are curvilinear. Mechanical log response usually reflects a progradational sequence, and the funnel-shaped SP or GR curve reflects a coarsening upward zone (Fig. 10.24D). Internal correlations are generally impossible over any significant area.

**Basin** Deep-water deposits of the step model (basin) and ramp model (outer) have many similarities. Both deposits can be subdivided into distal and proximal facies. Deposition in

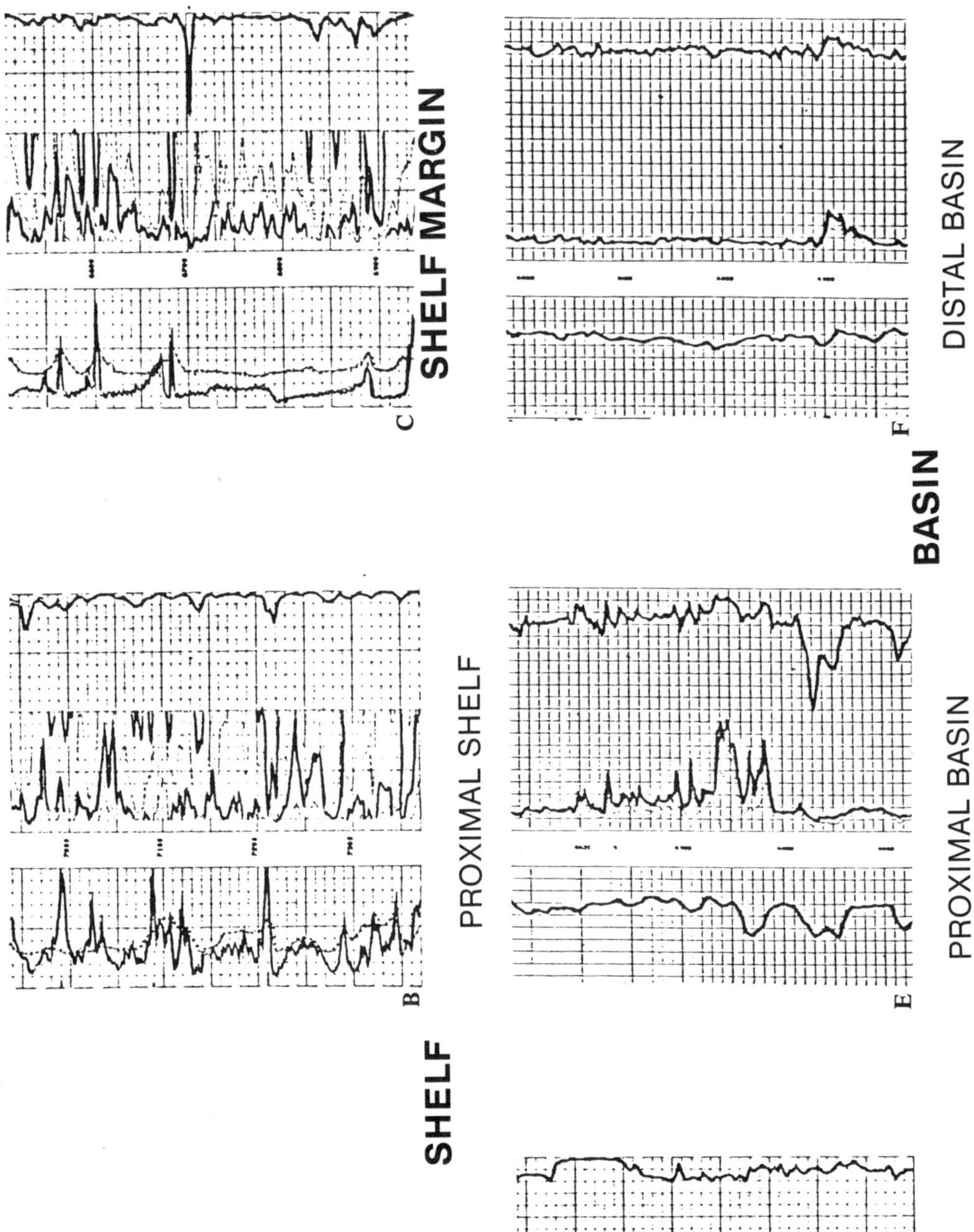

**Figure 10.24:** *Typical log response for shelf, shelf margin, slope, and basin facies.*

the proximal environment occurs in water depths shallow enough (but below wave base) to maintain well-oxygenated waters of normal marine salinities. If clastic influx is high, shale will be the dominant lithology; but if clastic influx is low, fossiliferous, highly-burrowed carbonates will occur. Deposition in the distal environment is characterized by water depths exceeding several hundred meters. During the Precambrian and Paleozoic, shales are the dominant deposit. In contrast, Mesozoic and Cenozoic deep water deposits are frequently chalks.

Submarine fan and turbidite deposits can occur in basinal deposits of step and platform depositional systems. They are absent in ramp deposits. Submarine fan and turbidite deposits contain the full spectrum of Dunham's textures. Sedimentary structures are identical to those described for sandstone turbidite and fan deposits. Fauna are represented by both displaced shelf and shelf-margin organisms as well as nektonic organisms.

Log response for distal and proximal nonsubmarine fan and turbidite deposits are depicted on Figure 10.24E and 10.24D. Correlations are generally straightforward except, of course, in the area of fan and turbidite deposits.

## EVOLUTION OF RAMPS TO STEPS

Carbonate production and, therefore, depositional rates are maximum along the transitional zone of a ramp where oxygen and nutrient supply is optimum. Therefore under certain conditions (slope angle, eustatic sea level changes, and/or rate of subsidence) ramp depositional patterns may construct a step depositional system (Fig. 10.25).

Degree of slope may be the most important controlling factor. If the slope is less than ½ degree or so (which is typical of the lower Paleozoic), the transitional zone may be tens of miles wide and hundreds of miles from the coastline. Although carbonate production is greater in this zone than in either the coastal or outer zones, the differential accumulation is spread over such a wide area that in order for a step to be constructed, the delicate balance between eustatic sea level and subsidence must exist over a significant span of geologic time (Fig. 10.25A & B). This situation rarely occurs.

In contrast, if the depositional slope is 3 degrees or more, ramp depositional patterns will "rapidly" construct a step depositional system (Fig. 10.25C & D). Maximum carbonate production within the transitional zone will produce a narrow belt of carbonates that parallels the coastline and that accretes to sea level. A shoreward shallow-water lagoonal environment will form on the once-active coastal zone where carbonate production was minimal. A deeper water environment will develop seaward of the belt on the outer zone because like the coastal zone, carbonate production is slow. During relative rise in sea level, the newly constructed shelf margin will depict onlap patterns. The amount of onlap will be determined by the rate of sea level rise and the gradient of the initial depositional surface (Fig. 10.25C). During relative sea level stillstand, the newly constructed shelf margin will prograde towards deeper water (Fig. 10.25D).

## EFFECT OF VARIATIONS OF SUBSIDENCE AND/OR SEA LEVEL ON DEPOSITIONAL PATTERNS

**Shelf and Shelf Margins** Rate of production of carbonate material is very sensitive to those factors that control biological processes which, in turn, are partly controlled by depth of water. Consequently, modification of carbonate depositional processes and patterns is highly influenced by eustatic sea level changes and differential rates of subsidence or uplift. An area of limited subsidence and a static sea level will be characterized by progradational deposition on the shelf margin and toplap on the shelf (Fig. 10.26A). Progradation will also occur in areas where carbonate production exceeds the rate of subsidence and/or sea level rise. Typically, the shelf margin becomes steeper and narrower with time. Onlap patterns will be replaced by toplap patterns when the shelf is equilibrated with sea level (Fig. 10.25B). Although an exact balance in nature is rare, it is possible for carbonate production to about equal subsidence and/or sea level rise. In these cases, the shelf and shelf margin will be characterized by vertical accretionary patterns (Fig. 10.26C).

Sea level falls are much more rapid than sea level rises *(Vail and others, 1977)*. Rarely can carbonate production keep pace which, in

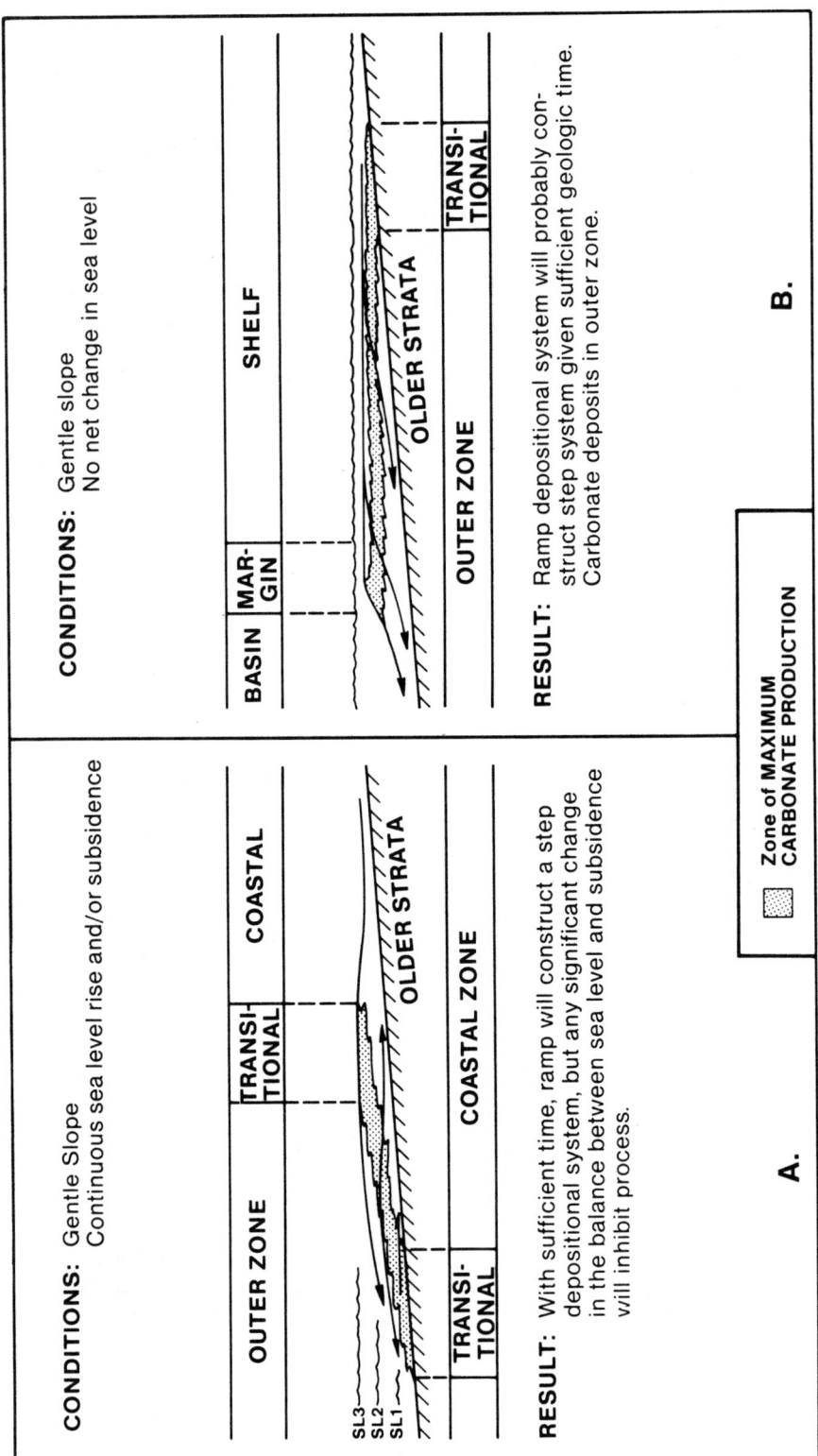

**Figure 10.25:** *Effect of slope and net sea level change on the evolution of ramp to step depositional systems.*

# CARBONATE DEPOSITIONAL ENVIRONMENTS AND STRATIGRAPHY

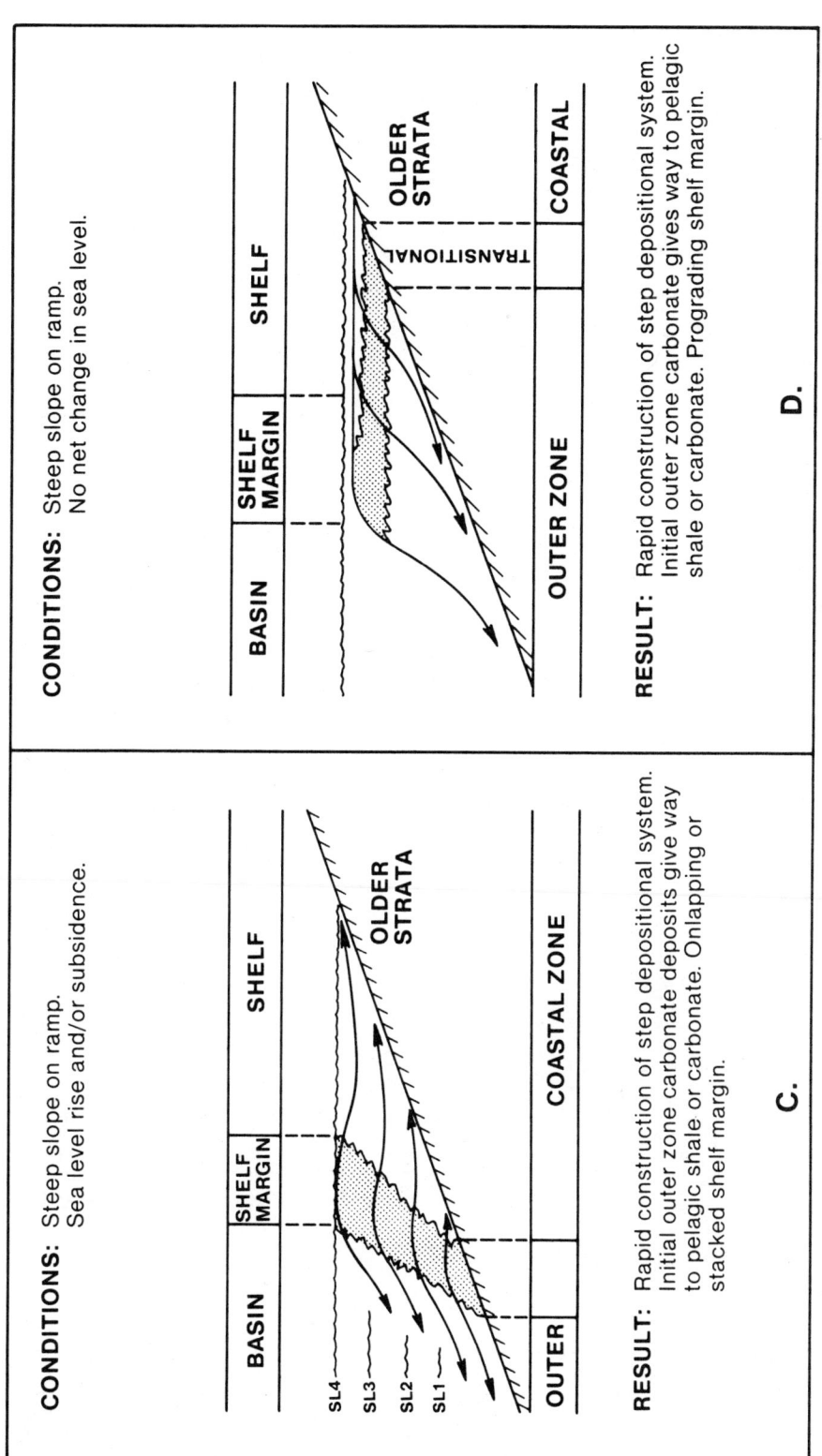

**Figure 10.25:** *Effect of slope and net sea level change on the evolution of ramp to step depositional systems.*

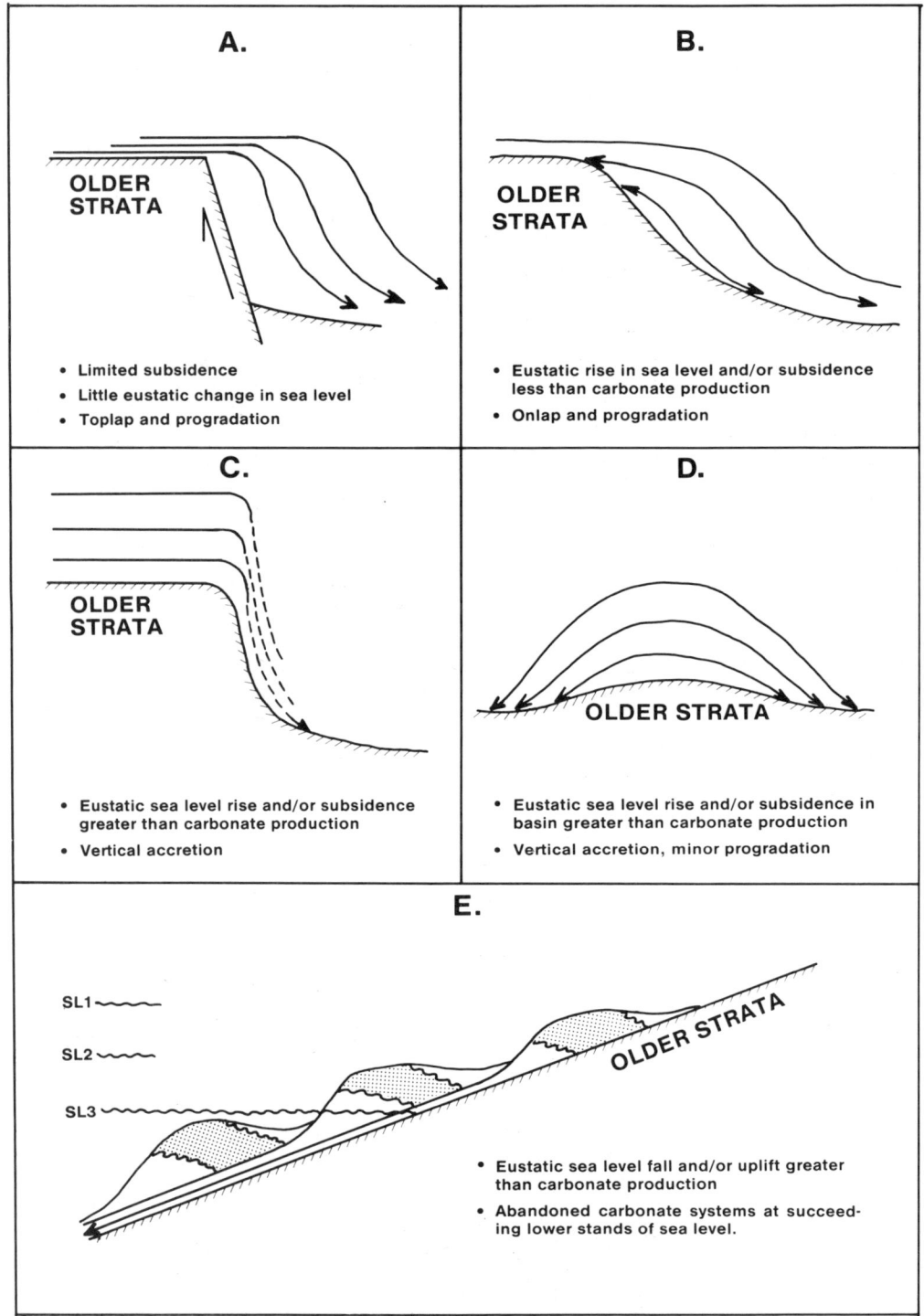

**Figure 10.26:** *Effect of eustatic sea level change and/or tectonics on carbonate depositional process.*

turn, results in stranded progradational depositional patterns at succeeding lower sea level stands within the basin (Fig. 10.26D).

**Basins** Typically basinal carbonate buildups initiated on a pre-carbonate high (*i.e.,* Enewetok, Marshall Islands, and Glovers Reef, Belize). These buildups are sensitive to differential rates of subsidence and/or sea level change. But in those cases where the buildups occur in deep water and they are "perched" well above the sea floor, progradational patterns are subordinate to vertical accretionary patterns. In contrast, where the buildup occurs in shallow water, progradational patterns are common (Fig. 10.26E).

## REGIONAL THICKNESS VARIATIONS

Regional isopach maps are useful to aid the explorationists to predict the distribution of potential reservoirs. There are numerous pitfalls in the construction and interpretation of isopach maps of carbonate rocks. Some of these include the following:
1. Boundaries of the isopach interval must approximate time surfaces. Selection of these surfaces is particularly difficult in those areas where changes in sea level and/or subsidence have produced several cycles of carbonate depositional patterns.
2. Correlation of shelf, shelf margin, and basin strata must be valid; otherwise the isopach map will depict erroneous patterns.
3. Submarine fan and turbidite deposits will be thickest in the basin, and it is possible (in areas of limited subsurface control) to mistake these thick areas as shelf margin deposits rather than basinal deposits.
4. Basinal structural "highs" may produce a platform on which thick carbonate accumulation will occur. Like the thick anomalies that were produced by submarine fan and turbidites, erroneous placement of the shelf margin is possible.

Diagrammatic isopach maps of step carbonate systems take on two general forms (Fig. 10.27). Thickness variation on the shelf is minimal and contour spacing is relatively constant. Minor closed thicks can occur on the shelf, particularly in those areas where the lagoon was several hundred feet deep. Contours are closely spaced along the shelf margin and, typically, they are more closely spaced across windward margins than across leeward margins. Local elongated closed contours along the shelf margin are common, and they may represent maximum carbonate production between major tidal channels or submarine canyons.

Isopach maps of carbonate ramp deposition are difficult to generalize as well as to construct. By definition, ramps are regionally-extensive, topographically-high surfaces with no break in slope. Minor changes in sea level or rates of subsidence may produce extensive discontinuities that are difficult to recognize in the subsurface. Furthermore, because of the large area involved, recognition of time surfaces from subsurface data may be impossible. Typically, ramp deposition will produce a wedge of sediment that will gradually thicken from the coastal and outer zones towards the transitional zone (Fig. 10.28A).

Isopach maps of carbonate platforms will depict an oval to elongated rock body (Fig. 10.28B). Contour spacing will be tighter on the windward side than on the leeward flank of the platform. Contour geometry in the center of the platform will be highly variable and dependent upon the nature of sea level changes, storms, and subsidence. If the platform was constructed on a structural high, it is probable that the center of the carbonate buildup will be thin rather than thick.

## SUBSURFACE STRATIGRAPHIC TECHNIQUES

**Assumptions** To date there is no clear-cut, single method to construct a valid stratigraphic framework for carbonate deposits. Each explorationist will be required to develop his or her own criteria to work out the physical stratigraphic relationships within the basin or area under investigation. And, unfortunately, the criteria developed for one area are not in total applicability to another area. However, the following assumptions and procedures have been useful to explore for hydrocarbons in carbonate rocks.
1. Sea level rises are rapid; sea level falls are abrupt; and the bulk of deposition occurs during periods of sea level stillstand *(Vail and others, 1977).*

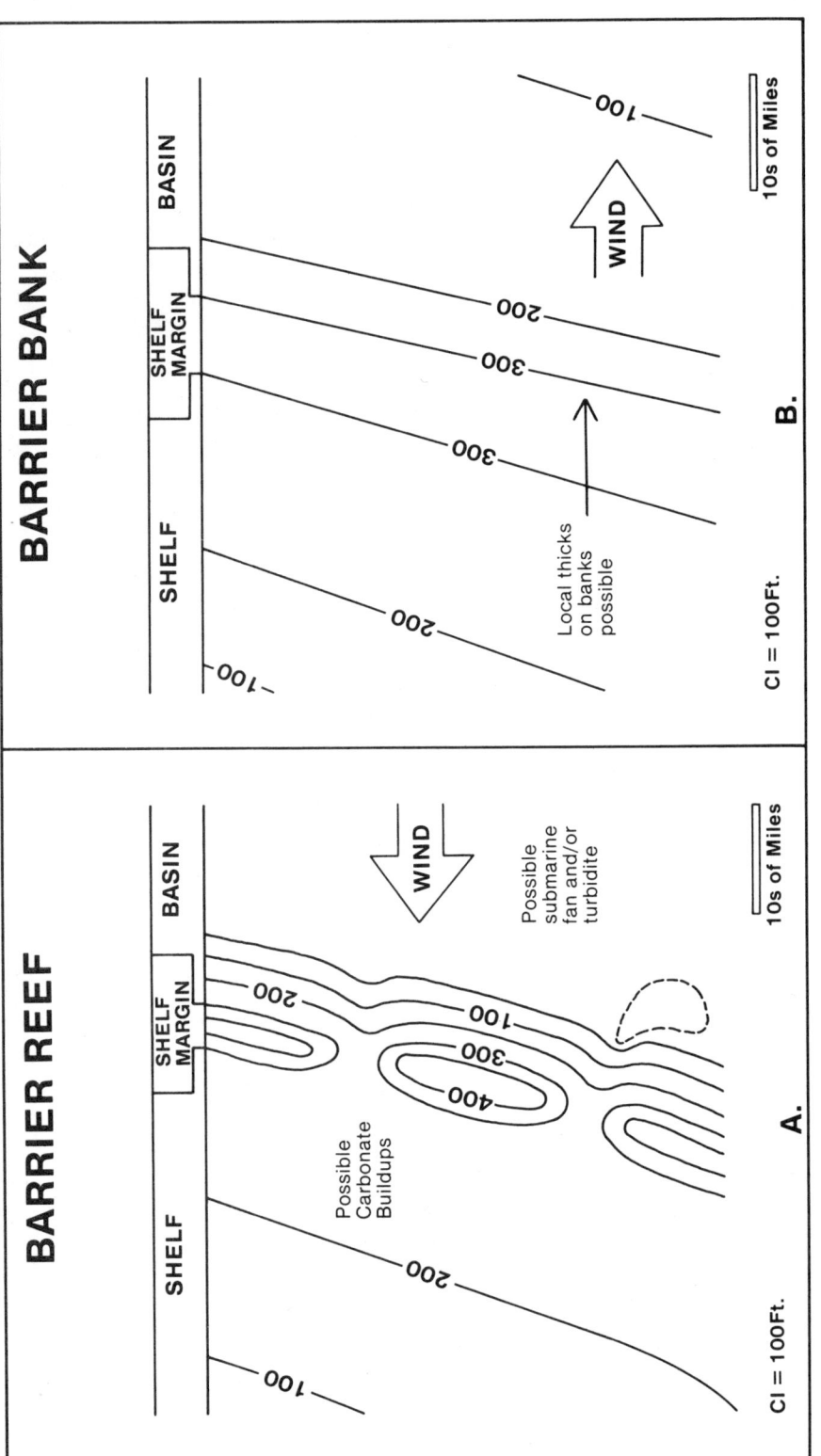

**Figure 10.27:** *Idealized isopach maps of step depositional systems.*

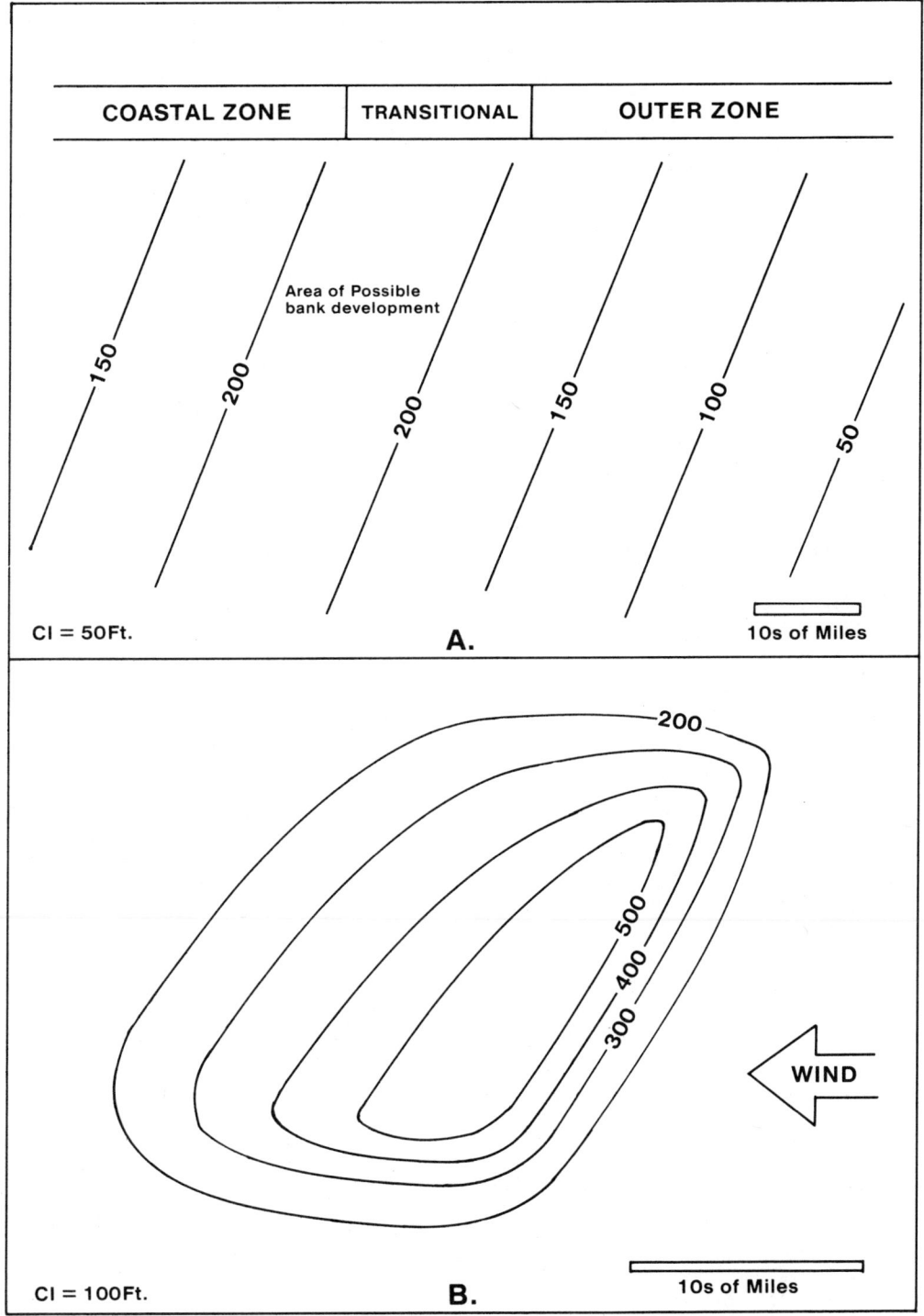

**Figure 10.28:** *Idealized isopach maps of A. Ramps and B. Platform.*

2. Marine transgression occurs over level surfaces except in those cases where the transgression occurred at the base of a sequence.
3. Carbonate production will equilibrate to sea level rapidly.
4. Time surfaces on the shelf and in the basin (or outer zone) will be more or less parallel to bedding surfaces.
5. Time surfaces on the shelf will be best manifested at the top of a shelf clastic that progrades across the shelf to the edge of the shelf margin either during sea level stillstand or sea level fall.
6. The rate of progradation across the shelf is more rapid than the rate of evolution of a single or group of species and, therefore, these progradational surfaces will correspond to excellent time surfaces.

**Suggested Technique** The following procedures have been useful for developing a stratigraphic framework for carbonate rocks.
1. Define sequence boundaries as outlined in Chapter Three. Each line of cross section should extend completely across the basin or area of investigation.
2. Construct a regional structural map on the basal unconformity of the sequence than contains the primary objective. Subtract out regional dip from the structure map in order to develop an appreciation for paleotopography.
3. Facies changes (*i.e.,* carbonate to shale) within the objective carbonate unit that were delineated by the regional cross sections should be plotted on a base map. This will produce a wide linear belt that will require further study. It may correspond to a break-in-slope depicted on the cross-plotted paleotopography map.
4. Construct a series of stratigraphic cross sections that trend normal to the facies change belt delineated in Step 3. These sections should be made using one inch equals 100-foot logs. All available sample, core, and DST data should be plotted on the mechanical logs. Use enough log interval to include the basal sequence unconformity and several marker beds above the carbonate interval under analysis.
5. Pattern correlate as many marker beds within the carbonate interval as practical. Typically, one or more of the markers will continue completely across the shelf.
6. Pattern correlate as many marker beds in the predominantly basinal clastic part of the cross sections.
7. Define lithogenetic rock units. Experience has shown that it is easiest to recognize the cyclic lithogenic units on the shelf, particularly in those cases where the basin facies is represented by shale. In those cases where the basin facies are represented by alternating clastics and carbonates, these cycles are a powerful clue to the number of cycles that will occur on the shelf.
8. The shelfward termination of markers within the basinal cycles coupled with the basinward termination of markers on the shelf cycles should permit the interpretation of prograding (or onlapping) coeval shelf margin cycles.
9. Cross check correlations throughout the other cross sections. Make sure, for example, that along depositional strike, you have not correlated a progradational unit with an onlapping unit without a reason. Construction of isopach maps of each chronostratigraphic unit will aid in checking correlation busts.
10. Construct facies maps for each chronostratigraphic unit.

## CONCLUSIONS

Unfortunately, there is no single or combination of methods that are universally applicable to develop a valid stratigraphic framework for cyclic carbonate depositional systems. But integration of all subsurface data within a sequence approach and careful observations of log response to relative sea level changes should produce effective results for a successful exploration program.

# 11
# Diagenesis of Potential Carbonate Reservoirs

## NATURE OF PROBLEM

Typically, modern grainstones and boundstones contain porosities in excess of 40% and permeabilities as high as 20,000 millidarcies. In contrast, ancient carbonate rocks rarely contain porosities above 3% and permeabilities greater than a few millidarcies. Figure 11.1 illustrates the relative frequency of carbonate reservoir porosity relative to porosities of Recent and Pleistocene rocks, and the dramatic pore occlusion that occurs in carbonate rocks over a very short period of geologic time. Early migration of hydrocarbons into a potential carbonate reservoir is almost mandatory if the high primary porosity and permeability is to be preserved.

The major processes of pore occlusion in carbonates include cementation, internal sedimentation, inversion, and neomorphism. Cementation is by far the most important process of pore destruction. A review of the early diagenetic processes of pore occlusion was presented by Longman *(1980)*.

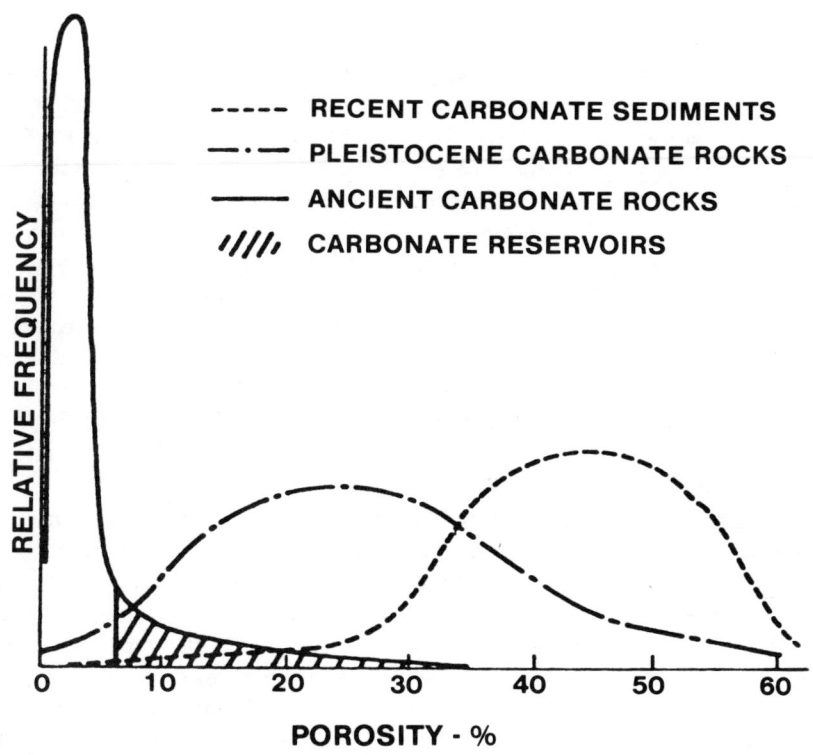

**Figure 11.1:** *Relative frequency of potential carbonate reservoirs plotted against their porosity.*

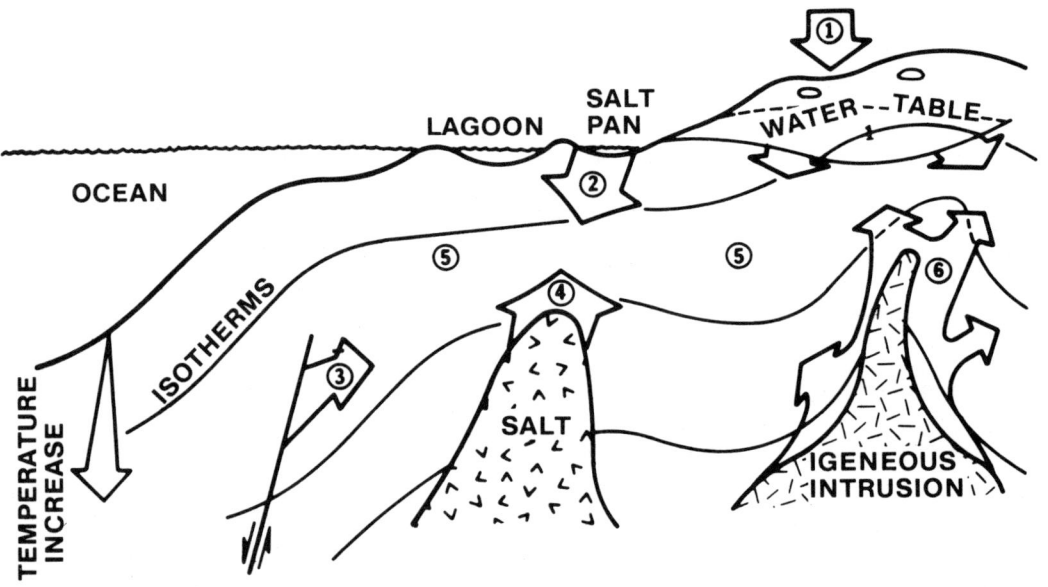

**Figure 11.2:** Sources of diagenetic water:
1. Meteoric  2. Hypersaline
3. Compaction  4. Dehydration
5. Connate  6. Hydrothermal

## PROCESSES OF PORE REDUCTION

### CEMENTATION

**Introduction** Carbonate cements are pore-filling calcite that may be composed of aragonite, Mg-calcite, calcite, or dolomite. The cement may contain varying amounts of $Fe$ (both $Fe^{+2}$ and $Fe^{+3}$), $Mn$, and $Sr$. The kind and amount of these trace elements are, in part, controlled by the source of pore water from which the cement precipitated. Figure 11.2 illustrates the various sources of diagenetic pore water that may migrate through carbonate rocks.

Controlling factors for pore occlusion by cementation include (1) chemistry of pore fluid, (2) rate of pore fluid migration, (3) rate of pore fluid recharge, and (4) duration that the rock is subjected to a given diagenetic environment. Cementation of carbonates can occur in each of the diagenetic environments (Fig. 8.2), and because each of these environments has a distinctive range in chemistry and textures, it is frequently possible for the explorationists to reconstruct the paragenetic sequence of cementation. This, in turn, permits the explorationist to interpret the relative timing of migration of hydrocarbons into the reservoir rock and, in many cases, use this as an exploration tool.

**Marine Cements** Unlike sandstones, cementation of potential carbonate reservoirs can occur almost contemporaneous to deposition. Figure 11.3 illustrates the textures of the most common marine cements. They are normally composed of aragonite or high Mg-calcite. Complete pore occlusion in the marine diagenetic environment is rare, but primary porosities can be reduced from 12% to 20%. Marine cements are most common in sediments produced in reefs or banks, and in sediments that accumulated in shelfal environments where sedimentation rates are slow.

**Phreatic Cements** The phreatic diagenetic environment is characterized by pore water that contains less than 1,000 milligrams per liter dissolved solids (a specific conductance of less than 2,000 micromhs/cm). Cementation in the phreatic environment is by far the most significant process of pore occlusion. These cements normally are blocky in texture (Fig. 11.4). Typically, they are separated from the framework grain by a marine cement, but solution of the latter may occur prior to

# DIAGENESIS OF POTENTIAL CARBONATE RESERVOIRS

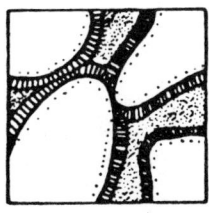
EVEN ("ISOPACHOUS") FIBROUS CEMENT

MUD CEMENT
• MATRIX
• PELLETS
• INTERNAL SEDIMENT

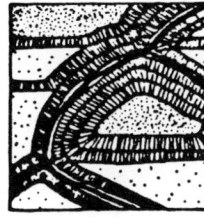
MULTIPLE GENERATIONS

**Figure 11.3:** *Early marine cements*

## INTERNAL SEDIMENTATION

Internal sedimentation of microcrystalline calcite in both intergranular and intragranular pores at or near the lithosphere surface can effectively reduce original pore volume. This process is especially important in supratidal, tidal flat, and reefal environments.

## CHEMICAL COMPACTION

The dissolution of carbonate particles and reprecipitation of the carbonate as a pore-filling cement results in the reduction of bulk volume. Net loss in porosity is equal to loss in bulk volume. Three general types of chemical compaction are illustrated in Figure 11.5.

Type I chemical compaction occurs early in the subaerial exposure of a carbonate rock or sediment. Meteoric water enters the rock or sediment column, dissolves carbonate, and transports it downward through the column. Grain to grain packing is not affected and

precipitation of the phreatic cement. Phreatic cements are low in Mg-calcite. Unless migration of hydrocarbons into the pores occurs prior to this stage of cementation, porosity will be reduced to less than 10%.

**Brackish Water Cements** The brackish water environment is characterized by pore water that contains greater than 1,000 milligrams per liter dissolved solids (a specific conductance of greater than 2,000 micrombs/cm). These typically later stage cements are equant, very coarse-grained, and fill the center of the pores. They contain varying amounts of *Fe* and *Sr,* but are always low Mg-calcite. Dolomite has been reported in the literature, but this is thought to be rare.

**Vadose Cements** Vadose cements are considered by this author as rare. When present, they are rhombic to needle fiber in texture and occur as a crust. They are low Mg-calcite and typically contain a great deal of *Mn*. Like brackish water dolomitic cements, vadose cements are not common.

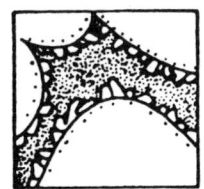
SINGLE LAYER DRUSE CEMENT

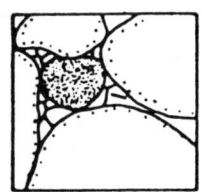
MENISCUS CEMENT

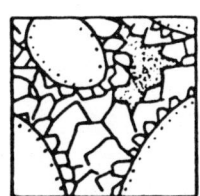
PORE FILLING CEMENT

OVERGROWTH ("SYNTAXIAL") CEMENT

**Figure 11.4:** *Phreatic blocky cements.*

### TYPE I: EXTERNAL EROSION OF SECTION

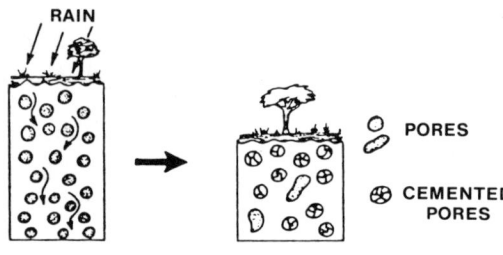

### TYPE II: INTERNAL EROSION OF SECTION

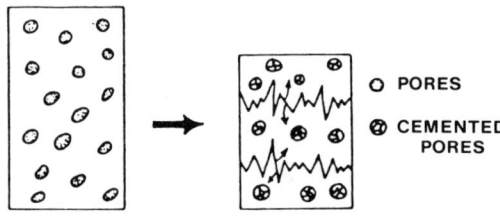

### MICROSCOPIC CHEMICAL COMPACTION

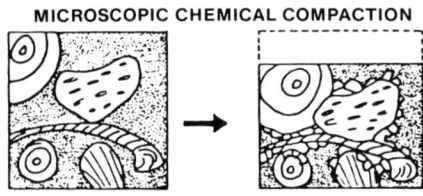

**Figure 11.5:** *Types of compaction processes.*

porosity normally deteriorates more rapidly than permeability. Normally, this process does not completely fill the pores, and the remaining porosity is typically concentrated in vugs or molds.

Type II chemical compaction occurs after relatively deep burial. The process is facilitated by clays, and the carbonates are transported by diffusion. The process is inhibited by early inversion of aragonite and Mg-calcite to the more stable form of calcite. Advanced stages of Type II compaction can result in extensive stylolitization, grain packing, and a completely occluded rock.

Type III chemical compaction results in pressure solution of grains at their point of contact. The process results in a closer packed rock and permeability is typically reduced more rapidly than porosity. The process is inhibited by early cementation and accelerated by mineral differences. The process is a self-inhibiting one and rarely is the reservoir potential totally destroyed by the process.

## INVERSION

Many organisms that constitute carbonate sediments are composed of varying amounts of aragonite and Mg-calcite. Although the aragonite and high Mg-calcite are at equilibrium while the organism is alive, once it expires, the minerals begin to invert (dissolve and be reprecipitated) to the more stable form, calcite. This process of inversion results in an increase in volume which produces a loss in porosity (Fig. 11.6).

## NEOMORPHISM

Neomorphism is the replacement of the original fabric by a mosaic of crystals. Normally, neomorphism produces a coarser-grained textured rock, but in some cases, grain-sized reduction may occur. Neomorphism does not involve significant change in chemical composition, but few carbonate rocks that have been completely recrystallized contain any effective porosity.

## PROCESSES OF POROSITY ENHANCEMENT

Like sandstones, carbonate secondary porosity and permeability may be formed even after complete pore occlusion, by various diagenetic processes. These processes include inversion, solution, fracturing, and dolomitization.

## INVERSION

Inversion of aragonite to calcite in certain framework organisms (corals) results in a reduction of porosity, but an increase in permeability, because it changes the pore aperture (pore throat) size during the change in fabric *(Pittman, 1974).* Research is currently underway to investigate this process in order to determine whether the process is applicable to non-framework organisms.

## SOLUTION - LEACHING

Carbonate minerals are extremely sensitive to changes in salinity, temperature, pressure, and pH. Consequently, any significant change in one or more of these solubility controlling factors may cause carbonate particles (includ-

# DIAGENESIS OF POTENTIAL CARBONATE RESERVOIRS

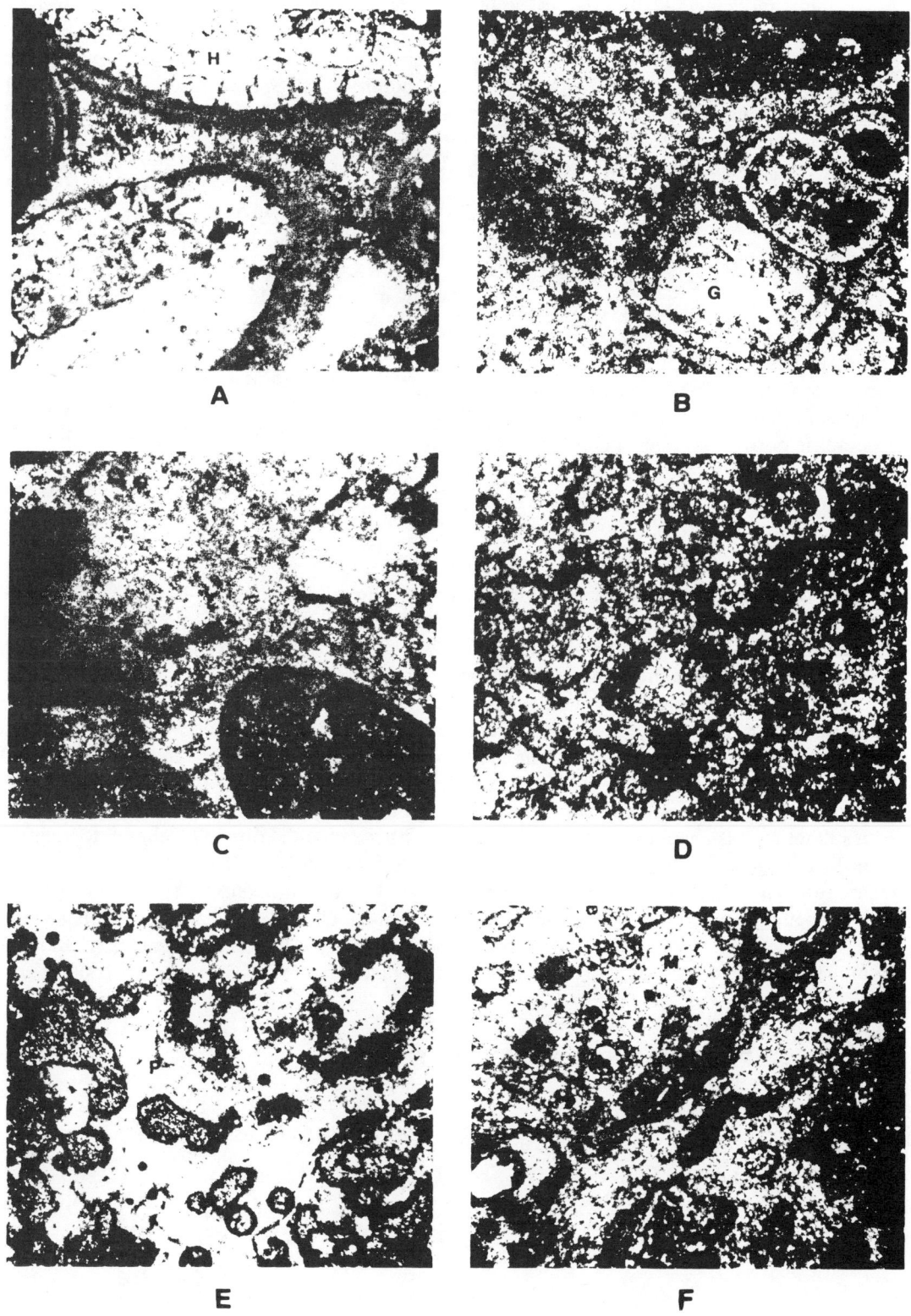

**Figure 11.6:** *Inversion results in pore occlusion, but in certain skeletal grains pore throats are enhanced up to a point prior to complete destruction of porosity.*

ing mud and cement) to go into solution. Because aragonite and high Mg-calcite are more soluble than calcite, particles that are composed of either of the former may be selectively leached, thus forming moldic porosity. Solution and leaching are most active when the sediment enters the phreatic environment from the marine environment, but it can occur in the marine and brackish environment.

## FRACTURING

Fracturing of completely cemented rock may produce new pathways for rapid migration of pore waters through the rock, and may produce a second stage of solution and leaching. Although fracturing, in itself, probably does not increase porosity, it does increase permeability.

## DOLOMITIZATION

**Introduction** Dolomitization is perhaps the most important diagenetic process that enhances porosity and permeability in carbonate rocks. Most dolomites are thought to be formed by replacement of magnesium for calcite in the calcite lattice. The $Mg$ ion is smaller than the $Ca$ ion, therefore, the replacement causes a 10% reduction in volume. This volume reduction is manifested in the form of intercrystalline porosity and fairly uniform permeability. The resultant rock is chemically stable and resists further chemical alteration. It will readily fracture and therefore, dolomites normally have better permeability than limestones.

The numerous processes of dolomitization may be classified according to the influence of facies on dolomitization patterns. *Concordant dolomites* are penecontemporaneous, they parallel bedding planes, and they are facies controlled. *Discordant dolomite* cuts bedding planes and they are only, in part, controlled by facies.

**Concordant Dolomites** Concordant dolomites can form by two very different processes, evaporation and biochemical sulfate reduction. During periods of supratidal exposure, sea water rises to the surface by capillary action where it is evaporated (Fig. 11.7). The upper layers of the supratidal flat become saturated with lime. Precipitation of $CaSO_4$ minerals from the brine concentrates $Mg$ ions. The excess brine and salts are dissolved during the next tidal flooding, and the $Mg$ ions replace $Ca$ ions in the calcite lattice. Dolomitization occurs layer by layer. Consequently, concordant dolomites have the characteristics mentioned above.

Microbial induced sulfate reduction appears to cause concordant dolomitization of Holocene muds in the northern Belize shelf lagoon *(Silver and Bloch, in preparation)*. Total organic material ranges from trace amounts to more than 5%. Sulfide was removed from the interstitial waters in the mud by sulfate reduction. Protodolomite ranged from trace amounts near the surface to 17% at depth and its abundance is directly related to degree of sulfate reduction (Fig. 11.8).

The initial results of the Silver and Rafalska-Bloch study are encouraging in light of the recently published experiments by Baker and Kastner *(1981)*. They demonstrated that the

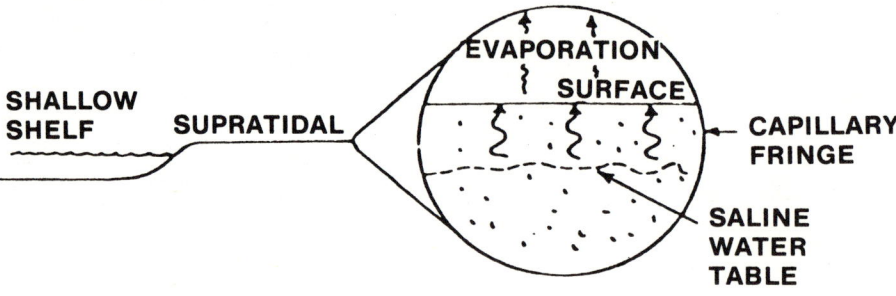

**Figure 11.7:** *Concordant dolomitization by supratidal evaporation.*

# DIAGENESIS OF POTENTIAL CARBONATE RESERVOIRS

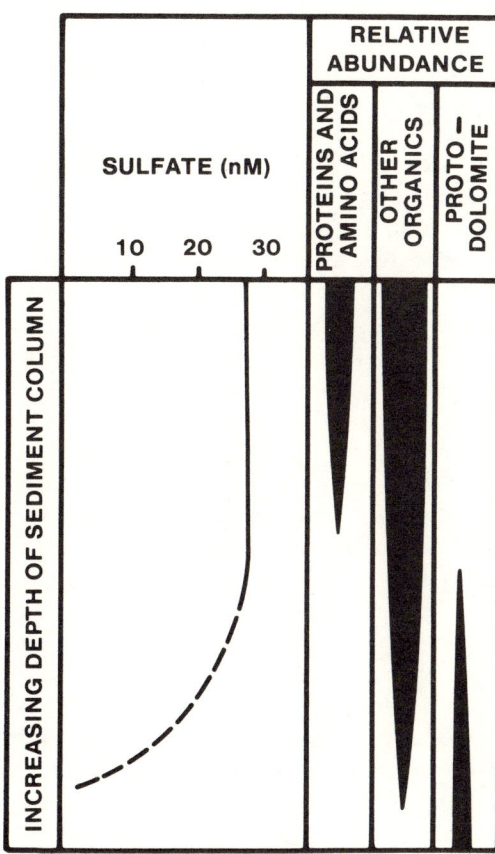

**Figure 11.8:** *Microbial induced sulfate reduction removes $SO_4^{2-}$ from pore water which, in turn, permits rapid dolomitization (Silver and Bloch, in preparation).*

addition of minor amounts of $SO_4^{2-}$ strongly inhibited the rate of calcite dolomitization. They further suggested that dolomite can form rapidly in nature only where $SO_4^{2-}$ concentrations are low. The probable correlation between abundant organics, microbial induced sulfate reduction, and the formation of dolomite suggests a strong cause-effect relationship *(Silver and Bloch, in preparation).*

**Discordant Dolomites** Discordant dolomites are thought to be formed by two different methods, seepage reflux and marine-phreatic mixing.

*Seepage Reflux:* Evaporation of sea water, either in a restricted lagoon or salinas adjacent to the lagoon, is enriched in brines. $CaSO_4$ precipitates, which, in turn, increases the Mg/Ca ratio. The dense Mg-rich brine flows downward and outward and dolomitizes the carbonate sediments in its path. Dolomitization thus cuts bedding planes, is controlled by primary porosity and permeability, and is frequently related to structure and/or unconformities (Fig. 11.9).

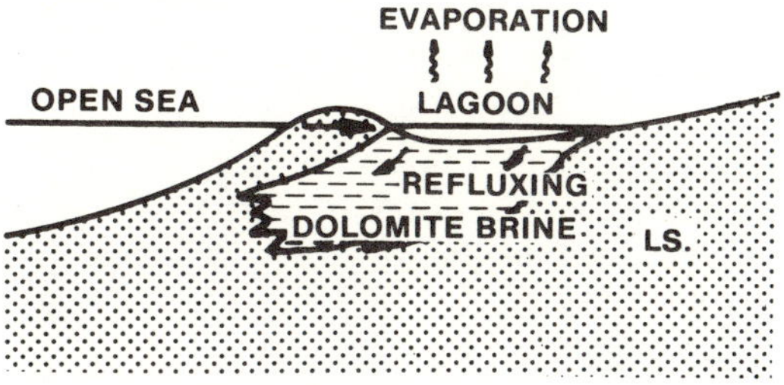

**REFLUX MODEL:**

**Figure 11.9:** *Discordant dolomitization by seepage reflux.*

*Mixing:* The mixing of carbonate-rich phreatic water (meteoric or groundwater) with sea water may produce a suitable chemical reaction for the replacement of *Mg* for *Ca*. *Mg* derived from inversion of aragonite or high Mg-calcite and/or sea water is dissolved in phreatic water that migrates up and down (as well as laterally) through the sediment column. The migration of phreatic waters is controlled by sea level variations, climate, distance from the ocean, and primary permeabilities. Like the reflux dolomites, dolomitization cuts across bedding planes and is often related to unconformities and structure (Fig. 11.10).

**MARINE PHREATIC MODEL:**

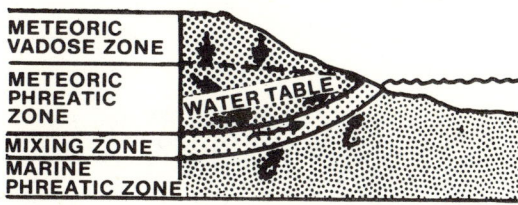

**Figure 11.10:** *Discordant dolomitization by marine phreatic mixing of pore waters.*

## A TECHNIQUE FOR RECONSTRUCTING THE DIAGENETIC HISTORY OF CARBONATES

Only in rare cases do explorationists have the luxury to study cores of potential carbonate reservoirs. If such an opportunity exists and the exploration play does not have a "short fuse," a detailed study of cores will be rewarding.

Perhaps the first step should be to refresh one's memory regarding the diagenetic processes and textures of carbonate rocks. A review of the literature would be helpful, and perhaps the first place to start would be with Bathurst *(1975, pp. 321-544)*. Longman *(1980)* presents a colorful review of the important early diagenetic process in carbonate sediments as well as a list of recent references. Papers that relate depth of burial to porosity include Schmoker and Halley *(1982)*, Wagner and Matthews *(1982)*, Moore and Druckman *(1981)*, and Mazzullo *(1981)*. Two other recent papers that should be reviewed are those by Pingitore, Jr. *(1982)* and Magaritz and others *(1979)*. A practical reference on techniques by Bebout and others *(1979)* would also be helpful.

The cores must be slabbed and most of the saw marks removed. This can be accomplished by spraying diluted HCl (1:10) on the core. The core must be systematically described using a binocular microscope. Thin sections and/or acetate peels may be required to describe detailed cement and grain textures, particularly during the initial stage of the study. The data must be recorded in a systematic manner that will permit interpretation. Appendix II contains a dictionary of symbols and a form for encoding data. Although the form is designed for key punching and electronic data processing, it can be colored like a strip log and the data interpreted by visual inspection.

The most important variables of the core descriptions for this unique study should be plotted on the entire suite of 5-inch logs from the cored well. Study the rock/log relations in order to determine if one or more of the logs can be used independent of core data to recognize the key diagenetic processes that produced the porosity and permeability.

Construct isopach maps that represent the thickness variations from the top of the carbonate unit to each succeeding unconformity. Search the isopach maps for clues that relate depth of burial to porosity enhancement (or preservation) and to generation, expulsion, and migration of hydrocarbons. The maps may also produce data on paleohydrostatic gradients and migration of phreatic lenses through the potential reservoir rock.

# 12
# Reservoirs and Carbonate Traps

## SHELF MARGIN TRAPS

### INTRODUCTION

Explorationists have been extremely successful in discovering major oil and gas reserves in shelf-margin deposits. This is, in part, attributed to the fact that shelf margins are typified by accumulation of porous and permeable rocks; *i.e.,* reefs and banks, and shelf margin deposits are normally linear belts that more or less parallel the margins of basins. Maximum carbonate production and accumulation occurs at, or near, the shelf margin and slightly downslope. Facies belts are normally parallel to the break in slope. They are wider and more variable where the slope is gentle and are narrow and regular where the slope is steep. Seals for the traps are normally shallow-water evaporites, tight limestones, or very fine-grained terrigenous sediments that were deposited across the shelf onto the shelf margin during lower stands of sea level.

### KEY EXPLORATION PROBLEMS

Development of a stratigraphic framework that permits valid correlation of shelf and basin strata and the selection of proper time surfaces for the construction of structural and isopach maps are major problems in exploration for potential shelf margin reservoirs. The Upper Wolfcampian through Guadalupian strata of the Permian Basin (Fig. 12.1) represents one of the best examples of rapid facies change from the shelf to the slope to the basin in North America. Topographic difference between coeval shelf and basin facies may have been as much as 1,800 feet.

Correct correlation of shelf, shelf margin, and basin strata in the Permian Basin is mandatory prior to risking exploration monies on drilling a wildcat. The rapid changes in facies, relatively steep depositional topography, and cyclic depositional patterns permit at least three very different correlations of shelf strata to basinal strata.

Cyclic changes in global sea level produced reciprocal sedimentation that alternately open and shut off transport of terrigenous clastics across the shelf into the basin. During high stand of sea level, carbonates were the dominant lithologies that accumulated on the shelf, shelf margin, and in the basin (Fig. 12.2). During succeedingly lower stands of sea level, terrigenous clastics prograded across the shelf, and initial clastic sedimentation occurred as they by-passed the slope and accumulated in the basin (Fig. 12.3 - 12.5). Time surfaces do not necessarily represent horizontal bedding planes (Fig. 12.6).

### EMPIRE ABO FIELD, SOUTHEAST NEW MEXICO

**History of Exploration Strategy**  The first Abo reef discovery was in 1951 at Lovington, located in southeastern New Mexico. The field is not large and did not produce much interest in the industry. Few geologists understood the significance of the discovery; that the Abo was a barrier reef that effectively divided the shelf from the basin facies.

Figure 12.7 shows the location of well control in 1957 along the Abo shelf margin prior to the discovery of the Empire Field. In 1956, geologists of a major oil company effectively presented the concept of a barrier reef trend to their management, and an exploratory well was drilled. It encountered water-bearing reef deposits in an envelope of basinal deposits. A second well was approved. It was drilled shelfward and was the discovery well. Again, not much "excitement" was generated by the discovery because the discovery well contained only 30 to 60 feet of net pay. However, later drilling resulted in about 750 feet of gross pay. To say the least, leases along

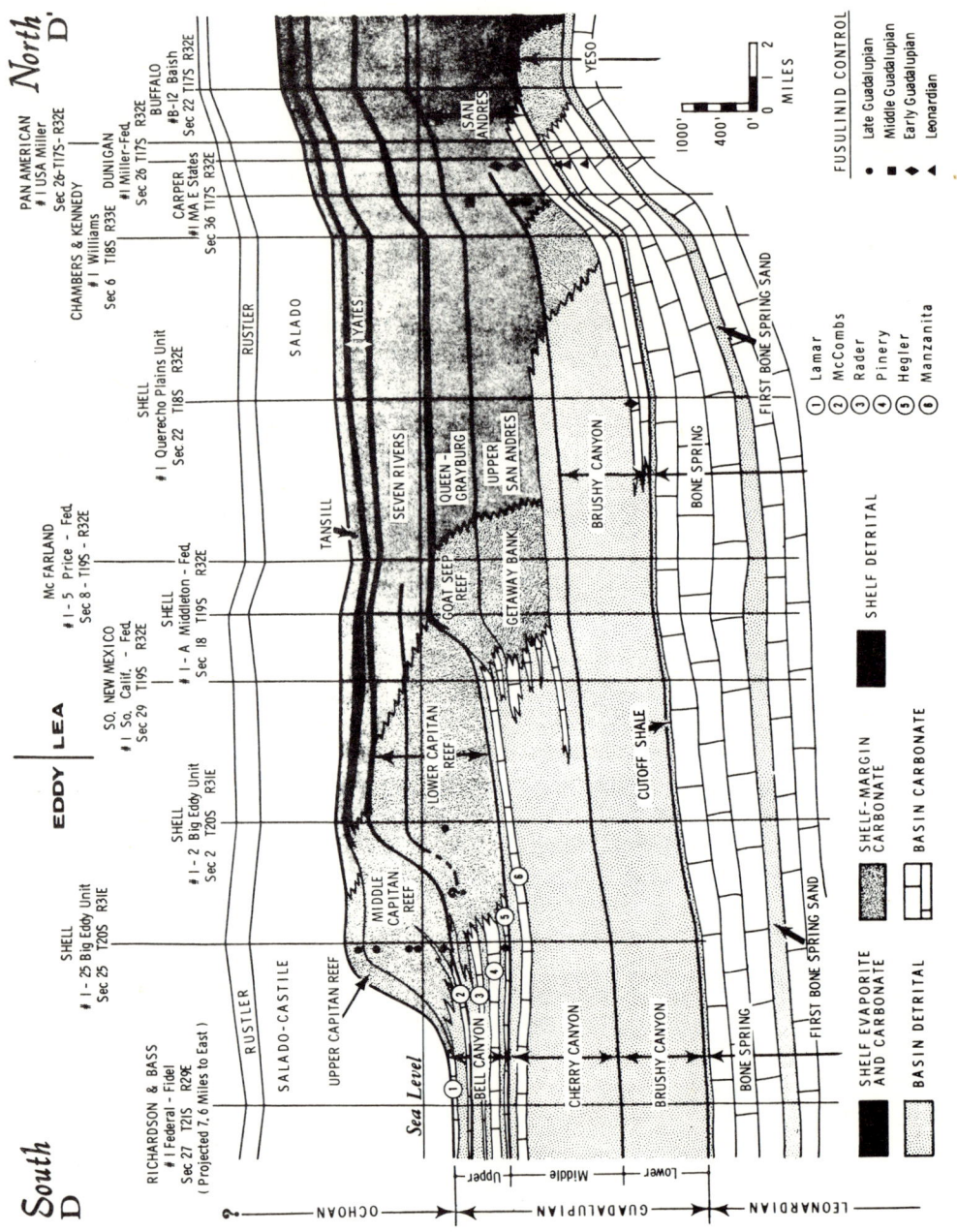

**Figure 12.1:** *Guadalupian physical stratigraphic framework, Delaware Basin, west Texas and southeastern New Mexico (Silver and Todd, 1969).*

# RESERVOIRS AND CARBONATE TRAPS

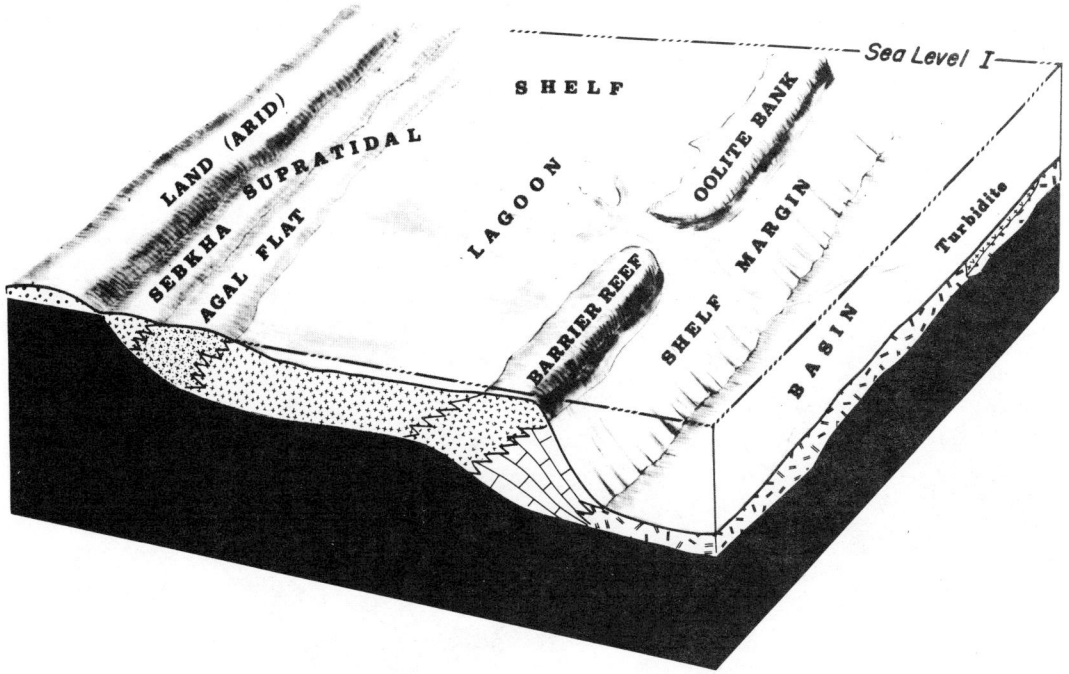

**Figure 12.2:** *High stand of sea level floods the shelf which, in turn, inhibits clastics from entering the basin. Carbonate accumulation occurs on the shelf, shelf margin, and basin (Silver and Todd, 1969).*

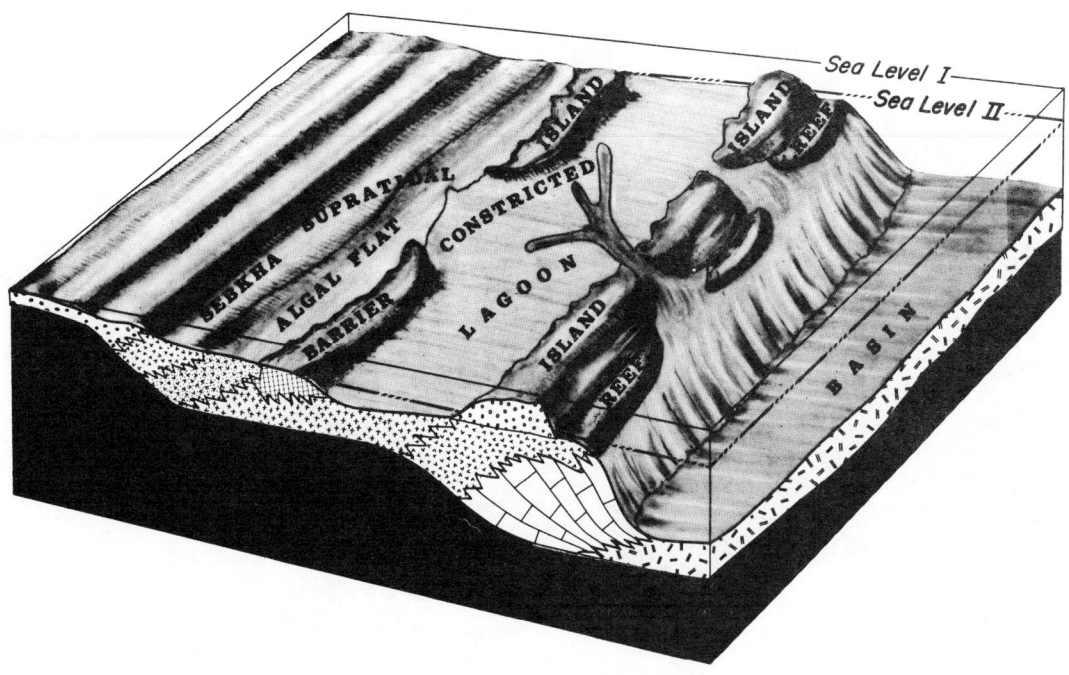

**Figure 12.3:** *Progradation of shelf and shelf margin facies occurs during a lower stand of sea level (Silver and Todd, 1969).*

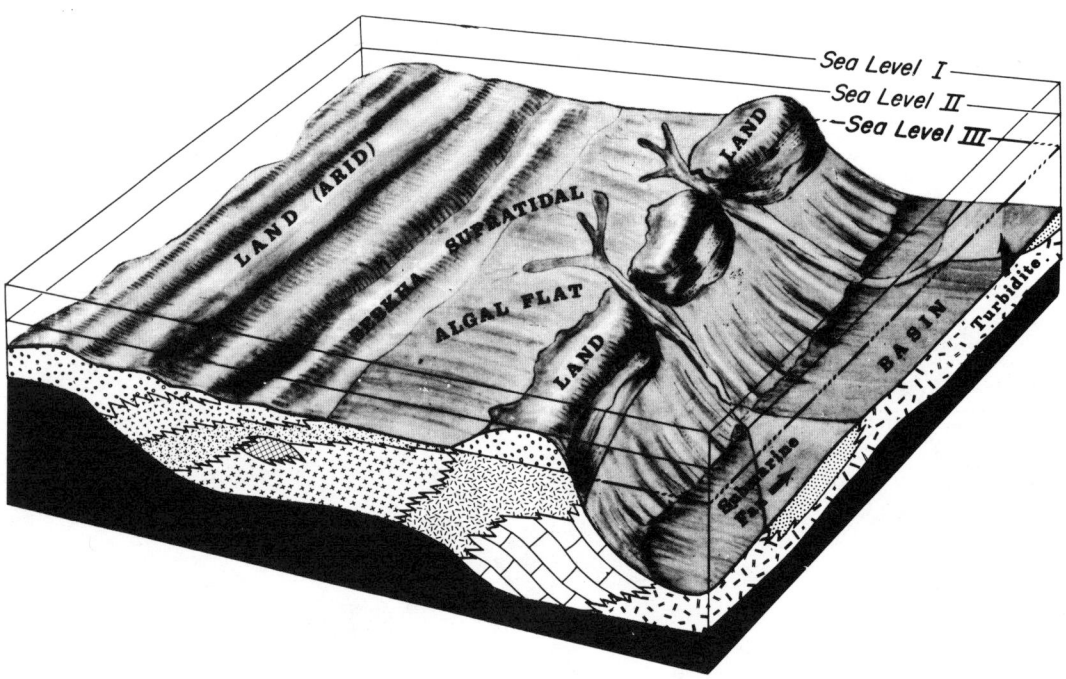

**Figure 12.4:** *Continued progradation of the shelf facies during a lower stand of sea level permits terrigenous clastics to enter submarine canyons. Extensive submarine clastic fan deposition gradually replaces carbonate deposition in the basin (Silver and Todd, 1969).*

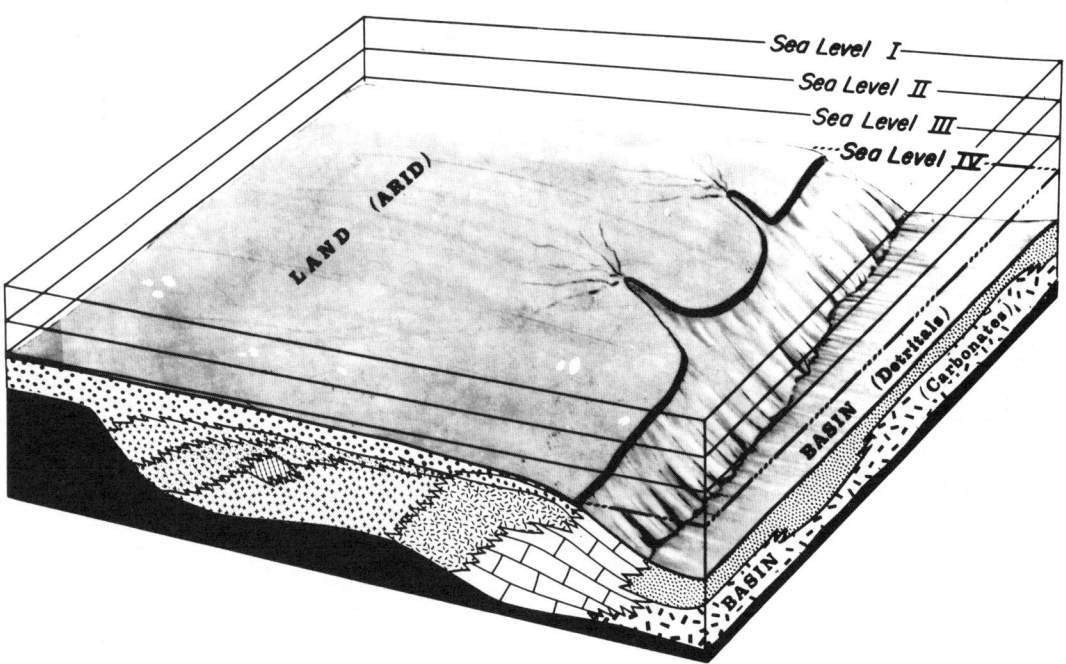

**Figure 12.5:** *At low stand of sea level, clastic debris progrades across the shelf to the edge of the shelf margin. Transport of clastics through the submarine canyons and via sediment by-pass across the shelf margin floods the basin with clastics (Silver and Todd, 1969).*

# RESERVOIRS AND CARBONATE TRAPS

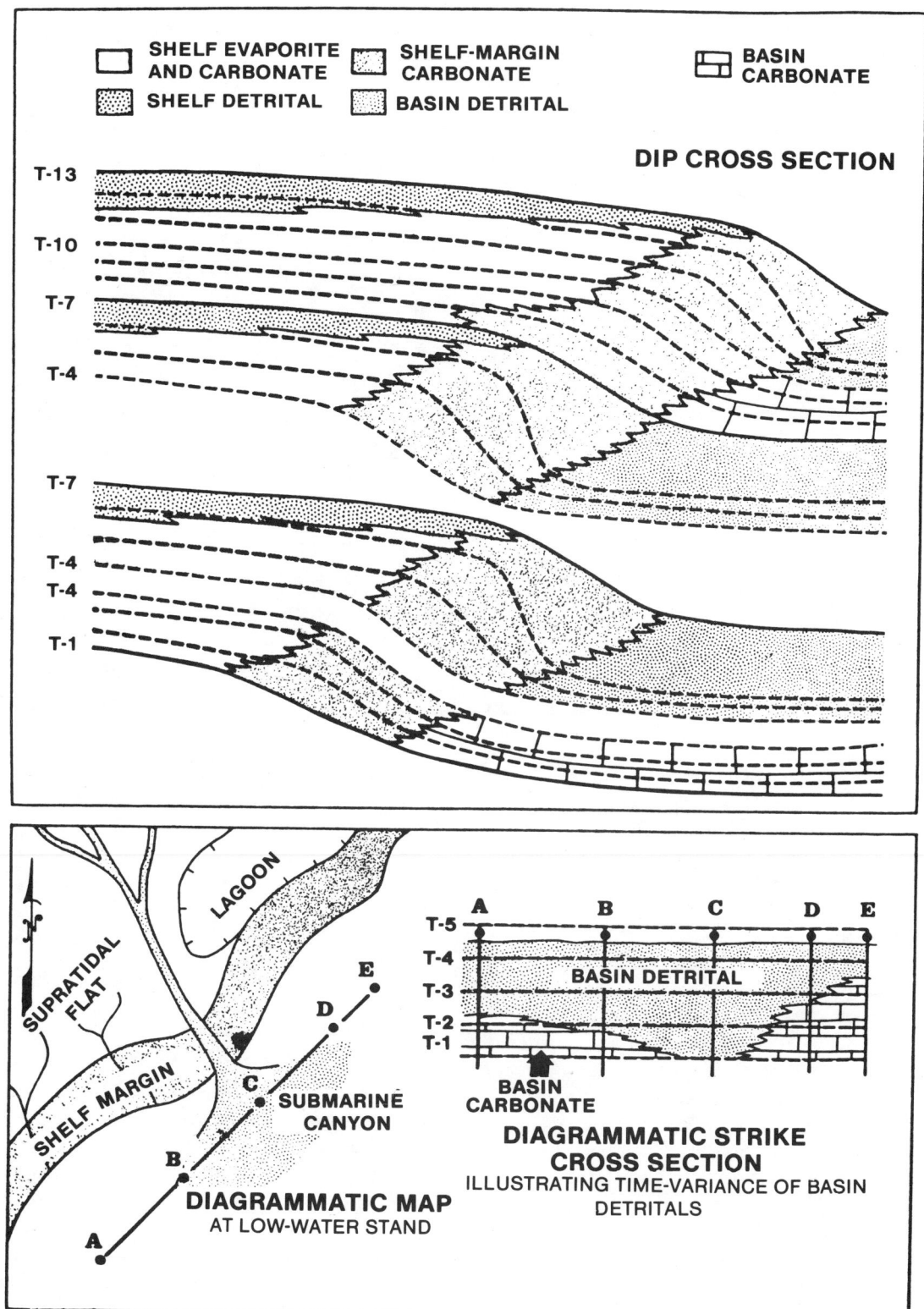

**Figure 12.6:** *Carbonate facies typically cross time surfaces (Silver and Todd, 1969).*

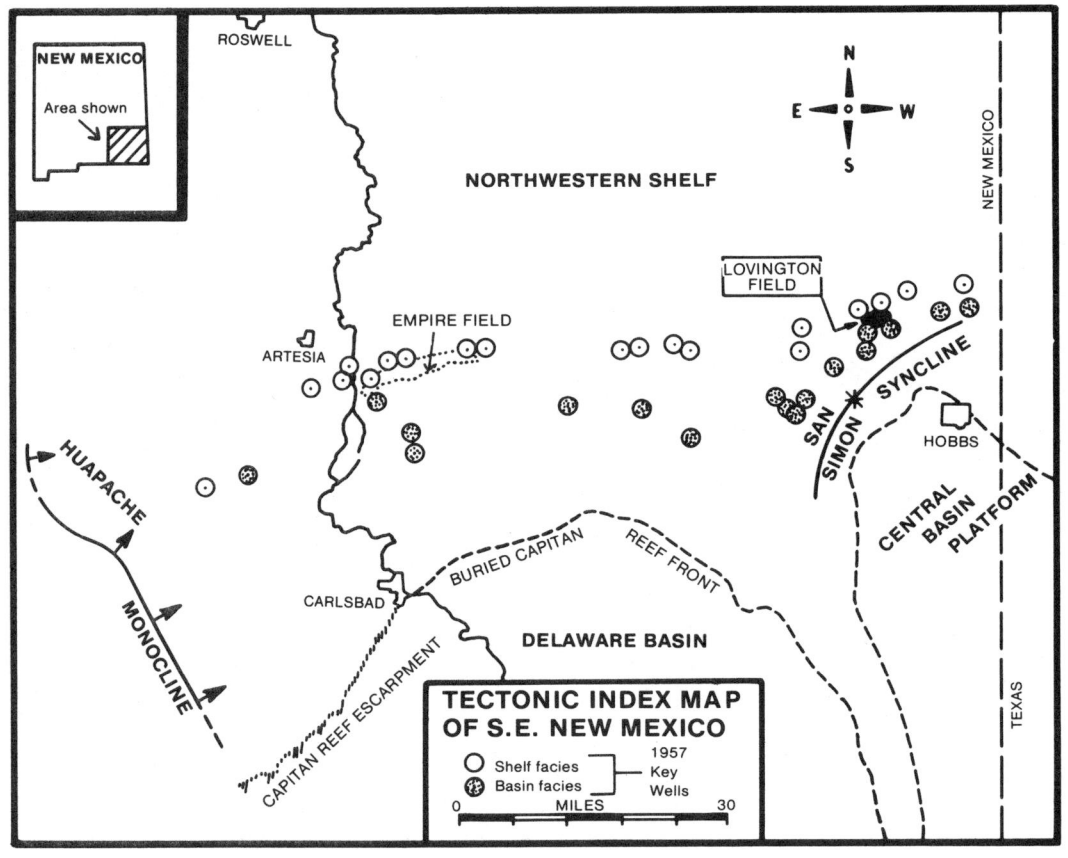

**Figure 12.7:** *Critical well control in 1957 along the Abo hinge line, prior to the discovery of Empire (LeMay, 1971).*

the Abo shelf margin were "consumed" rather rapidly. The Empire Field contains 250 million barrels of oil *(LeMay, 1972).*

**Trap** The trap is an updip pinchout of the porous and permeable reef deposits (Fig. 12.8). The loss of porosity and permeability corresponds to the facies change from reefal to intercalated terrigenous-derived green shales and shelf-deposited carbonates (Fig. 12.9). Southwest and eastern closure is attributed to a combination of structure and gradational reduction of porosity and permeability.

**Reservoir** Dolomitization by seepage reflux of the reef facies resulted in a fine to coarsely crystalline dolomite. Analysis of cores from the field demonstrates that most of the original reef material was dolomitized. In fact, some argument exists as to the original nature of the Abo. Was it indeed a reef, or was it a mound? The reservoir has well-developed intercrystalline porosity. Vugs (some filled with anhydrite) and fractures are common. Permeability is high.

## EXPLORATION AND EXPLOITATION STRATEGY

**Exploration Techniques** The first step the explorationist must take is to develop regional stratigraphic and depositional models for the stratigraphic interval in question. This study should be based on detailed descriptions of samples, cores, mechanical logs, and seismic data in the basin. All paleotological data should be incorporated into the analyses. Construction of stratigraphic cross sections, both strike and dip, will "tie" the correlations, which in turn will result in more precise maps.

The trend of the shelf margin may be mapped by isopaching the shelf margin, or

# RESERVOIRS AND CARBONATE TRAPS

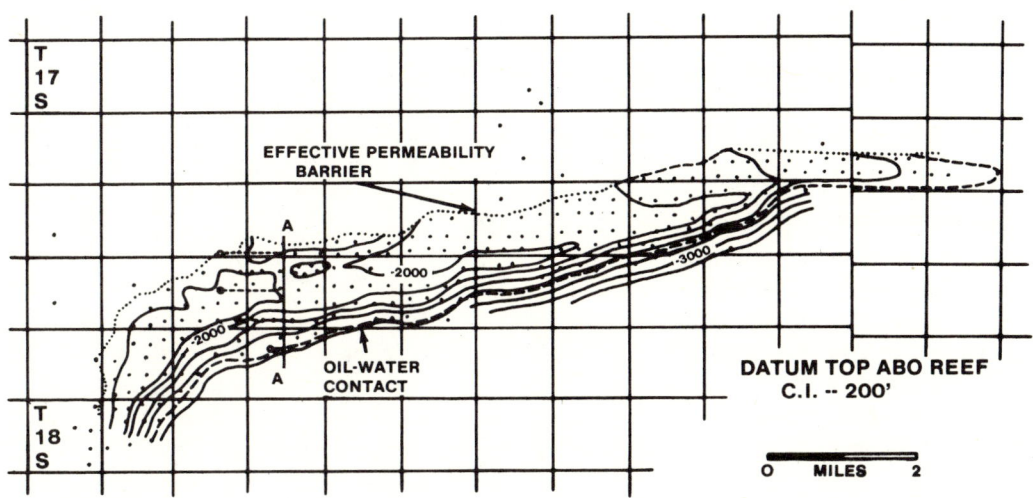

**Figure 12.8:** *Structural map on top of the Abo "reef" (LeMay, 1971).*

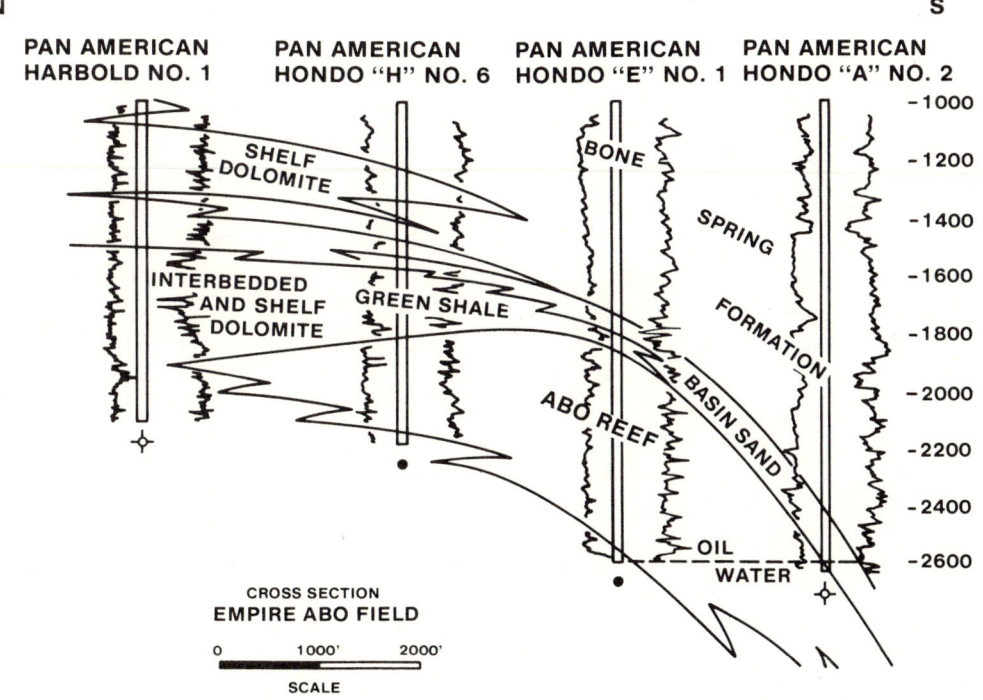

**Figure 12.9:** *Cross section across the Empire Field (LeMay, 1971).*

indirectly by isopaching the shelf and/or basin facies. Particularly in areas that are characterized by cyclic depositional patterns, it is necessary to document source rock and seal relationships.

**Exploitation Guideposts** The following maps must be kept up-to-date while a development program is in progress: (1) isopach gross reef (or bank) pay during development, (2) core wells when economically feasible and map facies changes, (3) maintain up-to-date isopach maps of shelf and basinal facies in order to use as a guide to predict reef (or bank) facies.

## SEISMIC CHARACTERISTICS OF SHELF MARGINS

**Midland Basin, West Texas** During Middle Wolfcampian through Early Guadalupian time, the Midland Basin of West Texas was characterized by carbonate shelf, shelf-margin, and basin deposition. Deposition on the shelf-margin was characterized by linear belts of barrier bank or barrier reef deposits. The back-barrier facies are characterized by supratidal evaporites and siltstones, and lagoonal dolomitized skeletal grainstones and packstones. The shelf-margin facies is composed of skeletal and oolitic grainstones with local buildups of boundstones. Basinal facies are represented by pelagic wackestones, mudstones, and siltstones.

High amplitude and continuous reflections on the shelf, grade into low amplitude, discontinuous reflections that are generated from the shelf-margin facies (Fig. 12.10). Reflections from the basinal facies are generally moderate amplitude and continuous. In response to cyclic sea level changes, basinal facies (*i.e.*, Sprayberry Sand) onlap shelf-margin facies (*i.e.*, Clear Fork). The basin facies do not interfinger with the shelf-margin facies as implied by Silver and Todd *(1969)*. Overall sea level drop is demonstrated by the progradation of the Abo through Leonardian shelf-margin facies (Fig. 12.10). All three environments produce hydrocarbons from both structural and stratigraphic traps. One critical variable to be considered in cyclic depositional patterns is the occurrence of a seal. Production does occur from the Clear Fork along line A (Fig. 12.10) but it is insufficient to drill a well;

it will pay for a workover. Although line A is only about three miles north of line B (Fig. 12.10), it appears that the character of the shelf-margin has changed significantly. This is attributed to the angle that the two lines cross the shelf-margins. Line A crosses the margin at about 35°, whereas line B crosses the margin at nearly a right angle.

**Gulf of Papua** The offshore Paupuan Basin is a segment of the northern slope of the Austrian continental margin during Tertiary time. The western portion of the Papuan Basin contains a barrier reef trend that was remarkably similiar in form, trend, and pattern to the modern Great Barrier Reef and platform of the Western Coral Sea *(Tallis, 1975)*. The barrier reef reached a thickness of over 4,000 feet. Shelf carbonates and patch reefs developed landward of the barrier and pelagic limestones and reef slope deposits accumulated on the seaward margin of the barrier reef.

One segment of the barrier reef was tested to be dry by the Borabi No. 1 well. The well penetrated two distinct reef cores that are separated by an off-reef limestone facies. The older reef facies is chalky *(Tallis, 1975)*. The chalk was probably produced in the vadose diagenetic environment. The younger reef displays no evidence of chalk, even though it was subaerially exposed for a considerable length of time as indicated by Late Pliocene carbonate muds that overlie the top of the reef. The upper part of this buildup is dolomitized.

Seismic reflections in the lower Miocene section of the Gulf of Papua indicate carbonate step sedimentation. Abrupt changes in the slope of reflectors depicts the back-reef and fore-reef facies (Fig. 12.11). Internal reflectors in the shelf facies (back reef-lagoon) are moderate in amplitude, continuous, and parallel. These grade seaward into discontinuous, nearly reflection-free shelf-margin facies (reef). Shelf-margin slope facies are characterized by convergent cycles of high amplitude.

## CARBONATE BUILDUPS

### INTRODUCTION

For practical purposes, the wide variety of carbonate buildups will be grouped into four categories: pinnacles, patches, atolls, and

# RESERVOIRS AND CARBONATE TRAPS

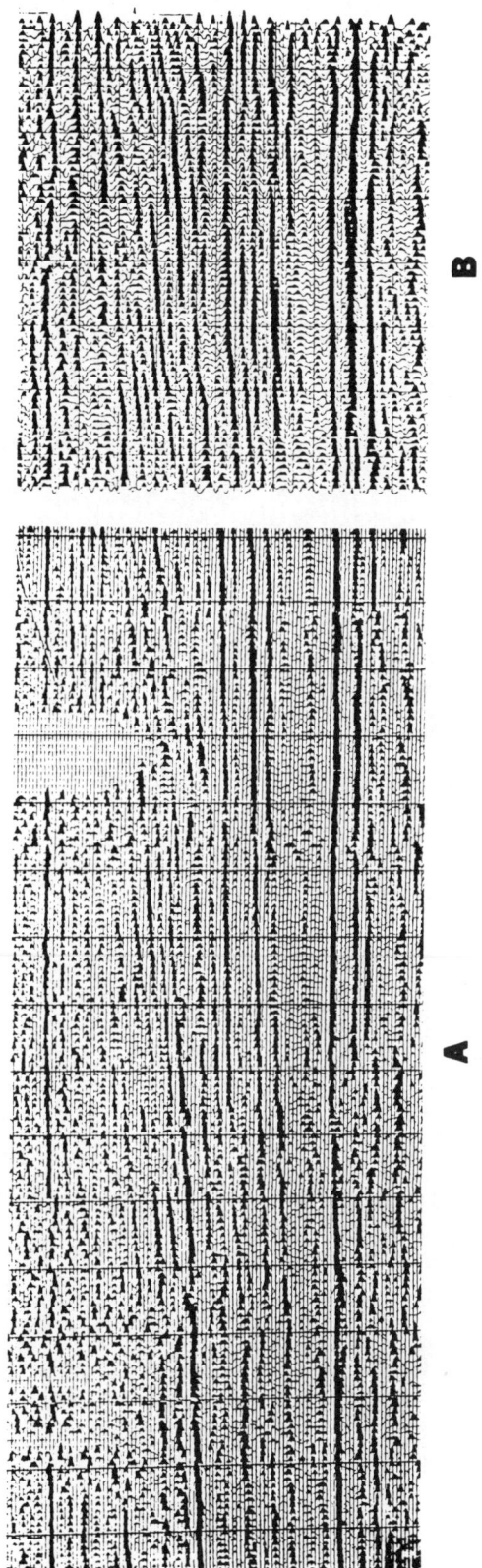

**Figure 12.10:** *Parts of two 24-fold, CDP lines that cross the shelf margins of the Midland Basin. Line A is perpendicular to the margin; line B is oblique to the shelf margin (courtesy of Olympic Exploration & Production Co.).*

Figure 12.11: *Seismic line of shelf margin reef deposits, Gulf of Papua (Bubb and Hatlelid, 1977).*

mounds. Again, don't permit complex or varied usage of carbonate terminology to inhibit your understanding of carbonate reservoir rocks.

## PINNACLE BUILDUPS

**Definition** Pinnacle buildups are roughly equi-dimensional and are normally surrounded by relatively deep water. Most pinnacle buildups do not have a well-defined lagoon. Dominant framework organisms are corals, stromatoporoids, or rudstids. Algae and associated mollusks are important constituents of most pinnacle buildups.

**Key Exploration Problems** Pinnacles are normally developed along a trend, but in many petroleum basins they occur almost randomly in deep water or in a shelf environment. Most productive Paleozoic pinnacles occur seaward of the shelf margin, whereas most productive Mesozoic pinnacles occur shelfward of the shelf margin. The pinnacles are normally small and very difficult to successfully explore for with only subsurface geologic data. To compound the exploration problem, not all pinnacles are porous.

### Michigan Basin Pinnacle Fields

*History of Exploration:* The Michigan Basin was ringed by a barrier reef of Silurian Age (Fig. 12.12). The barrier reef is 450 to 600 feet thick and is composed of coarse grainstones, stromatoporoids, and tabulate corals.

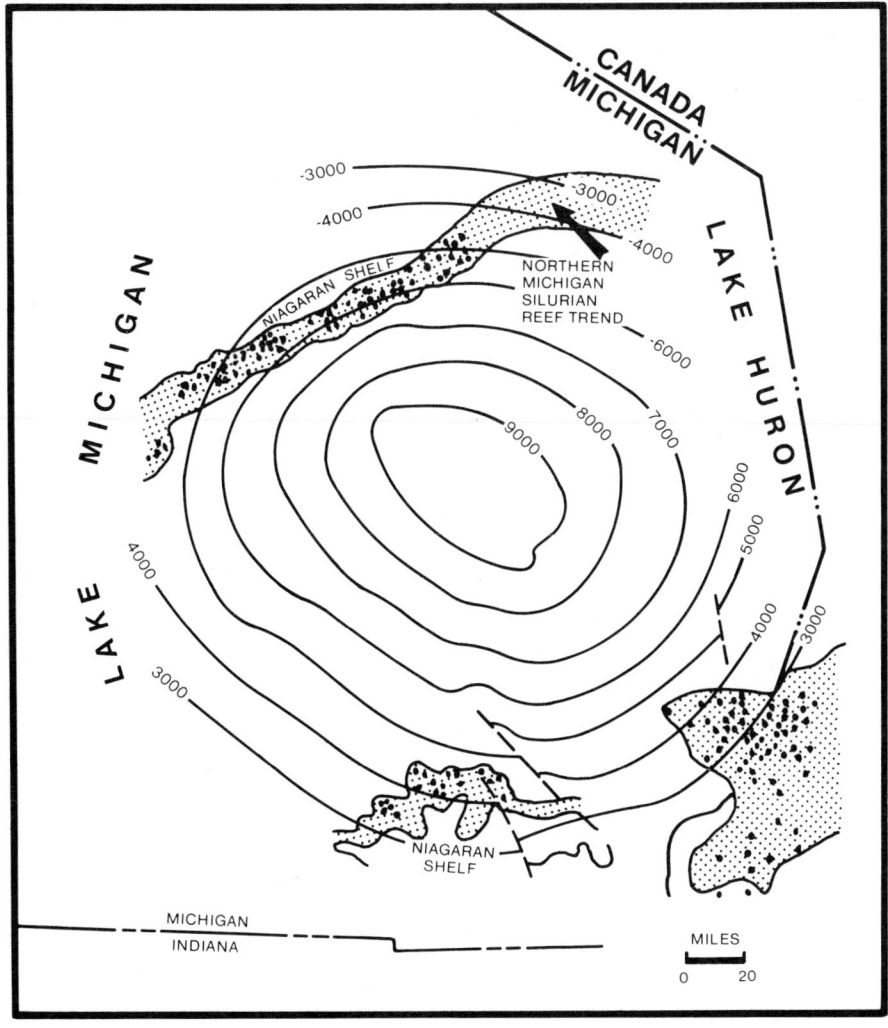

**Figure 12.12:** *Index map of the Michigan Basin (Caughlin and others, 1976).*

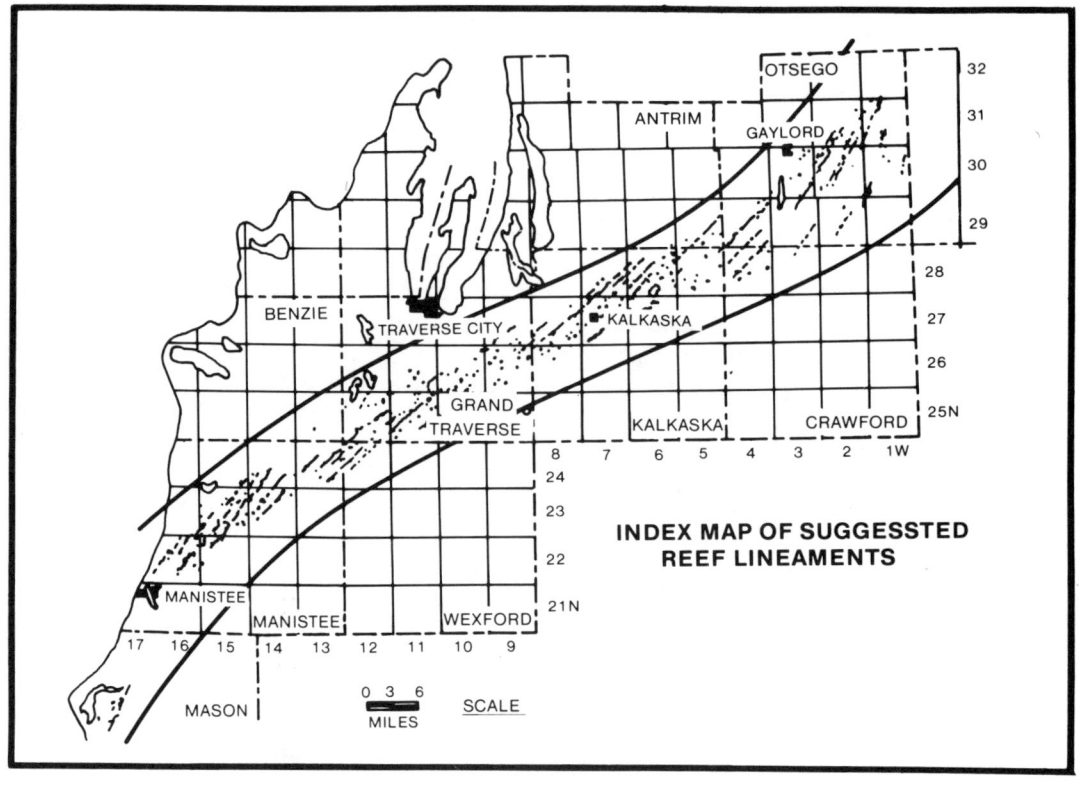

**Figure 12.13:** *Reef lineaments in the northern Michigan Basin (Mantech, 1976).*

The barrier reef is non-productive to date.

Two trends of pinnacle reefs occur in the basin (Fig. 12.12). The southern trend was extensively developed during the 1940-1960s and the northern trend during the 1950-1970s. Development of both trends continue and success rates exceed 80%. Mantech *(1976)* has demonstrated the possibility that there is a subtle association of the pinnacle reef development and reactivated Grenville shear fractures in the northern Fairway (Fig. 12.13). These lineations trend 15-17° to the strike of the major reef trend. He suggested that these lineations represent low relief, en echelon, tension fracturing, and normal faulting that was produced by basin subsidence and left-lateral shearing. The pinnacles would tend to initiate on the seaward edges of the topographic high.

*Traps:* The pinnacle reefs occur along a belt 5-10 miles wide and 160 miles long. Typically the fields are not more than 1½ to 2 miles wide, and thickness of the individual reefs increases basinward (Fig. 12.14). The seal is either nonporous mudstone or evaporites.

*Reservoirs:* The reservoir rock is composed of a complex sequence of grainstones, packstones, and some boundstones. From base to top the sequence is (1) crinoid, bryozoan mudstone, (2) stromatoporoid (commonly algal encrusted), coral packstone to boundstone, (3) laminar stromatoporoid and algal stromatolite packstone, and (4) evaporites. The model preferred by most explorationists working in the Michigan Basin is illustrated by Figure 12.15.

Many of the pinnacles are dolomitized. Dolomitization is not necessary for production but it does enhance production rates and ultimate recovery. Porosity ranges from 3-16%, normally they average from 6-10%. Permeabilities range from 5 to 50md, they average around 15md. Total reserves may be in excess of 500 million barrels of oil and 5 TCF of gas.

*Exploration and Exploitation Strategy:* An integrated geologic and geophysical ap-

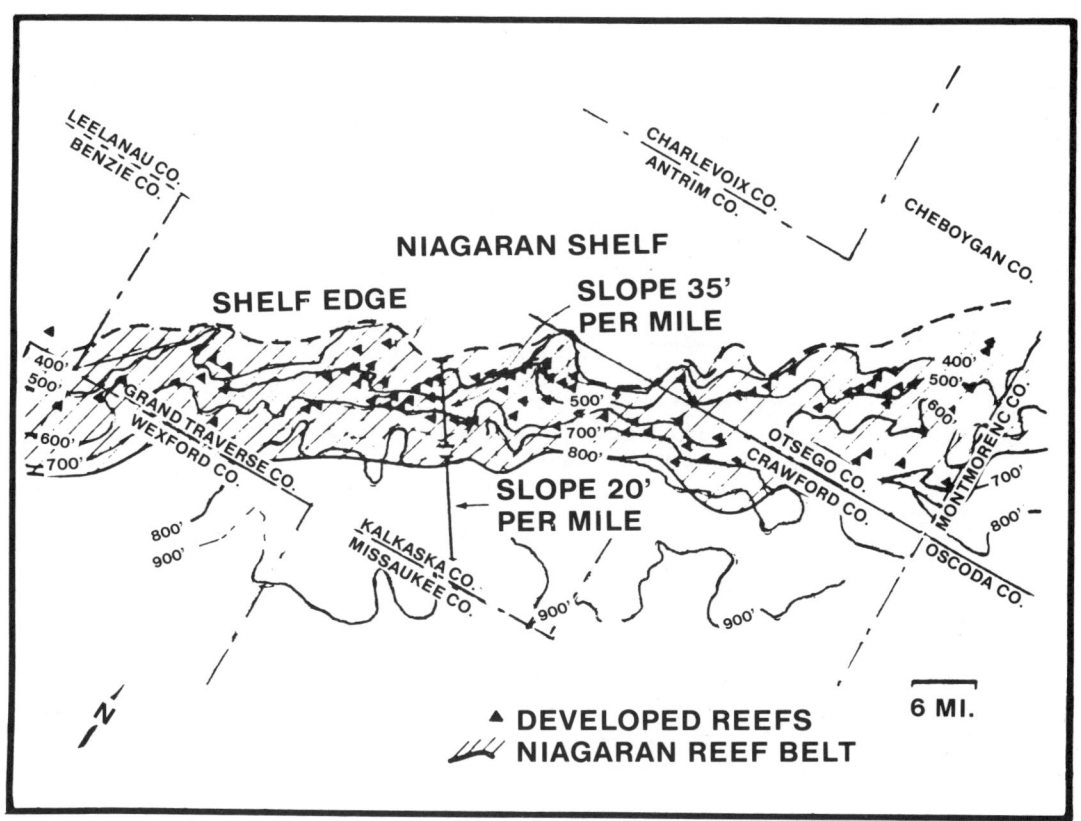

**Figure 12.14:** *Slope on which pinnacles formed in the Michigan Basin (Caughlin, et al., 1976).*

proach is necessary for a successful exploration program in the Michigan Basin. Regional structural maps on the Niagarian permit the exploration team to delineate anomalous "wrinkles" on the basin floor. Regional isotime maps built from the base of the A-2 carbonate to the top of the Niagaran are useful to isolate specific areas to shoot seismic lines on a one mile grid spacing.

The pinnacle reefs in the Michigan Basin have three seismic characteristics (Fig. 12.16). In many cases, an isotime anomaly that is due to thinning between the Dundee and the A-2 Carbonate denotes a reef. In other situations, the Niagara event is disrupted by the presence of a reef. A third seismic model is characterized by continuous reflections from both the Niagaran and A-2 Carbonate and the latter denotes topography on top of the pinnacle *(Laughlin and others, 1975).*

Dip meter and borehole gravitometer data has been used both in exploitation and exploration programs. The borehole gravitometer is particularly useful if a wildcat has "missed" the reef. In many cases, in-hole-gravity data has supported the decision to plug back, set a whipstock, and slant hole the well into the reef.

### Nisku Pinnacle Reef Fields, Alberta

*History of Exploration:* The initial discovery of oil in Nisku pinnacle reefs was made in early 1977. The exploration program was based on over 1,300 miles of 12-fold seismic data that was obtained prior to the spud date of the discovery well. The numerous discoveries since 1977 are an important addition of reserves in an area that was considered to be a mature exploration province.

*Trap:* The trapping mechanism for Nisku reef production is totally stratigraphic. The pinnacle reefs are located seaward of the Nisku shelf margin on regional dip (Fig. 12.17).

*Reservoir:* The hydrocarbon bearing pinnacle reefs occur in Zeta Lake Member of the Nisku Formation *(Chevron Exploration Staff,*

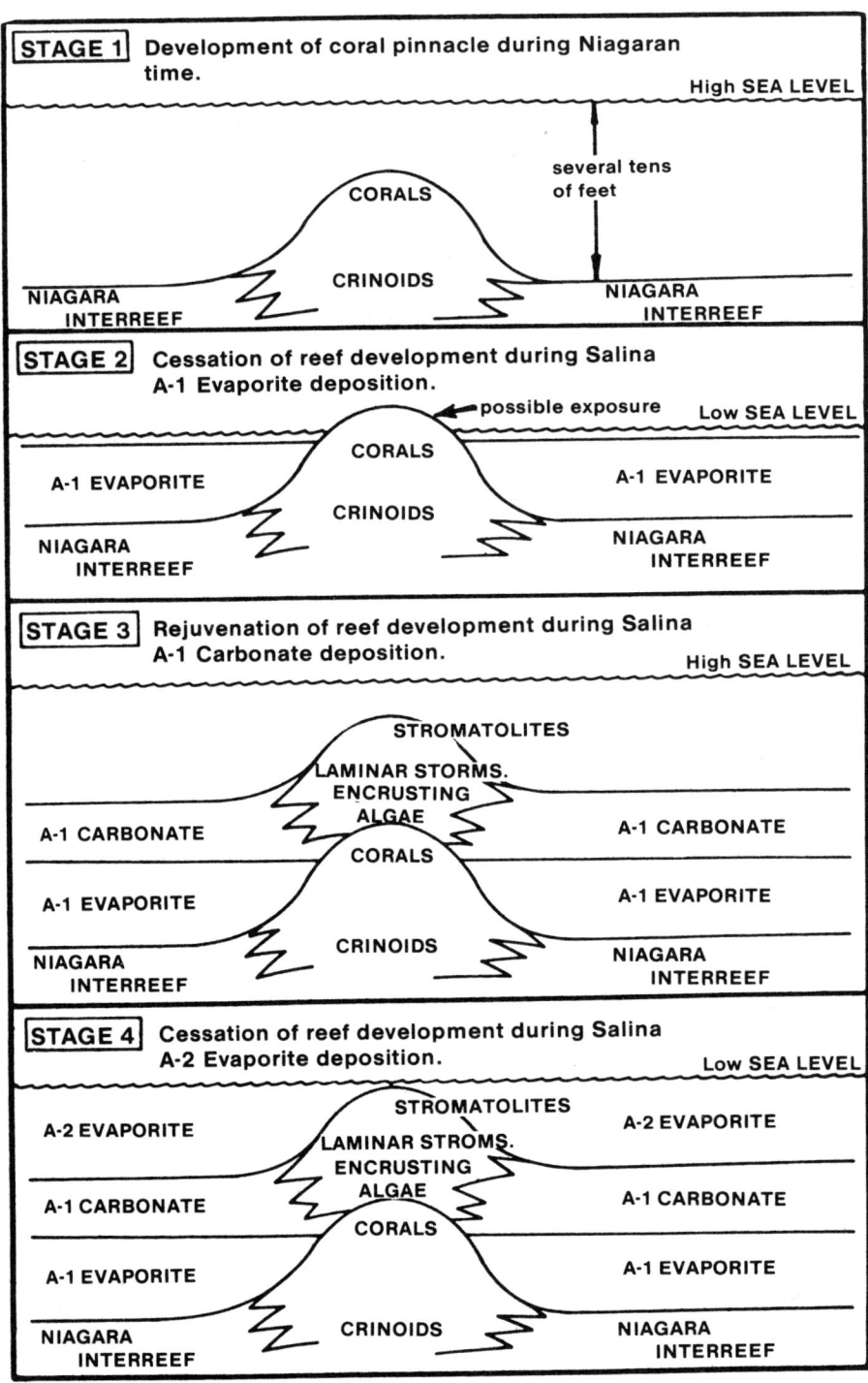

**Figure 12.15:** *Model to explain the evolution of Michigan Basin pinnacle reefs (Messolella and others, 1974).*

# RESERVOIRS AND CARBONATE TRAPS

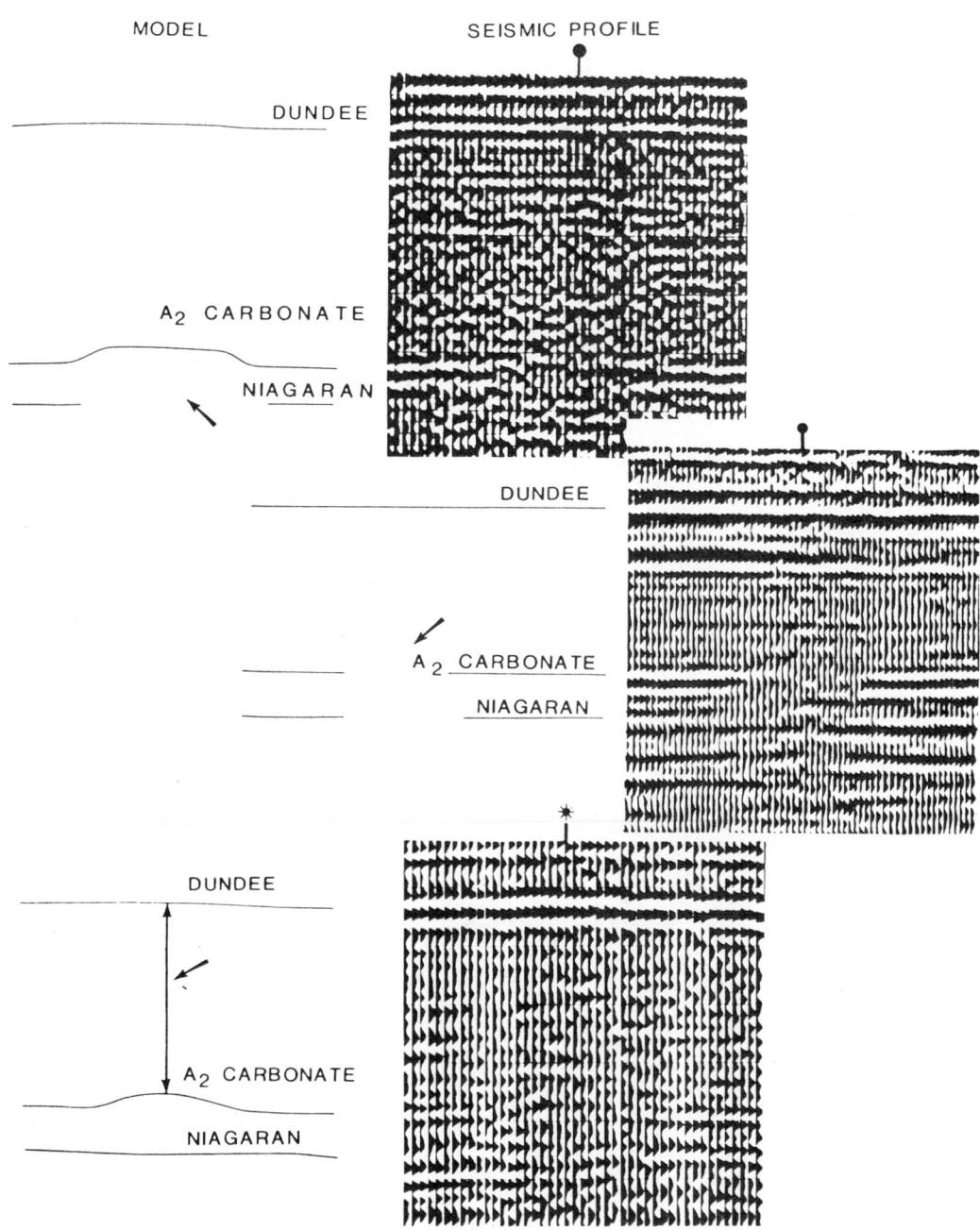

**Figure 12.16:** *Exploration seismic models for pinnacle reef definition (Caughlin, et al., 1976).*

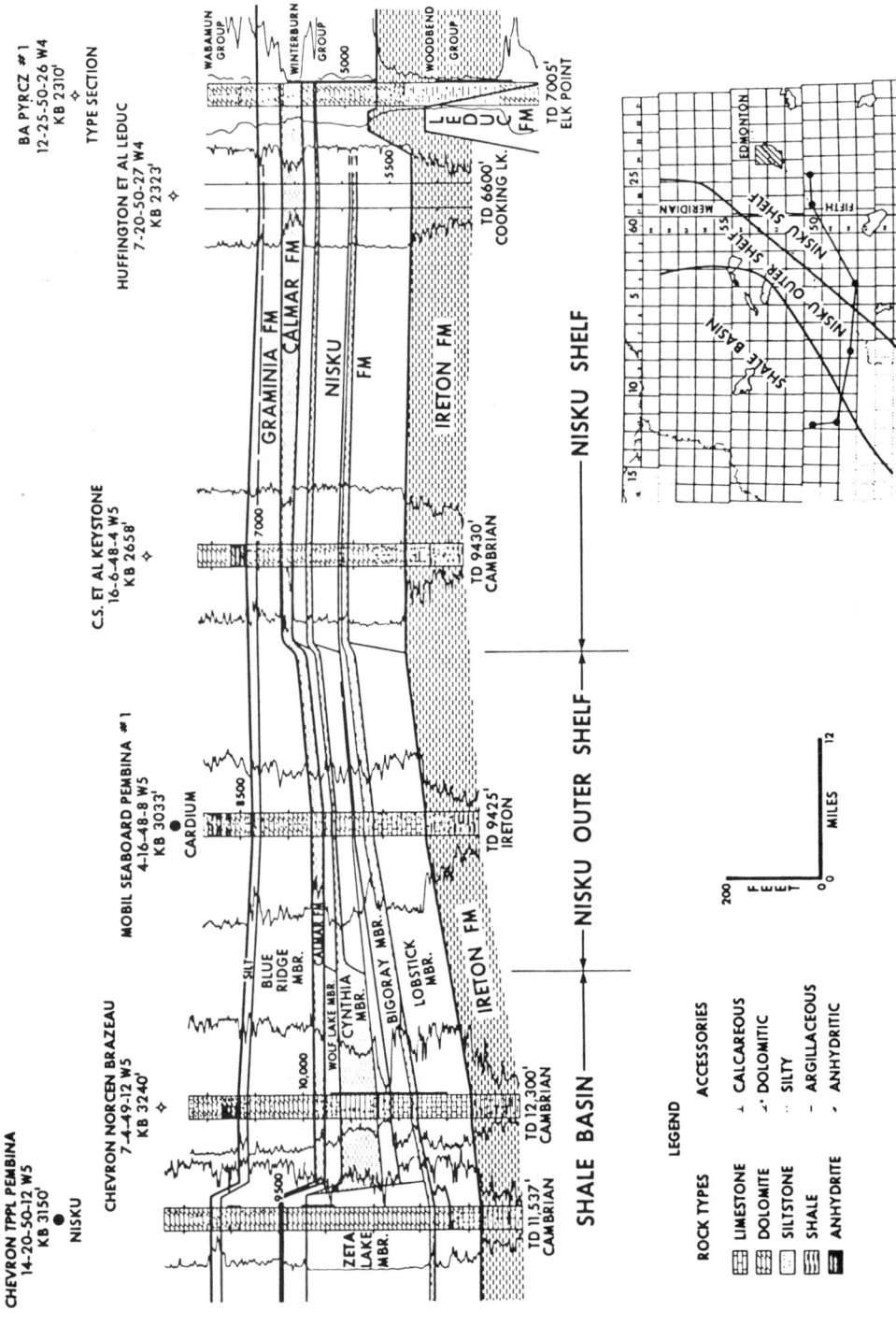

**Figure 12.17:** *Regional cross section showing relationship between Zeta Lake pinnacle reef and Nisku (Devonian) shelf margin, Pembina Area, Alberta, Canada (Chevron Exploration Staff, 1979).*

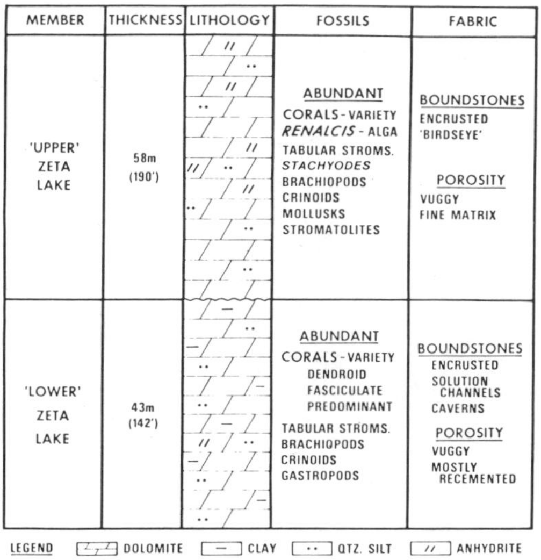

by the Chevron Exploration Staff (1979). The velocity log of the off-reef deposits shows three distinct reflectors: top of Cynthia, base of Cynthia, and top of Ireton (Fig. 12.20A & B). The Nisku event occurs in the sum of the trace. The first peak can be considered as the Cynthia; the following trough, Nisku; and the second peak, top of the Ireton. Figure 12.20C depicts a typical Zeta Lake anomaly. Note that the first and second peaks are stronger in the reef interval because the former represents more porous carbonates than the non-reef deposits and the latter depicts the Niksu which is higher velocity than off-reef deposits.

At least three additional seismic anomalies have been evaluated. Anomaly A is character-

Figure 12.18: *Lithology of typical Zeta Lake pinnacle reef, Alberta, Canada (Chevron Exploration Staff, 1979).*

1979). Basal deposits of the Zeta Lake Member are gray, slightly argillaceous and silty dolomite. Framework organisms are dominated by dendroid fasciculated rugose corals. Tabular stromatoporoids and encrusting corals are a minor framework component. Typically, this unit is about 140 feet thick (Fig. 12.18).

The upper zone in the Zeta Lake Member consists of gray to brown, fossiliferous dolomite. Rugose and tabulate corals, a few fasciculate corals, and stromatoporoids occur as boundstones. Lime mud and pellets are the dominant matrix components.

Diagenetic processes have extensively altered the depositional texture of the pinnacle reef facies. Dolomite is very common and at some localities the entire pinnacle facies have been dolomitized. Abundant vuggy porosity has been formed by leaching of all or portions of the non-dolomitized fossils matrix. Anhydrite occurs in intervals throughout the reefs, but they are most common in the upper part of the pinnacles. These anhydrite zones as well as the dolomitization are thought to have been formed during two low stands of sea level (Fig. 12.19).

*Exploration Strategy:* Once the geologic setting was established, a tight network of seismic lines were shot normal to the Nisku shelf margin. A simple three-reflector model for reef and off-reef positions was developed

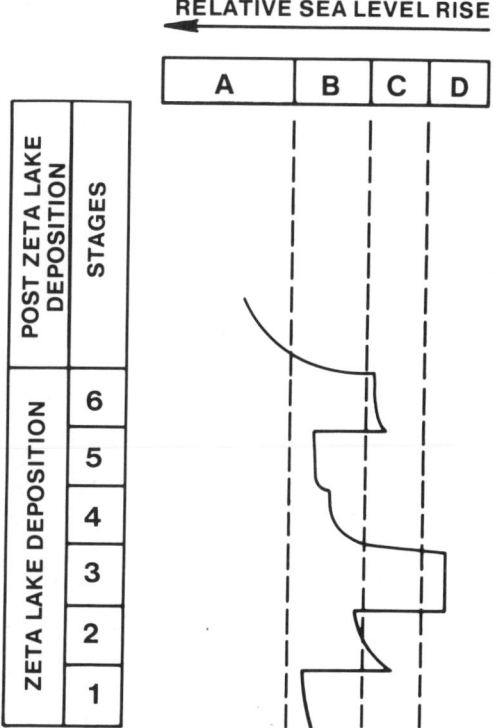

**EXPLANATION**

A Water too deep for optimum carbonate production
B Optimum carbonate production
C Subtidal, progradation, & shoaling
D Subaerial exposure

Figure 12.19: *Interpreted relative sea level changes, Zeta Lake pinnacle reefs, Pembina Area, Alberta, Canada (adapted from Chevron Exploration Staff, 1979).*

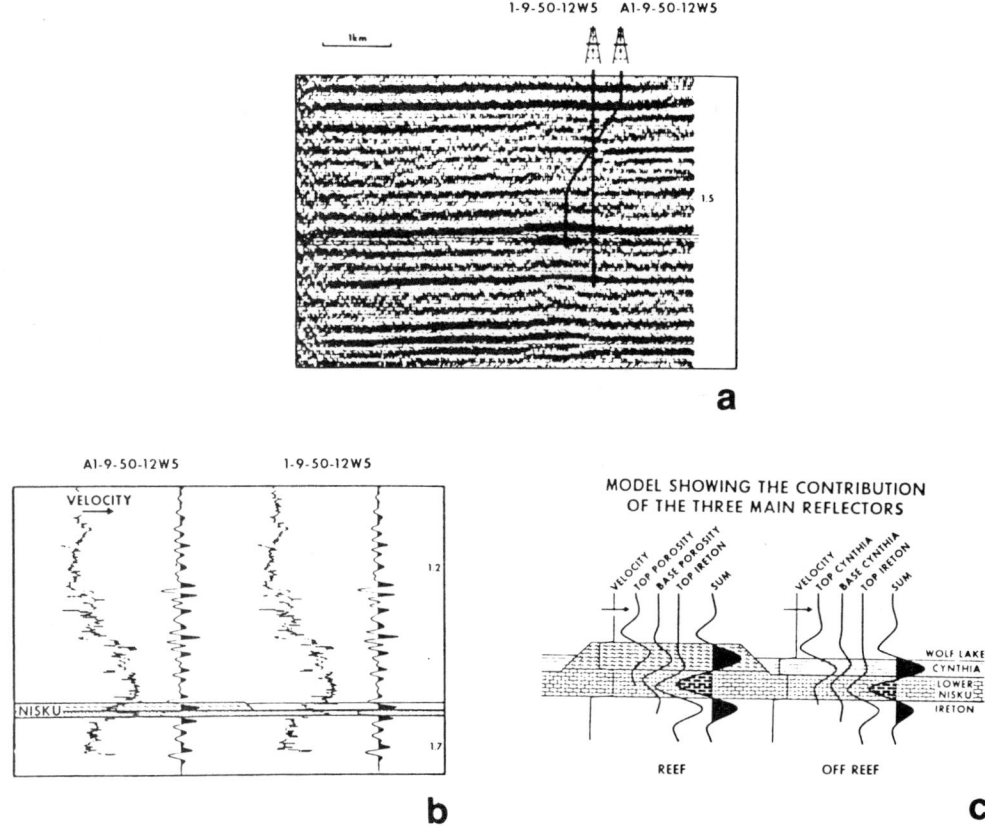

**Figure 12.20:** *Seismic section and model of typical Zeta Lake pinnacle reef, Alberta, Canada (Chevron Exploration Staff, 1979).*

ized by a very low-amplitude first Nisku peak because no distinct boundary occurs at the top of the pinnacle reef and the velocity of the lower Nisku is similar to that of the porous reef (Fig. 12.21A). Anomaly B has also been drilled. In this case the anomaly also has a weak first peak in the Nisku event (Fig. 12.21B). But in this case the character is due to relatively low-velocity Blue-Ridge above a high-velocity, nonporous Nisku (Fig. 12.21C). Anomaly C is the reverse of the typical anomaly in that the pinnacle has a higher velocity than the beds above and below.

## PATCH BUILDUPS

**Introduction** Patch buildups can occur on the deep shelf of a step carbonate system, the transitional zone of a ramp system as well as on topographic highs within the ramp outer zone of deposition. In each case, the fauna, depositional relief, and geometry of the patch buildups are varied and few generalities can be made. An excellent review of reefs is presented by James *(1979)*.

### Smackover Reefs, Gulf Coast

*Reservoir:* Ramp-type depositional patterns characterize Smackover (Jurassic) deposits along the United States Gulf Coast. The Smackover is separated into a lower member consisting of dark colored mudstones, calcareous shales, and/or sands; and an upper member of predominantly non-skeletal grainstones through mudstones *(Bishop, 1968)*. The lower member is considered to have been deposited in a deep ramp environment and the upper member accumulated in the transitional and coastal zones of a ramp. In all probability, the members cross time surfaces and are facies of one another.

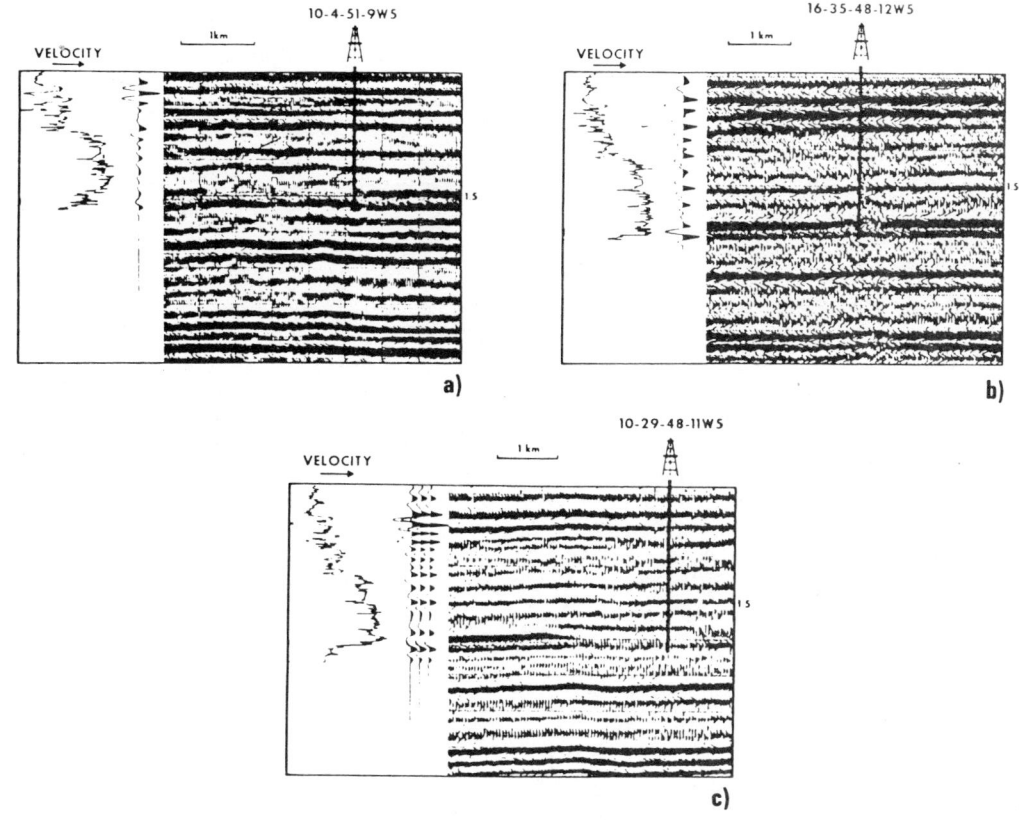

**Figure 12.21:** *Seismic anomalies produced by Zeta Lake pinnacle reefs, Alberta, Canada (Chevron Exploration Staff, 1979).*

The patch buildups occur on subtle topographic highs seaward of the high-energy transitional zone of the Smackover ramp *(Baria and others, 1982)*. The reefs occur on basement ridges, upthrown basement fault blocks, and upthrown salt-cored fault blocks (Fig. 12.22). The reefs are constructed of digitate and branching "stromatolitic" blue-green, algal, *Tubiphytes* sp., and marine cements. Corals, skeletal algae, sponges, bryozoans, and hydrozoans are locally important constituents of the buildups.

Unfortunately most of the Smackover patch reefs drilled to date have limited reservoir qualities. The original interframework and interparticle porosities of Walker Creek (Arkansas) and Melvin fields (Alabama), for example, were reduced to 5-15% by early marine cementation *(Baria and others, 1982)*. Compaction, styolization, and late cementation further occluded the primary pores at Walker Creek and present porosities range from 1-5%. These late stage cements appear to be the most damaging pore occlusion process. For example, the boundstones in Melvin Field are much more porous than those at Walker Creek and the former lack the late-stage cements *(Baria and others, 1982)*.

*Exploration Strategy:* Because regional uplifting, faulting, or salt core structures produced the topographic highs on which Smackover-type reefs occur, seismic data and regional structural maps may be the best tools for predicting their occurrence. Typically the reefs will be indicated on a seismic line by an amplitude anomaly on the high side of a faulted structure (Fig. 12.23). Models to predict the occurrence of porosity within the buildups have not been published to date. However, even in those cases where the reefs

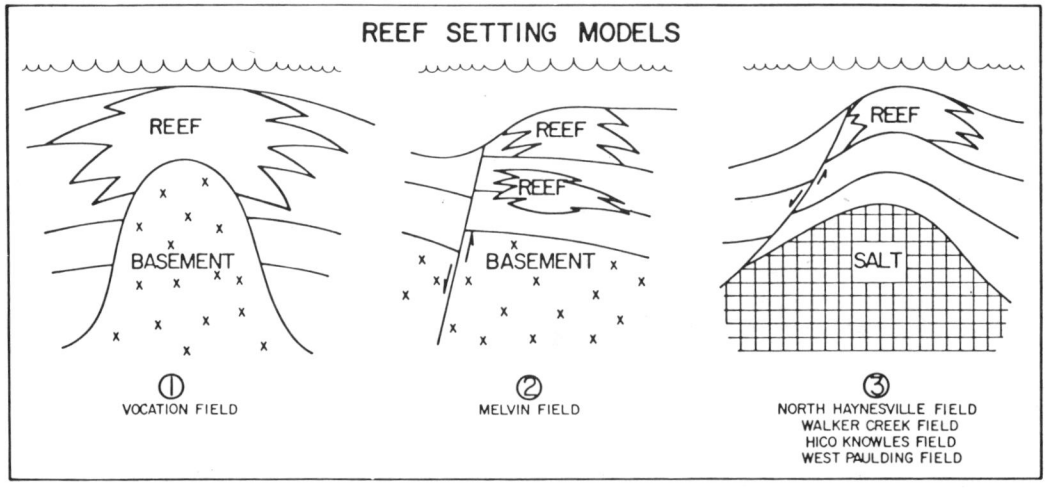

**Figure 12.22:** *Schematic diagram showing the structural setting of patch reefs in the Smackover Formation, Gulf Coast, U.S.A. (Baria and others, 1982).*

have been cemented by early marine processes, the topography on the reef has influenced subsequent depositional patterns. Frequently the reefs are capped by dolomitized grainstones that are production (Fig. 12.24 and Fig. 12.25).

**Seismic Characteristics of Pinnacles**
*Southwest Africa:* Pinnacle reefs are common in Southwest Africa. Two such pinnacle reefs are illustrated in Figure 12.26. Excellent high amplitude reflections outline the top and base of the buildups. Depositional topography on the buildups is clearly illustrated by two cycles of onlap along the flank of the reefs. Indirect evidence of the reef include drape and a negative velocity anomaly within the reefs. Internal reflectors are surprisingly continuous.

*Zama, Alberta:* Zama Field is located in the northwest corner of Alberta, and the geology of the area has been reported in detail by McCamis and Griffith *(1968).* Porosities in the field range up to 11% and the productive zone ranges up to 160 feet. The field was discovered by five-fold CDP vibroseis data *(Evans, 1972).* The original data was at best difficult to interpret (Fig. 12.27). However, using a deconvolution program developed by Conoco Research, the reprocessed data is not only useful for exploration, but also exploitation (Fig. 12.28).

## ATOLL AND MOUND BUILDUPS

Atolls are more or less circular carbonate buildups that encircle a lagoon. They may occur on a deep shelf, *i.e.,* Belize, but more frequently they develop seaward of the shelf margin in deep oceanic basins. Both geologic and geophysical exploration techniques have been successfully applied to prospecting for atoll-accumulated reservoirs.

Middle and Late Devonian marks the beginning of major expansion of corals and stromatoporoids. Vast hydrocarbon accumulation is the resultant carbonate buildup occurs in Alberta and British Columbia (Fig. 12.29). The major reserves are located in Upper Devonian Leduc buildups (5 BBO and 10 TCF gas) and middle to late Devonian Beaverhill Lake buildups (6 BBO and 10 TCF gas).

### Judy Creek Field, Alberta, Canada

*History of Exploration:* Six discrete reef complexes (Snipe Lake, Goose Lake, Kaybob, Virginia Hills, House Mountain, and Swan Hills) provide the backbone for the Canadian petroleum industry. In 1957, several discoveries at Swan Hills, Kaybob, and Virginia Hills suggested the presence of an extensive continuous porous Devonian reef. A flurry of drilling activity followed which proved that the disco-

# RESERVOIRS AND CARBONATE TRAPS

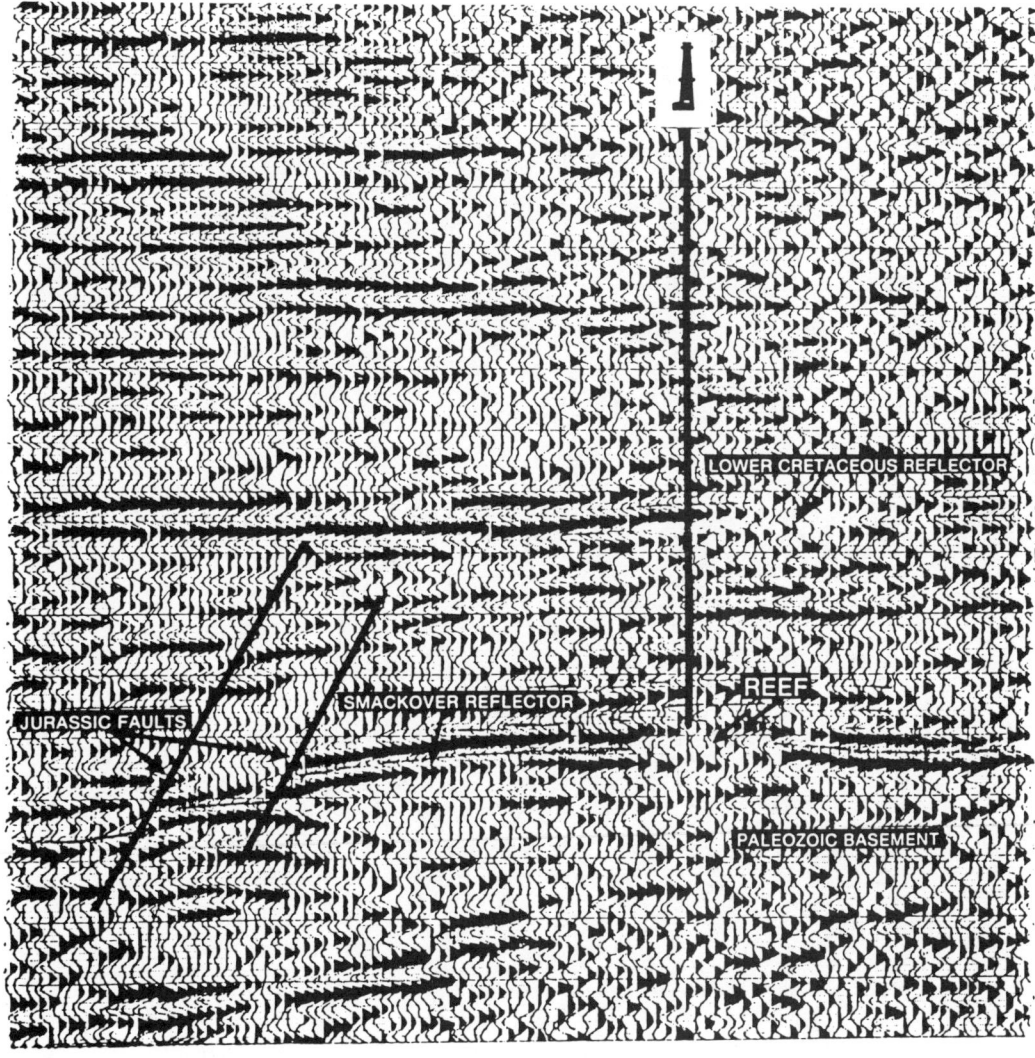

**Figure 12.23:** *Porous Smackover reef is located on the upthrown block of a down-to-basin fault and is depicted on the seismic line as an amplitude anomaly (Baria and others, 1982).*

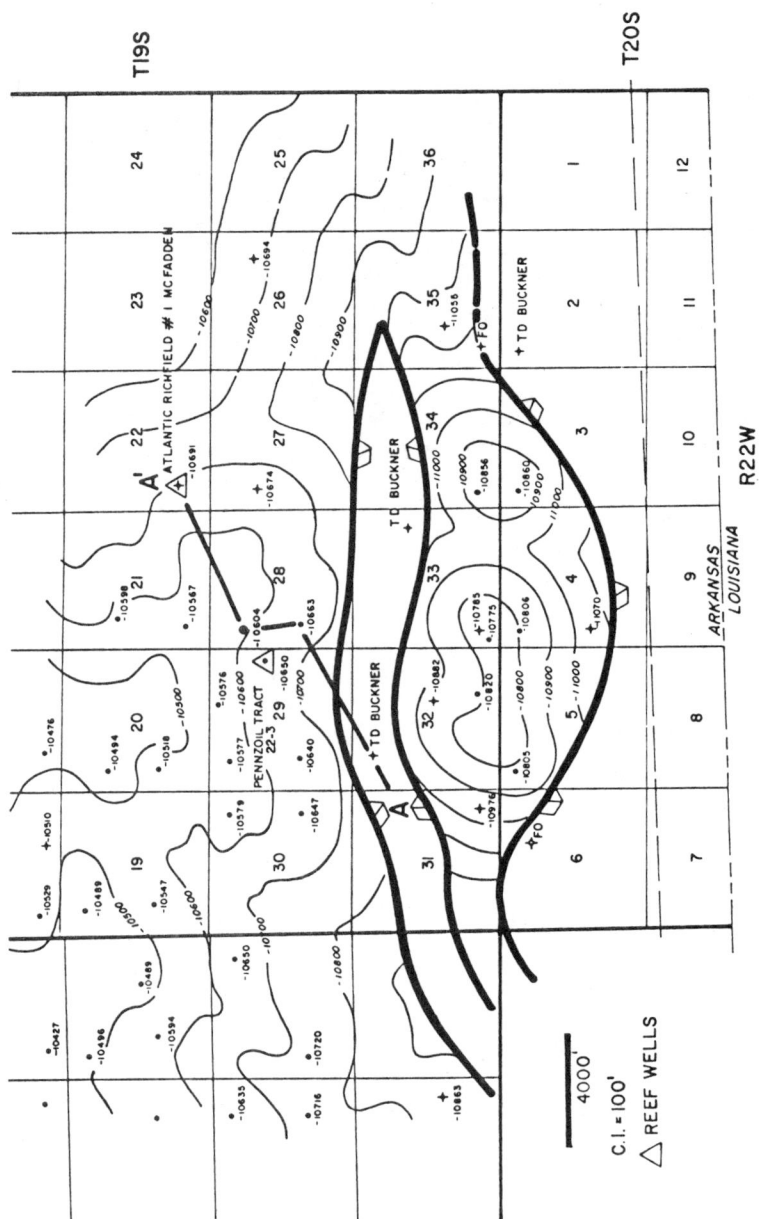

**Figure 12.24:** *Smackover structure map of Walker Creek and Welcome Fields, southern Arkansas (Baria and others, 1982).*

# RESERVOIRS AND CARBONATE TRAPS

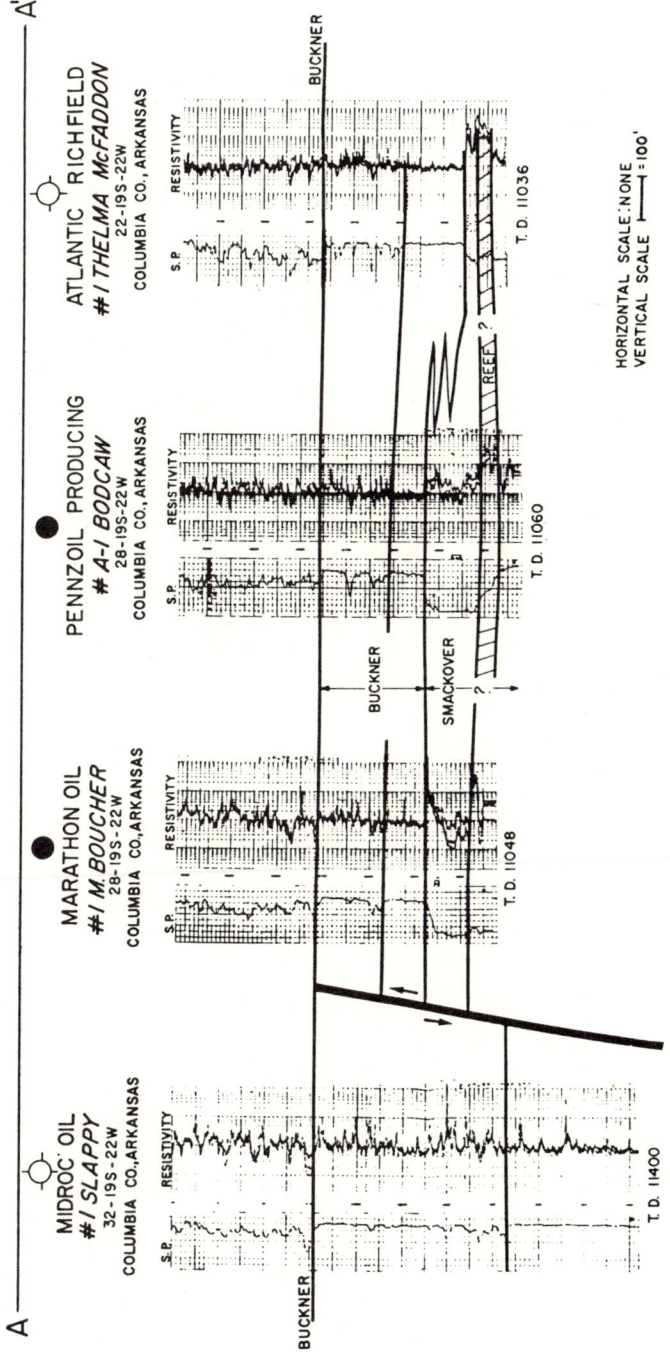

**Figure 12.25:** *Northeast-southwest cross section of Walker Creek Field, Arkansas. Location of section depicted on Figure 12.24 (Baria and others, 1982).*

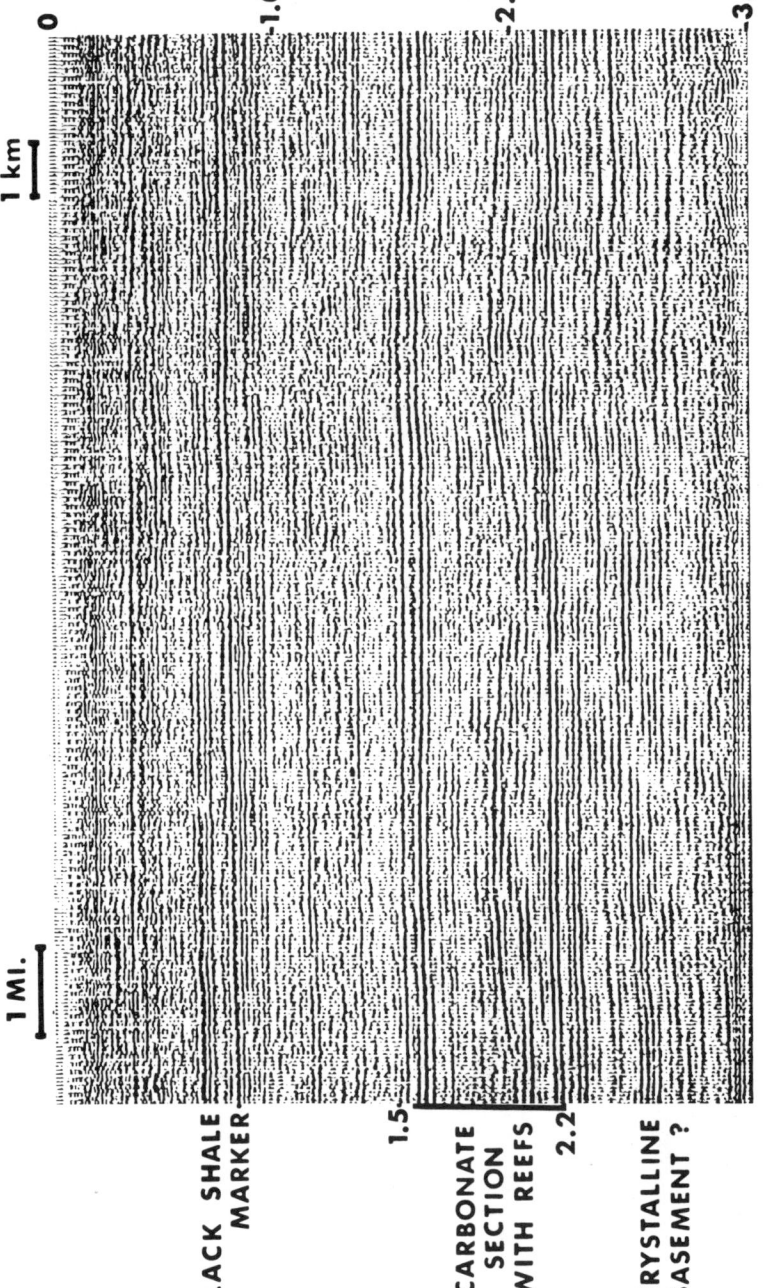

Figure 12.26: *Seismic line across two untested pinnacle reefs, Etosha Basin, southwest Africa (Laing, 1972).*

# RESERVOIRS AND CARBONATE TRAPS

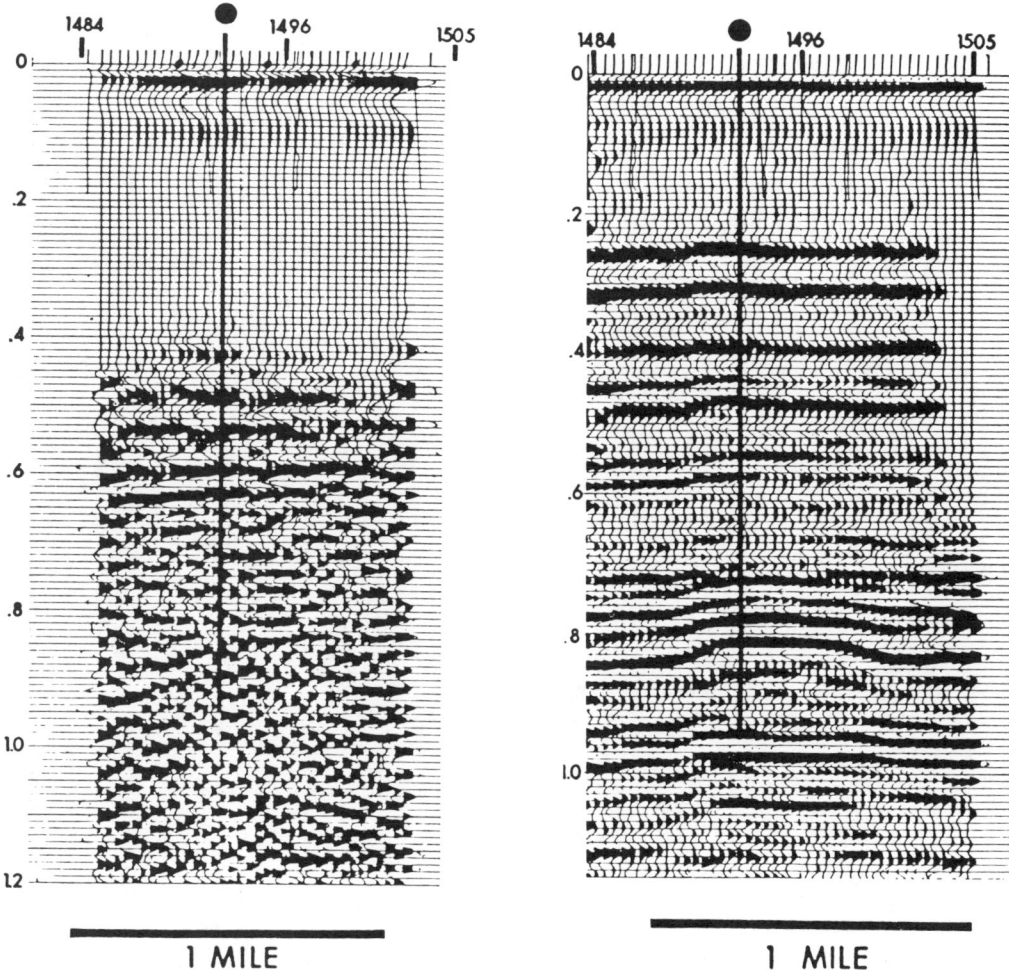

Figure 12.27: *Zama line 116-1, Alberta, Canada (Evans, 1972).*

Figure 12.28: *Zama line 116-1 reprocessed by a deconvolution program. This data is sufficient in quality to be used for development drilling (Evans, 1972).*

veries were on local buildups, rather than on a continuous barrier buildup.

*Trap:* Production at Judy Creek (Fig. 12.30) is from a carbonate buildup of probable early Devonian age. Reserves are estimated at 800 million barrels of 41° API oil. The trap is totally stratigraphic and related to the distribution of porosity and permeability within the buildup (Fig. 12.31).

*Reservoir:* Limestones in the Judy Creek Field are composed of (1) organic framebuilders — dominantly massive stromatoporoids, (2) nonframebuilders — namely tabular *Amphipora* and *Stachyodes,* solenporid algae, and tabulate corals, (3) grains — mostly skeletal, but pellets, intraclasts, and algal-coated grains are locally abundant, (4) micrite,

(5) various spar cements, and (6) pores. Figure 12.32 illustrates the geometry and facies distribution of the carbonate buildup.

The depositional history of a typical Beaverhill Lake reef can be described in six stages *(Hemphill and others, 1970).* The Beaverhill Lake sea transgressed westward and produced widespread shoal conditions (Stage I, Fig. 12.33). These conditions were favorable for carbonate production by organisms that produced extensive shallow water environments. Stromatoporids occupied the northeastern margins of the shoal environments where energy levels existed. Stage II was characterized by a persistent relative rise in sea level

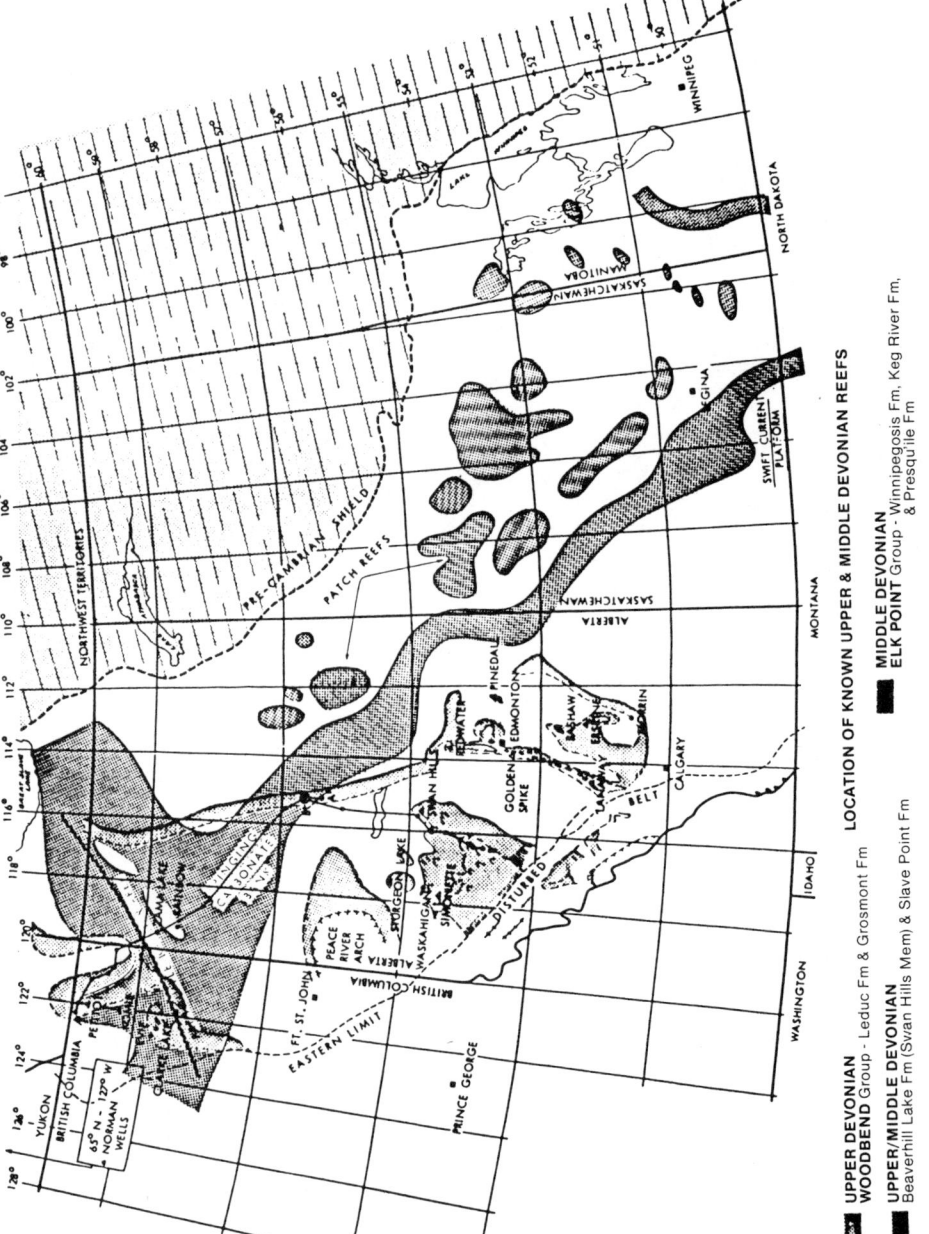

**Figure 12.29:** *Distribution of Devonian reefs in southcentral Canada.*

RESERVOIRS AND CARBONATE TRAPS 259

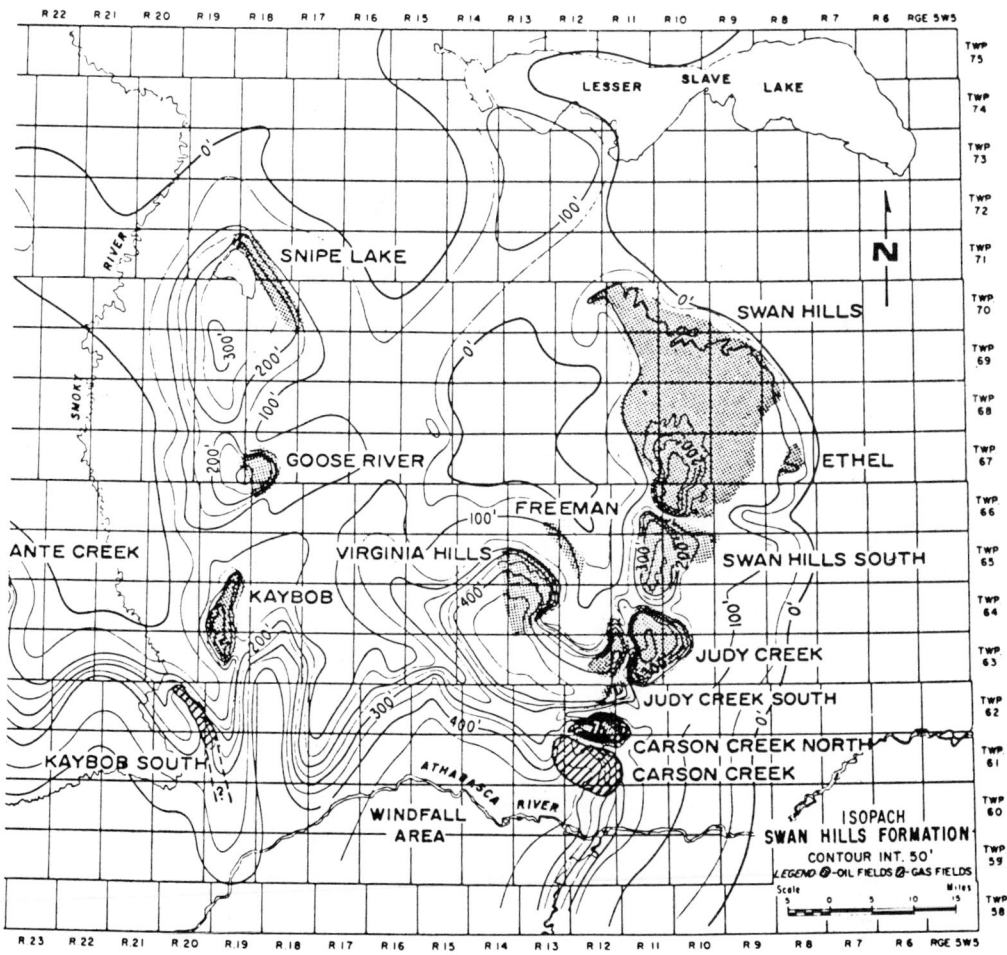

**Figure 12.30:** *Isopach map of Swan Hills Formation with locations of the major fields, Alberta, Canada (Hemphill and others, 1970).*

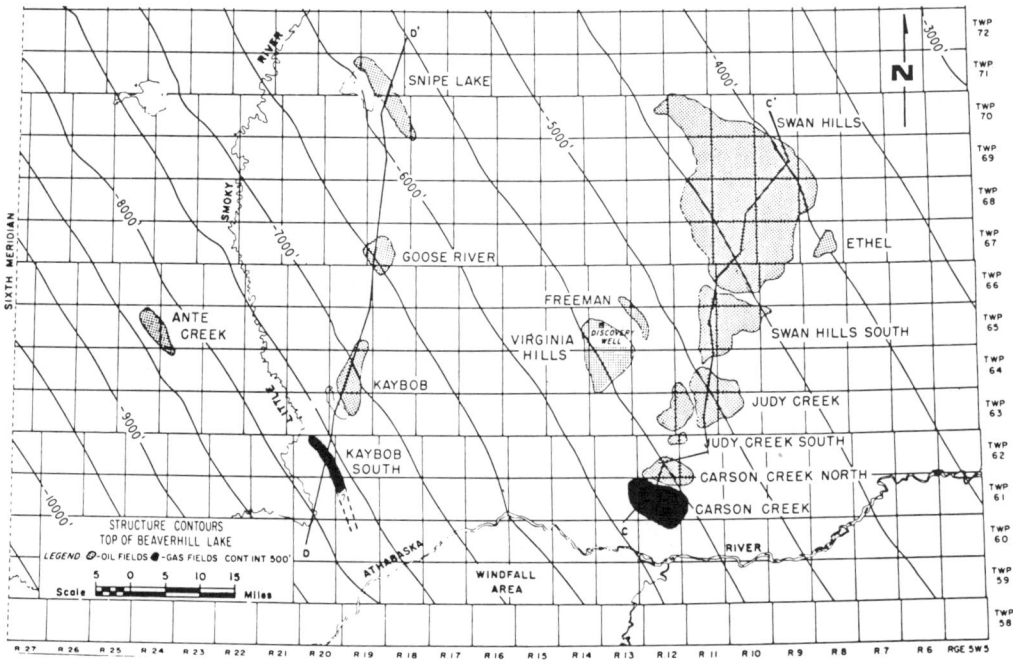

**Figure 12.31:** *Structural contour map on top of the Beaverhill Lake depicts southwesterly regional dip (Hemphill and others, 1970).*

that may have been produced by subsidence. This permitted continuous stromatoporoid reef growth on the northeast rim and produced a broad shelf lagoon. With continued subsidence, rigid stromatoporoid reef growth continued while *Amphipora*-rich deposits accumulated in the lagoon. Slight eastward tilting of the platform was too rapid for the slow carbonate producing Stromatoporoids to keep pace, and the reef development ceased. At the same time, the western margin was briefly exposed and reworked (Stage III).

Accumulation of *Amphipora* fragments in the quiet water, submerged platform-dominated Stage IV. These grainstones and packstones were reworked by worms to produce local bank buildups that provided the necessary topography for reef growth to re-initiate Stage V. The "Table reef" which developed on the once emergent shoals of Stage IV was restricted by a significant stillstand and developed laterally. Slow but steady subsidence resulted in the development of a lagoon encircled by a fringing stromatoporoid reef.

The final stage was marked by a return to tidal and supratidal conditions. This re-initiated extensive *Amphipora*, pelletoid, and mud to accumulate across Stage V lagoonal and fringing reef deposits. Stage VI was climaxed by rapid subsidence and/or sea level rise which drowned the reef complex.

### Horseshoe Atoll, Texas

*History of Exploration:* The late Paleozoic (Strawn-lower Wolfcampian) Horseshoe Atoll is a sinuous chain of carbonate mounds that extends for a distance of 175 miles in the subsurface of the northern part of the Midland Basin, west Texas. The first field to be discovered on the Atoll was made in 1948 and an extremely active exploration and exploitation program was conducted in the area from 1949-1953. More than 5,500 wells have been drilled on 40-acre spacing and recoverable reserves are estimated to be 2.54BBO *(Vest Jr., 1970)*.

*Trap:* Production at the Horseshoe Atoll comes from fifteen fields. Each field is totally stratigraphic. Regional dip averages about 30 feet per mile to the west. Although a map on top of the Horseshoe reef complex appears to contain structural closures, inspection of an

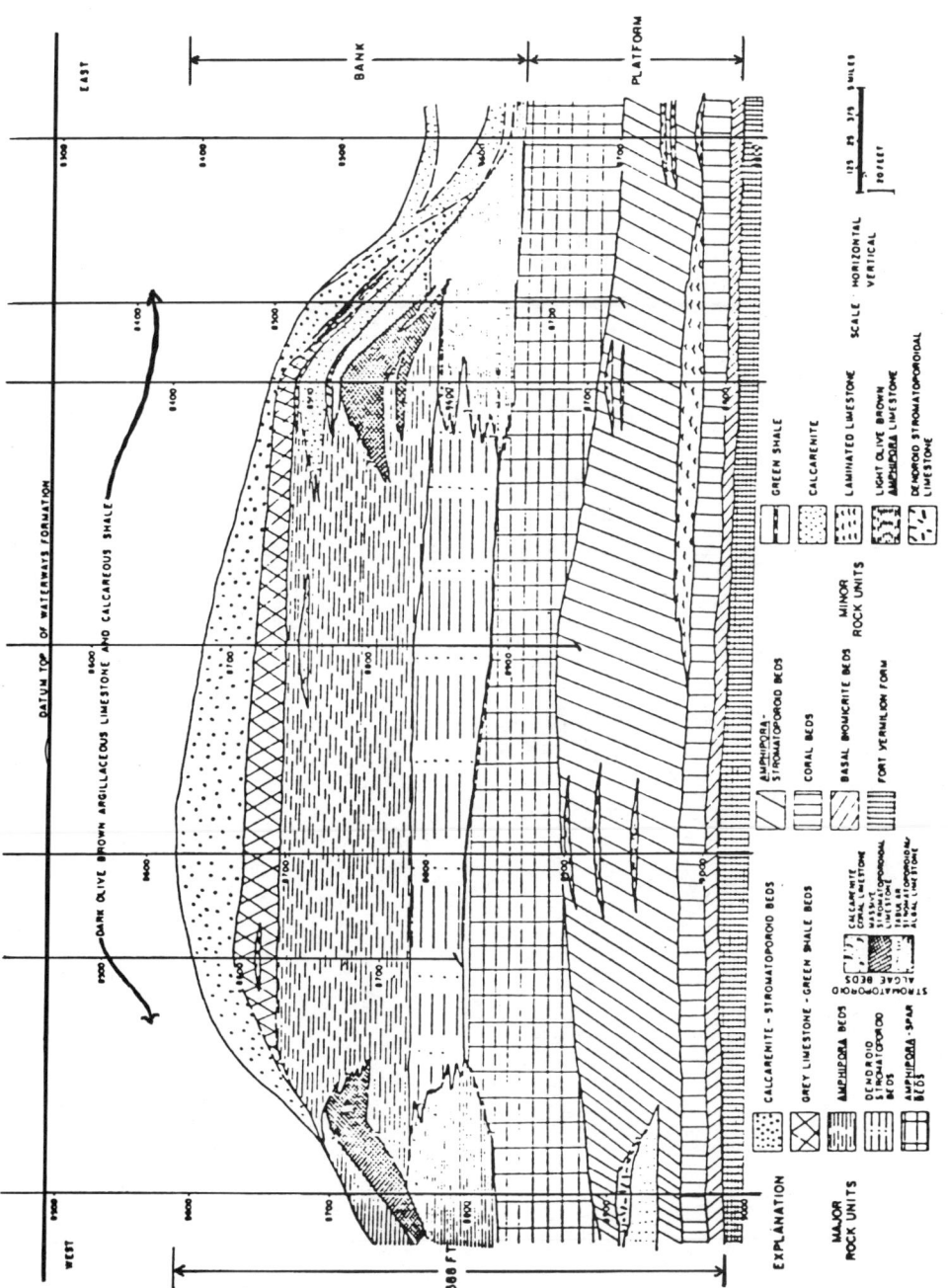

**Figure 12.32:** *Cross section illustrating lateral and vertical facies relationships in Judy Creek Field (Murray, 1966).*

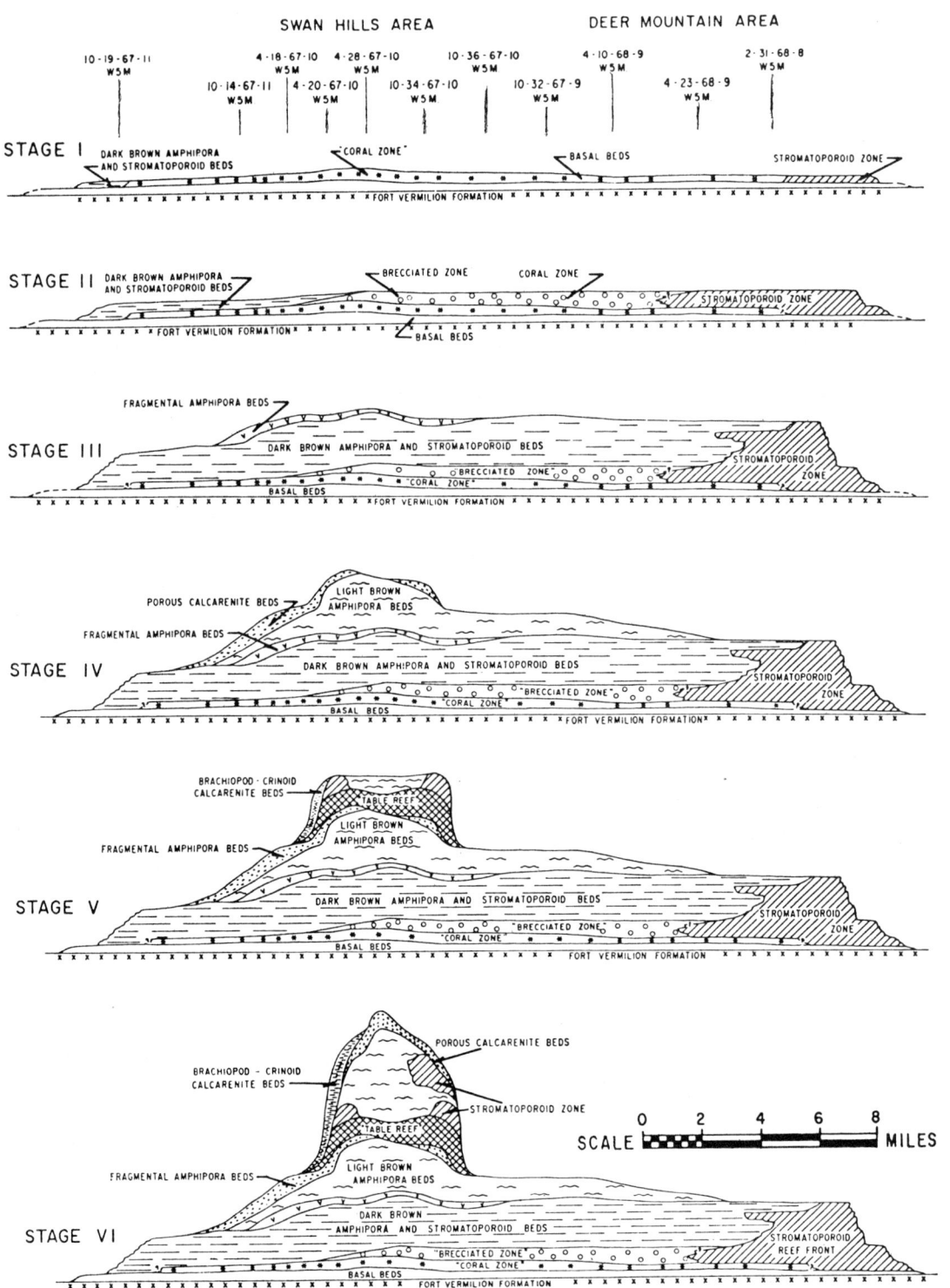

**Figure 12.33:** *Stages of typical Swan Hills reef development, Alberta, Canada (Hemphill and others, 1970).*

isopach map of the carbonate will reveal that the "structural" contours reflect depositional topography rather than structural closures (Fig. 12.34 and Fig. 12.35). Sedimentary-dip averages are generally less than 8° but locally exceed 20° *(Vest Jr., 1970)*.

*Reservoir:* The Horseshoe reef reservoirs consist of Strawn, Canyon, Cisco (Pennsylvanian), and lower Wolfcampian (Permian) limestones. Typically these carbonates are massively bedded grainstones and packstones. No boundstones have been reported; however, most probably various algae, bryozoans, and specialized brachiopods provided a framework that may have been locally rigid. The section is laced with interbeds of shales that range in thickness from a few feet to nearly 20 feet. Although the shale beds can be correlated over large areas, they did not prevent vertical as well as lateral migration of hydrocarbons. Maximum thickness of the reef complex occurs along its western margin where it reaches a thickness of over 1,700 feet (Fig. 12.31). The overall westerly thickening of the reef complex is probably due to differential subsidence that began in late Cisco time.

**Exploration and Exploitation Strategy**

*Exploration:* Regional isopach and facies maps of stratigraphic targets aid in prediction of atoll and mound buildups. Core and sample analysis provide the basis for facies analyses and rock-log calibrations are essential. Mechanical logs and seismic data should be integrated for the construction of critical isopach maps. Synthetic seismic lines should be constructed in order to provide a guide for interpretation of the field area.

*Exploitation:* Time surfaces are difficult to document in most carbonate buildups. Therefore, arbitrary slice maps have been found to be adequate to map porosity. These

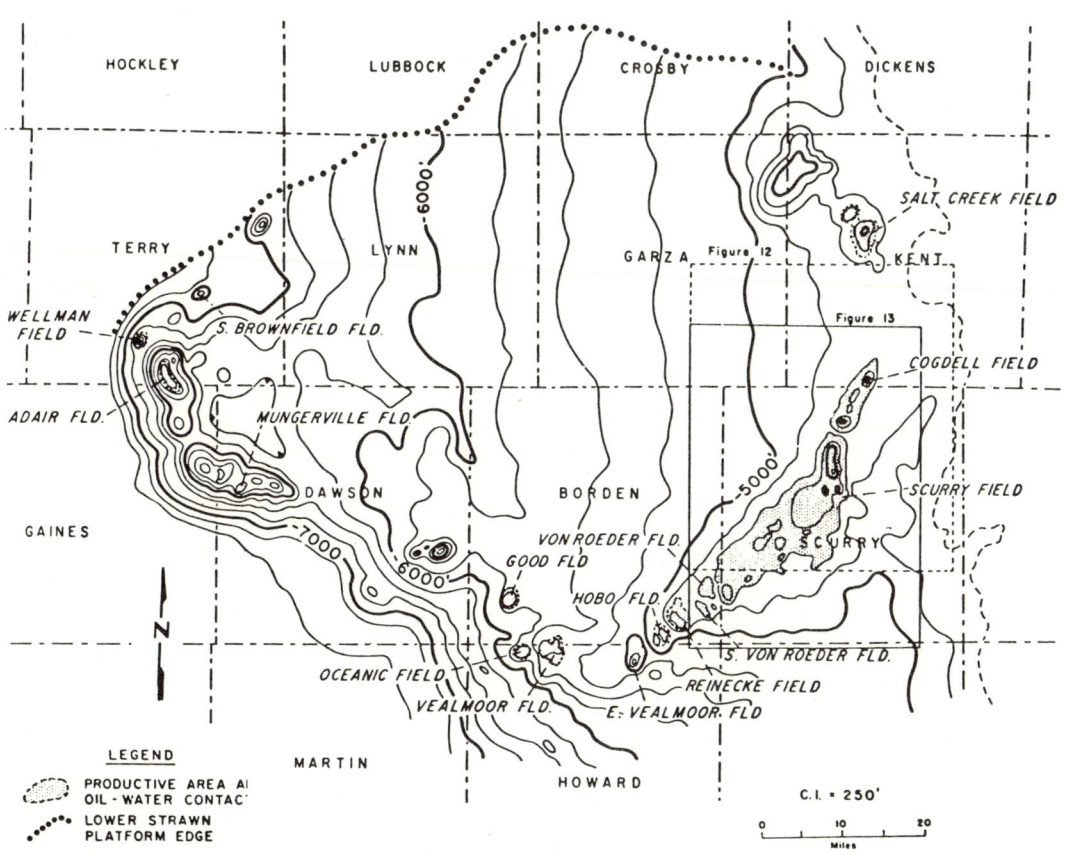

**Figure 12.34:** *Structural contour map on top of the Horseshoe reef complex, Midland Basin, Texas (Vest, Jr., 1970).*

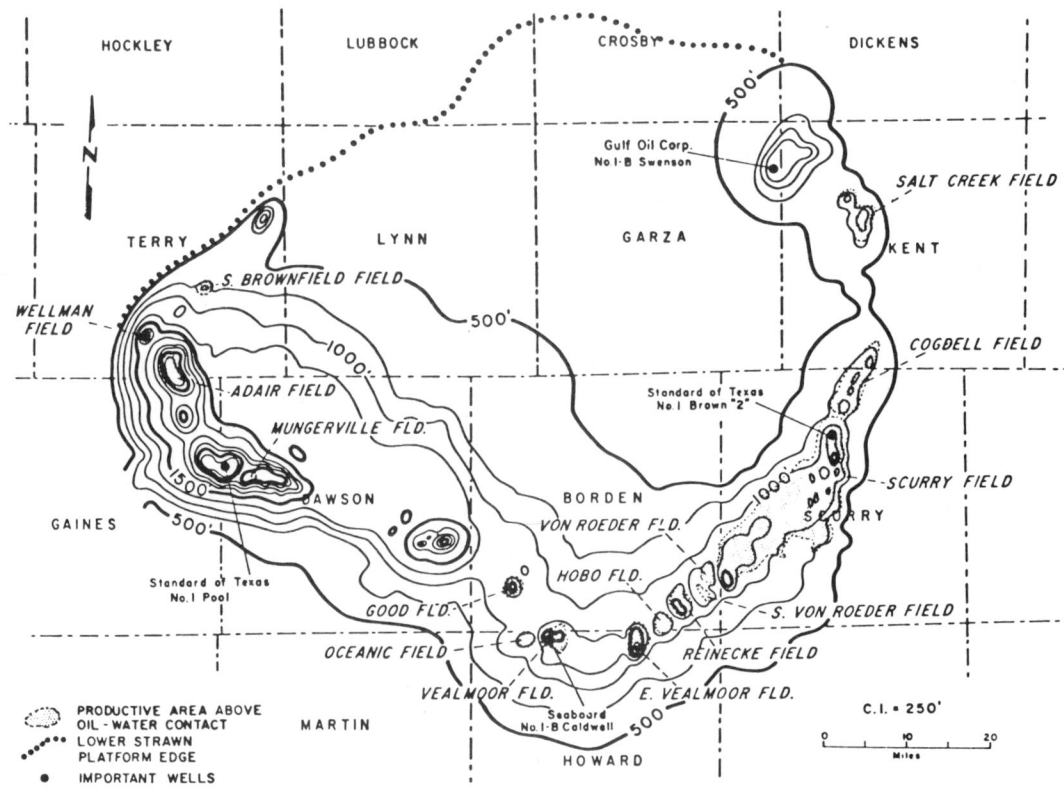

**Figure 12.35:** *Isopach map of the Horseshoe reef complex, Midland Basin, Texas (Vest, Jr., 1970).*

maps should be updated during the development program. An understanding of the facies distribution will add to the knowledge of pore geometry, which is useful during pressure maintenance, and enhanced recovery programs.

**Seismic Characteristics of Atolls & Mounds**
Figure 12.36 illustrates a seismic line across the Horseshoe Atoll of West Texas. An excellent high amplitude and continuous reflector depicts the top and depositional slope of the carbonate buildup. Two cycles of continuous reflectors onlap the flanks of the buildup. Reflections within the buildup are typically discontinuous. Drape over the buildup is minimum because the "basinal" facies are composed of moderately cemented siltstones and sandstones.

## SPREAD-OUTS

### INTRODUCTION

Significant hydrocarbon reserves have been discovered in three types of carbonate "spread-outs." They are grainstone shoals, tidal flat deposits, and chalks. Chalk reservoirs are not included in this discussion of case histories because these reservoirs typically require structure as the trapping mechanism.

### GRAINSTONE SHOALS

**Introduction** Like terrigenous sands, carbonate grains are winnowed and sorted in response to waves and currents, but unlike terrigenous sands, carbonate grains can grow larger by accretion. The resultant grains are called ooliths, pisolites, or oncolites, depending on their size. The grainstone shoals can form along the transitional zone of a carbonate ramp or along the leeward side of a shelf margin.

**Key Exploration Problems** Grainstone shoals that accumulate on the lee side of a shelf margin or along the coastline of a carbonate ramp are not too difficult to map. However, those shoals that accumulate in the transitional zone of an epeiric sea (ramp) are

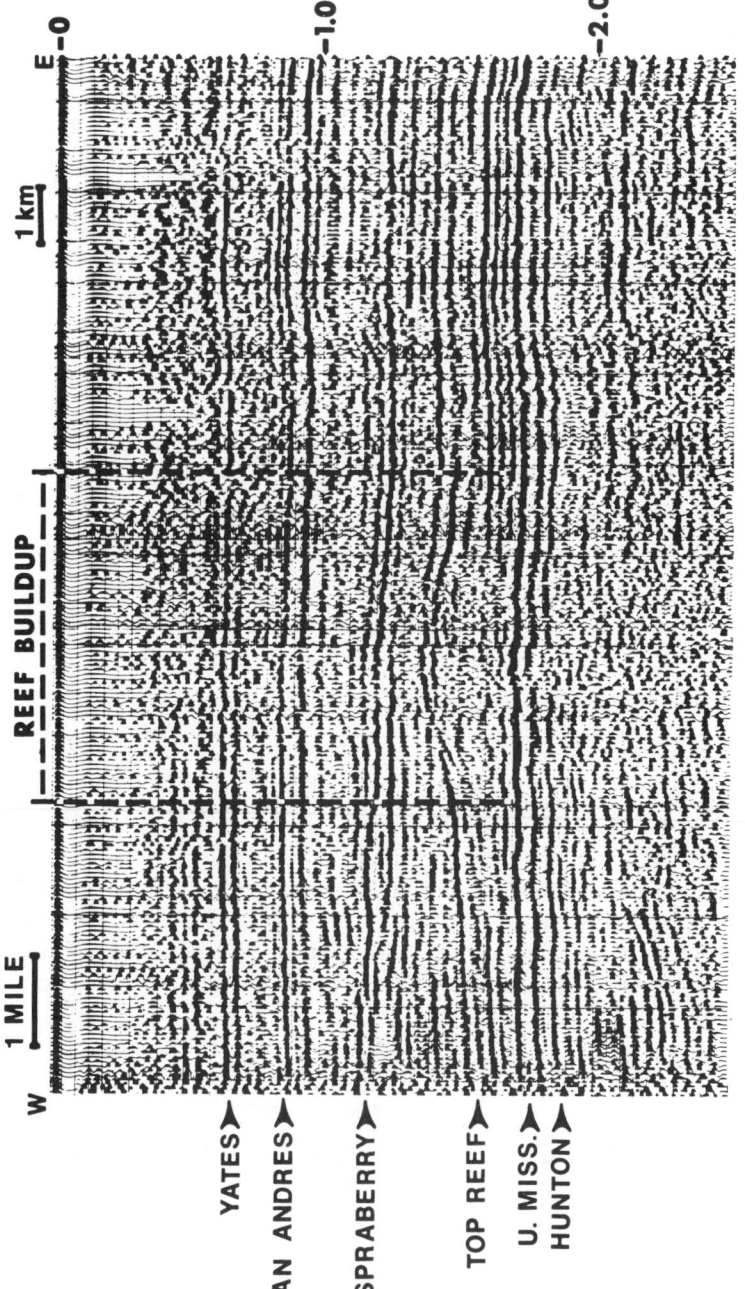

Figure 12.36: *Seismic line across Horseshoe reef complex, Midland Basin, Texas (Laing, 1972).*

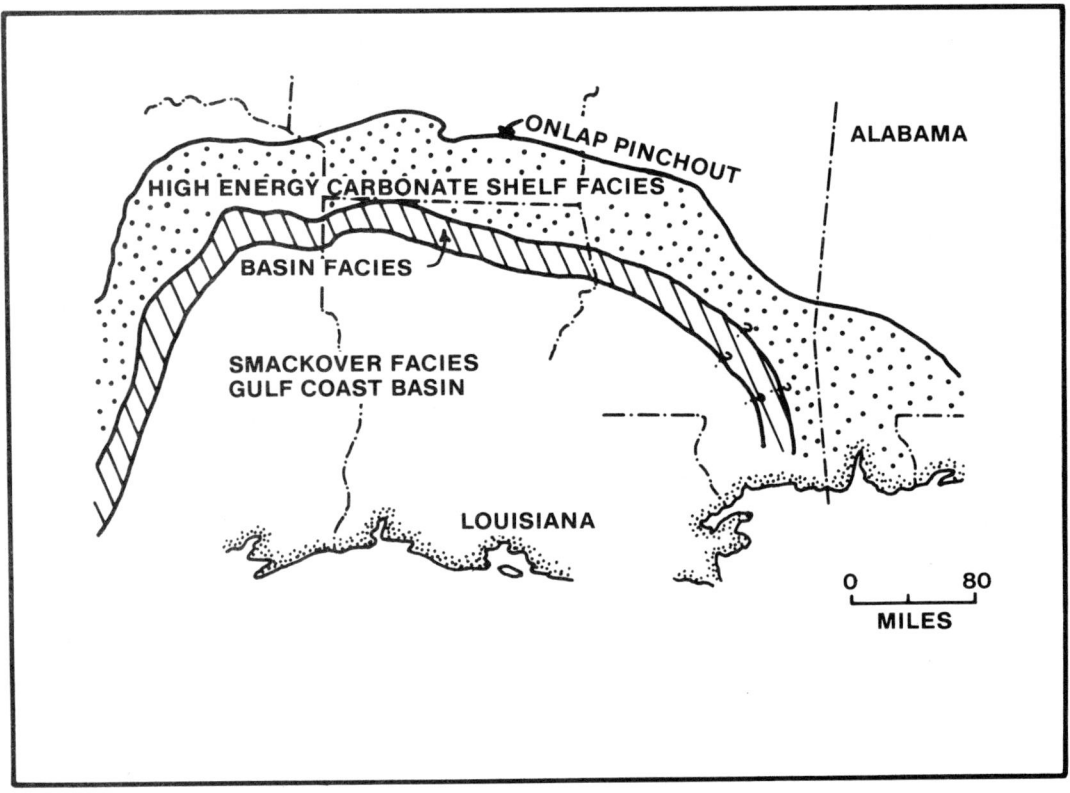

Figure 12.37: *Regional distribution of the Smackover facies in the Gulf Coast, U.S.A. (Ottman and others, 1973).*

generally randomly developed and provide a very difficult challenge to the explorationist. Furthermore, seismic data is not often useful to aid in the delineation of carbonate shoals because they are generally thin and do not result in a buildup. Some grainstone shoals have been mapped by amplitude changes, bright and/or dim spots.

**Jay Field**

*History of Exploration:* The Jurassic Smackover represents a 200 to 400-foot thick belt of high energy deposits (Fig. 12.37). The ramp-deposited rocks form an accurate trend from Mexico to Florida. Basal strata of the Smackover onlaps the eroded roots of the Ouachita Mountain chain, and these mountains appear to control the accumulation of carbonate sand.

Between 1937 and 1950, about 250 million barrels of oil were discovered in 34 fields located in southern Arkansas and northern Louisiana. Smackover production occurs between 6,000 and 12,000 feet and was discovered with the aid of conventional seismic data. No additional major reservoirs were discovered in the Smackover until 1960 when the then new CDP seismic technique was introduced. This permitted definition of deeper structures, and exploration shifted into south Texas and into southern Mississippi where 38 new fields were discovered. From 1967, the present Smackover exploration was extended into Florida. Jay and Big Excambia Creek Fields were discovered during this interval.

*Trap:* Jay Field is a combination stratigraphic-structural trap (Fig. 12.38). The trap contains about 400 feet of oil column that extends over a 21 square mile area. Recoverable reserves are in excess of 400 million barrels of oil.

*Reservoir:* The Smackover in the Jay area may be subdivided into three parts (Fig. 12.39). The upper part is commonly an oolite and pelletoid line grainstone that is locally

# RESERVOIRS AND CARBONATE TRAPS

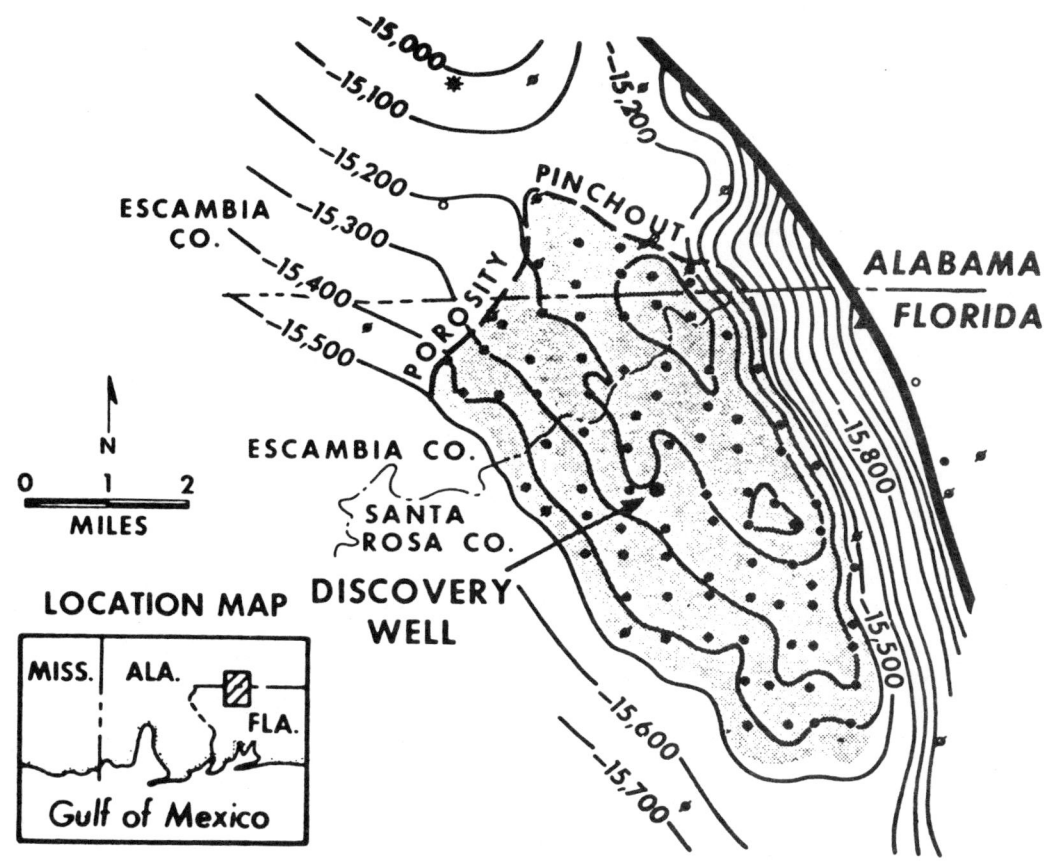

**Figure 12.38:** *Structure map on top of the Smackover, Jay Field, Florida and Alabama, U.S.A. (Ottman and others, 1973).*

dolomitized. Regionally, this is the main reservoir zone. A brown mudstone and pelletoid wackestone characterizes the middle zone of the Smackover. The lower Smackover is normally a dark microlaminated limestone-dolomite that contains silt and organic matter. Pelagic fauna are abundant, and it is this zone that is thought to be the source rock. Figure 12.40 depicts the inferred environments of deposition of the Smackover.

In Jay Field the reservoir is composed of pelletal-oncolitic micrite, pelletal grainstones, pelletal-oolitice grainstones, and algal stromatolitic wackestones. Figure 12.41 illustrates the three pore types: (1) grain moldic dolomite, (2) intercrystalline dolomite, and (3) leached matrix.

Vertical distribution of the porosity is illustrated by Figure. 12.39. The major pay zone is the hardened pellet grainstone facies in the upper part of the Smackover. This zone of moldic porosity was produced by invasion of phreatic pore water into the dolomitized pellet grainstone during subaerial exposure of the pellet-oolitic shoals. Porosity values are high for a limestone reservoir (8-18%), and maximum permeability is about 300md.

## EXPLORATION AND EXPLOITATION STRATEGY

**Exploration** A successful exploration program for Jay Field type discoveries must be based on a sound understanding of the regional facies, influence of pre-existing topography, salt solution, and changes in sea level on carbonate accumulation, and complete integration of subsurface geology and geophysics.

**Exploitation** Where economics permit, each well should be cored and a complete suite

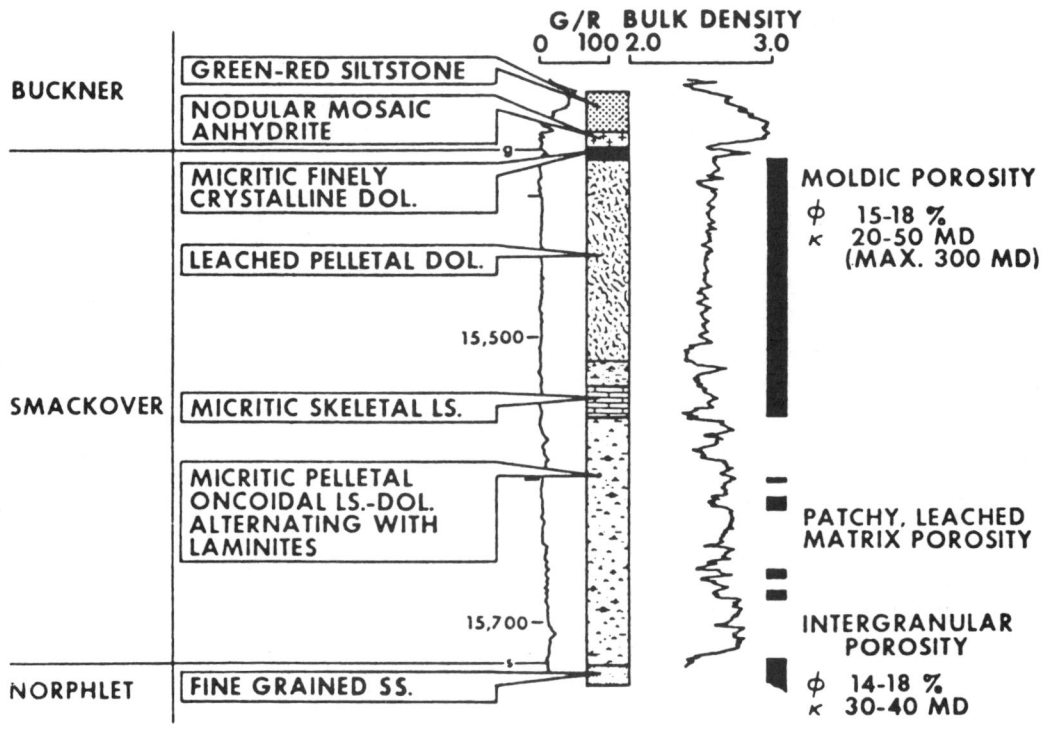

**Figure 12.39:** *Vertical relationship of facies and pore types, Jay Field, Florida (Ottman and others, 1973).*

of porosity logs run, *i.e.,* sonic and density. Detailed facies analysis of the cores should be made prior to spudding the next development well. Analysis of pore casts will be useful to aid the enhanced recovery program.

### Hanford Field, Texas

*History of Exploration:* Hanford Field was discovered in 1977 by seismic and surface mapping *(Kumar and Foster, 1982).* The field is located on the southern edge of the San Simon Channel along the northeastern corner of the Central Basin Platform (Fig. 12.42). It has thirty-one producing wells in a San Andros (Permian) carbonate shoal deposit with and EUR of 6.3 million barrels of oil.

*Trap:* A structural map on top of the San Andros Formation would not depict a closure over the field area. However, an isopach map of the producing interval shows a thick anomaly over the productive area of the field and the top of the productive zone does have closure (Fig. 12.43).

*Reservoir:* The productive interval (informally called the "green zone") consists almost entirely of dolomitized grainstones, packstones, and wackestones with interspersed anhydrite. Kumar and Foster *(1982)* interpreted the lithofacies to represent numerous cyclic units that range from 10-20 feet thick (Fig. 12.44.) Each cycle records a shoaling upward from subtidal to supratidal environments. Maximum porosity and permeability occurs in the subtidal grainstone facies of each cycle.

The high-energy shoaling conditions in the field area are thought to be related to a pre-San Andros mound that occurs about 1,000 feet below the field (Fig. 12.45). The mound was drowned during transgression and was later gradually buried by prograding, cyclic subtidal to supratidal carbonates and anhydrites. Sufficient topography over the mound existed during the deposition of the reservoir to place the field area in a shoal environment within a deeper subtidal system. The shoal was an area of rapid carbonate production which in turn produced about 140 feet of depositional relief on top of the shoal *(Kumar and Foster, 1982).*

*Exploration Strategy:* The model suggested by Kumar and Foster *(1982)* implies that in

# RESERVOIRS AND CARBONATE TRAPS

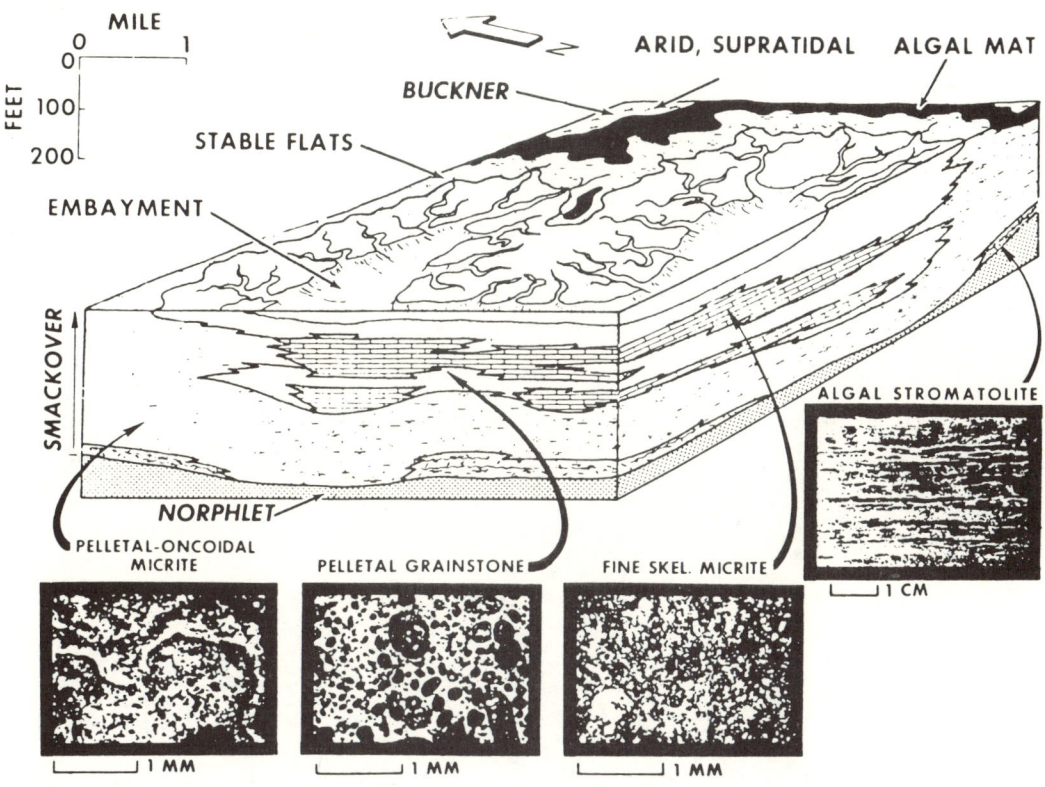

**Figure 12.40:** *Reconstruction of depositional environments, Jay Field area (Ottman and others, 1973).*

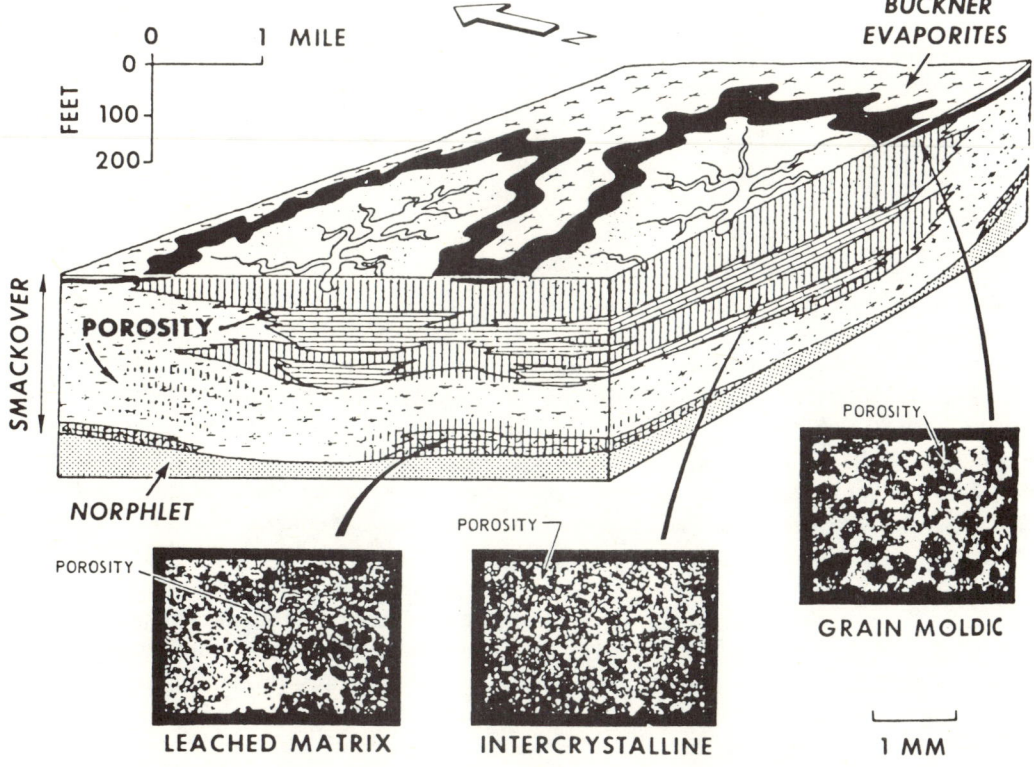

**Figure 12.41:** *Pore types, Jay Field, Florida and Alabama (Ottman and others, 1973).*

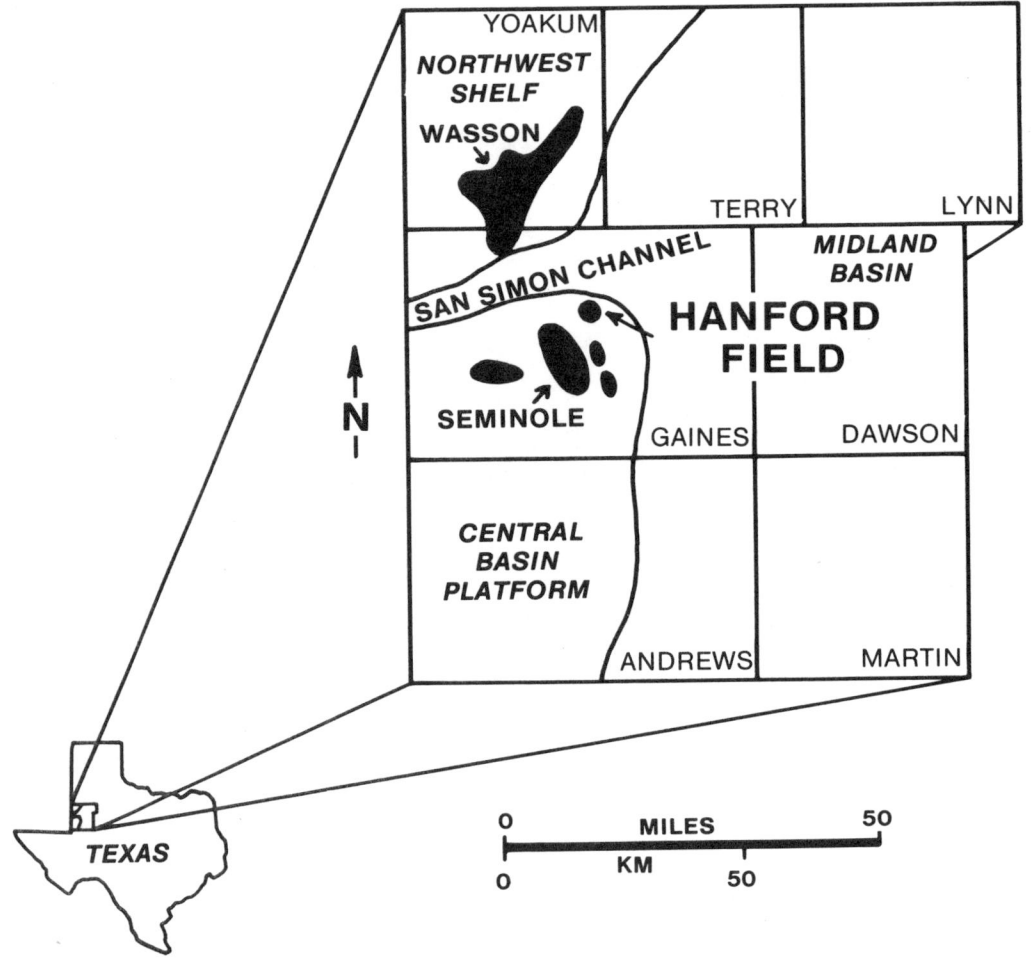

**Figure 12.42:** *Hanford Field is located on the northeastern edge of the Central Basin Platform, west Texas (Kumar and Foster, 1982).*

areas where pre-San Andros mounds occur beneath subtidal San Andros deposits, the later would be excellent sites for San Andros exploration. The earlier mounds can be mapped seismically, and the regional distribution of the mounds are known to occur around the shelf edges of the Permian Basin, Texas. Regional facies maps of the San Andros in conjunction with trend maps of the pre-San Andros mounds would outline a prospective trend.

## TIDAL FLAT DEPOSITS

**Introduction** Tidal flat deposits occur in both carbonate ramp and step models of deposition. They represent the distal member of the ramp or step marine carbonate package. Typically tidal flats have little depositional topography, therefore they are highly influenced by minor sea-level changes, rate of subsidence, influx of terrigenous clastics and marine and/or wind generated tides. Whether the tidal flat environment is located in a humid climatic belt (Belize) or an arid climate (Trucial Coast) the restricted nature of the tidal flat produces a carbonate facies that typically includes dolomites and evaporites.

Ancient and modern tidal flat deposits contain more similarities than any other ancient-modern carbonate analog. This is in part due to the restricted nature of the environment which typically produces hypersaline conditions. The fauna is usually limited to extensive mats of blue-green algae that may produce stromatilitic structures. The mats are normally covered by grazing, pellet-producing

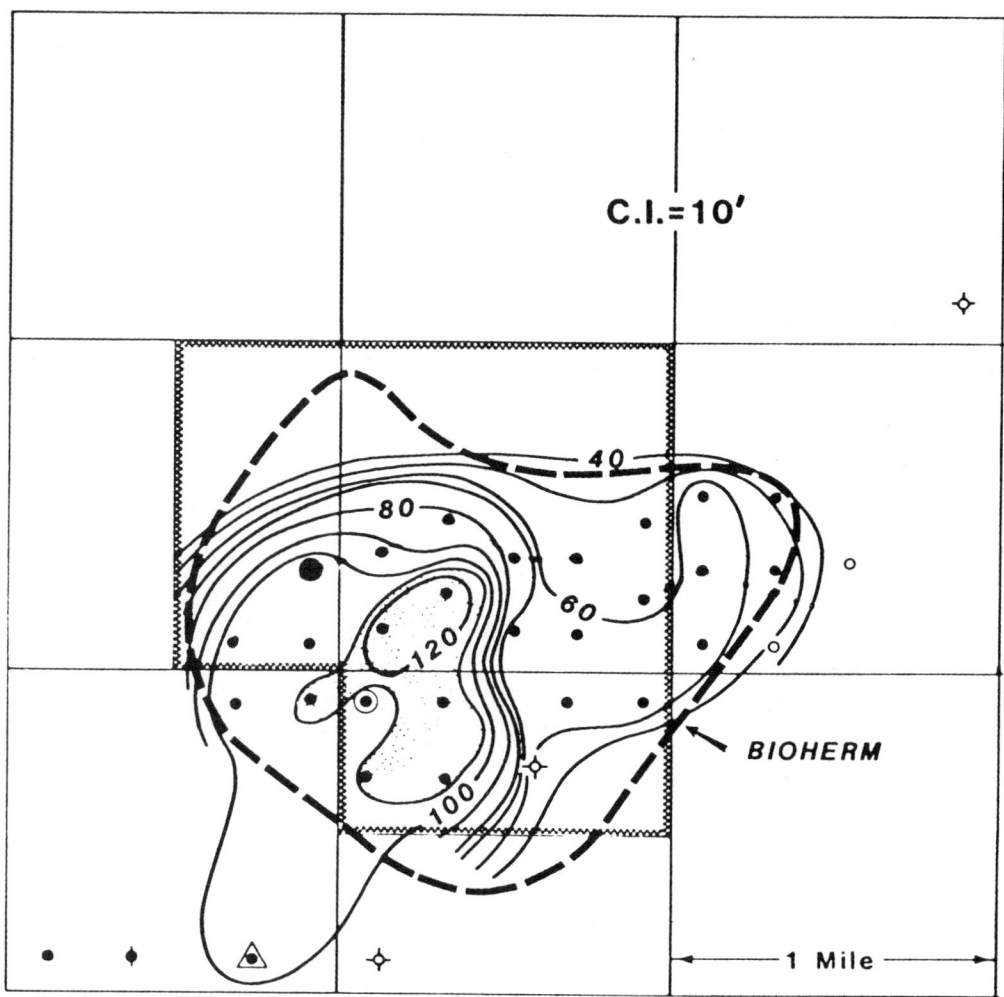

**Figure 12.43:** *Isopach map of the producing interval at Hanford Field, west Texas. Dashed line represents limits of pre-San Andros bioherm (Kumar and Foster, 1982).*

mollusks. The mat is dissected by tidal channels and although winnowing currents are normally weak, they effectively rework the pellets into a porous and permeable facies. These subtidal environments are typically bioturbated to such an extent that primary sedimentary structures are totally destroyed. Further landward, the algal-mat facies is bordered by extensive laminated facies that may contain nodular evaporites, or if the climate is sufficiently arid, bedded evaporites. Concordant and discordant dolomitization patterns are commonly associated with this facies belt.

**Key Exploration Problems** Primary porosity and permeability of non-dolomitized tidal flat deposits are normally low. Therefore, not only must the explorationist successfully predict the location of the deposit, he or she must be able to successfully predict the dolomitization patterns that produce secondary porosity and permeability. These deposits are typically thin, and because bathemtry is such a critical variable, depositional strike can change abruptly.

**Levelland-Slaughter Field**
*History of Exploration:* An abnormally thick sequence of alternating dolomite, limestone, fine-grained terrigenous clastics, and evaporites occurs on the northwestern shelf of the Delaware and Midland Basins (Fig. 2.13).

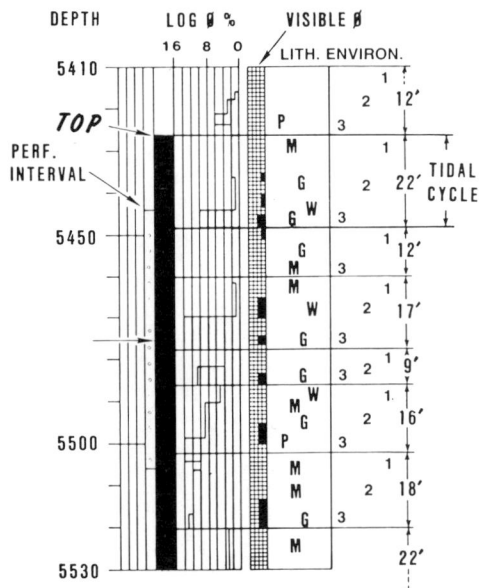

**Figure 12.44:** *Influence of cyclic depositional patterns on reservoir quality, Hanford Field, west Texas (Kumar and Foster, 1982).*

Drilling what was considered to be structures, discoveries by "accident" were made along this stratigraphic trend that date back into the 1930s. Levelland-Slaughter is one of the largest fields located in West Texas and Southeastern New Mexico.

*Trap:* The trap is an updip porosity pinchout of porous and permeable dolomite into tight dolomite and/or dense evaporite (Fig. 12.46). Some lateral facies changes control reservoir quality. There is some subtle structural control on oil-water contacts that are typically associated with minor changes in reservoir capillarity.

*Reservoir:* The reservoir is a microcrystalline dolomite. Average porosity ranges from 10-14%. Permeabilities are low (1-10 md), but because the pore throats are equidimensional, capillary properties are good to excellent. The porosity zone is completely dolomitized. The reservoir is a result of cyclic sedimentation that is similar to that which occurs along the Trucial Coast (Figs. 10.20 and 12.47).

**Exploration Strategy** The explorationist must map the thickness and distribution of supratidal deposits with the knowledge that where the supratidal deposits wedge out, the dolomite should wedge out. The best porosity in the dolomite should be located farthest from the supratidal rocks. It should be continuous from the interface of the dolomitized limestone-supratidal terrigenous and/or evaporite boundary down to the dolomitized limestone-nondolomitized boundary (Fig. 12.48).

## EXPLORATION STRATEGY FOR CARBONATE RESERVOIR ROCKS

Potential carbonate reservoir rocks are highly sensitive to global changes in sea level, differential rates of subsidence, climate conditions, and diagenetic environments. Moreover, carbonate reservoir rocks are frequently source and seal, cyclic, reflect significant depositional topography, and facies are randomly developed. It is therefore mandatory that the exploration team follow a series of systematic steps to predict the depositional and post depositional history of carbonate rocks.

1. A valid stratigraphic framework within an unconformity bounded sequence must be developed. This requires pattern correlation of a network of stratigraphic cross sections and/or seismic lines. Where control permits, the lines of section should be constructed parallel and perpendicular to depositional strike. All wells between the lines of section should be projected along strike into the closest line of section for correlation.

2. Sample and/or core data should be plotted on the appropriate logs. This data should be integrated into all seismic lines.

3. Preliminary facies maps should be constructed in light of the modern step or ramp model of carbonate deposition. The modern models must be used only as a guideline for initial interpretations. Data derived from rock calibrated logs, pattern correlations, and seismic reflection character (Fig. 12.49) will significantly alter the initial interpretations that are based on the modern generalized models.

4. If production occurs from carbonate reservoirs in the area under investigation,

# RESERVOIRS AND CARBONATE TRAPS

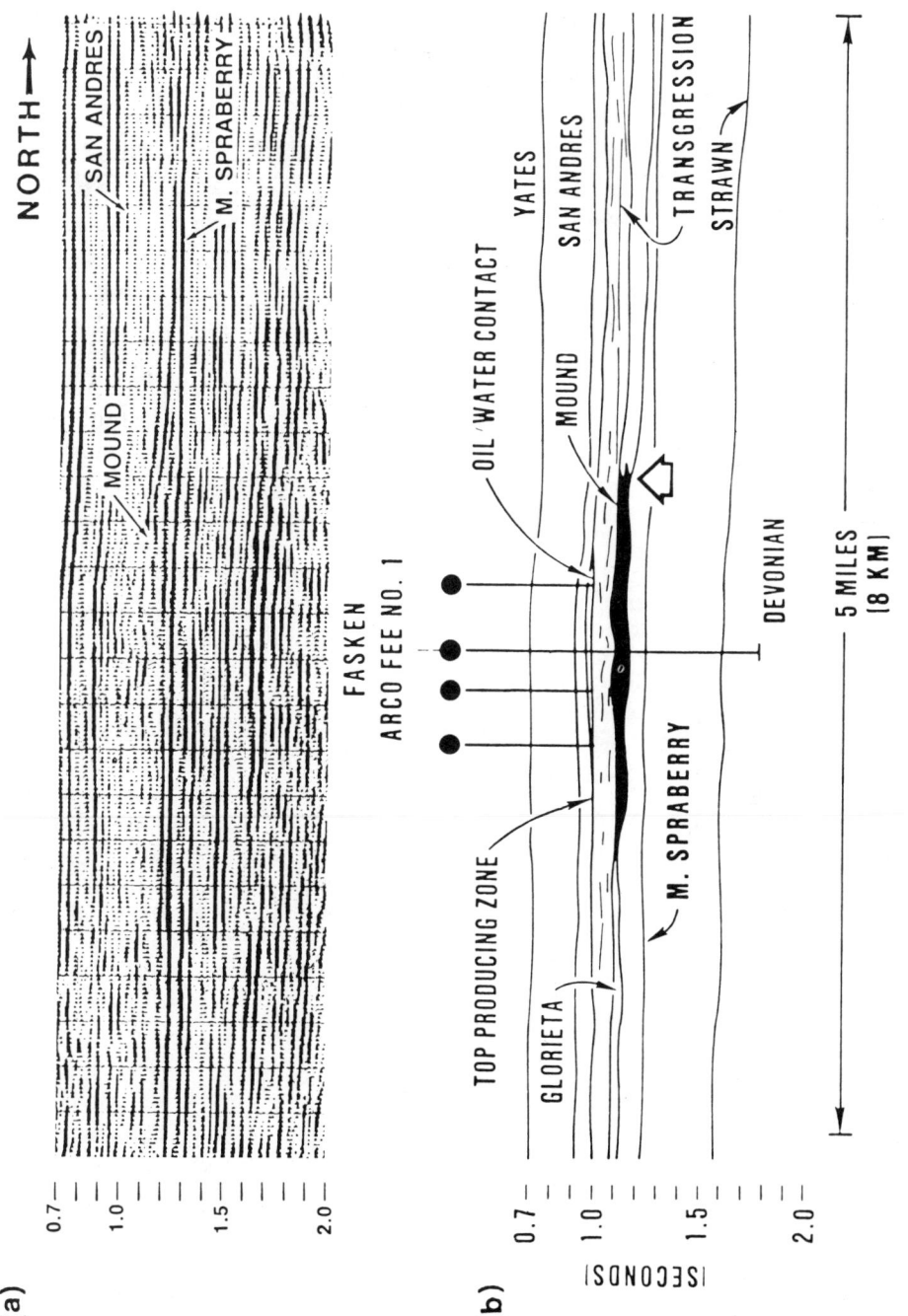

**Figure 12.45:** Seismic line and interpreted cross section depict the influence of a pre-San Andros mound on the depositional environment of the overlying San Andros (Kumar and Foster, 1982).

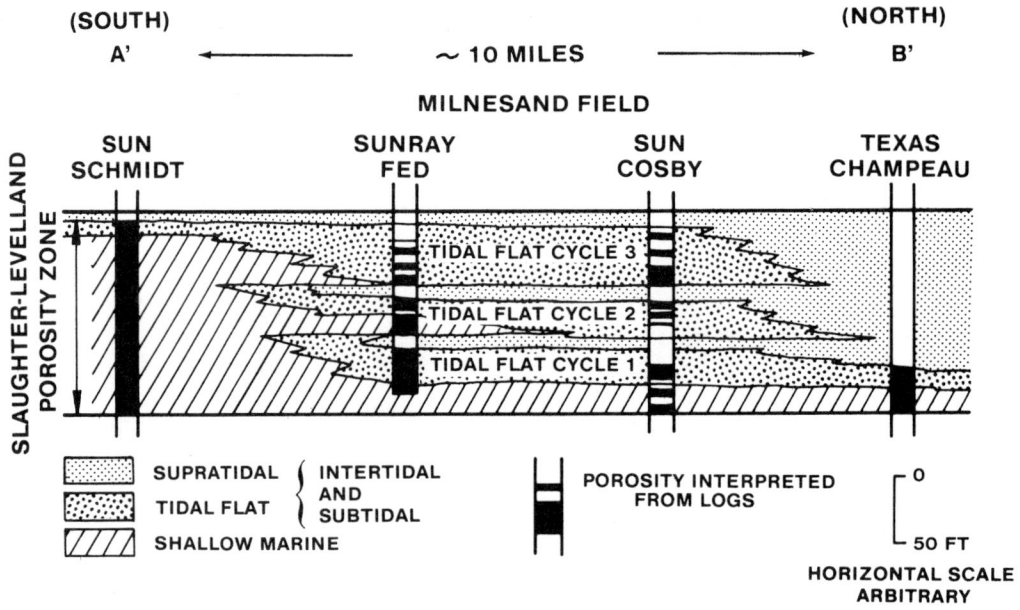

Figure 12.46: *North-south cross section showing pinchout of the Slaughter-Levelland porosity zone, southeast New Mexico.*

complete a study of each field, evaluate the trapping mechanism and source of petroleum. Classify each field as to trap type. Determine what map or set of maps would indicate the presence of each trap type prior to its discovery.

5. Refinement of the preliminary facies maps can be made by selecting proper stratigraphic intervals to isopach in order to highlight shelf margins, buildups, shoals, or tidal flat facies. Use the information derived from the field study to determine what maps should be constructed.

6. Depending upon the availability of seismic data and/or budget constraints, recommend a seismic program that will cross at right angles the predicted facies belt(s) that is considered to be the target horizon(s).

7. Refine maps and cross sections in light of the new seismic data in order to make the final decision — to drill or not to drill.

# RESERVOIRS AND CARBONATE TRAPS

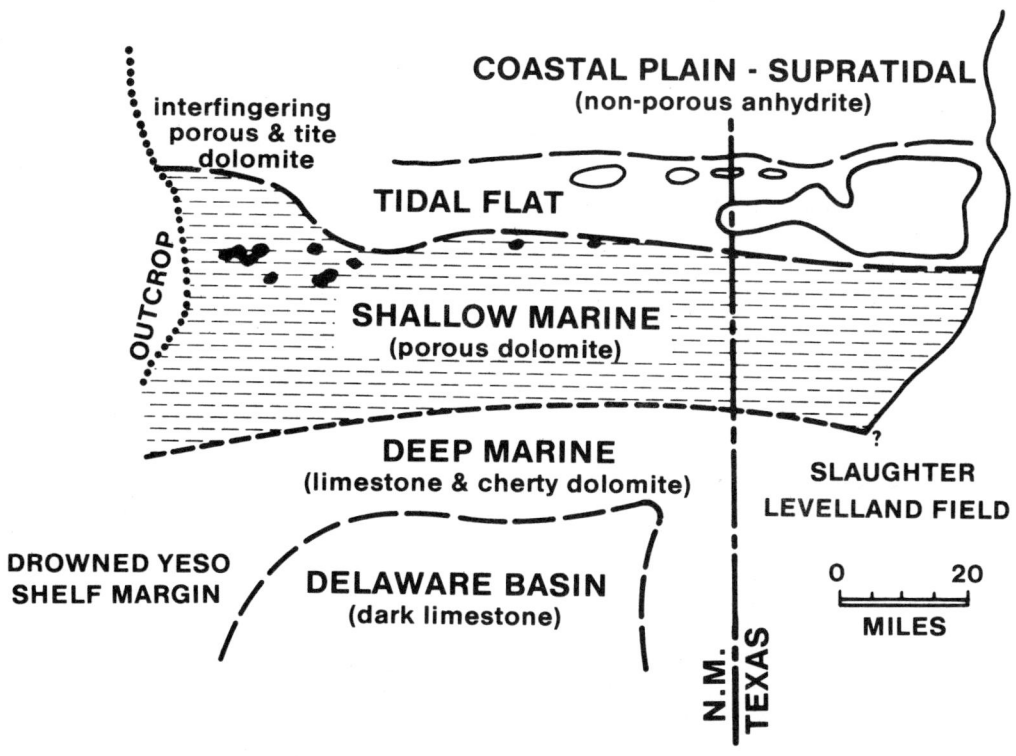

**Figure 12.47:** *Depositional environments of the Slaughter-Levelland (Middle San Andres time), southeast New Mexico and west Texas.*

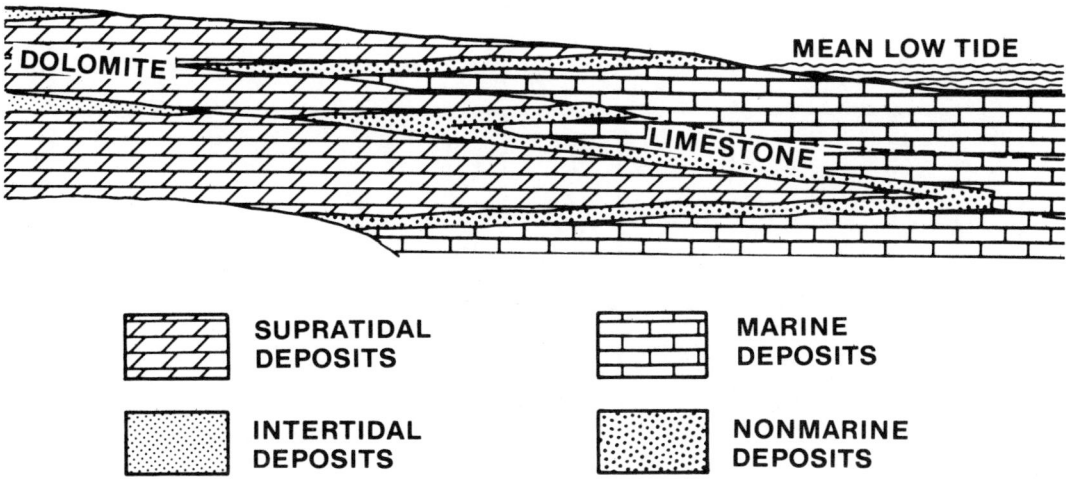

**Figure 12.48:** *Relationship between tidal flat sedimentation and reflux dolomitization.*

| ENVIRON-MENT | DIRECT EVIDENCE | | | | INDIRECT EVIDENCE | | | |
|---|---|---|---|---|---|---|---|---|
| | DEPOSITIONAL TOPOGRAPHY | SHAPE | ONLAP | AMPLITUDE | CONTINUITY | DRAPE | VELOCITY | SPURIOUS EVENTS |
| SHELF MARGIN |  | Linear | Yes | Yes | Discontinuous | Basinal side | Slope facies higher than shelf or basin | Yes |
| BARRIER |  | Linear | Yes | Yes | Discontinuous | Both sides | Barrier is high velocity | Yes |
| PINNACLE |  | Oval | Yes | Yes | Discontinuous | All Sides | Pinnacle may be high | Yes |
| PATCH |  | Oval | Normally none | Normally none | Normally continuous | Normally insignificant | Generally none | Yes |
| ATOLL |  | Oval to elongate | Yes | Yes | Discontinuous | All sides | Reef core higher than lagoon or basin | Yes |
| TIDAL FLAT |  | Linear to sheet | Very Gentle | High to low | Discontinuous | No | Evaporites higher than dolomites | No |
| CARBONATE SHOAL |  | Linear | Very Gentle | High to low | Discontinuous | No | Porous Slower than non-porous | No |

**Figure 12.49:** *Summary of carbonate reservoir seismic facies.*

# REFERENCES AND ADDITIONAL READING

# References

Allen, J. R., *Physical Processes of Sedimentation,* New York: Elsevier Publ. Co., Inc., (1969).

———, "Identification of Sediments—Their Depositional Environment and Degree of Compaction—From Well Log Curves," *Developments in Sedimentology 18A, Compaction of Coarse-Grained Sediments I,* New York: Elsevier Publ. Co., Inc., (1975), pp. 349-401.

Almon, D. R., and A. L. Schultz, "Electric Log Detection of Diagenetically Altered Reservoirs and Diagenetic Traps," *Gulf Coast Association of Geol. Soc. Trans.,* v. 24 (1979), pp. 126-137.

Anderson, F. M., "Origin of California Petroleum," AAPG *Bulletin,* v. 37 (1926), pp. 585-614.

Anderson, R. Y., and D. W. Kirkland, "Origin, Varves, and Cycles of Jurassic Todilto Formation, New Mexico," AAPG *Bulletin,* v. 44 (1960), pp. 37-52.

Asquith, D. O., "Depositional Topography and Major Marine Environments, Late Cretaceous, Wyoming," AAPG *Bulletin,* v. 54 (1970), pp. 1184-1226.

Atwater, G. I., and E. E. Miller, "Unpublished Data from D. L. Ziegler and J. H. Spotts," AAPG *Bulletin,* v. 62 (1965), p. 814.

Baria, L. R., D. L. Stoudt, P. M. Harris, and P. D. Crevello, "Upper Jurassic Reefs of Smackover Formation, United States Gulf Coast," AAPG *Bulletin,* v. 66 (1982), pp. 1449-1482.

Bathurst, R. G. C., *Carbonate Sediments and Their Diagenesis,* New York: Elsevier Scientific Publishing Co., (1975), 658 pp.

Bebout, D., G., Davies, C. Moore, P. Scholle, and N. Wardlaw, *Geology of Carbonate Porosity,* AAPG Continuing Education Course Note Series 11 (1979), 248 pp.

Beers, Roland F., "Radioactivity and Organic Content of Some Paleozoic Shales," AAPG *Bulletin,* v. 29 (1945), pp. 1-22.

Berg, R. R., "Depositional Environment of Upper Cretaceous Sussex Sandstone, House Creek Field, Wyoming," AAPG *Bulletin,* v. 58 (1975), pp. 2099-2110.

———, "Capillary Pressure in Stratigraphic Traps," AAPG *Bulletin,* v. 59 (1975), pp. 939-956.

———, "Point-Bar Origin of Fall River Sandstone Reservoirs, Northeastern Wyoming," AAPG *Bulletin,* v. 52 (1968), pp. 2116-2122.

———, "Seismic Detection and Evaluation of Delta and Turbidite Sequences: Their Application to Exploration for Subtle Trap," AAPG *Bulletin,* v. 66 (1982), pp. 1271-1288.

———, *Exploration for Sandstone Stratigraphic Traps,* AAPG Continuing Education Course Note Series No. 3 (1978), pp. 69-83.

———, "Hilight Muddy Field—Lower Cretaceous Transgressive Deposits in the Powder River Basin, Wyoming," *The Mountain Geologist,* v. 13 (1976), pp. 33-45.

Berven, R. J., "Cardium Sandstone Bodies, Crossfield—Garrington Area, Alberta," *Bulletin* Canadian Petrol. Geol., v. 14 (1966), pp. 208-240.

Bishop, William F., "Petrology of Upper Smackover Limestone in North Haynesville Field, Claiborne Parish, Louisiana," AAPG *Bulletin,* v. 52 (1968), pp. 92-128.

———, "Geology of a Smackover Stratigraphic Trap," AAPG *Bulletin,* v. 55 (1971), pp. 51-63.

———, "Geology of Upper Member of Buckner Formation, Haynesville Field Area, Claiborne Parish, Louisiana," AAPG *Bulletin,* v. 55 (1971), pp. 566-580.

Blixt, J. E., "Cut Bank Oil and Gas Field, Glacier County, Montana," *Stratigraphic Type Oil Fields,* AAPG (1941), pp. 327-381.

Borger, H. D., "Case History of Quirquire Field, Venezuela," AAPG *Bulletin,* v. 36 (1952), pp. 2291-2330.

Brown, L. F., Jr., A. W. Cleaves II, and A. W. Erxleben, *Pennsylvanian Depositional Systems in North-Central Texas: A Guide for Interpreting Terrigenous Clastic Facies in a Cratonic Basin,* Bureau of Econ. Geol., The University of Texas at Austin, Guidebook No. 14 (1973), p. 129.

Bubb, J. N., and W. G. Hatlelid, "Seismic Stratigraphy and Global Changes in Sea Level, Part Ten," *Seismic Stratigraphy—Applications to Hydrocarbon Exploration,* Tulsa: AAPG Memoir 26 (1977), pp. 185-204.

Busch, D. A., "Genetic Units in Delta Prospecting," AAPG *Bulletin,* v. 55 (1971), pp. 1137-1154.

Caughlin, W. G., et al., "The Detection and Development of Silurian Reefs in Northern Michigan," Geophysics *Bulletin,* v. 41 (1976), pp. 646-658.

Chapman, R. E., "Primary Migration of Petroleum from Clay Source Rocks," AAPG *Bulletin,* v. 56 (1972), pp. 2185-2191.

Choquette, P. W., and L. C. Pray, "Geologic Nomenclature and Classification of Porosity in Sedimentary Carbonates," AAPG *Bulletin,* v. 54 (1970), pp. 207-250.

Chuber, S., and E. Rodgers, *Relationships of Oil Composition and Stratigraphy of Pennsylvanian and Wolfcampian Reservoir,* Report by the West Texas Geological Society Subcommittee, (1968).

Clark, F. R., "Origin and Accumulation of Oil," *Problems of Petroleum Geology,* Tulsa: AAPG (1934).

Cluff, R. M., and Z. Lasemi, "Paleochannel Across Louden Anticline, Fayette County, Illinois," Illinois State Geological Survey, *Ill. Petrol.* 119 (1980), p. 21

Cook, C. W., "Study of Capillary Relationships of Oil and Water," *Econ. Geology,* v. 18 (1923), pp. 162-182.

Cordell, Robert J., "Depths of Oil Origin and Primary Migration: A Review and Critique," AAPG *Bulletin,* v. 56 (1973), pp. 2029-2067.

Couch, R. F., et al., "Drilling Operations on Enewetak Atoll During Project EXPOE," *AFTR,* pp. 75-216.

Crevello, P. D., "Sedimentary Record of Longitudinal Transport in Late Pleistocene—Holocene of Bahamian Trough," AAPG *Bulletin,* v. 63 (1978), pp. 507-508.

Curry, W. H., and W. H. Curry III, "South Glenrock Field, Wyoming: Prediscovery Thinking and Postdiscovery Description," *Stratigraphic Oil and Gas Fields—Classification, Exploration Methods and Case Histories,* R. E. King, ed., AAPG Memoir 16, SEG Special Publication No. 10 (1972), pp. 415-527.

# REFERENCES

Davies, D. K., and R. R. Berg, "Sedimentary Characteristics of Muddy Barrier-Bar Reservoir and Lagoonal Trap at Bell Creek Field," *The Economic Geology of Eastern Montana and Adjacent Areas,* Eastern Montana Symposium, 20th Annual Conference (1969), pp. 97-105.

Davies, D. K., F. G. Ethridge, and R. R. Berg, "Recognition of Barrier Environments," AAPG *Bulletin,* v. 55 (1971), pp. 550-565.

Davies, J. L., *Geomorphic Variations in Coastal Development,* Hafner Publishing Co., New York (1973), 204 p.

Dillion, W. P., and J. G. Vedder, "Structure and Development of the Continental Margin of British Honduras," Soc. America *Bulletin,* v. 84 (1973), pp. 2713-2732.

Douglas, T. R., and T. A. Oliver, "Environments of Deposition of the Borden Island Gas Zone in the Subsurface of the Sabine Peninsula Area, Melville Island, Arctic Archipelago," *Bulletin of Canadian Petroleum Geology,* v. 27, No. 3 (1979), pp. 273-313.

Dunham, R. J., "Classification of Carbonate Rocks According to Depositional Structure," *Classification of Carbonate Rocks,* Tulsa: AAPG Memoir 26 (1962), pp. 108-121.

Dutton, S. P., "Pennsylvanian Fan-Delta and Carbonate Deposition, Mobeetle Field, Texas Panhandle," AAPG *Bulletin,* v. 66 (1982), pp. 389-407.

Evans, H., "Zama—A Geophysical Case History," *Stratigraphic Oil and Gas Fields—Classification, Exploration Methods, and Case Histories,* AAPG Memoir 16 (1972), pp. 440-452.

Evans, W. E., "Imbricate Linear Sandstone Bodies of Viking Formation in Dodsland—Hoosier Area of Southwestern Saskatchewan, Canada," AAPG *Bulletin,* v. 54 (1970), pp. 469-486.

Fash, Ralph H., "Theory of Origin and Accumulation of Petroleum," AAPG *Bulletin,* v. 28 (1944), pp. 1510-1518.

Field, M. E., "Sandbodies on Coastal Plain Shelves: Holocene Record of the U. S. Atlantic Inner Shelf off Maryland," *Journal of Sedimentary Petrology,* v. 50 (1980), pp. 505-528.

Fisher, W. L., L. F. Brown, Jr., A. J. Scott, and J. H. McGowen, *Delta Systems in the Exploration for Oil and Gas,* Texas Bureau Econ. Geol. (1969).

Fisher, W. L., and others, *Environmental Geologic Atlas of the Texas Coastal Zone: Galveston—Houston Area,* Bureau of Econ. Geol., The University of Texas at Austin (1973), 91 p.

Fisher, W. L., and J. H. McGowen, "Depositional Systems in the Wilcox Group (Eocene) of Texas and Their Relation to Occurrence of Oil and Gas," AAPG *Bulletin,* v. 53 (1969), pp. 30-54.

Fisk, H. N., "Bar-finger Sands of Mississippi Delta," *Geometry of Sandstone Bodies,* Tulsa: AAPG (1961).

Flores, R. M., and F. G. Ethridge, "Nonmarine Deposits and the Search for Energy Resources and Minerals," *Recent and Ancient Nonmarine Depositional Environments: Models for Exploration,* F. G. Ethridge and R. M. Flores, eds., SEPM Special Publication No. 31 (1982), pp. 1-17.

Folk, R. L., "Spectral Subdivision of Limestone Types," *Classification of Carbonate Rocks,* Tulsa: AAPG Memoir 1 (1962), pp. 62-84.

Friedman, M., and D. W. Stearns, "Relations between Stresses Inferred from Calcite Twin Lamellae and Microfractures, Teton Anticline, Montana," G.S.A. *Bulletin,* v. 82 (1971), pp. 3151-3161.

Fryberger, S. G., "Eolian-Fluviatile (Continental) Origin of Ancient Stratigraphic Trap for Petroleum in Weber Sandstone, Rangely Oil Field, Colorado," *The Mountain Geologist,* v. 16 (1979), pp. 1-36.

Fuchtbauer, Hans, "Zur Diagenese Fluviatiler Sandsteine," *Geol. Rundschau,* v. 63 (1974), pp. 904-925.

Gatlin, C., *Petroleum Engineering,* New Jersey: Prentice-Hall Inc., (1960).

Gehman, H. M., Jr., "Organic Matter in Limestones," *Geochim. et Cosmochim Acta,* v. 26 (1962), pp. 885-897.

Hackford, J. E., "The Chemistry of the Conversion of Algae into Bitumen and Petroleum and of the Fucosite-Petroleum Cycle," *Journal Institute Petroleum Technology,* v. 18 (1932), pp. 74-114.

Halbouty, Michael T., "Rationale for Deliberate Pursuit of Stratigraphic, Unconformity, and Paleogeomorphic Traps," AAPG *Bulletin,* v. 56 (1972), pp. 537-541.

Harms, J. C., "Brae Field Area, North Sea, 5 (abst.)," AAPG *Bulletin,* v. 64 (1980), p. 718

————, "Stratigraphic Traps in a Valley Fill, Western Nebraska," AAPG *Bulletin,* v. 50 (1966), pp. 2119-2149.

Haun, John D., "Training the Petroleum Geologist," *Subsurface Geology in Petroleum Exploration,* Eds. J. D. Haun and L. W. Leroy, Golden, Colorado: Colorado School of Mines (1958).

Hayes, John B., "Sandstone Diagenesis—The Hole Truth," *Aspects of Diagenesis,* S.E.P.M., Spec. Publication, v. 26 (1979), pp. 127-140.

————, "Geologic Aspects of Origin of Petroleum," AAPG *Bulletin,* v. 48 (1964), pp. 1755-1803.

Hayes, M. O., "Transitional Coastal Environments—Introduction," *Terrigenous Clastic Depositional Environments: Some Modern Examples,* Hayes and Kona, eds., AAPG Field Course (1977), pp. I-32 – I-40.

Hemphill, C. R., R. L. Smith, and F. Szabo, "Geology of Beaverhill Lake Reefs, Swan Hills Area, Alberta, *Geology of Giant Oil Fields,* AAPG Memoir 14 (1970), pp. 50-90.

Hodgson, Gordon W., *et al., Geochemistry of Porphyrins: Fundamental Aspects of Petroleum Geochemistry,* New York: Elsevier Publ. Co. (1967).

Holmes, A., *Principles of Physical Geology,* New York: Ronald Press Co. (1965), 1288 p.

Hriskevich, Michael E., "Middle Devonian Reef Production, Rainbow Area, Alberta, Canada," AAPG *Bulletin,* v. 54 (1970), pp. 2260-2281.

Hunt, J. M., "The Origin of Petroleum in Carbonate Rocks," *Carbonate Rocks, Developments in Sedimentology 9B,* C. V. Chiliinger, J. J. Bissel, and R. W. Fairbanks, eds., New York: Elsevier Publ. Co. (1967), pp. 225-251.

Hunt, T. S., "Notes on the History of Petroleum or Rock Oil," *Annual Report for 1861,* Washington, D. C.: Smithsonian Institute, (1861).

Hsu, J. K., K. Kelts, and J. W. Valentine, "Resedimented Facies in Ventura Basin, California, and Model of Longitudinal Transport of Turbidity Currents," AAPG *Bulletin,* v. 64 (1980), pp. 1034-1051.

Ingels, J. J. C., "Geometry, Paleontology, and Petrography of Thornton Reef Complex, Silurian of Northeastern Illinois," AAPG *Bulletin,* v. 47 (1963), pp. 405-440.

Inman, D. L., and C. E. Norstrom, "On the Tectonic and Morphologic Classification of Coasts," *Journal Geology,* v. 79 (1971), pp. 1-21.

Jageler, A. H., and D. R. Matusak, "Use of Well Logs and Dipmeters in Stratigraphic Trap Exploration," *Stratigraphic Oil and Gas Fields—Classification, Exploration Methods, and Case Histories,* R. E. King, ed., AAPG Memoir 16 (1972), pp. 107-135.

James, N. P., and R. N. Ginsburg, *The Seaward Margin of Belize Barrier and Stoll Reefs,* International Association of Sedimentologists, Special Publication, v. 3 (1979).

James, N. P., "Facies Models 10-Reefs," *Geoscience Canada,* v. 5 (1979), p. 16-26.

Jenik, A. J., and J. F. Lerbekmo, "Facies and Geometry of Swan Hills Reef Member of Beaverhill Lake Formation (Upper Devonian), Goose River Field, Alberta, Canada," AAPG *Bulletin,* v. 52 (1968), pp. 21-56.

Jonas, E. C., and E. F. McBride, *Diagenesis of Sandstone and Shale: Application to Exploration,* Austin, Texas: University of Texas, Dept. of Geol. Science, Continuing Education Program, Publication No. 1 (1977).

Jones, H. P., and R. G. Speers, "Permo-Triassic Reservoirs in Prudhoe Bay Field, North Slope, Alaska," *North American Oil and Gas Fields,* J. Braunstein, ed., AAPG Memoir 24 (1976), pp. 23-50.

Kirkpatrick, C. V., "Formation Testing," *The Petroleum Engineer,* Oct., B-139.

Kraft, J. C., and J. J. Chacko, "Lateral and Vertical Facies Relations of Transgressive Barriers," AAPG *Bulletin,* v. 63 (1979), pp. 2145-2163.

Kraft, J. C., *A Guide to the Geology of Delaware's Coastal Environments,* College of Marine Sciences, Dept. of Del., Newark, Delaware (1971), 220 pp.

Kumar, N., and J. D. Foster, "Effect of an Underlying Bioherm on the San Andres (Permian) Reservoir and Trap, Hanford Field, Gaines County, Texas," AAPG *Bulletin,* v. 66 (1982), pp. 2571-2583.

La Fon, A. N., "Offshore Bar Deposits of Semilla Sandstone Member of Mancos Shale (Upper Cretaceous) San Juan Basin, New Mexico," AAPG *Bulletin,* v. 65 (1981), pp. 706-721.

Laing, W. E., *The "Vibroseis" System,* Society Exploration Geophysics and Continental Oil Co., (1972).

Land, C. B., and R. J. Weimer, "Peoria Field, Denver Basin, Colorado—J Sandstone Distributary Channel Reservoir," *Energy Resources of the Denver Basin,* Pruit, et al., eds., Rocky Mountain Association Geol. (1978), pp. 81-104.

Langton, J. Roger, and George E. Chin, "Rainbow Member Facies and Related Reservoir Properties, Rainbow Lake, Alberta," AAPG *Bulletin,* v. 52 (1968), pp. 1925-1955.

Lawson, D. E., and C. W. Crowson, *Geology of Arch Unit and Adjacent Areas, Sweetwater County, Wyoming,* Wyoming Geological Association Guidebook, 16th Annual Conference (1961), pp. 280-299.

LeMay, W. J., *Empire Abo Field, Southeastern New Mexico,* Tulsa: AAPG Memoir 16 (1972), pp. 472-480.

Leighton, M. W., and C. Pendexter, "Carbonate Rock Types," *Classification of Carbonate Rocks—A Symposium,* W. E. Ham, ed., Tulsa: AAPG Memoir 1 (1962), pp. 33-61.

Leopold, L. B., et al., *Fluvial Processes in Geomorphology,* San Francisco: Freeman, (1964).

Levandowski, D. W., M. E. Kaley, S. R. Silverman, and R. G. Smalley, "Cementation in Lyons Sandstone and Its Role in Oil Accumulation, Denver Basin, Colorado," AAPG *Bulletin,* v. 57 (1973), pp. 2217-2244.

Levorsen, A. I., *Geology of Petroleum,* San Francisco: W. H. Freeman and Co., (1956).

Leythaeuser, D. V., R. G. Schaefer, and A. Yukler, "Role of Diffusion in Primary Migration," *AAPG Bulletin,* v. 66 (1982), pp. 408-429.

Longman, M. W., "Carbonate Diagenetic Textures from Near-Surface Diagenetic Environments," *AAPG Bulletin,* v. 64 (1980), pp. 461-487.

Lynn, J.R., *Cut Bank Oil and Gas Field, Glacier County, Montana,* Billings Geological Society Guidebook, 6th Annual Field Conference (1955), pp. 195-197.

McPherson, B. A., "Sedimentation and Trapping Mechanism in Upper Miocene Stevens and Older Turbidite Fans of Southeastern San Joaquin Valley, California," *AAPG Bulletin,* v. 62 (1978), pp. 2243-2274.

Magaritz, M., E. Gavish, N. Bakler, and U. Kafi, "Carbon and Oxygen Isotope Composition-Indicators of Cementation Environment in Recent, Holocene, and Pleistocene Sediments along the Coast of Israel," *Journal of Sedimentary Petrology,* v. 49 (1979), pp. 401-412.

Malek-Aslani, Morad, "Lower Wolfcampian Reef in Kemnity Field, Lea County, New Mexico," *AAPG Bulletin,* v. 54 (1970), pp. 2317-2335.

Mann, A., *The Economic Importance of the Diatom,* Washington, D. C.: Smithsonian Institute (1916).

Masters, J. A., "Deep Basin Gas Trap, Western Canada," *AAPG Bulletin,* v. 63 (1979), pp. 152-181.

Mather, K. F., and M. C. Mehl, "The Importance of Drainage in Estimating the Possibilities of Petroleum Production from an Anticlinal Structure," Dennison Univ. Science Lab. *Bulletin,* v. 19 (1919), pp. 143-146.

Matuszczak, R. A., "Wattenberg Field, Denver Basin, Colorado," *North American Oil and Gas Fields,* Jules Braunstein, ed., AAPG Memoir 24 (1976), pp. 136-144.

Mazzullo, S. J., "Stratigraphy and Depositional Mosaics of Lower Clearfork and Wichita Groups (Permian), Northern Midland Basin, Texas," *AAPG Bulletin,* v. 66 (1982), pp. 210-227.

―――, "Facies and Burial Diagenesis of a Carbonate Reservoir: Capman Deep (Atoka) Field, Delaware Basin, Texas," *AAPG Bulletin,* v. 65 (1981), pp. 850-865.

McAuliffe, C. D., "Role of Solubility in Migration of Petroleum from Source," *AAPG Bulletin,* v. 62 (1978), pp. 541-542.

McCamis, J. G., and L. S. Griffith, "Middle Devonian Facies Relations, Zama Area, Alberta," *AAPG Bulletin,* v. 52 (1968), pp. 1899-1924.

McCubbin, D. G., "Cretaceous Strike-Valley Sandstone Reservoirs, Northwestern New Mexico," *AAPG Bulletin,* v. 53 (1969), pp. 2114-2140.

McGregor, A. A., and C. A. Biggs, "Bell Creek Field, Montana: A Rich Stratigraphic Trap," *AAPG Bulletin,* v. 52 (1968), pp. 1869-1887.

McKee, E. D., *Significance of Tree Climbing-Ripple Structure,* Washington, D. C., U.S. Geological Survey, Prof. Paper No. 550-D (1966), D94-D103.

―――, et al., *Paleotectonic Investigations of the Permian System in the United States,* Washington, D. C., U. S. Geological Survey, Prof. Paper No. 515 (1967).

McLean, J.R., "Regional Considerations of the Elmworth Field and the Deep Basin," *Bulletin,* Canadian Petroleum Geology, v. 27 (1979), pp. 53-62.

Mesollella, K. J., *et al.,* "Cyclic Deposition of Silurian Carbonates and Evaporites in Michigan Basin," *AAPG Bulletin,* v. 58 (1974), pp. 34-62.

Michaelis, E. R., and G. Dixon, "Interpretation of Depositional Processes from Sedimentary Structures in the Cardium Sand," *Bulletin,* Canadian Petroleum Geology, v. 17 (1969), pp. 410-443.

Middleton, G. V., and M. A. Hampton, "Sediment Gravity Flows: Mechanics of Flow and Deposition," *Turbidites Deep-Water Sedimentation: Pacific Sect,* Middleton and Bouma, eds., Soc. Econ. Paleon. and Mineral., Short Course Notes (1973), pp. 119-157.

Mitchum, R. M., et al., "Part Two: The Depositional Sequence as a Basic Unit for Stratigraphic Analysis," *Seismic Stratigraphy—Applications to Hydrocarbon Exploration,* Charles Payton, ed., AAPG Memoir 26 (1977), pp. 53-62.

―――――, "Regional Seismic Interpretation Using Sequences and Eustatic Cycles," AAPG *Bulletin,* v. 60 (1976), p. 699.

Moore, C. H., and Y. Druckman, "Burial Diagenesis and Porosity Evolution, Upper Jurassic Smackover, Arkansas and Louisiana," AAPG *Bulletin,* v. 65 (1981), pp. 597-628.

Moore, G. T., "Lodgepole Limestone Facies in Southwestern Montana," AAPG *Bulletin,* v. 57 (1973), pp. 1703-1713.

Morgan, J. T., J. T. Cordiner, and A. R. Livingston, "Tensleep Reservoir, Oregon Basin Field, Wyoming," AAPG *Bulletin,* v. 62 (1978), pp. 609-632.

Morgridge, D. L., and W. B. Smith, Jr., "Geology and Discovery of Prudhoe Bay Field, Eastern Arctic Slope, Alaska," *Stratigraphic Oil and Gas Fields—Classification, Exploration, and Case Histories,* AAPG Memoir 16 (1972), pp. 409-501.

Murray, J. W., "An Oil-Producing Reef Fringed Carbonate Bank in the Upper Devonian Swan Hills Member, Judy Creek, Alberta," Canadian Petroleum Geol. *Bulletin,* v. 14 (1966), pp. 1-103.

Murray, R. C., "Origin of Porosity in Carbonate Rocks," *Journal Sedimentary Petrology,* v. 30 (1960), pp. 59-84.

Mutti, E., and F. Ricci Lucchi, *Le torbiditi dell'Appenio settentroinale: Introduzoine all'analisi di facies,* Mem. Soc. Geol. Italina, v. 11 (1972), pp. 161-199.

Nanz, R. H., Jr., "Genesis of Oligocene Sandstone Reservoir, Seeligson Field, Jim Wells and Kleberg Counties, Texas," AAPG *Bulletin,* v. 38 (1954), pp. 96-117.

Natland, M. L., "Paleoecology of West Coast, Tertiary Sediments," *Treatise on Marine Ecology and Paleoecology, V. 2,* Geol. Soc. America Memoir 67 (1957), pp. 543-572.

Nielsen, N. R., "Cardium Stratigraphy of the Pembina Field," *Journal Alberta Petrol Geo.,* v. 5 (1957), pp. 64-72.

Oakwood, T. S., "Optical Activity of Petroleum," *Ind. and Eng. Chemistry,* v. 44 (1952), pp. 2568-2570.

Ottmann, Robert P., et al., "Jay Field, Florida: Jurassic Stratigraphic Trap," AAPG *Bulletin,* v. 57 (1973), pp. 798-799.

Peterson, James A., "Stratigraphic vs. Structural Controls on Carbonate-Mound Hydrocarbon Accumulation, Aneth Area, Paradox Basin," AAPG *Bulletin,* v. 50 (1968), pp. 2068-2081.

Peterson, R. A., et al., "The Synthesis of Seismographs from Well Log Data," *Geophysics,* v. 20 (1955), pp. 516-538.

Pettijohn, F. J., P. E. Potter, and R. Siever, *Sand and Sandstone,* New York: Springer, (1972).

Philippi, G. T., "On the Depth, Time, and Mechanism of Petroleum Generation," *Geochim, Cosmochim.,* ACTA 29 (1965), pp. 1021-1049.

Phinney, A. J., *The Natural Gas Field of Indiana,* Washington, D. C.: U. S. Geological Survey, Ann. Dept. 11, Pt. 1 (1891), pp. 579-742.

Pittman, E. D., "Porosity and Permeability Change During Diagenesis of Pleistocene Corals, Barbados, West Indies," G.S.A. *Bulletin,* v. 85 (1974), pp. 1811-1820.

―――――, "Porosity, Diagenesis, and Productive Capability of Sandstone Reservoirs," *Aspects of Diagenesis,* Eds. Scholle and Schluger, S.E.P.M., Special Publication v. 26 (1979), pp. 159-174.

―――――, "Effects of Cementation on Physical Properties of Sandstones," AAPG *Bulletin,* v. 64 (1980), p. 766 (abst.).

Powell, T. G., D. M. McKirdy, "Geologic Factors Controlling Crude Oil Composition in Australia and Papua, New Guinea," AAPG *Bulletin,* v. 59 (1976), pp. 1176-1197.

Powell, T. G., and L. R. Snowdon, "Geochemistry of Oils and Condensates from the Mackenzie Delta Basin, N. W. T.," *Canadian Geological Survey,* Paper No. 76-1, Part C (1976), pp. 41-43.

Pray, L. C. and J. L. Wray, "Porous Algal Facies (Pennsylvanian) Honaker Trail, San Juan Canyon, Utah," *Shelf Carbonates of the Paradox Basin,* Ed. R. O. Bass, Four Corners Geological Society, (1963).

Price, L. D., "Aqueous Solubility of Petroleum as Applied to Its Origin and Primary Migration," AAPG *Bulletin,* v. 60 (1976), pp. 213-244.

Purdy, E. G., "Recent Calcium Carbonate Facies of the Great Bahama Bank, Part I: Petrography and Reaction Groups," *Journal Geology,* v. 71 (1963), pp. 334-355.

―――――, "Recent Calicum Carbonate Facies of the Great Bahama Bank, Part II: Sedimentary Facies," *Journal Geology,* v. 71 (1963), pp. 472-497.

―――――, "Karst-Determined Facies Patterns in British Honduras: Holocene Carbonate Model," AAPG *Bulletin,* v. 58, (1974), pp. 825-855.

Purdy, E. G., W. C. Pusey, III, and K. F. Wantland, "Continental Shelf of Belize—Regional Shelf Attributes," *Belize Shelf-Carbonate Sediments, Clastic Sediments, and Ecology,* AAPG, Studies in Geology, No. 2 (1975), pp. 1-52.

Putnam, P. E., and T. A. Oliver, "Stratigraphic Traps in Channel Sandstones in the Upper Mannville (Albian) of East-Central Alberta," *Bulletin,* Canadian Petroleum Geology, v. 28 (1980), pp. 489-508.

Rainwater, E. H., "Look for Ancient Deltas in Your Search for Oil," *Oil and Gas Journal,* v. 1 (1964), pp. 110-116.

Redfield, A. C., "Preludes to the Entrapment of Organic Matter in Sediments of Lake Maracaibo, in Habitat of Oil," AAPG *Bulletin,* v. 42 (1958), pp. 968-981.

Rice, D. D., "Coastal and Deltaic Sedimentation of Upper Cretaceous Eagle Sandstone: Relations to Shallow Gas Accumulations, North-Central Montana," AAPG *Bulletin,* v. 64 (1980), pp. 316-338.

Rich, J. L., "Function of Carrier Beds in Long-Distance Migration," AAPG *Bulletin,* v. 15 (1939), pp. 918-924.

Rittenhouse, G., "Pore Space Reduction by Solution and Sedimentation," AAPG *Bulletin,* v. 55 (1971), pp. 80-91.

―――――, "Stratigraphic Trap Classification," *Stratigraphic Oil and Gas Fields—Classification, Exploration Methods, and Case Histories,* AAPG Memoir 16 and SEPM Special Publication No. 11 (1972), pp. 14-28.

Roberts, W. H., *Design and Function of Oil and Gas Traps,* AAPG Memoir 32 (1980), pp. 217-241.

Sabins, F. F., "Anatomy of Stratigraphic Trap, Bisti Field, New Mexico," AAPG *Bulletin,* v. 47 (1963), pp. 193-228.

Sangree, J. B., and J. M. Widmier, *Interpretation of Depositional Facies from Seismic Data,* Continuing Education Symposium, Houston: Geophysical Society of Houston, (1974).

―――, "Seismic Stratigraphy and Global Changes of Sea Level, Part 9," *Seismic Stratigraphy-Applications to Hydrocarbon Exploration,* Tulsa: AAPG Memoir 26 (1977), pp. 165-184.

Savkevich, S. S., "Relation of Secondary Porosity Caused by Leaching to the Main Phase of Petroleum Formation," *Akas. Nauk SSSR., IZV., Ser. Geol.,* v. 6 (1971), pp. 70-78.

Sawyer, K. C., "Preservational Patterns of Organic Material in a Carbonate Shelf Environment, Southwest Puerto Rico," *Diss. University of Oklahoma,* (1980).

Schmidt, V., and D. A. McDonald, "The Role of Secondary Porosity in the Course of Sandstone Diagenesis," *Aspects of Diagenesis,* S.E.P.M., Special Publication 26 (1979), pp. 175-208.

Schmoker, J. W., and R. B. Halley, "Carbonate Porosity versus Depth: Aperdectable Relation for South Florida," AAPG *Bulletin,* v. 66 (1982), pp. 2561-2570.

Schowalter, T. T., "The Mechanics of Secondary Hydrocarbon Migration and Entrapment," Wyoming Geol. Assoc. Earth Science *Bulletin,* v. 9 (1976), pp. 1-43.

Seeling, A., "The Shannon Sandstone, A Further Look at the Environment of Deposition at Heldt Draw Field, Wyoming," *The Mountain Geologist,* v. 15 (1978), pp. 133-144.

Shannon, J. P., Jr., and A. R. Dahl, "Deltaic Stratigraphic Traps in West Tuscola Field, Taylor County, Texas," AAPG *Bulletin,* v. 55 (1971), pp. 1194-1205.

Shelton, J. W., "Stratigraphic Models and General Criteria for Recognition of Alluvial, Barrier-Bar, and Turbidity-Current Sand Deposits," AAPG *Bulletin,* v. 51 (1967), pp. 2441-2461.

Sheriff, R. E., "Limitations of Resolution of Seismic Reflections and Geologic Detail Derivable from Them," *Seismic Stratigraphy Applications to Hydrocarbon Exploration,* Ed. Charles Payton, Tulsa: AAPG Memoir 26 (1977), pp. 3-14.

Shirley, M. L., and J. A. Ragsdale, *Deltas,* Houston: Houston Geological Society, (1966).

Silver, B. A., *Exploration Geology,* Tulsa: IED Exploration, Inc., (1980).

Silver, B. A., and R. G. Todd, "Permian Cyclic Strata, Northern Midland and Delaware Basins, West Texas and Southeastern New Mexico," AAPG *Bulletin,* v. 53 (1969), pp. 2223-2251.

Silver, B. A., R. Goldhammer, D. Neese, and G. Stewart, *The Belize Carbonate Complex: Shelf-Margin Model for Hydrocarbon Exploration,* Tulsa: IED Exploration, Inc., (1980).

Silver, B. A., and D. E. Wermeil, "Diagenetic History of Mississippian Carbonate Rocks, Pediagosa Basin," AAPG *Bulletin,* v. 60 (1976), p. 723.

Sloss, L. L., "Sequences in the Ciatonic Interior of North America," G.S.A. *Bulletin,* v. 74 (1963), pp. 93-114.

Stapor, F. W., Jr., "Origin of the Todilto Gypsum Mounds," *Ghost Ranch Area, North Central New Mexico,* The Mountain Geologist, v. 9 (1972), pp. 59-64.

Stewart, G. S., "Influence of Tectonics of Patch Reef Development, Southwestern Puerto Rico," *Diss. University of Oklahoma,* (1982).

Stoddart, D. R., "The Reefs and Sand Cays of British Honduras: Cambridge Exped. British Honduras, 1959-1960," *Gen. Rep.* (1960), pp. 16-22.

Stubblefield, W. L., "Sediment Response to the Present Hydraulic Regime on the Central New Jersey Shelf," *Journal of Sedimentary Petrology,* v. 45 (1975), pp. 337-358.

Sullwold, H. H., Jr., "Turbidites in Oil Exploration," *Geometry of Sandstone Bodies,* J. A. Peterson and J. C. Osmond, eds., AAPG (1961), pp. 63-81.

Sundborg, A., "The River Klaraluen: A Study of Fluvial Processes," *Geografiska Hnnaler,* v. 38 (1956), pp. 125-316.

Tallis, N. C., "Development of the Tertiary Offshore Papuan Basin," *The APEA Journal,* (1975), pp. 55-60.

Tanner, W. F., "Triassic—Jurassic Lakes in New Mexico," *The Mountain Geologist,* v. 7 (1970), pp. 281-289.

―――, "History of Mesozoic Lakes in Northern New Mexico," *Ghost Ranch, Central— Northern New Mexico,* New Mexico Geological Society, 25th Field Conference Guidebook (1974), pp. 219-223.

Tizzard, P. G., and J. F. Lerbekmo, "Depositional History of the Viking Formation, Suffield Area, Alberta, Canada," *Bulletin,* Canadian Petroleum Geology, v. 23, pp. 715-752.

Treibs, A., "Porphyrin in Bituminosen Gesteinen and Erdol-Kohlen: Zur Entstehung des Erdols," *Angew. Chem.,* v. 44 (1936), p. 551.

Vail, P. R., R. M. Mitchum, Jr., *et al.,* "Seismic Stratigraphy and Global Changes of Sea Level, Parts 1-11," *Seismic Stratigraphy Applications to Hydrocarbon Exploration,* Charles Payton, ed., Tulsa: AAPG Memoir 26 (1977), pp. 49-212.

Vest, E. L., Jr., "Oil Fields of Pennsylvania-Permian Horseshoe Atoll, West Texas," *Giant Oil Fields,* AAPG Memoir 14 (1970), pp. 185-203.

Vincelette, R. R., and W. E. Chittum, "Exploration for Oil Accumulations in Entrada Sandstone, San Juan Basin, New Mexico," *AAPG Bulletin,* v. 65 (1981), pp. 2546-2570.

Wahner, P. D., and R. K. Matthews, "Porosity Preservation in the Upper Smackover (Jurassic) Carbonate Grainstone, Walker Creek Field, Arkansas: Response of Paleophreatic Lenses to Burial Processes," *Journal Sedimentary Petrology,* v. 52 (1982), pp. 3-18.

Walker, R. G., "Deep-Water Sandstone Facies and Ancient Submarine Fans: Model for Exploration for Stratigraphic Trap," *AAPG Bulletin,* v. 62 (1978), pp. 932-966.

―――, "Facies Models of Shallow Marine Sands," *Facies Models,* Walker, ed., Geoscience Canada Reprint Series I (1979), pp. 75-89.

Wardlaw, N. C., and G. E. Reinson, "Carbonate and Evaporite Deposition and Diagenesis, Middle Devonian Winnipegosis, and Prairie Evaporite Formations of South-Central Saskatchewan," *AAPG Bulletin,* v. 55 (1971), pp. 1759-1786.

Waugh, B., "Petrology, Provenance, and Silica Diagenesis of the Penrith Sandstone (Lower Permian) of Northwest England," *J.S.P.* v. 40 (1970), pp. 1226-1240.

Webb, G. W., "Stevens and Earlier Miocene Turbidite Sandstones, Southern San Joaquin Valley, California," *AAPG Bulletin,* v. 65 (1981), pp. 438-465.

Weeks, L. G., "Factors of Sedimentary Basin Development that Control Oil Occurrences," *AAPG Bulletin,* v. 36 (1952), p. 2103.

Wegemann, C. H., "The Salt Creek Oil Field, Natrona Co., Wyoming," U. S. Geological Survey *Bulletin,* v. 452 (1911), pp. 37-83.

Weimer, R. J., "Time-Stratigraphic Analysis and Petroleum Accumulations, Patrick Draw Field, Sweetwater County, Wyoming," AAPG *Bulletin,* v. 50 (1966), pp. 2150-2175.

Westcott, W. A., and F. G. Ethridge, "Fan-Delta Sedimentology and Tectonic Setting—Yallahs Fan Delta, Southeast Jamaica," AAPG *Bulletin,* v. 64 (1980), pp. 374-399.

Wilde, P., W. R. Normark, and T. E. Chase, "Channels Sands and Petroleum Potential of Monterey Deep-Sea Fan, California," AAPG *Bulletin,* v. 62 (1978), pp. 967-983.

Williams, G. D., and C. R. Stelck, "Speculations on the Cretaceous Paleogeography of North America," *The Cretaceous System in the Western Interior of North America,* Geol. Assoc. Canada Spec. Paper 12 (1975), pp. 1-20.

Wilson, J. L. "Characteristics of Carbonate Platform Margins," AAPG *Bulletin,* v. 58 (1974), pp. 810-824.

Wonless, H. R., "Sea Level Rising—So What?" *Journal of Sedimentary Petrology,* v. 52 (1982), pp. 1051-1054.

Woolnough, W. G., "Sedimentation in Barred Basins, and Source Rocks of Oil," AAPG *Bulletin,* v. 21 (1937), pp. 1101-1157.

Wuenschel, P. C., "Seismogram Synthesis Including Multiples and Transmission Coefficients," *Geophysics,* v. 25 (1960), pp. 106-129.

# Additional Reading

Ahr, W. M., "The Carbonate Ramp: An Alternative to the Shelf Model," *Gulf Coast Association Geological Soc. Trans.,* v. 23 (1973), pp. 221-225.

Allen, J. R. L., "A Review of the Origin and Characteristics of Recent Alluvial Sediments," *Sedimentology,* v. 5 (1965), pp. 89-191.

———, "Packing and Resistance to Compaction of Shells," *Sedimentology,* v. 21 (1965), pp. 89-191.

Aoyagi, Koichi, "Petrophysical Approach to Origin of Porosity of Carbonate Rocks in Middle Carboniferous Windsor Group, Nova Scotia, Canada," *AAPG Bulletin,* v. 57 (1973), pp. 1692-1702.

Ball, M. M., "Carbonate Sand Bodies of Florida and the Bahamas," *Journal Sedimentary Petrology,* v. 37 (1967), pp. 556-591.

Bass, R. O., ed., *Shelf Carbonates of the Paradox Basin—A Symposium,* Durango, Colorado: Four Corners Geological Society (1963).

Bouma, A. H., *Sedimentology of Some Flysch Deposits: A Graphic Approach to Facies Interpretation,* Amsterdam: El Sevier Publ. (1962).

Bull, W. B., "The Alluvial-Fan Environment," *Progress in Physical Geography,* v. 1 (1977), pp. 222-270.

Bush, D. A., "Genetic Units in Delta Prospecting," *AAPG Bulletin,* v. 55 (1971), pp. 566-580.

Choquette, P. W., and J. D. Traut, *Pennsylvanian Carbonate Reservoirs, Ismay Field, Utah and Colorado: Shelf Carbonates of the Paradox Basin,* R. O. Bass, ed., Durango, Colorado: Four Corners Geological Society (1963).

Coleman, J. M., and S. M. Gagliano, "Cyclic Sedimentation in the Mississippi River Deltaic Plain,"*Gulf Coast Association Geological Society Trans.,* v. 14 (1964), pp. 67-80.

Coogan, A. H., B. G. Debout, and C. Maggio, "Depositional Environments and Geologic History of Golden Land and Poza Rica Trend, Mexico, An Alternative View," *AAPG Bulletin,* v. 56 (1972), pp. 419-1447.

Cooke, R. U., and A. Warren, *Geomorphology in Deserts,* London: B. T. Batsford, Ltd. (1973).

Crandall, K. H., "Putting Exploration Back into Focus," *AAPG Bulletin,* v. 53 (1969), pp. 2055-2061.

Davies, D. K., F. G. Ethridge, and R. R. Berg, "Recognition of Barrier Environments," *AAPG Bulletin,* v. 55 (1971), pp. 550-565.

Davies, H. K., *Geographical Variation in Coastal Development,* Edinburgh: Oliver and Boyd, (1973).

Dean, W. E., "Shall-Water versus Deep-Water Evaporites," *AAPG Bulletin,* v. 59 (1975), pp. 534-535.

Dott, R. H., and M. J. Reynolds, *Sourcebook for Petroleum Geology,* Tulsa: AAPG Memoir 5 (1969).

Exploration Staff, Chevron Standard Limited, "The Geology, Geophysics and Significance of the Nisku Reef Discoveries, West Pembina Area, Alberta, Canada," *Bulletin of Canadian Petroleum Geology,* v. 27 (1979), pp. 325-359.

Fisk, H. N., *Sand Facies of Recent Mississippi Delta Deposits,* Report by the Fourth World Petroleum Congr., Rome: Proc. Sec. 1/C (1955), pp. 377-398.

Ginsburg, R. N., ed., "South Florida Carbonate Sediments," *Geological Society Am. Guidebook,* Report by the Geologial Society Am., Miami Convention, Miami, (1964).

Ginsburg, R. N., et al., "Shallow-Water Carbonate Sediments," *The Sea,* v. 3, Hill, ed., London: Pergamon Press, (1963).

Ginsburg, R. N., "Introduction to Comparative Sedimentology of Carbonates," AAPG *Bulletin,* v. 58 (1974), p. 781.

Halbouty, Michael T., "Hidden Trends and Subtle Traps in Gulf Coast," AAPG *Bulletin,* v. 53 (1969), pp. 3-29.

———, "Stratigraphic-Trap Possibilities in Upper Jurassic Rocks, San Marcos Arch, Texas," AAPG *Bulletin,* v. 50 (1966), pp. 3-24.

Hallana, A., ed., "Depth Indicators in Marine Sedimentary Environments," *Marine Geology,* v. 5, Special Issue (1967), pp. 329-567.

Imbrie, J., and N. D. Newell, eds., *Approaches to Paleoecology,* John Wiley & Sons (1964).

Kinsman, David J. J., "Modes of Formation, Sedimentary Associations, and Diagnostic Features of Shallow-Water," AAPG *Bulletin,* v. 53 (1969), pp. 330-840.

Land, L. S., F. T. MacKenzie, and S. J. Gould, "Pleistocene History of Bermuda," Geological Society Am. *Bulletin,* v. 78 (1967), pp. 993-1006.

LeBlanc, R. J., "Geometry of Sandstone Reservoir Bodies," *Underground Waste Management and Environmental Implications,* Tulsa: AAPG Memoir 18 (1972), pp. 133-190.

Margaritz, M., "Lithification of Chalky Limestone: A Case Study in Senonian Rocks from Israel," *Journal Sedimentary Petrology,* v. 44 (1974), pp. 947-954.

Markevich, V. P., "The Concept of Facies," *International Geological Review,* v. 2 (1960), pp. 376-379.

Middleton, G. V., "Johannes Walther's Law of the Correlation of Facies," Geological Society of America *Bulletin,* v. 84 (1973), pp. 979-988.

Moore, R. C., *Meaning of Facies,* Geological Society of America Memoir 39 (1949), pp. 1-34.

Potter, P. E., "Sand Bodies and Sedimentary Environments: A Review," AAPG *Bulletin,* v. 51 (1967), pp. 337-365.

Purdy, E. G., and J. Imbrie, "Carbonate Sediments, Great Bahama Bank," *Geological Society of America Guidebook No. 2,* Report by the Geological Society of America, Miami Convention, Miami (1964).

Roehl, Perry O., "Stony Mountain (Ordovician) and Interlake (Silurian) Facies Analogs of Recent Low-Energy Marine and Subaerial Carbonates, Bahamas," AAPG *Bulletin,* v. 51 (1967), pp. 1979-2032.

Shaw, A. B., *Time in Stratigraphy,* New York: McGraw-Hill Book Co. (1964).

Shelton, J. W., "Stratigraphic Models and General Criteria for Recognition of Alluvial, Barrier-Bar, and Turbidity-Current Sand Deposits," AAPG *Bulletin,* v. 51 (1967), pp. 2441-2461.

Swift, D. J. P., D. J. Stanley, and O. H. Pileky, eds., *Shelf Sediment Transport: Process and Pattern,* Stroudsberg, Pennsylvania: Dowden, Hutchinson, and Ross (1972).

Teichert, C., "Concept of Facies," AAPG *Bulletin,* v. 42 (1958), pp. 2718-2744.

Wheeler, H. E., "Time Stratigraphy," AAPG *Bulletin,* v. 42 (1958), pp. 1047-1063.

———, "Baselevel Transit Cycle," *Symposium on Cyclic Sedimentation,* D. F. Merriam, ed., 2 vols., Kansas Geological Survey (1964).

Wheeler, H. E., and V. S. Mallory, "Factors in Litho-Stratigraphy," AAPG *Bulletin,* v. 40 (1956), pp. 2711-2723.

Wheller, J. M., *Stratigraphic Principles and Practice,* New York: Harper & Brothers, Publishers (1960).

Wright, L. D., "Sediment Transport and Deposition at River Mouths: A Synthesis," Geological Society of America *Bulletin,* v. 88 (1977), pp. 857-868.

# APPENDIX I

## Electronic Data Format for Clastics

# GENERAL INSTRUCTIONS
## CLASTIC EDP FORMAT

### INTRODUCTION

The EDP format contained in the appendix is designed for computer processing. The format is also adaptable to a strip-log approach in that the original description can be copied and various columns grouped, spliced, and colored to facilitate visual analysis and correlation.

### INSTRUCTIONS

1. Columns 1-5:     Reserved for encoding the project
2. Columns 6-10:    Initials of describer
3. Columns 11-15:   Formation code
4. Columns 16-20:   Footage (right-hand justify)
5. Column 21:       Color, may revise to fit standard color scheme
6. Columns 22-32:   Each column represents 10% by volume. If describing cuttings, use gamma ray E/or SP logs to make interpretative rather than a percentage log.
7. Remaining part of dictionary self explanable.

### RECOMMENDATION

Revise the general clastic dictionary to fit the suite of rocks under study. It is possible to combine the clastic and carbonate formats (Appendix II) by using a two-card system. Reserve one of the columns for encoding card 1 or 2, depending upon which rock type is being described.

# APPENDIX I
## Clastic EDP Dictionary

# CLASTIC EDP DICTIONARY

**1. COLOR** (21)

    9 = colorless to white
    8 = cream to tan
    7 = light greys and browns
    6 = intermediate greys and browns
    5 = dark browns and blacks
    4 = greens
    3 = red, orange, pink
    1 = mottled
    0 = remarks for others

**2. COMPOSITION** (22-31)

    S = sandstone
    L = limestone
    D = dolomite
    A = anhydrite
    G = gypsum
    H = halite
    C = chert
    I = igneous
    M = metamorphic
    B = bentonite
    O = coal
    X = other
    R = recrystallized grains
    E = lithic
    F = feldspathic

**3. PERCENT GRAINS** (32)

    0 =   0 - 10
    1 = 11 - 20
    2 = 21 - 30
    3 = 31 - 40
    4 = 41 - 50
    5 = 51 - 60
    6 = 61 - 70
    7 = 71 - 80
    8 = 81 - 90
    9 = 91 - 100

| COLOR | ROCK TYPE COMPOSITION | | | | | | | | | | PERCENT GRAINS |
|---|---|---|---|---|---|---|---|---|---|---|---|
| 21 | 22 | 23 | 24 | 25 | 26 | 27 | 28 | 29 | 30 | 31 | 32 |

## 4. GRAIN SIZE RANGE (33 and 34)

    1 = mud   .03mm.
    2 = silt (0.03 - .0625)
    4 = f   (.125 - .25)
    5 = m  (.25 - .5)
    6 = c   (.5 - 1.0)
    7 = vc (1.0 - 2.0)
    8 = p  (2.0 - 0.0)
    9 = b  (greater than 10)

## 5. DOMINANT SIZE (36)

    See 4.

## 6. GRAIN SUPPORT (36)

    W = wackestone, grains not in contact
    P = packstone, grains in contact
    G = grainstone, grains 90% of rock

## 7. GRAIN SHAPE (37)

    1 = angular
    2 = subangular
    3 = subrounded
    4 = rounded

## 8. TRACE MINERALS (39 and 40)

    T = glauconite
    P = pyrite
    N = siderite
    E = phosphatic
    F = plant fragments
    Z = sulfur
    K = mica
    Q = feldspars
    J = hematite
    T = limonite
    L = fossils

## 9. PERCENT TRACE MINERALS (41)

    See 3.

| | SANDGRAINS | | | | | | TRACE MIN. | |
|---|---|---|---|---|---|---|---|---|
| | | GRAINS | | | | | | |
| PERCENT GRAINS | SIZE RANGE | DOM. SIZE | SUPPORT | MAT-CEM. | SHAPE | | TYPE | PERCENT |
| 32 | 33 | 34 | 35 | 36 | 37 | 38 | 39 | 40 | 41 |

# CLASTIC EDP DICTIONARY

## 10. SHALE COMPOSITION (42)

See 4.

## 11. SHALE GRAIN TYPE (43)

1 = detrital clay
2 = authegentic clay
3 = silt
4 = detrital clay and silt

## 12. SEDIMENTARY STRUCTURES (44 and 45)

A = even parallel laminae less than 3 mm. thick
B = even parallel laminae greater than 3 mm. thick
C = wavy laminae
D = small scale cross laminae
E = large scale cross laminae
F = graded bedding
G = flame structure
H = groove casts
I = convalute bedding
J = burrowed
K = contorted bedding
L = flute casts
M = load casts
N = geopetal structures
O = stromatolitic
P = birdseye
Q = breccia
U = scoured surface
R = modular
S = small scale festoon cross beds
T = large scale festoon cross beds
U = current ripples
V = ossilation ripples
W = small scale wedge cross beds
X = large scale wedge cross beds
Y = small scale planar cross beds
Z = large scale planar cross beds

## 13. SHALE MODIFIERS (46)

A = mudrock
B = mudstone
C = mudite
D = clay rock
E = clay stone
F = clayite
G = silt rock
H = silt stone
I = siltite

| SHALE | | | | |
|---|---|---|---|---|
| COMPOSITION | TYPE GRAINS | SEDIMENTARY STRUCTURES | | MODIFIERS |
| 42 | 43 | 44 | 45 | 46 |

## 14. CEMENT TYPE (47)

    C = sparry calcite
    D = dolomite
    S = silica
    A = anhydrite
    F = friable
    B = calcite and anhydrite
    E = dolomite and silica
    I = calcite and dolomite
    J = calcite and silica
    K = dolomite and anhydrite
    G = gypsum
    H = hematite
    M = micrite

## 15. CEMENT OCCURRENCE (48)

    I = intergranular
    A = intragranular
    O = organic
    B = fenestral
    M = moldic
    X = intercrystalline
    V = vug
    F = fracture
    U = cavity of unknown origin

## 16. CEMENT PERCENT (49)

    1 = 1 - 5%
    2 = 6 - 10%
    3 = 11 - 15%
    4 = 16 - 20%
    5 = 21 - 25%
    6 = 26 - 30%
    7 = 31 - 35%
    8 = 36 - 40%
    9 = 41 - 45%
    0 = 46 - 50%

## 17. MATRIX COMPOSITION (50)

    A = clay
    B = silt
    C = vf sand

## 18. MATRIX PERCENT (51)

    See 16.

# CLASTIC EDP DICTIONARY

**19. POROSITY TYPE** (52 and 53)

See 15.

**20. POROSITY PERCENT** (54)

See 16.

**21. SEDIMENTARY STRUCTURES** (55, 57, and 59)

See 12.

**22. FREQUENCY OF SEDIMENTARY STRUCTURES** (56, 58, and 60)

1 = abundant
2 = common
3 = rare

**23. CONTACTS** (61 and 62)

S = sharp
G = gradational
U = unconformable
R = scour
X = other

**24. GEOMETRY** (63)

T = tabular
W = wedge
S = sheet
L = lenticular
C = crescent

| POR. | | SEDIMENTARY STRUCTURES | | | | | | CON-TACTS | | |
|---|---|---|---|---|---|---|---|---|---|---|
| TYPE | PERCENT | TYPE | ABUNDANCE | TYPE | ABUNDANCE | TYPE | ABUNDANCE | UPPER | BASAL | SAND GEOMETRY |
| 52 | 53 | 54 | 55 | 56 | 57 | 58 | 59 | 60 | 61 | 62 | 63 |

## 25. FOSSILS (needs expansion and refinement) (64 - 69)
### (if no fossils, put 0 in first abundance column)

Be sure to use 2 columns for each fossil, third column abundance—abundance indicated by:

```
5 = very abundant   = 75%
4 = abundant        = 25 - 75%
3 = common          = 5 - 25%
2 = rare            = 1 - 5%
1 = trace           = 1%
```

  — = unidentified (Note that this can be easily expanded)
1. A- = *algae*
2. AS = stromatolitic algae          AU = cunephycus
3. AP = platy, blade, phylloid type  AK = komia
4. AL = lace (cunephycus type)       AI = Ivanovia
5. AE = encrusting
6. AC = club (dasyclad)
7. AO = (ottonosia - types, algal biscuits)

8. B- = *Bryozoans*
9. BF = fenestrate
10. BI = fistuliporid
11. BR = ramose
12. BA = archimedids

13. I- = *Brachiopods*
14. IL = linguloid
15. IP = productid
16. IS = spiriferid
17. IR = pentamarids         IC = conchicium

18. C- = *Corals*            CH = Halysites
19. CT = tabulates           CF = Favosites
20. CR = rugose and tetracorals  CL = Heliolites
21. CE = hexacorals

22. K- = *Crinoids* (all parts)

23. N- = *Conodonts*

24. X- = *Calcispheres*

25. MA = *Ammonoids*
26. M- = Molluscan
27. MD = Dentalium
28. MN = Nautiloids

29. G- = *Gastropods*
30. GT = Turritellid type
31. GE = Ecphorid type
32. GB = Busycon type
33. GO = Ovid type
34. GU = Euophalid type
35. GB = Bellerophonid type

| | FOSSILS | | | | |
|---|---|---|---|---|---|
| TYPE | ABUNDANCE | TYPE | ABUNDANCE | TYPE | ABUNDANCE |
| 64 | 65 | 66 | 67 | 68 | 69 |
| | | | | | |
| | | | | | |
| | | | | | |
| | | | | | |
| | | | | | |
| | | | | | |
| | | | | | |
| | | | | | |
| | | | | | |
| | | | | | |
| | | | | | |
| | | | | | |
| | | | | | |
| | | | | | |
| | | | | | |
| | | | | | |
| | | | | | |
| | | | | | |

36. O-  = *Ostracods*

37. L-  = *plant remains*
38. LL = leaves
39. LR = roots and stems

40. F-  = *Foraminifera*
41. FF = Fusulinid
42. FM = Miliolids
43. FP = Paleotextularid
44. FC = Calcitornellid
45. FB = Buliminids
46. FE = Endothyrids
47. FS = Saccamminids
48. FK = Coskinollinids
49. FL = Lituolids
50. FO = Orbitoids
51. FT = Textularids
52. FR = Orbitolinids
53. FG = Globogerinids
54. FD = Globorotalids
55. FZ = Rotalids
56. FQ = Trochamminids
57. FA = Ammobaculitids
58. FY = Bathysiphonids = Hedburgella

59. H-  = *Insoluble fossils*
60. HC = chitinozoans
61. HH = hystrichosphraerids
62. HS = scolecodonts
63. HG = siculae
64. HP = spores and pollen
65. HL = scilerites

66. E-  = *Echinoids*
67. ES = echinoids spines

68. GR = *Graptolites*

69. P-  = *Pelecypods*
70. PE = Exogyrids
71. PG = gryphae type
72. PI = inoceramids
73. PA = alectryonids
74. PV = venericardids
75. PL = glycimirids
76. PN = nuclanids
77. PP = pectinids
78. PU = unids

79. Q-  = *Radiolarids*

80. S-  = *Stromatoporids*
81. SM = massive
82. SW = lenticular, wafer
83. SF = finger *(Amphipora, Stachyodes)*

84. X-  = *Sponges*
85. SX = sponge, spicules
86. XC = sponge, calcareous
87. XL = sponge, siliceous

88. T-  = *Trilobites*

89. U-  = *Tubiphytes*

90. J-  = *Holothurians*
91. JS = Holothurians scelerites

92. R-  = *Rudistids*
93. RT = rudistids, toucasids
94. RM = rudistids, monopleurids
95. RC = rudistids, caprinids
96. RR = rudistids, radiolitds

97. W-  = *fish remains*
98. WO = otoliths
99. WS = scales
100. WK = teeth

101. V-  = *vertebrates*

102. Y-  = *annelids    trace fossils*
103. YA = worms
104. YT = trace fossils
105. YC = connularids
106. YN = tentaculitids

107. Z  = other - see remarks

## 26. DIAGENESIS TYPE (70 and 71)

- D = desiccation
- B = burrowing
- M = mixing
- A = accretion
- G = aggregation
- C = corrosion
- L = leaching and solution
- R = recrystallization
- N = replacement
- I = mech. internal sedimentation
- V = inversion
- X = cementation
- Y = decementation
- O = compaction
- T = tectonic fractures
- P = pressure solution

## 27. DEGREE OF ALTERATION (72)

- 1 = 0 - 10% or textures easily observed
- 2 = 10 - 50% or textures partially observed
- 3 = 50 - 90% or textures considerably altered
- 4 = 90 - 100% or textures completely obliterated

## 28. LOCATION OF DIAGENESIS (73)

- L = lithosphere surface
- V = vadose
- M = marine
- X = meteoric
- P = phreatic
- B = brackish

## 29. PROCESS OF DEPOSITION (74)

- F = fluvial
- A = alluvial
- E = eolian
- M = marine
- T = turbidity flow
- P = mass transport
- L = landslide
- W = mud flow

## 30. DEPOSITIONAL SITE (75 - 76)

- C- = *continental*
- CE = eolian
- CB = braided stream
- CP = point bar
- CA = abandoned channel
- CC = colluvial
- CF = alluvial fan
- CP = alluvial plain
- CI = levee
- DP = *deltaic plain*
- DM = marsh - swamp
- DI = inner distributary
- DC = crevasse fan
- DT = tidal channel
- DC = distributary channel
- DF = deltaic front
- DS = stream mouth bar
- DM = delta margin island
- DB = beach ridge
- DX = interchannel mouth bar
- DA = bay
- PD = *pro delta*
- I- = *interdeltaic*
- IB = beach
- IL = lagoonal
- IB = bay
- IF = tidal flat
- IM = mud flat
- IS = shoreface
- IR = foreshore
- II = barrier island
- M- = deep marine
- MT = turbidite
- MF = submarine fan
- MD = fan delta
- Mt = terrigenous shales and/or silt

| DEP. ENV. | | | | |
|---|---|---|---|---|
| PROCESS | SITE | ENERGY | REMARKS | |
| 74 | 75 | 76 | 77 | 78 | 79 |
|  |  |  |  |  |  |
|  |  |  |  |  |  |
|  |  |  |  |  |  |
|  |  |  |  |  |  |

## 31. DEPOSITIONAL ENERGY (77)

- 4 = high
- 3 = moderate
- 2 = low
- 1 = very low

## 32. REMARKS (including formational and paleo markers) (78 - 80)

# APPENDIX II

## Electronic Data Format for Carbonates

# GENERAL INSTRUCTIONS
# CARBONATE EDP FORMAT

### INTRODUCTION

The EDP format contained in the appendix is designed for computer processing. The format is also adaptable to a strip-log approach because the original descriptions can be copied and various columns grouped, spliced, and colored to facilitate visual analysis and correlation.

### INSTRUCTIONS

1. Columns 1-5:    Project identification (right-hand justify)
2. Columns 6-10:    Initials of describer
3. Columns 11-15:    Formation code
4. Columns 16-20:    Footage (right-hand justify)
5. Column 21:    Color, may revise to fit standard color system or add specific colors characteristic of the rocks under study.
6. Columns 22-31:    Each column represents 10% of the rock by volume. If describing cuttings, use GR or SP logs to aid in descriptions.
7. Remaining part of dictionary self explanable.

### RECOMMENDATION

Revise the carbonate dictionary to fit the suite of rocks under study. It is possible to combine the carbonate and clastic formats (Appendix I) by using a two-card system. Reserve one of the columns for encoding card 1 or 2, depending upon which rock type is being described.

# APPENDIX II

**Carbonate EDP Dictionary**

# CARBONATE EDP DICTIONARY

### 1. COLOR (21)

    9 = colorless to white
    8 = cream to tan
    7 = light greys and browns
    6 = intermediate greys and browns
    5 = dark browns and blacks
    4 = greens, etc.
    3 = red, orange, pink
    2 = variegated (interbedded)
    1 = mottled (throughout)
    0 = remarks for others

### 2. CARBONATE GRAIN TYPE
(relict if dolomite) (22-31)

    M = micrite
    G = unidentified grains
    S = skeletal
    O = oolite (2 or more laminae)
    C = coated grains (including pseudo
         oolite - one coat)
    B = algal-coated (oncolite)
    P = pellet
    L = lump
    D = detrital (non-carbonate grains)
    I = intraclast
    R = recrystallized (dolomite crystals in
         which no relict textures are seen,
         possibly interpreted as grains)
    A = organic
    E = caliche pisolite
    X = algae

### 3. GRAIN PERCENT (applies to grain carbonate shown in 2 - micrite = 0) (32)

    0 =  0 - 10
    1 = 11 - 20
    2 = 21 - 30
    3 = 31 - 40
    4 = 41 - 50
    5 = 51 - 60
    6 = 61 - 70
    7 = 71 - 80
    8 = 81 - 90
    9 = 91 - 100

**4. GRAIN SIZE RANGE** (show minimum and
                            maximum size—a
                            measure of sorting) (33 - 34)

    1 = mud < .03 mm.
    2 = silt (0.03 - .0625)
    3 = vf (.0625 - .125)
    4 = f   (.125 - .25)
    5 = m (.25 - .5)
    6 = c  (.5 - 1.0)
    7 = vc (1.0 - 2.0)
    8 = p  (2.0 - 10.0)
    9 = b  (greater than 10)

**5. DOMINANT GRAIN SIZE** (see above) (35)

**6. INPLACE OR TRANSPORTED GRAINS** (36)

    1 = inplace
    2 = transported
    3 = transported, mechanically rounded

**7. GRAIN FABRIC** (37)

    1 = grain supported
    2 = in situ growth position
    3 = grains supported in spar or micrite

**8. DOLOMITE MODIFIER** (38 - 39)
    ROCK TYPE (if interbedded show percentage of each type)

    − = shale (silica particles less than .03 mm)
        (argillaceous if modifier) (39)
    F = sandstone (silica particles greater than .03 mm)
    G = limestone (calcareous if modifier)
    H = dolomite
    J = anhydrite
    K = gypsum
    N = halite
    Q = chert
    T = igneous
    U = metamorphic
    V = bentonite
    W = coal, including lignite
    Y = marl (defined limestone)
    Z = quartz

| SIZE RANGE | | DOM. SIZE | SUPPORT | MAT-CEM. | SHAPE | DOMINANT MODIFIER | |
|---|---|---|---|---|---|---|---|
| 33 | 34 | 35 | 36 | 37 | 38 | 39 | 40 |

## 9. TRACE ROCKS OR MINERALS (less than 10%) (40)

See above - plus
T = glauconite
P = pyrite
N = siderite
E = phosphatic
F = plant fragments including spores and pollen
Z = sulphur
K = mica
Q = feldspathic
J = hematite or iron staining
X = other - see remarks
╪ = limonite
/ = quartz
w = clay

## 10. CRYSTAL SIZE RANGE (41 - 42)

See 4.

## 11. DOMINANT CRYSTAL SIZE (43)

## 12. CRYSTAL PERCENT
(% crystals larger than 0.125 mm) (44)

1 = 0 - 10%
2 = 10 - 50%
3 = 50 - 90%
4 = 90 - 100%

## 13. CHERT COLOR (45)

See 1.

## 14. CHERT TYPE (46)

G = glassy
N = novaculite
P = porcellanite
T = tripolitic
X = other - see remarks
D = glassy with dolomite rhombs

## 15. CHERT RELICT TEXTURE (47)

See 3.

| DOLOMITE | | | | | | |
|---|---|---|---|---|---|---|
| CRYSTAL SIZE | | | | CHERT | | |
| RANGE | DOM. SIZE | % CRYSTALS | COLOR | TYPE | REL. TEX. | |
| 41 | 42 | 43 | 44 | 45 | 46 | 47 |

## 16. DIAGENETIC PROCESSES (48 - 50)

- D = desiccation
- B = burrowing
- X = mixing
- A = accretion
- G = aggregation
- O = corrosion
- L = leaching
- S = solution
- R = recrystallization
- E = replacement
- T = mech. internal sedimentation
- I = inversion
- M = micritization
- C = cementation
- N = decementation
- W = compaction
- F = tectonic fracture
- P = pressure solution

## 17. DOMINATE DIAGENETIC PROCESS (51)

See 16.

## 18. ENVIRONMENT OF DIAGENESIS (52 and 53)

- C = continental
- N = nearshore
- M = marine
- V = vadose
- P = phreatic
- S = brackish
- L = lakes and streams
- U = supratidal
- T = tidal
- K = sabkha
- D = deep marine (sea floor)
- G = marine groundwater lens

# CARBONATE EDP DICTIONARY

## 19. DOMINATE DIAGENETIC ENVIRONMENT (54)

See 18.

## 20. CEMENT MINERALOGY (55 and 56)

C = calcite
A = argonite
L = low mag. calcite
H = high mag. calcite
D = dolomite
S = silica
Y = anhydrite
B = calcite and anhydrite
E = dolomite and silica
I = calcite and dolomite
J = calcite and silica
K = dolomite and anhydrite
G = gypsum
M = hematite

## 21. CEMENT OCCURRENCE (57)

I = intergranular
A = intragranular
O = organic
B = fenestral (birdseye)
M = moldic
X = intercrystalline
V = vug
F = fracture
U = cavity of unknown origin

## 22. CEMENT TEXTURE (58)

E = equant
B = blocky
F = fribrans
M = meniscus
R = micrite
S = spar

## 23. CEMENT ENVIRONMENT (59)

See 18.

## 24. PERCENT CEMENT (of total rock) (60)

1 = 1 - 5%
2 = 6 - 10%
3 = 11 - 15%
4 = 16 - 20%
5 = 21 - 25%
6 = 26 - 30%
7 = 31 - 35%
etc.

## 25. POROSITY TYPE (Most important to left) (61 and 62)

See 21.

## 26. POROSITY DEGREE (% of rock) (63)

1 = 1 - 5
2 = 6 - 10
3 = 11 - 15
4 = 16 - 20
5 = 21 - 25
6 = 26 - 30
7 = 31 - 35
8 = > 35
Blank = no effective porosity

## 27. DEGREE OF ALTERATION
(either dolomite or limestone) (64)

1 = 0 - 10% or textures easily observed
2 = 10 - 50% or textures partially obliterated
3 = 50 - 90% or textures considerably altered
4 = 90 - 100% or textures completely obliterated

## 28. SHAPE OF CRYSTALS (65)

A = anhedral
S = subhedral
E = euhedral
F = fibrous
B = bladed
X = other - see remarks

## 29. MINERALIZATION (66)

P = pyrite           C = calcite
H = sphalerite       Q = quartz
G = galena           S = silica
D = dolomite         K = hydrocarbons

| DIAGENESIS | | | | | |
|---|---|---|---|---|---|
| PERCENT | POROSITY | | ALT. | | MINERALIZATION |
| | TYPE | PERCENT | DEGREE | SHAPE | |
| 60 | 61 | 62 | 63 | 64 | 65 | 66 |

**30. FOSSILS** (needs expansion and refinement) (67 - 75)
    (if no fossils, put 0 in first abundance column)

Be sure to use 2 columns for each fossil, third column abundance—abundance indicated by:

    5 = very abundant  = > 75%
    4 = abundant       = 25 - 75%
    3 = common         = 5 - 25%
    2 = rare           = 1 - 5%
    1 = trace          = < 1%

    – = unidentified   (Note that this can be easily expanded)
1. A-  = *algae*
2. AS  = stromatolitic algae      AU = cunephycus
3. AP  = platy, blade, phylloid type    AK = komia
4. AL  = lace (cunephycus type)    AI = Ivanovia
5. AE  = encrusting
6. AC  = club (dasyclad)
7. AO  = (ottonosia - types, algal biscuits)

8. B-  = *Bryozoans*
9. BF  = fenestrate
10. BI = fistuliporid
11. BR = ramose
12. BA = archimedids

13. I-  = *Brachiopods*
14. IL  = linguloid
15. IP  = productid
16. IS  = spiriferid
17. IR  = pentamarids       IC = Conchicium

18. C-  = *Corals*           CH = Halysites
19. CT  = tabulates          CF = Favosites
20. CR  = rugose and tetracorals   CL = Heliolites
21. CE  = hexacorals

22. K-  = *Crinoids* (all parts)

23. N-  = *Conodonts*

24. X-  = *Calcispheres*

25. MA = *Ammonoids*
26. M-  = Molluscan
27. MD = Dentalium
28. MN = Nautiloids

29. G-  = *Gastropods*
30. GT  = Turritellid type
31. GE  = Ecphorid type
32. GB  = Busycon type
33. GO  = Ovid type
34. GU  = Euophalid type
35. GB  = Bellerophonid type

36. O-  = *Ostracods*

37. L-  = *plant remains*
38. LL = leaves
39. LR = roots and stems

40. F-  = *Foraminifera*
41. FF = Fusulinid
42. FM = Miliolids
43. FP = Paleotextularid
44. FC = Calcitornellid
45. FB = Buliminids
46. FE = Endothyrids
47. FS = Saccamminids
48. FK = Coskinollinids
49. FL = Lituolids
50. FO = Orbitoids
51. FT = Textularids
52. FR = Orbitolinids
53. FG = Globogerinids
54. FD = Globorotalids
55. FZ = Rotalids
56. FQ = Trochamminids
57. FA = Ammobaculitids
58. FY = Bathysiphonids        = Hedburgella

59. H-  = *Insoluble fossils*
60. HC = chitinozoans
61. HH = hystrichosphraerids
62. HS = scolecodonts
63. HG = siculae
64. HP = spores and pollen
65. HL = scilerites

66. E-  = *Echinoids*
67. ES = echinoids spines

68. GR = *Graptolites*

69. P-  = *Pelecypods*
70. PE = Exogyrids
71. PG = gryphae type
72. PI = inoceramids
73. PA = alectryonids
74. PV = venericardids
75. PL = glycimirids
76. PN = nuclanids
77. PP = pectinids
78. PU = unids

79. Q-  = *Radiolarids*

| FOSSILS | | | | | | | | |
|---|---|---|---|---|---|---|---|---|
| KIND | | ABUNDANCE | KIND | | ABUNDANCE | KIND | | ABUNDANCE |
| 67 | 68 | 69 | 70 | 71 | 72 | 73 | 74 | 75 |

80. S-  = *Stromatoporids*
81. SM = massive
82. SW = lenticular, wafer
83. SF  = finger *(Amphipora, Stachyodes)*

84. X-  = *Sponges*
85. SX = sponge, spicules
86. XC = sponge, calcareous
87. XL = sponge, siliceous

88. T-  = *Trilobites*

89. U-  = *Tubiphytes*

90. J-  = *Holothurians*
91. JS = Holothurians scelerites

92. R-  = *Rudistids*
93. RT = rudistids, toucasids
94. RM = rudistids, monopleurids
95. RC = rudistids, caprinids
96. RR = rudistids, radiolitids

97. W-  = *fish remains*
98. WO = otoliths
99. WS = scales
100. WK = teeth

101. V-  = *vertebrates*

102. Y-  = *annelids - trace fossils*
103. YA = worms
104. YT = trace fossils
105. YC = connularids
106. YN = tentaculitids

107. Z   = other

## 31. SEDIMENTARY STRUCTURES (needs expansion) (76 and 77)

- A = even parallel laminae less than 3 mm thick
- B = even parallel laminae greater than 3 mm thick
- C = wavy laminae
- D = small-scale cross laminae
- E = large-scale cross laminae
- F = graded bedding
- G = flame structure
- H = groove casts
- I = convolute bedding
- J = burrowed or churned by organisms
- K = contorted bedding
- L = flute casts
- M = load casts
- N = geopetal structures
- O = stromatrolitic
- P = birdseye
- R = breccia
- S = scoured surface
- U = nodular
- V = bedded
- W = ripple marks
- X = cross bedded
- Y = desiccation cracks

## 32. LITHOFACIES (set up dictionary for each project) (78)

## 33. CARBONATE ENVIRONMENTS (79 - 80)

STEP

Shelf

- SA = algal flat
- ST = tidal flat
- SS = supratidal
- SI = intertidal
- SE = subtidal
- SF = shoreface
- SL = lagoon-bay
- SG = restricted lagoon
- SP = patch reef
- SU = reef flank
- SJ = interreef
- SO = shoal
- SK = mound
- SH = pinnacle
- SD = atoll
- SC = linear buildup

| SEDIMENTARY STRUCTURES | | LITHOFACIES | DEPOSITIONAL ENVIRONMENT | |
|---|---|---|---|---|
| 76 | 77 | 78 | 79 | 80 |
| | | | | |
| | | | | |
| | | | | |
| | | | | |
| | | | | |
| | | | | |
| | | | | |
| | | | | |
| | | | | |
| | | | | |
| | | | | |
| | | | | |
| | | | | |
| | | | | |
| | | | | |
| | | | | |
| | | | | |
| | | | | |
| | | | | |
| | | | | |

# CARBONATE EDP DICTIONARY

Shelf Margin

SB = bank
SR = reef
SY = forereef slope
SZ = backreef
SW = forebank slope
SX = backbank

Basin

SM = pelagic
SV = turbidite
SN = submarine fan
SQ = submarine channel

## RAMP

Coastal

RA = algal flat
RT = tidal flat
RS = supratidal
RE = subtidal
RH = beach
RF = shoreface
RC = channel
RL = lagoon-bay
RG = restricted lagoon
RP = patch reef
RM = mound
RU = reef flank
RJ = interreef
RO = shoal
RK = intershoal

Transitional

RB = bank
RW = forebank
RX = backbank
RV = interbank
RR = buildup
RY = fore buildup
RX = back buildup

Outer

RV = inner neritic
RQ = middle neritic
RN = outer neritic

# Index

## A

abandoned channels, 74
Abo, 235, 238
acropora, 197, 201
  palmata, 204, 207
A. cervicornis, 205, 207
acoustic logs, 30
aerating currents, 23
Alaska, 37
  east coast fan deltas, 107
  Prudhoe Bay Field, 61, 80 113-115
algae, 21, 197, 201
algal
  encrusted grains, 182
  kerogen, 22
  mat, 269
  stromatolite, 212, 269
  stromatolite packstone, 244
Allen, *cited* (1969) (1975), 9, 65
allochemical rocks, 183
alluvial fans, 50-51, 67, 69-71, 84, 129
alluvial plain, 68
Almond formation, 124, 137, 138
alopods, 215
Amazon, 91
Ambergris Cay, 194
American Commission of Stratigraphic
  Nomenclature, 40
amplitude anomaly, 253
ammonites, 212
Amphipora, 260
Amoseas, 37
Anadarko Basin, 115
anaerobic conditions, 23
anatexis, 166
Anderson (1926), 21
Anderson and Kirkland, *cited* (1960), 87
Andros, 200, 203
Angelina-Caldwell flexure, 35
anhydrites, 29, 50, 268
anthracene, 38
antidune plane beds, 65
apparent anomalies, 52

aquatic plants, 21
aqueous solubility, 35
Arabia, 211
aragonite, 181
  crystals, 201
Arbuckle limestone, 35
Arkoma Basin, Oklahoma, 109
Arctic Slope Province, 113
aromatics, 24, 36
  benzene
  naphthalene
Ash Falls, 40
Asphalt, 21-22, 24
Asquith (1970), 154
Atlantic Shelf, 142
atolls, 187
  carbonate reservoirs, 198
attached dunes, 71
Atwater and Miller, *cited* (1965), 170
authigenic cement, 175
authigenic minerals, 8, 9
authigenic replacement, 175
autochthonous reef rocks, 183

## B

back reef, 24, 27
  bank, 180
bacteria, 21
bacterial oxidation, 23
Bahamian Platform, 199-205
Baker and Kastner, *cited* (1981), 230
Bakersfield Arch, 159
bank, 54, 187
Baltic Sea, 23
barchans, 71
Baria and Others, *cited* (1982), 251-254
barred basin, 23
barrier
  bank, 51
  island sands, 131-132
  reef, 24, 43, 51, 54, 187, 192, 195
Barrow Arch, 113

bar sand, 50
basal, 42
base-concordance, 43
baselap, 43
   onlap
   downlap
base level, 142
"basinal" facies, 264
basin
   carbonate, 234
   detrital, 234, 237
   facies, 235
   floor, 204
bathymetric, 92
bathymetry zones, 191
Bathurst, cited (1967) (1975), 203, 232
beaches, 40
   barrier island complex, 129, 130
   facies, 199
beach deposits, 125
beach-ridge sequence, 106
beach-ridge reservoirs, 101
beach sequence, 123
beach-shoreface reservoir, 140
Beaverhill Lake reef, 257, 260
Bebout and Others, cited (1979), 232
bedding plane, 15
   time surfaces, 13
bed form, 63-64
   low flow (nipples, sandwaves, dunes)
   upper flow (planes, antidunes, chutespools)
Belize, Yucatan Peninsula, 191-197
Bell Creek Field, Montana, 61, 134-137
Beluga formation, 80
Beers, cited (1945), 24
Bend Arch, 115
Benezene concentration, 37
benthonic, 21, 212
benzofluorene, 38
benzo pyrene, 38
Berg, cited (1968) (1975) (1976) (1978) (1982), 61, 80, 108-109, 134, 148, 151, 153, 159
Berry Islands, 201-202
Berven (1966), 154
bifurcation, 92
bioclastic drift debris, 212
biodegraded, 38
biofacies, 179
bioherms, 188
biostratigraphic boundary, 41
biostratigraphic facies, 45
biostratigraphic unit, 41-42

biota, 205, 215
biotic growth, 23
biotite, 171
bioturbation, 101
Bimini, 201, 203
Bishop (1968), 250
Bisti Field, San Juan Basin, N.M., 134
Blackadore Cay, 194
Blixt (1941), 80, 87
blowouts, 71
Bogan, 198
Bone Springs sands, 159
Boothroyd and Nummdal (1978), 107
Borabi No. 1 well, 240
Borden Island Gas Field, 134, 140
Borger (1952), 84-85
boring, 186
borehold, 30
Borregas Field, West Texas, 35, 80
bottom set, 188
boundstones, 184, 215, 244
Brachiopod, 179, 215
brackish, 165
brackish water cements, 227
braided river sediments, 64
braided stream, 68
Brae Field, 113
breccia, 186
"bright-spot", 81
Bristol Channel, England, 104
brittle grains, 173
Brown and Others (1973), 66
bryozoan mudstone, 244
bryozoans, 215, 251
Bubb and Hatlelid, cited (1977), 242
burial diagenesis, 171
burial relationship, 32
burrow, 74, 186
Busch, cited (1971), 109
butylbenzenes, 38
by-pass slope margin, 195

## C

Cabbage Ridge area, 192, 194
calcareous mud, 194
caliper log, 30
Cambrian, 22, 30, 47
Cambrian Age reservoirs, 30
Canadian Arctic Archipelago, 140
Canyon Field, Valverde Basin, Texas, 159
capillary fringe, 230
capillary pressures, 31, 61

INDEX

carbon, 21
carbonates, 23-24, 33
　anomalies, 52
　bioherms, 33
　buildup, 187, 189
　deposits, 214, 220
　geology, 179
　mud, 184, 193
　sediment, 33, 45
　turbidites, 179
carbonate reservoirs, 177-189
　facies, 237
　rocks, 272
　sands, 211
　sedimentation
　seismic facies summary, 276
Carboniferous, 169
carbohydrates, 21, 22
carbonization, 28
　scale, 29
carbon isotope value, 33
carbon number, 38
Cardium, Cross Field, Alberta, 146
Cardium, Pembina Field, Alberta, 146
Cardium formation, 153
Cardium sandstone, 154, 158
Caribbean Plate, 191
carrier bed, 31, 33, 35, 40
Carya translucency, 29
Caspian Sea, 23
catalytic reactions, 24
Caughlin and Others, *cited* (1976), 243, 245, 247
cavern, 186
caving bank, 73
Cay Corker, 194
Cayman Ridge, 192
Cayman trough, 191-192
cays, 194
cementation, 168, 182
　aragonite, 226
　calcite, 168
　dolomite, 227
　Mg-calcite, 226
　silica, 168
Cenozoic, 217
　reservoirs, 123
　sediments, 25
Central Basin Platform, 33
chalkify, 212
chalks, 212
channel, 73, 186
　sand, 50

channel-mouth sand bar, 49, 98
　sequence, 103
Chapman (1973), 32
Chetumal Bay, 192
Chevron Exploration Staff, *cited* (1979), 245, 248-250
chlorophyll, 21
cholesterin, 21
Choquette and Pray, *cited* (1970), 185-186
chronostratigraphic unit, 224
Chuber and Rodgers, *cited* (1968), 33
Cincinnati Arch, 35
CITCO Publication (1979), 28
Clark, *cited* (1924), 21
clastic anomalies, 50
clastic rocks, 9, 33
clay
　drape, 74
　grains, 32
Clear Fork, 204, 240
Cliff and Lasemi, *cited* (1980), 134
climate, 62
cline form, 188
clinoform sections, 109
coal, 22
coalification, 29
coastal
　deposits, 46, 91, 188
　onlap, 43-44
　toplap, 44
coast lines, 91
　collision
　macrotidal
　marginal sea
　mesotidal
　microtidal
　trailing edge
coated grains, 182
cobbles, 198
coccolith algae, 212
coeval
　markers, 50-51
　shales, 50
　shelf, 235
Cole Creek Field, 80
collodial, 36
　aggregates, 36
Colorado, 91
Columbia, 92
compaction, 31, 172, 266
　curves, 31
　processes, 228
combination pattern, 62

concentration, 37
Concho Arch, 115
concordant dolomites, 230
condensate-gas, 30
Congo, 92
connate, 226
Conoco research, 252
contemporaneous structure, 56
continental facies, 78
continental margin, 42
continental sands, 64
    alluvial channels, fans
    eolian
continental sandstone reservoirs, 61
continental shelf, 24
continuous phase migration, 35
continuous velocity logs (CVL), 13
Cook, *cited* (1923), 31
Cook Inlet Basin, Alaska, 80
corals, 21
coral-algal sand, 194
coralgal facies, 200
coralgal sand, 202
coraline algae, 207
coraline facies, 201
coral packstone, 244
coral rank, 29
Cordell, *cited* (1973), 36
core analyses, 3, 5
cores, 3
corozal, 192, 194
correlation
    checks, 50
    geologic method, 44-48
    geophysical method, 48
    stratigraphic 48-50
Couch and Others, *cited* (1977), 199
Craton, 42
crescentic dunes, 71
Cretaceous, 47
    Cut Bank, 80, 87, 89-90
    Dakota, 35, 112
    deltaic strata, 109
    Edwards limestone, 36, 148, 204
    Fall River, 80
    Kootenai formation, 87
    Lewis, 112
    limestone, 25
    Mesa Verde, 112
    Tuscaloosa delta, 148
Crevasse fan sequence, 100
Crevello (1978), 204-205
crinoids, 63, 244

critical well control, 238
cross time surfaces, 45
crude oil, 24
crytocrystalline grains, 194
Curry and Curry (1972), 80
Cut Bank Sand, Montana, 80, 87, 89-90
Cynthia, 249

# D

Dakota Group, 35
Danube, 21, 91
data distribution, 63
Davies, *cited* (1973), 91
Davies and Berg, *cited* (1969), 61
Davies and Others, *cited* (1971), 130
deep marine, 275
    debris-flow deposits, 69
    debris flows, 146
    deep marine sands, 145
dehydration, 226
Delaware Basin, 189, 234, 275
delta, 23, 92
    classification, 94
    complexes, 96-97
    deposits, 23
    development, 92-97
    size, 93
deltaic front reservoir, 64, 95, 118
    beach ridge sands
    channel mouth
    delta margin islands, 101, 105
deltaic plain reservoirs, 64, 95-98
    braided stream
    crevasse fan
    distributary sand deposits
    point bar
deltaic sedimentary patterns, 133
deltaic swamp, 48
density, 30
Denver Basin, Colorado, 109
depositional energy, 22
depositional topography, 61, 180, 187
deposits, 23
    floodplain
    marsh
    swamp
detrital (intraclasts), 182
Devonian, 22, 47
    Beaverhill Lake buildups, 252
    reef, 253, 258
diagenesis, 30, 63, 165
diagenetic clay, 8-9
diagenetic kerogen, 28

diagenetic pattern, 61
Diatoms, 21
dictyota, 24
differential
   entrapment, 38-39
   topography, 45
Dillon and Vedder, *cited* (1973), 191-192
dimethylbenzenes, 38
dimethylnaphthalenes, 38
dipcross section, 237
Diplora sp, 205, 207
dip reversal, 35
discordant dolomites, 231
disequilibrium pattern, 62
dissolution, 174-175
distal facies, 215
drillstem tests, 3, 7
dolomite, 181
dolomitization
   concordant, 230
   discordant, 231
   processes, 230-231
Douglas and Oliver, (1979), 134, 140-141
downlap, 56
drainage area, 35
Drake Point Field, 140-141
draping, 56
Druse cement, 227
Duane and Others, *cited* (1972), 142
ductile grains, 173
Dundee, 247
Dune reservoir, 85
Dunham (1962) Classification, 184
Dutton, *cited* (1982), 113, 119-120
dythatic hydrocarbons, 24

## E

Early Guadalupian, 234
Early Oligocene time, 191
Easterly Trade Winds, 199
Eastern Venezuela Basin, 84
East Texas Field, 35
Echinoderms, 181, 198, 215
Edwards
   limestone, 148
   reef, 111, 204
   reef trend, 36
electric log correlations, 14
elemental analyses, 30
Eleuthera Island, 206
Ellenburger crude, 33
Elmworth Field, Alberta, 113
Empire Abo Field, southeast N. M., 233, 238

Enewetak Atoll, 198, 211
Enmedio reef, 24, 27
Entrada sandstone, 85-87
Eolian
   depositional processes, 80
   dunes
   reservoirs, 71
Epeiric seas, 206
   nearshore
   middle
   outer
   Persian Gulf, 211-212
   tidal flat-shoreline, 211
epeirogeny, 42
equilibrium
   depth, 31
   pattern, 62
era, 40
erosional truncation, 43
estuarine environment, 24
Etosha Basin, southwest Africa, 256
eustatic
   changes, 42, 44-45, 190, 220
evaporites, 23, 33, 244
   anomalies, 52
   Buckner, 269
   dolomites, 211
Evans, *cited* (1970) (1972), 153, 252, 257
expulsion, 31, 32
exsolution mechanisms, 36, 39
Exuma Sound, 204

## F

fabric, 61, 182
facies
   belts, 213
   change, 61
   equivalent, 15
      lithostratigraphic units, 15
   lines, 15
   pinchouts, 71
Fahler member, 113
Fairway Field, 244
Fall River Sandstone, 80
fan
   apex, 69
   delta, 104, 113
   surface, 69
   toe, 69
farmers whole oil, 37
Fash, *cited* (1944), 24
fault communication, 33

faunal
  anomalies, 33
  marker, 48
  unit, 48
fenestral, 186
festoon cross beds, 68, 74-76
Field, *cited* (1980), 143
final flow pressure (FLP), 3
final shut-in pressure (FSP), 3
fish, 21
  Scales Marker, 82
Fisher, *et al., cited* (1973), 124
Fisher and McGowen, *cited* (1969), 109
Fisher and Others, *cited* (1969), 94
Fisk, *cited,* (1961), 102
flame structures, 145
flood plain, 68
  silts, 33
Flores and Ethridge, *cited* (1981), 80
Florida Straits, 199
flow regime, 63
fluid density, 9
fluidized flows, 146
fluid saturation, 9
flushing, 32
flute casts, 145
fluvial deposits, 40
fluvial-dominated deltas, 108, 110
  complex, sigmoid oblique
  oblique (tangential)
  shingled
  sigmoid
Folk, *cited* (1962), 183
fondoform, 188
foraminifera, 21, 181, 212
forams, 215
foredune ridges, 71
fore reef, 24, 27, 195
foreset slope, 188
formation, 41-42
  boundary
fossils, 42, 63, 183
  biomicrite
  biosparite
  fauna
  flora
fracture, 174, 186, 238
  igneous rocks, 33
fracturing, 174, 230
Freidman and Sanders, *cited* (1978), 107
Freidman and Stearns, *cited* (1971), 174
fresh water, 165
Fresnal Zone, 12

fringing reef, 187
Fruitale Field, 159
Fryberger, *cited* (1979), 80
Fulchtbauer, *cited* (1974), 169

## G

gamma ray logs (GR), 9, 30
gas cap, 38
gastropod, 179
gathering area, 35
Gehman, *cited* (1962), 23
geic acid, 21
generation stage, 30, 31
geochronology, 40
geologic
  history, 91
  reef, 187
geometric interpretations, 9
  bed thickness
  facies
  lithologies sequence
  shape of potential reservoir bodies
  structure
geostatic pressure, 31
Ghyben-Herzberg lens, 165
Gilbert-type deltas, 69
Ginsburg and James, *cited* (1979), 189, 194-195
global
  sea level changes, 40, 63
  tectonics, 40, 91
Glovers Atoll, Belize, 192, 195, 197, 221
Goose Lake, Alberta, 252
graded zone, 204
grain flows, 146
grain moldic, 269
grains, 257
grainstone, 184, 215, 243
grainstone shoals, 211, 214, 264, 266, 268
grain type, 61, 183, 214
grapestone facies, 200
graphite, 29
Grayson, *cited* (1972), 29
Great Bahama Bank, 199, 202
Greely area, San Joaquin Basin, CA, 159-160
Green River Basin, 19, 72, 109, 112, 133
Grenville shear fractures, 244
gross time-stratigraphic correlations, 45
groundwater, 36
group, 42
growth framework, 186

Guadalupian strata, 235
 Permian Basin
Gulf
 Coast, 32, 109, 124
 of California, 123, 129
 of Honduras, 191-192
 of Kara Bogaz, 23
 of Mexico
 of Papua, 240
 of Riga, 23
gullied slope, 204
Gum Hollow, 107

# H

Halbouty, *cited* (1972), 80
Halimeda, 24
 sands, 194-195, 201
halophytic vegetation, 212
Handford, *cited* (1980), 120
Handford Field, Texas, 268, 270
Harms, *cited* (1966) (1980), 80, 113
Hayes, *cited* (1977) (1979), 91, 123, 172, 176
Hayes and Others, *cited* (1976), 80
Hedberg, *cited* (1959), 32
Helca Point Field, Canada, 140-141
Heldt Draw Field, WY, 153
hemin, 21
Hemphill and Others, *cited* (1970), 257, 259-260, 262
Hertz Pulse, 11
heterogeneity, 33
Hicks, *cited* 194
Hilight Field, WY, 134
Himus, *cited* (1951), 29
Hobson and Others, *cited* (1982), 153, 155-157
Holmes, *cited* (1965), 104, 107
Holocene
 Coast of N. Carolina and Georgia, 123
 fan deltas, 107
 U.S.A. Gulf Coast, 123
Holocene Epoch, 124, 142
homocline, 137
Horseshoe Atoll, Texas, 260, 265
 canyon limestone
 cisco limestone
 lower Wolfcampian limestone
 reservoirs, 263
 strawn limestone
Horseshoe Field, San Juan Basin, N.M., 134
Hough area, Oklahoma, 104
House Creek Field, WY, 153

House Mountain, 252
Hsu, *cited* (1977), 160, 163
Hsu and Others, *cited* (1980), 160
humic acid, 21
humid climatic belts, 23
Hunt, *cited* (1862), 35
Hunton limestone, 42
Hutton, James, *cited* (1795), 22
hydrocarbon
 expulsion, 31
 fraction, 24
 gas analysis, 30
 generating stage, 29
 generation, 28, 30
 migration, 33, 38
 molecules, 32
hydrodynamic gradients, 35
hydrographic regime, 91
hydrologic gradients, 31
hydrothermal, 226
hydrozoans, 251
hypersaline, 226

# I

igneous intrusion, 226
illite mud, 193
impurities in oil, 20-22
 $CO_2$
 $H_2O$
 $H_2S$
 N
 $O_2$
 sulphur
Ingels, *cited*, (1963), 180
inland lakes, 24
Inman and Norstrom, *cited* (1971), 91
inorganic carbon, 30
in-place organic structure, 182
insular shelf, 24
intercrystalline, 269
interdeltaic sands, 64, 123
 barrier island, 126-132, 134-140
 beach, 123-124, 140-141
 tidal channel, 126
intercrystal, 186
interfluvial, 23
interior range, 84
internal sediments, 227
interparticle, 186
interreef trends, 24
interstitial
 porosity, 31
 sulfate, 24

intertinite, 22
intraclasts, 183
  intramicrite
  intrasparite
intraparticle, 186
intraplatform basin, 204
inversion, 228
Ireton shale, 249
isopach maps, 50
isopachous fibrous cement, 227
isotherms, 220
isotopic composition, 26-27

## J

Jageler and Matuszak, *cited* (1972), 9
Jamaica, southeastern, 107
James and Ginsburg, *cited,* (1979), 95
Jay Field, 266, 268
  depositional environments, 269
Jonas and McBride, *cited* (1977), 173
Jones and Speers, *cited* (1976), 61
Judy Creek Field, Alberta, 252, 261
Jurassic, 47, 49, 85, 109, 113
  Smackover, 266

## K

Kaolinite mud, 193
Karst topography, 11
Kaskaskia rocks, 42
Kaybob, 252
kerogen
  amorphous, 30
  fraction, 30
  marine, 22
  organic matter, 28
  terrigenous, 30
  translucency, 29
Koch, *cited* (1973), 196
Kraft, *cited* (1971), 124
Kraft and Chacko, *cited* (1979), 123
Krueger, *cited* (1963), 192
Kumar and Foster, *cited* (1982), 268, 270, 273

## L

lacustrine, 104
lagoon lithofacies, 199
Laing, *cited,* (1972), 79, 110, 112, 133, 151, 256, 265
Lake Maracaibo, Venezuela, 24
laminae, 24
laminar stromatoporoid, 244
laminations, 74

land, 26
Land and Weimer, *cited* (1978), 109
Las Piedras formation, 84
L.A. State, 37
Late Guadalupian, 234
lateral distribution, 40
Laughlin and Others, *cited* (1975), 245
Lawrencia, 24
Lawson and Crawson, *cited* (1961), 140
layering, 40
leached matrix, 269
leaching, 228
"leased plays", 215
LeMay, *cited* (1971), 238-239
Leonardian, 234
Leopold and Others, *cited* (1964), 74, 77
Levelland-Slaughter Field, TX, 271
Levandoski and Others, *cited* (1973), 61
Levorsen, *cited* (1958), 35
Lewis formation, 137
Leythaeuser and Others, *cited* (1982), 32
Lighthouse Reef, Belize, 192, 195
lignins, 21
lignite, 29
limestones, 28
linear sand ridges, 143
liquid hydrocarbon, 29, 32
lipid material, 24, 26
lithofacies, 179
lithogenetic facies, 45
lithogenetic rock units, 224
lithogenetic unit or bed, 48
lithographic characteristics, 40
lithosphere surface, 31, 36
littoral deposits, 46
load casts, 145
Lobate delta, 155
Long Cay, 194
longitudinal dunes, 71
Longman (1980), 232, 255
long migration, 33
Louden Field, Illinois Basin, 134
Louisiana cores (south), 170
Lovington Field, N.M., 235
lumps, 182
Lynmouth delta, 104
Lynn, *cited,* (1955), 87
Lyons Sandstone, 61

## M

Mackenzie River Delta, 91, 93
MacPherson, *cited* (1978), 159
macrotidal coasts, 123

Magaritz and Others, *cited* (1972), 232
Mancos shale, 146
Mann, *cited* (1916), 21
Manteck (1976), 244
Manville Group, central Alberta, 80, 82
mappable units, 53
    sequences, 44
margin reef deposits, 242
marine, 21, 46, 64, 144, 165, 226
    cements
    deep marine fan
    deposits
    fungi
    lens
    offshore shallow
    sands
    sandstone reservoirs
    surface
marine-phreatic, 231
    mixing
    model
marker beds, 45, 48
Marshall Islands, 198, 221
marsh shales, 48
Martin Hills Field, Alberta, 61
mass transport, 146
    rabbit model
    turtle model
Masters, *cited* (1979), 113
Mather and Mehl, *cited* (1919), 35
Matuszak and Jageler, *cited,* (1972), 9
matrix, 227
maxima cement, 174
maximum carbonate production, 218
Maya Mountains, Belize, 192
Mazullo, *cited* (1981) (1982), 159, 232
McCubbin, *cited* (1969), 134
McDonald and Lewis, *cited* (1973), 107
McGowan, *cited* (1970), 107
McGregor and Biggs, *cited* (1968), 134-137
McKee, *et al, cited* (1967), 34
McLean, *cited* (1979), 113
meandering stream, 73, 77
mechanical logs, 8-9
    response, 10, 30, 216
Media Field, N.M., 85-86, 88
mega fauna, 21
megapore, 186
MeKong Delta, 91
Melville Island, Canada, 140
Melvin Fields, Alabama, 251
member, 42
meniscus cement, 227

mesopore, 186
mesotidal coasts, 123
Mesozoic, 217, 243
    pinnacles
Messolella and Others, *cited* (1974), 246
metamorphism, 166
metamorphosed, 28
meteoric, 226
methane, 32
    carbon dioxide water, 29
methylnaphthalenes, 38
micaceous silty sands, 169
micelles, 36
Michaelis and Dixon, *cited* (1969), 153
Michigan Basin Pinnacle Fields, 243, 246
micrite, 257
micritic fluorapatite, 24
microcrystalline, 272
    calcite
    dolomite
micro fauna, 21
microorganisms, 21, 24
micropore, 186
microsolution desiccation fabrics, 212
microtidal coasts, 123
mid-cretaceous time, 24
Middle Devonian Elk Point, 258
Middle Guadalupian, 234
Middleton and Hampton, *cited* (1973), 147
Midland Basin, W. Texas, 33, 159, 240-241
migration, 31, 33
    paths
Millapora, 201
Miller Creek Field, Montana, 80
milliolids, 194
    mud
mineral density, 9
mineralogy, 168
Miocene
    en echelon, 205
    gas, 30
    La Pica formation, 84
    mudstone, 207
Mio-Pliocene time, 84
Mississippi, 47, 91, 93, 102, 112, 165
Mitchum and Others, *cited* (1977), 42-43, 108-109
mg-calcite, 212
Mobeetie Field, TX, 113, 115, 119, 121-122
modern carbonate sediments, 181
moldic, 186
mollusks, 21, 201, 215
Monastria, 201, 205, 207

Montaqua-Cayman Trench, 191
Montastrea, 197
Montmorillonite mud, 193
Moore and Druckman, cited (1981), 232
Morgan and Others, cited (1978), 80
Morgridge and Smith, cited (1972), 80, 113-115
mound, 54, 187
Muddy sandstone, 134-137
mud logs, 3, 6
mudstone, 184
"multipay" fields, 33
multiple generations, 227
Murray, cited (1966), 261
Mutti and Ricci Lucchi, cited (1972), 146

## N

n-alkanes, 36
Nanz, cited (1954), 80
naphthalene, 36, 38
Napierian Log, 31
Natland, cited (1957), 160
natural levee
   reservoirs, 98
   sequence, 99
Naval Petroleum Reserve No. 4, 113
Neilsen, cited (1957), 158
neomorphic process, 182
neomorphism, 228
neutron logs, 9
New Province Channel, 202
Niagaran
   reef belt, 245
   shelf, 245
   Time, 246
Niger River, 126
Niger-Venue, 93
Nile, 91
Nisku (Devonian) shelf margin, 248
Nisku Pinnacle Reef Fields, Alberta, 245
nitrogen, 21, 24
nodules, 24
noise, 11
   harmonic distortions
   microseisms
   shot-generated
   tape-modulation
nonframebuilders, 257
   amphypora
   solenporoid algae
   stachyodes
   tabulate corals
non-platy grains, 173

norphlets, 269
North Alazan Field, 35
Northern River, 194
North Fish Cay, 202
North Slope, Alaska, 108, 110
nutrient supply, 61

## O

odd-carbon ratio, 24
offlap sediments, 42
offshore barrier island reservoir, 126
offshore facies, 212
offshore marine sandstone reservoirs, 142
oil-condensate-gas, 30
oil generation, 28-31
oil migration, 235
oil seeps, 33, 84
onlap, 42, 43, 56
oolite facies, 200, 203, 214, 267
oolites, 24, 43, 182-183
   oomicrite
   oosparite
opaque kerogen, 30
Ordovician crude, 33
Oregon Basin Field, 80
organic carbon, 24, 30
organic framebuilders, 257
   stromatoporoids
organic matter, 28
organic molecules, 36
organic-rich shales, 30
organic solvents, 24
orogeny, 42
orthochemical rocks, 183
Ostracode-oolite, 194
Ottman and Others (1973), 226, 268-269
Ouachita Mountain chain, 266
Outer Basement Ridge, 192
outer shelf environments, 24
oxidation-reduction, 22
oxidized, 38

## P

paleo axis, 35
paleobiologist, 28
paleography, 51
paleohydrostatic gradients, 232
paleomorphic map, 137
paleontology, 40
paleoslope, 53
paleotopography, 53

paleozoic, 42, 217
    crinoid biofacies, 179
    sequences
    strata, 213
Paloma test, 159
pan-continental, 42
paraffin-napthene value, 33
parallel laminae, 76, 101
particles, 182
patch reefs, 187, 197
Patrick Draw Field, WY, 134, 137-140
pattern correlation, 48
Paupuan Basin, 240
"pay zone", 84
Peake and Hodgson, cited (1967), 36
peak generation curve, 30
peat, 29
pebble imbrication, 69
pebbly sands, 204
pelagic, 181
pelagic chalk, 212
    carbonates, 213
pelecypods, 212
pellet mud facies, 200
pelletoids, 193
pellets, 24, 182-183
    pelmicrite
    pelsparite
Pembina area, Alberta, 248
Pembina Field, Alberta, 153, 158
Peneroplid sand, 194
penesoplidae, 201
Pennsylvanian, 33, 47, 115
    Red Rock Sand, 83
    river-dominated delta, 109
Peoria Field, Alaska, 113
Peoria Field, Colorado, 109
period, 40, 47
peri-platform mode, 204
permeabilities, 33
Permian, 47
    Basin, 34
    Jurassic faulting, 114
    Lower Triassic, 113
perrenial, 195
    hemipelagic
    pelagic
Persian Gulf, 206, 211-212
Peruvian Gulf, 24
    pore volume, 31
Peterson and Others, cited (1955), 15
petroleum, 20
    cycloparaffins, 20

cyclobutane
cyclopropane
genesis, 23
impurities, 20
napthenes
reservoirs, 33
Pettijohn and Others, cited (1972), 166, 171, 173
Phanerozoic time, 44, 47
phenatherene, 38
Philippi, cited (1965), 32
Phinney, cited (1891), 35
phosphorites, 24
photosynthesis, 21
phreatic, 165
    cements, 226
    lenses, 232
physiochemical processes, 176
Pierre shale, 153
"pigeon holes", 182
pigmentation, 28
Pingetore, Jr. (1982), 232
pinnacle, 187
    buildups, 243
    reef definition, 247
pisolites, 182
Pittman, (1974) (1979) (1980), 9, 174, 185, 228
plane beds, 65
plankton, 23
plankton forms, 21
planktonic lime, 195
plant cuticle, 22
plant debris, 29
platforms, 223
    carbonate reservoirs, 199
Platte River, 67
platy grains, 173
Pleistocene sea cliffs, 129
    Pleistocene rocks, 202, 225
Pliocene, 22, 23
    pinnacle reef, 54
    Repetto formation, 159
    sands, 159
point bar, 73-75, 78, 87
    deposits, 33
pollen, 22
pores, 257
    filling cement, 227
    geometry, 32
    occlusion, 168, 229
    size reduction, 31
    space classification, 184

pore wall, 32
Porities, 201, 205, 207
porosity, 31
porphyrin pigments, 21
postdepositional anomalies, 42
Post Mississippi Unconformity, 83
Powder River Basin, 134, 153
Powell and Snowden, *cited* (1976), 24
Precambrian, 33, 47, 217
pregeneration, 30
pressure solution, 173
Price, *cited* (1976), 35, 37
prodeltaic shales, 48
progradation, 56, 92, 235-236
protopetroleum, 31
Providence channel, 202
proximal facies, 215
Prudhoe Bay Field, Alaska, 61, 80, 113-115
Puerto Rico, 24, 207
Punta Gorda, 192
Purdy, *cited* (1963) (1974), 192, 200
Purdy, *et al., cited* (1975), 193-194
Putnam and Oliver, *cited* (1980), 80-82
Pyrene, 38
pyromorphite, 24

## Q

Qatar Peninsula, 211
quartz arenite, 171
quartz grains, 32
Quirquire Field, Venezuela, 71, 84-85, 113

## R

radioactive bombardment, 24
Rainwater, *cited* (1963), 23
ramp carbonate buildups, 207
    carbonate reservoirs, 206
    outer, 215
    ramps, 187, 223
    ramp to step depositional systems, 218
Rangely Field, Colorado, 80
rapid sedimentation, 22-23
rapid transport, 146
    rabbit model
recent rocks, 225
recrystallization, 172
Redfield, *cited* (1958), 24
Red River, 67
Reedy Creek, 36
reef facies, 198
    backreef (reef flat)
    forereef
    reef crest

reefs, 23
    apron, 24, 27
    crest, 24, 27
    flat sediments, 24, 27
    lineaments, 244
    trends, 24
reflections, 13, 18
reflux dolomitization, 275
Regili, 198
regional
    anomalies, 52
    layering, 53
    tilting, 38
    transect, 26
relative oil potential, 22
remobilization, 38
Repetto formation, 160, 163
replacement, 172
reservoir, 238
    Age, 31
    rocks, 23, 40, 61
        location
        quality
    trap potential, 95
reworked silt, 212
Rhone River Delta, 92
Rice, *cited* (1980), 109
Rich, *cited* (1939), 35
Rierdon formation, 87
rip up clasts, 145
Rittenhouse, *cited* (1971) (1972), 80, 172, 174
Roberts, *cited* (1980), 80
rock interpretations, 9
    lithology type
    reservoir potential
rock strata, 42
rock-stratigraphic units, 40
Rocky Mountains, 134, 137
Roja, 198
Rosedale Ranch, 159
rubble sands, 204
rudstids, 243
Rugose corals, 249

## S

Sabine
    Peninsula, 140
    Uplift, 35
    water table, 230
Sabins, *cited* (1963), 134
Sadlerochit formation, 113
Sahl, *cited* (1970), 115

salinities, 22
salinity, 35, 37
salt
　lens, 55
　pan, 225
　River, 67
　solution, 55
San Andros, 268, 273
San Bernard River, 67
sand grains, 32
sandstone, 137, 140, 165
　biotite
　chert
　chlorite
　feldspar
　musconite
　quartz
　zircon
sandstone geology, 179
sandstone interdeltaic reservoirs, 130
sandstone porosity, 175
sand waves, 143
San Felipe, Baja, CA, 129
Sangree and Widmier (1974), 48
San Joaquin Basin, CA, 151
San Juan Basin, N.M., 85
San Simon channel, 268
satellite reefs, 180
Sauk rocks, 42
Savkevic, cited (1971), 176
Sawyer, cited (1980), 21, 23-27
scale, 62
Schmidt and McDonald (1979), 168, 171, 175-176
Schmoker and Halley (1982), 232
Schowater (1976), 80
Scott (1966), 193
scout tickets, 17, 19
sea level, 26, 236
　movements, 91
sea margins, 23
secondary migration, 33, 38
secondary porosity formation, 174
sedimentary patterns, 44
sediment hydrocarbon, 24
sediment movement, 63
sedimentology, 40
sediment supply, 91
　source
sediment-water interface, 22-23
Seeling, cited (1978), 153
seepage reflux, 231
seismic data, 9, 40

profile, 53
reflections, 13
section, 15
sediments, 25
sequence, 48
tract
seismic profiles
　basin buildups, 247, 250-251, 257, 265
　continental sands, 79, 81, 83, 88
　deltaic sands, 110-112, 114
　general, 18, 45
　interdeltaic sands, 133
　marine sands, 151, 152
　shelf carbonates, 208-210, 253, 256, 273
　shelf margin carbonates, 241, 242
seismic reflection pattern (SRP), 108
seismic signature, 208-210
Selley, cited (1979), 149
SEM, 213
semi anthracite, 29
sequence, 42
serpentine, 25
Seven Oaks gas field, 109
shale, 24
　grains, 32
shale lens, 51
shallow marine reservoirs, 142
shallow marine sands, 144
shallow marine sandstone reservoirs, 153
shallow water sedimentary carbonates, 179-180, 191-195, 199-206, 211-212
shamal winds, 211
Shannon and Dahl, cited (1971), 115-118
Shannon Sandstone, 153
sheet sands, 35
shelf, 187
shelf buildups
　Bahama, 201
　Belize, 191-193
　Hanford Field, 268
　Puerto Rico, 204-206
　Smackover, 250-252
shelf detrital, 236
shelf lithofacies, 192, 201
　carbonate grain
　carbonate mud
　grapestone
　island
　mixed-terrigenous-carbonate mud
　oolitic
　pelletal mud
　reef
　skeletal mud

terrigenous
shelf margin, 188
shelf margin lithofacies, 201, 214-222
   back bank, 201
   back reef, 195
   bank, 187, 201
   forebank, 202
   fore reef, 196
   reef, 187, 195, 201, 233-243
shelter, 186
Shelton, *cited* (1967), 89-90
Sheriff, *cited* (1977), 11-12
shingled seismic reflections, 111
Shipstern Lagoon, 194
Shirely and Ragsdale (1966), 92
shoal, 54
shoal retreat massifs, 142
"shoestring", 144
short migration, 33
shrinkage, 174, 186
   voids
Sigmoid reflections, 110
silica precipitate, 171
Silled Basins, 23
Silurian, 47
   age, 243
   Thronton reef complex, 179
Silver, *cited* (1980), 61, 75, 103
Silver and Bloch, in preparation, *cited*, 230
Silver, and Todd, *cited* (1969), 45, 159, 234-237, 240
Silver and Wermiel, *cited* (1976), 167-168
Silver, *et al, cited* (1981) 191, 193-194
Simpson Group, 35
skeletal
   algae, 251
   content, 61
   fragments, 24
   grains, 182
   mud facies, 200
Slaughter-Levelland Field, 274
Sloss, *cited* (1963), 42
slump, 212
Smackover
   deposits, 250
   ramps, 251
   reefs, Gulf Coast, 250, 252-253, 266
snapshot in time, 62
Snipe Lake, 252
soft sediment formation, 145
solution, 228
Solution Uplift, 35
Sonic logs, 9

source
   bed, 23, 30
   rock, 22-23, 28, 35, 40
South Glenrock Field, Wyoming, 80
Southwest Africa, 24, 252
spar cements, 257
Sparky, 82
sparry calcite cement, 183
spill point, 38
Spirit River, 113
sponges, 215
spore coloration, 28
spores, 22
Sprayberry sands, 159, 240
"spreadouts", 212, 264
stable bonds, 28
stable flats, 269
Stapor, *cited* (1972), 87
step
   carbonate reservoirs, 191
   depositional systems, 222
step model basin, 215
Sterling Formation, 80
Stevens, sandstone, 159
Stewart and Silver, in preparation, *cited,* 204-207, 209-210
Stoddart (1960), 195
Straits of Hormuz, 211
stranded beach reservoirs, 126-127
Stratigraphic
   Code, Articles 29 and 37, 40
   framework, 13, 40
   horizons, 40
   trap, 67, 71, 78, 80-90, 95, 98, 101, 104, 115-121, 126, 130, 134-141, 146, 148, 153-164, 234-275
   units, 40
strawn, 115
stream power velocity, 65
stress equilibrium, 31
stromatoporoids, 179-180, 215, 243
Stubblefield and Others (1975), 143
structured kerogen, 30
structured reversal, 35
styolization, 251
subaqueous sediment gravity flow, 147
sub biluminous, 29
submarine canyon channel fill deposits, 159
submarine fan reservoir
   fans, 146, 149, 150, 154
submarine sand, 50
subsidence, 23
subsurface exploration stratigraphy, 40

subzone, 42
Suffield area, Alberta, 153
Sullwald (1961), 159-160
suprafan lobes, 150
supratidal, 188, 212
    evaporation model, 230
surface information, 40
surface sediment, 26
Sussex sandstone, 153, 155-157
Sverdrup Basin, northwest Canada, 140
swamp or salt marsh, 212
Swan Hills, 252, 259, 262
Swanson, *cited* (1972), 78
Swift, *cited* (1974), 142
Swift and Others, *cited* (1975), 143
Swift and Sears, *cited* (1974), 142
Swift formation, 87
synchronous surfaces, 40
syntaxial cement, 227
synthetic section, 14, 15, 17

**T**

tabulate corals, 243, 249
Tallis, *cited* (1975), 240
Tanner, *cited* (1970) (1974), 87
tectonics, 62
temperature, 37
    time relationship, 28
Tensleep Group, 35
    sandstone, 80
terrigenous
    analogues, 183
    clastics, 43, 45-46
    sand, 194
tertiary, 47
    basins, 159
    Fort Union, 112
Thermal Basin, 31
thermal cracking reactions, 28
thermal expansion, 32
thermochemical generation, 28, 30
thin bedding, 74
Thompson, *cited* (1968), 107
tidal, 188
tidal flat
    deposits, 270
    reservoirs, 126, 128
    sedimentation, 275
time line, 40
time stratigraphic unit, 40, 42 45
Tippecanoe strata, 42
Tizzard and Lerbekmo, *cited* (1975), 153
Todelto formation, 87

toe-of-slope, 204
Tongue of the Ocean, 199
topset interior, 188
Toulvene, 38
trailing edge coast, 91
    afro-trailing
    amero-trailing
    neo-trailing
transformation process, 28
transgressive margin, 72
transitional lithofacies, 211
translucent, 29
transparency, 28
transverse dune ridges, 71
trap, 33, 38, 238
Trenton limestone, 35
Triassic, 47
Trimethylbenzenes, 38
Trucial Coast, 211, 270
truncation anomalies, 52
Tubiphytes, 251
turbidites, 145, 151, 160
    reservoirs, 146
turbidity flows, 46
Turneffe Island, Belize, 192
Tuscallosa deltaic system, Texas, 151
tying loops, 50

**U**

ulmic acid, 21
unconformity, 13, 41, 61
undaform, 188
Union, 37
uplift, 38
Upper Devonian Leduc buildups, 252
Upper Devonian Woodbend, 258
Upper/Middle Devonian, 258
upper patterns, 43
    concordance
    toplap
Upper Wolfcampian, 233, 234
upward migration, 39
Uruguay, 21
U.S.A. Gulf Coast, 123
U-shaped dunes, 71

**V**

Vadose, 165
    cements, 227
    zone, 11
Vail and Others, *cited* (1977), 11, 13-16, 18, 44-47, 217, 221
Vaqueros, Miocene, 146

varves, 40
velocity cross section, 16
velocity relations, 13
Ventura anticline, 163
Ventura fields, 159-160, 163
vertical
　distribution, 40
　migration, 33
Vest, Jr. (1970), 263
Viking Sandstone, Alberta, 153
Vincelette and Chittum (1981), 85-88
Virginia Hills, 252
vitrinite, 28
visual kerogen analysis, 30
volatiles, 28-29
Volga, 91, 93
volume, 31
Vug, 186, 238

## W

Wackestone, 184, 195
Wagner and Matthews, *cited* (1982), 232
Wallace and Schaferman, *cited* (1977), 197
Walker, *cited* (1978) (1979), 142-143, 150, 159
Walker Creek, Arkansas, 251, 254
Walther's law, 48
Wamsutter Arch, 134
Wanless, *cited* (1982), 190
Washburn Ranch Field, La Salle County, Texas, 113
wave-dominated delta, 109, 111, 123
wave energy, 203
wavefronts, 12
wavelength, 11
Webb (1981), 159-160
Weber sandstone, 80
Weeks (1952), 23
Weimer (1966), 134, 137-138, 140
Wegemann (1911), 35

Welcome Fields, 254
well
　bore, 33
　cuttings, 3
Westcott and Ethridge (1980), 107, 120
Western Canadian Arctic Archipelago, 134, 140
Western Coral Sea, 240
Western Great Plains of North America, 137
West Tuscola Field, Texas, 115-117
Whale Cay, 202
Wheeler, *cited* (1963), 42
Wilcox, 109, 113
Williams and Stelck (1975), 154
Wilson, *cited* (1974) (1975), 181, 213
Winchel, Alexander (1865), 22
wind direction, 61
windward shelf margins, 215
wireline log, 9, 40
Wisconsin Age, 91
Wolfcampian, 33
woodbine, 36
Woodbine prograded delta, 111
woody debris, 22
Woolnough, *cited* (1937), 23
Wright, *et al, cited* (1950), 192
"wrinkles", 212
Wuenshel, *cited* (1960), 15

## Y

Yallahs fan delta, 107
Yukon, 93

## Z

Zama, Alberta, 252
Zama line, 257
Zeta Lake pinnacle reef, 248
zone, 42
zonule, 42
Zooanthus Sociatus, 24